Natural Science in Archaeology

Series editors

Günther A. Wagner
Christopher E. Miller
Holger Schutkowski

More information about this series at http://www.springer.com/series/3703

Laure Dussubieux • Mark Golitko •
Bernard Gratuze
Editors

Recent Advances in Laser Ablation ICP-MS for Archaeology

Springer

Editors
Laure Dussubieux
Integrative Research Center, Elemental Analysis Facility
Field Museum of Natural History
Chicago, IL, USA

Mark Golitko
Department of Anthropology
University of Notre Dame
Notre Dame, IN, USA

Bernard Gratuze
Institut de Recherche sur les Archéomatériaux
Centre Ernest Babelon
CNRS/Université d'Orléans
Orléans, France

ISSN 1613-9712
Natural Science in Archaeology
ISBN 978-3-662-49892-7 ISBN 978-3-662-49894-1 (eBook)
DOI 10.1007/978-3-662-49894-1

Library of Congress Control Number: 2016947426

Printed on acid-free paper

This Springer imprint is published by Springer Nature
The registered company is Springer-Verlag GmbH Berlin Heidelberg

Prologue

Since its inception as a modern discipline, archaeology has strived to produce more quantifiable data to test its theories on how human cultures change and transform. One particularly effective application for charting the transformation of objects of human ingenuity or of human beings themselves has been through the analysis of the chemical composition of material culture. I refer to both provenance studies that seek to identify the geological source material of artifacts or ecofacts and to chemical characterization studies focused on the alteration of a material through biological, environmental, or manufacturing processes. The former include obsidian sourcing, clay sourcing, metallic ore sourcing, or biological sourcing through isotopic signatures, for example. The latter include applications such as identifying diagenic processes on human bone, archaeochemical evidence for metallurgical activities, or measuring heavy element contaminants in archaeobotanic materials. Advances in technology in the twentieth century have propelled the ability of archaeologists to measure and report with much greater precision than ever before on these transformations, creating the discipline of archaeometry in the process.

Technical advances in instrumentation and techniques have always been a driver in archaeometry. This is certainly the case with Laser Ablation Inductively Coupled Plasma-Mass Spectrometry (LA-ICP-MS). Since its development toward archaeological applications in the 1990s, LA-ICP-MS has come into its own as one of the premier archaeological tools for nearly nondestructive multielement compositional analysis of objects (Giussani et al. 2009; Gratuze et al. 1993: Resano et al. 2010; Speakman and Neff 2005). It builds on earlier high resolution multielement techniques like Instrumental Neutron Activation Analysis (INAA), with the same multielement capabilities and low detection limits in the parts per billion or, in the case of solution, parts per trillion range. It has high throughput capabilities of tens of samples processed per day, and unlike INAA, it does not produce dangerous long-term radioactive waste. It is cost-effective and has minimal impact on the object of study.

LA-ICP-MS is not without its challenges, of course. Technique specific issues include the stability of the instrument's detector, accounting for doubly charged and oxide species, interferences, and limits of detection (see Pollard et al. 2007 for detailed descriptions of these problems). Issues with replicability and reproducibility can create problems for quantitative analysis, especially in heterogeneous materials where difficulty in matrix matching between the sample and standard may introduce errors. Likewise, the analysis produces large amounts of data that may have varying degrees of accuracy and precision across elements and matrices. The data must be evaluated for quality, and unreliable data may be omitted from the analysis. There is a need for curating large amounts of data long term and making them accessible to other researchers.

Despite the challenges, LA-ICP-MS has become the most widely utilized high resolution technique for multielement characterization in twenty-first century archaeology. Numerous university and museum labs dedicated to archaeological applications of LA-ICP-MS have sprung into being, and many other multidisciplinary labs are in existence. There are certain advantages to having laboratories dedicated to archaeological LA-ICP-MS, including specialized method development, prioritization of archaeological sampling, and the development of staff with technological know-how around archaeological materials. One example of such a lab is The Field Museum's Elemental Analysis Facility, founded in 2005. Equipped with a quadrupole LA-ICP-MS, a standard 213 nm laser, and an experimental adaptable chamber 266 nm laser, the EAF serves archaeologists and collections from around the world as well as the museum's own vast archaeological collections housed in the same building. Several of the chapters in this volume derive from EAF-based research.

This volume highlights these advances in LA-ICP-MS applications in archaeology, with reviews of how the technology works (Chap. 1) and innovations in sample introduction including new adaptable laser cell and profiling technologies (Chaps. 2–5). The latter chapters are dedicated to exploring the application of the technique to a variety of material types, from non-vitreous materials primarily of metallic origin (Chaps. 6–8) to vitreous materials including glass and obsidian (Chaps. 9–14). The final chapters explore the expansion of LA-ICP-MS to materials including slag, garnet, stone, mineralized tissue, and lead glazes (Chaps. 15–19). This work highlights the results of a 20-year history of Laser Ablation ICP-MS in archaeology and its potential for future growth. Given the state of the discipline, it is clear that LA-ICP-MS will continue to revolutionize archaeology as the next generation of archaeologists takes it to new frontiers.

Integrative Research Center, Social Sciences
Field Museum of Natural History
Chicago, IL, USA

Patrick Ryan Williams

References

Giussani B, Monticelli D, Rampazzi L (2009) Role of laser ablation – inductively coupled plasma – mass spectrometry in cultural heritage research: a review. Anal Chim Acta 635:6–21

Gratuze B, Giovagnoli A, Barrandon J-N, Telouk P, Imbert J-L (1993) Apport de la méthode ICP-MS couplée à l'ablation laser pour la caractérisation des archéomatériaux. Revue d'Archéométrie 17:89–104

Pollard M, Batt C, Stern B, Young SMM (2007) Analytical chemistry in archaeology. Cambridge University Press, Cambridge

Resano M, Garcia-Riuz E, Vanhaecke F (2010) Laser ablation – inductively coupled plasma mass spectrometry in archaeometric research. Mass Spectrom Rev 29:55–78

Speakman RJ, Neff H (eds) (2005) Laser ablation-ICP-MS in archaeological research. University of New Mexico Press, Albuquerque, NM

Contents

1 **Instrumentation, Fundamentals, and Application of Laser Ablation-Inductively Coupled Plasma-Mass Spectrometry** 1
Mattias B. Fricker and Detlef Günther

Part I Sample Introduction

2 **Introduction to Solid Sampling Strategies** 23
Mark Golitko

3 **Open-Cell Ablation of Killke and Inka Pottery from the Cuzco Area: Museum Collections as Repositories of Provenience Information** 27
Mark Golitko, Nicola Sharratt, and Patrick Ryan Williams

4 **Optimization of 2D LA-ICP-MS Mapping of Glass with Decorative Colored Features: Application to Analysis of a Polychrome Vessel Fragment from the Iron Age** 53
Johannes T. van Elteren, Serena Panighello, Vid S. Šelih, and Emilio F. Orsega

5 **LA-ICP-MS Analysis of Ancient Silver Coins Using Concentration Profiles** 73
Guillaume Sarah and Bernard Gratuze

Part II Application to Non-siliceous Materials

6 **Analysis of Non-siliceous Archaeological Materials by LA-ICP-MS** 91
Laure Dussubieux

7 **Precise and Accurate Analysis of Gold Alloys: Varna, the Earliest Gold of Mankind—A Case Study** 95
Verena Leusch, Michael Brauns, and Ernst Pernicka

8 **LA-ICP-MS Analysis of Prehistoric Copper and Bronze Metalwork from Armenia** 115
David L. Peterson, John V. Dudgeon, Monica Tromp, and Arsen Bobokhyan

Part III Application to Vitreous Materials

9 Analysis of Vitreous Archaeological Materials by LA-ICP-MS 137
Bernard Gratuze

10 Contributions of LA-ICP-MS to Obsidian Sourcing in the Pacific 141
Christian Reepmeyer, Wallace Ambrose, and Geoffrey Clark

11 The Role of LA-ICP-MS in the Investigation of Archaeological Glass 163
Simone Cagno, Kevin Hellemans, and Koen Janssens

12 Glass Characterization Using Laser Ablation-Inductively Coupled Plasma-Mass Spectrometry Methods 179
Bernard Gratuze

13 A Proficiency Test for LA-ICP-MS Laboratories Analyzing Ancient Glass 197
Laure Dussubieux

14 Characterization of Slag Inclusions in Iron Objects 213
Maxime L'Héritier, Stéphanie Leroy, Philippe Dillmann, and Bernard Gratuze

Part IV Expanding the Application of LA-ICP-MS in Archaeology

15 Expanded Applications of Laser Ablation-ICP-MS in Archaeology 231
Mark Golitko

16 Determining the Provenience of Garnet Beads Using LA-ICP-MS 235
Alison K. Carter

17 Application of LA-ICP-MS to Black Stone Objects Used During the Iron Age in Celtic Europe 267
Anne Baron and Bernard Gratuze

18 Investigating Biogenic Versus Diagenetic Trace Element Incorporation in Archaeological Mineralized Tissues with LA-ICP-MS 323
John V. Dudgeon, Monica Tromp, Bryan K. Hanks, and Andrei V. Epimakhov

19 Pb Isotopic Composition of Panamanian Colonial Majolica by LA-ICP-MS 343
Javier G. Iñañez, Jeremy J. Bellucci, Juan Guillermo Martín, Richard Ash, William F. McDonough, and Robert J. Speakman

Contributing Authors

Wallace Ambrose Archaeology and Natural History, School of Culture, History, and Language, Australian National University, Canberra, Australia

Richard Ash Department of Geology, University of Maryland, College Park, MD, USA

Anne Baron Institut de Recherche sur les Archéomatériaux, Centre Ernest Babelon, CNRS/Université d'Orléans, Orléans, France

Jeremy J. Bellucci Department of Geology, University of Maryland, College Park, MD, USA

Department of Geosciences, Swedish Museum of Natural History, Stockholm, Sweden

Arsen Bobokhyan Institute for Archaeology and Ethnography, Armenian National Academy of Sciences, Yerevan, Republic of Armenia

Michael Brauns Curt-Engelhorn-Zentrum Archäometrie, Mannheim, Germany

Simone Cagno Department of Chemistry, University of Antwerp, Antwerp, Belgium

Alison K. Carter Department of Anthropology, University of Illinois at Champaign-Urbana, Urbana, IL, USA

Geoffrey Clark Archaeology and Natural History, School of Culture, History, and Language, Australian National University, Canberra, Australia

Philippe Dillmann Laboratoire Archéomatériaux et Prévision de l'Altération (LAPA), IRAMAT LMC, UMR 5060 CNRS and NIMBE UMR 3685 CEA/CNRS, CEA/Saclay, Gif-sur-Yvette Cedex, France

John V. Dudgeon Department of Anthropology, Center for Archaeology, Materials, and Applied Spectroscopy, Idaho State University, Pocatello, ID, USA

Laure Dussubieux Integrative Research Center, Elemental Analysis Facility, Field Museum of Natural History, Chicago, IL, USA

Andrei V. Epimakhov Institute of History and Archaeology RAS (South Ural Department), South Ural State University, Chelyabinsk, Russia

Mattias B. Fricker Department of Chemistry and Applied Biosciences, Laboratory of Inorganic Chemistry, ETH Zurich, Zurich, Switzerland

Mark Golitko Department of Anthropology, University of Notre Dame, Notre Dame, IN, USA

Bernard Gratuze Institut de Recherche sur les Archéomatériaux, Centre Ernest Babelon, CNRS/Université d'Orléans, Orléans, France

Detlef Günther Department of Chemistry and Applied Biosciences, Laboratory of Inorganic Chemistry, ETH Zurich, Zurich, Switzerland

Bryan K. Hanks Department of Anthropology, University of Pittsburgh, Pittsburgh, PA, USA

Kevin Hellemans Department of Chemistry, University of Antwerp, Antwerp, Belgium

Javier G. Iñañez Grupo de Investigación en Patrimonio Construido (GPAC), Universidad del País Vasco/Euskal Herriko Unibertsitatea (UPV/EHU), CIEA Lascaray, Vitoria-Gasteiz, Spain

IKERBASQUE, Basque Foundation for Science, Bilbao, Spain

Koen Janssens Department of Chemistry, University of Antwerp, Antwerp, Belgium

Maxime L'Héritier Department of History, Université Paris 8, Saint Denis, France

Stéphanie Leroy Laboratoire Archéomatériaux et Prévision de l'Altération (LAPA), IRAMAT LMC, UMR 5060 CNRS and NIMBE UMR 3685 CEA/CNRS, CEA/Saclay, Gif-sur-Yvette Cedex, France

Verena Leusch Curt-Engelhorn-Zentrum Archäometrie, Mannheim, Germany

Juan Guillermo Martín Universidad del Norte, Barranquilla, Colombia

William F. McDonough Department of Geology, University of Maryland, College Park, MD, USA

Emilio F. Orsega Department of Molecular Sciences and Nanosystems, University Ca'Foscari di Venezia, Venice, Italy

Serena Panighello Department of Molecular Sciences and Nanosystems, University Ca'Foscari di Venezia, Venice, Italy

Ernst Pernicka Curt-Engelhorn-Zentrum Archäometrie, Mannheim, Germany

David L. Peterson Department of Anthropology, Center for Archaeology, Materials and Applied Spectroscopy, Idaho State University, Pocatello, ID, USA

Christian Reepmeyer College of Society, Culture and Education, James Cook University, Cairns, Australia

Guillaume Sarah Institut de Recherche sur les Archéomatériaux, Centre Ernest Babelon, CNRS/Université d'Orléans, Orléans, France

Vid S. Šelih Laboratory for Analytical Chemistry, National Institute of Chemistry, Ljubljana, Slovenia

Nicola Sharratt Department of Anthropology, Georgia State University, Atlanta, GA, USA

Robert J. Speakman Center for Applied Isotope Studies, University of Georgia, Athens, GA, USA

Monica Tromp Department of Anatomy, University of Otago, Dunedin, New Zealand

Johannes T. van Elteren Laboratory for Analytical Chemistry, National Institute of Chemistry, Ljubljana, Slovenia

Patrick Ryan Williams Integrative Research Center, Social Sciences, Field Museum of Natural History, Chicago, IL, USA

Abbreviations

ATW	Atmospheric thin window
BG	Background
BSED	Backscattered electron detector
CA	Correspondence analysis
CCD	Charge-coupled device
CCT	Collision cell technology
CMG	Corning Museum of Glass
cps	Counts per second
CRM	Certified reference material
DC	Direct current
DP	Depth profile
DT	Dwell time
EDTA	Ethylenediaminetetraacetic acid
EF	Elemental fractionation
EPMA	Electron probe microanalysis
ESA	Electrostatic sector analyzer
ESEM	Environmental scanning electron microscope
ESR	Electron spin resonance
FI	Fractionation index
FTIR	Fourier transformed infrared
GC	Gas chromatography
GCMS	Gas chromatography-mass spectrometry
GED	Gas exchange device
GF-AAS	Graphite furnace-atomic absorption spectrometry
HEAD	High efficiency aerosol dispersion
HFSE	High field strength element
HIMT	High iron manganese titanium
HR-ICP-MS	High resolution inductively coupled plasma-mass spectrometry
ICP-MS	Inductively coupled plasma-mass spectrometry
ICP-MS-AES	Inductively coupled plasma-atomic emission spectrometry
ICP-OES	Inductively coupled plasma-optical emission spectrometry

ICP-Q-MS	Inductively coupled plasma-quadrupole-mass spectrometry
ICPS	Intensity counts per second
INAA	Instrumental neutron activation analysis
IR	Infrared
ISI	Internal standard independent
LA	Laser ablation
LA-ICP-MS	Laser ablation inductively coupled plasma-mass spectrometry
LAMMA	Laser microprobe mass analysis
LIBS	Laser-induced breakdown spectroscopy
LILE	Large-ion lithophile element
LIP	Late intermediate period
LMG	Low magnesium glass
LMHK	Low magnesium high potassium
LOD	Limits of detection
LREE	Light rare earth elements
MC-ICP-MS	Multi-collector inductively coupled plasma-mass spectrometry
MSA	Magnetic sector analyzer
NAA	Neutron activation analysis
ND	Normal distribution
Nd:YAG	Neodymium-doped yttrium aluminum garnet
NIST	National Institute of Standards and Technology
NORC	New Ohio Red clay
NRC	Non-reduced compounds
OIB	Oceanic island basalt
PCA	Principal components analysis
PCR	Principal component regression
PD	Poisson distribution
PGE	Platinum group elements
PIXE-PIGE (or PIGME)	Particle-induced X-ray emission—proton-induced gamma-ray emission
PKT	Phu Khao Thong
pLA	Portable laser ablation
PNG	Papua New Guinea
ppb	Parts per billion
ppm	Parts per million
PXRF	Portable X-ray fluorescence analysis
Py-GC/MS	Pyrolysis gas chromatography mass spectrometry
QA/QC	Quality assurance/quality control
REE	Rare earth element
RF	Radio frequency
RGB	Red, Green, and Blue
RSD	Relative standard deviation

SD	Standard deviation
SDD-EDS	Silicon drift detector-energy dispersive spectrometry
SEA	Southeast Asia
SEM-EDS (or EDX)	Scanning electron microscopy with energy dispersive X-ray spectrometry
SG	Signal
SI	Slag inclusion
SIMS	Secondary ion mass spectrometry
SN	Sum normalization
SRM	Standard reference material
STD	Standard
TEM	Transmission electron microscopy
TIMS	Thermal ionization mass spectrometry
TOF-MS	Time of flight mass spectrometry
TRA	Time-resolved analysis
UHP	Ultra high purity
USGS	United States Geological Survey
UV	Ultraviolet
VOLM	Volume-optional low memory
VUV	Vacuum ultraviolet
WDS or WDX	Wavelength dispersive X-ray spectroscopy
WNB	West New Britain
XRF	X-ray fluorescence
μ-XRF	Micro-X-ray fluorescence

Instrumentation, Fundamentals, and Application of Laser Ablation-Inductively Coupled Plasma-Mass Spectrometry

1

Mattias B. Fricker and Detlef Günther

Abstract
Laser ablation-inductively coupled plasma-mass spectrometry (LA-ICP-MS) is becoming a versatile and mature analytical technique for quantitative major, minor, and trace element analysis and isotope ratio determinations. A wide variety of solid and liquid samples can be analyzed. Besides the fast growing application of this method as an imaging technique, geology still is and has been the major driving force for the development of LA-ICP-MS. In this review, the method, instrumentation, fundamental observations and quantification procedures are explained. In addition, pros and cons of the technique are discussed and some trends and new sampling strategies, with special focus on archaeological applications, are summarized.

1.1 Introduction

Laser Ablation-Inductively Coupled Plasma-Mass Spectrometry (LA-ICP-MS) is currently the technique of choice for trace elemental analysis of solid samples. In the 29 years since the introduction of LA-ICP-MS in 1985 (Gray 1985), the fundamental principle of the technique has not changed. In addition to many other parameters, the performance of LA-ICP-MS depends upon laser parameters (wavelength, pulse width), the design of the ablation cell and related aerosol transport, and the ICP-MS operating conditions. The ablation cell is an air-tight container, where samples and calibration reference materials are placed. It incorporates a window made of a material that is transparent at the wavelength of the laser being used, through which the laser beam is focused onto the sample surface, typically with diameters of tens to hundreds of µm. Each pulse of the high-energy laser beam ejects particles of the sample and the resulting laser generated aerosol is transported to the ICP by a carrier gas flow (commonly He, although Ar is also used). During the short residence time (µs-ms) in the plasma at over 6000 K, the solid aerosol is vaporized, converted to atoms, and ionized. The ion beam is then extracted into the mass spectrometer interface,

M.B. Fricker • D. Günther (✉)
Department of Chemistry and Applied Biosciences, Laboratory of Inorganic Chemistry, ETH Zurich, Vladimir-Prelog-Weg 1, 8093 Zurich, Switzerland
e-mail: guenther@inorg.chem.ethz.ch

L. Dussubieux et al. (eds.), *Recent Advances in Laser Ablation ICP-MS for Archaeology*, Natural Science in Archaeology, DOI 10.1007/978-3-662-49894-1_1

where the pressure is reduced from ambient to a moderately high vacuum in two stages, by passing through a sampler cone and a skimmer cone. In the high-vacuum chamber the ion beam is shaped by ion optics, and photons as well as negatively-charged and neutral particles are separated by a photon stop or similarly functioning ion optics. Positive ions pass through various kinds of analyzers, where ions with differing mass to charge (m/z) ratios are separated and then collected on a detector, where the ions are either counted or a generated current is measured.

The purpose of LA sampling is to generate an aerosol that can be transported, vaporized, atomized and ionized within the ICP. The ions thus formed are then extracted into the interface, separated by the analyzer and recorded by the detector and associated electronics. For bulk analysis of homogeneous samples the recorded transient signal can be integrated for quantification. When performing a line scan along a heterogeneous sample (such as a stalagmite, paint layers, or bronze objects, for instance), the transient signal can be used to show variations which correspond to concentration changes in the sample and provide insights into growth or the manufacturing procedure. Parallel line scans can be performed across an area of the sample yielding the distribution of elements in two dimensions, i.e. elemental imaging or mapping, which is today one of the fastest growing applications of LA-ICP-MS. Recording these variations in transient signals requires an ablation cell with a fast washout of the aerosol to minimize mixing of aerosol from different positions of the sample, allowing higher spatial resolution. Most commercially available ablation cells are relatively small, requiring cutting large samples into smaller pieces in order for the sample piece to fit into the ablation cell. For many samples, cutting and resulting destruction is not possible, thus a large ablation cell able to contain larger samples in their entirety, is required. Analysis without any ablation cell would be most ideal for many applications, and some promising approaches are discussed later in this chapter.

Research and development of LA-ICP-MS during the last decade has produced a large number of applications, which is largely due to the fact that using this technique, samples are analyzed under atmospheric pressure, which makes LA-ICP-MS easier to apply when compared to secondary ion mass spectrometry (SIMS), laser microprobe mass analysis (LAMMA), or electron probe micro-analysis (EPMA). All of these techniques require low-pressure vacuum conditions, which limits their applicability to some samples. However, it must be mentioned that some of the figures of merit for these techniques (limits of detection, spatial resolution) are superior compared to LA-ICP-MS. Furthermore, EPMA is often a complementary technique to LA-ICP-MS, as it allows the determination of major elements very accurately, a prerequisite of trace element determinations (e.g. Dussubieux et al. 2008). The ability to obtain quantitative determinations of major and trace element composition in any solid sample along with the ability to determine isotope ratios with high accuracy and high precision are some of the key reasons for the success of LA-ICP-MS. The technique is widely used in many areas, but especially in earth sciences, gemology, material sciences, and industrial quality control applications (Durrant 1999). Apart from bulk- and microanalysis of archeological objects, e.g. glass and ceramics (Barca et al. 2007; Dussubieux et al. 2007; Gratuze 1999), and depth profiling (Pisonero et al. 2007; Plotnikov et al. 2001), more advanced applications such as analyses or fingerprinting of gemstones and archeological objects such as pottery and ceramics (Arnold et al. 2012; Fontaine et al. 2010; Vaughn and Dussubieux 2011), isotope ratio determinations (Cottle et al. 2009) and elemental imaging and/or line scans (Bi et al. 2000; Campbell and Humayun 1999; Raab et al. 2009; Sinclair et al. 1998; Treble et al. 2003; Wirth et al. 2009), profiling (Mattey et al. 2008), and mapping (Becker et al. 2005; Treble et al. 2005; Woodhead et al. 2007) are rapidly growing applications, and provide valuable information for different fields of research.

1.2 Instrumentation

1.2.1 Inductively Coupled Plasma-Mass Spectrometry

The most common sample introduction systems used with ICP as an ion source are schematically shown in Fig. 1.1. The two basic systems are used for liquid samples or for solid samples or combinations thereof. The six most common mass analyzers used for ICP are also shown schematically, and will be explained in more detail in the following sections. Soon after ICP-MS was introduced, reports of elemental analysis were published with solution nebulization remaining the dominant sample introduction device (Jenner et al. 1990). The first coupling to laser ablation (Gray 1985), and later overviews (Durrant 1999; Hattendorf et al. 2003; Jackson et al. 1992; Russo et al. 2002a) gave some indications of the versatility of this ionization source. ICP-MS has several advantages in comparison to the longer established method of ICP-Optical Emission Spectrometry (ICP-OES), including lower limits of detection, a wider linear dynamic range, access to isotopic information, and simpler spectra. Due to the sensitivity of this technique, analyte concentrations can be diluted in solution ICP-MS (Longerich 2012), which reduces matrix effects and contributes to the wide range of applications reported in the literature.

The ion source, an argon plasma, vaporizes introduced particles, converts molecules to atoms, and ionizes these atoms, regardless of whether sample enters as a gas, liquid, or solid. The ion beam is then extracted through an interface where the sample passes through a sampler cone and a skimmer cone in two or three stages, dropping from atmospheric pressure into a moderate vacuum. A photon stop or equivalent device, removes negatively charged ions, neutrals and photons from the ion beam, which is then shaped by the ion optics before entering the mass analyzer. The mass filter (or analyzer) separates these ions by their mass to charge ratios, which are then counted by a single detector or multi collector.

The three most common types of mass analyzers in inorganic mass spectrometry are the quadrupole, the magnetic sector, and the time of flight mass spectrometer (TOF-MS), as shown in Fig. 1.1. The quadrupole is a low resolution, fast sequential mass filter consisting of four parallel rods, connected pair-wise to radio frequency (RF) voltage and direct current (DC) voltage. Depending on the applied voltages, ions of only one mass to charge ratio pass through the filter and are focused onto the detector. All types of ICP-MS suffer from interferences, as the mass analysers separate ions based on their mass to charge ratios, most often singly charged ions of an isotope of an element. Several isobaric interferences occur due to overlapping ions or the formation of polyatomic ions with a similar mass per charge ratio (Hattendorf and Günther 2001; Houk et al. 1980). A selection of some more common interference types are listed in Table 1.1, which occur at low mass resolution ($m/\Delta m=300$, quadrupole ICP-MS). These interferences also occur during LA, however, oxide formation is significantly lower with dry aerosol introduction.

If alternative isotopes that do not suffer from interference are not available, high resolution (HR) ICP-MS can overcome most interferences. The most widely used HR-ICP-MS instrumentation is based on various combinations of a magnetic field analyzer (MSA), and an electrostatic analyzer (ESA, a kinetic energy analyzer) with single, multiple or array detectors as shown in Fig. 1.1. The sector field ICP-MS with a single detector usually uses a magnetic field prior to the ESA, which is known as reverse Nier-Johnson geometry. Sector field ICP-MS with several detectors (multi collector or MC) ICP-MS (Nier-Johnson geometry) has also been used for more precise isotope ratio measurements. This instrument type records isotope signals simultaneously with high precision, and has found application in geochronology and other analyses requiring precise measurement of isotope ratios (Halliday et al. 1998). Developments have been made towards simultaneous detection of all elements (from m/z 7 to 250) using a Mattauch-

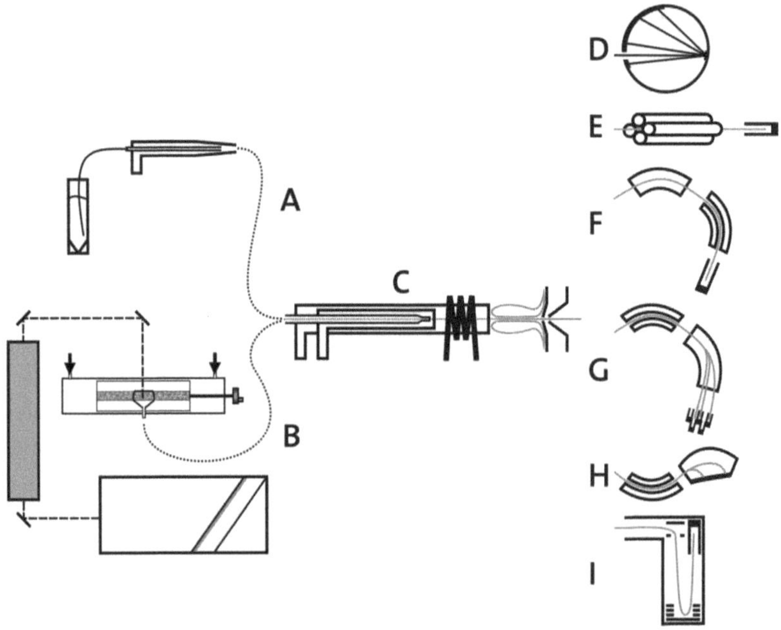

Fig. 1.1 Schematic representation (Fricker 2012) of the most common optical emission and mass spectrometer configurations used in combination with ICP as the photon emission or ion source: torch, load coil, plasma, and interface (C). Aerosols generated from solution nebulization (A) or from laser ablation (B) are transported (*dotted line*) to the ion source (C). The ions (*grey solid line*) can be separated using several types of analyzers (E–I) before being converted to a suitable electrical signal by various detectors. The quadrupole (E) and the single collector (reverse Nier-Johnson) sector field (F) are sequential MS, while G (multi collector, Nier-Johnson geometry) and H (Mattauch-Herzog geometry) are simultaneous MS. The time-of-flight (I) is a pulsed MS and while not exhibiting simultaneous detection does simultaneously sample the ions. ICP optical emission spectrometry is represented by D which shows a multichannel system

Table 1.1 List of selected common interferences and backgrounds in ICP-MS analysis (Fricker 2012) and suggested alternative isotopes

Interference form	Interference example	Isotope of interest	Alternative
An^{++}	$^{138}Ba^+$	$^{69}Ga^+$	$^{71}Ga^+$
$AnAr^+$	$^{23}Na^{40}Ar^+$	$^{63}Cu^+$	$^{65}Cu^+$
AnO^+	$^{40}Ar^{16}O^+$	$^{56}Fe^+$	$^{57}Fe^+$
O_2^+	$^{16}O_2^+$	$^{32}S^+$	$^{34}S^+$
An^+	$^{50}Cr^+$	$^{50}Ti^+$	$^{48}Ti^+$
$AnCl^+$	$^{40}Ar^{35}Cl^+$	$^{75}As^+$	HR-ICP-MS
Ar^+, high abundance	$^{40}Ar^+$, $^{80}Ar^+$	$^{40}Ca^+$, $^{80}Se^+$	$^{42}Ca^+$, $^{78}Se^+$

The formation of interferences needs to be evaluated for each analysis, as interferences heavily depend on matrix composition. Analyte (An) can be any element to be detected

Herzog ESA-MSA sector field geometry (Mattauch and Herzog 1934) in combination with an array detector (Cromwell and Arrowsmith 1996; Felton et al. 2011; Fricker 2012; Resano et al. 2012; Schilling et al. 2009). The coupling of TOF-MS to the ICP ion source is complicated due to the continuous ion beam delivered to the ICP and the pulsed nature of the TOF analyzer. However, reports using ICP-TOF-MS have been published (Borovinskaya et al. 2013; Feldmann et al. 2002; Gonzalez et al. 2012; Guilhaus 2000; Mahoney et al. 1996; Myers and Hieftje 1993; Tanner and Günther 2006) and commercially available and custom built instruments are currently both in use. The TOF-MS configuration is potentially very interesting for the simultaneous sampling of all isotopes and therefore very interesting for direct solid analysis by laser ablation. However, early TOF-MS instruments had significantly lower sensitivity when compared to available quadrupole or sector field instruments. Recent developments in TOF-MS have achieved limits of detection comparable with quadrupole ICP-MS (Borovinskaya et al. 2013).

1.2.2 Laser Ablation

The availability of ruby lasers first led to studies which introduced solid aerosols instead of liquid droplets into an ICP ion source. The early development of laser ablation was carried out by coupling LA to an ICP-Optical Emission Spectrometer (Carr and Horlick 1982; Ishizuka and Uwamino 1983; Thompson et al. 1981), which was soon followed by LA-ICP-MS (Arrowsmith 1987; Arrowsmith and Hughes 1988; Gray 1985). The laser is commonly used as a micro-sampling device, with the laser beam focused onto the sample surface using spot sizes of between ~ten to a hundred or more μm diameters. The high-energy laser pulse ablates a small amount of sample, which is transported by a carrier gas as a solid aerosol or gas phase via transport tube into the ICP. In contrast to all other laser-based techniques, in LA-ICP-MS, sampling and aerosol ionization are physically separated from each other. This is one of the major advantages of this technique as 100 % aerosol transport and 100 % vaporization within the ICP allows quantitative analysis without using matrix matched calibration materials.

The types of lasers used for LA-ICP-MS are categorized (Table 1.2) by lasing source, the wavelength of the emitted energy (157–1064 nm), and the pulse duration (approximately nanosecond to femtosecond range). All lasers used for ablation are pulsed lasers. The most common types are solid state lasers (Nd:YAG) and gas lasers (ArF excimer) both of which have ns pulse widths. More recently IR/UV fs Ti:sapphire solid state lasers have also been applied for various applications.

Most solid state lasers, such as the Nd:YAG, emit light in the near infrared (1064 nm) which can be frequency doubled (532 nm), tripled, quadrupled (266 nm), or quintupled (213 nm). However, each step of frequency multiplication results in the loss of energy. Despite the lower energy output, short wavelength lasers are now most often used, as it has been shown that shorter wavelength LA generates aerosols with nm particle sizes favorable for ICP-MS due to more complete vaporization within the plasma (Guillong and Günther 2002). Several reports of the superior qualities of UV LA compared to IR LA have been published (Geertsen et al. 1994; Jeffries et al. 1995, 1998; Shuttleworth 1996). The output energy of an ArF excimer laser is sufficiently high (240 mJ) to enable the use of beam homogenizing optics, leading to flat top laser beam profiles that result in more controlled, homogeneous ablation, and flat bottomed ablation craters. Despite the loss of energy through the additional optical components used, fluencies of more than 40 J/cm^2 can be reached. The use of a homogenized beam improved precision and accuracy in various applications, particularly on highly transparent samples (Günther et al. 1997; Heinrich et al. 2003). Another advantage of the imaging optics is that the beam energy density is independent of the crater size. The schematic optical beam path of a 193 nm ArF laser ablation system is shown in Fig. 1.2. Performing flat top ablation leads to improved signal stability over time, whereas a Gaussian profile crater quickly reaches a critical depth from which little or no aerosol leaves the crater site. The influence of the aspect ratio (depth/diameter) and its significant influence on the representative aerosol sampling has been studied in detail by Mank and Mason (1999).

The influence of carrier gas composition on laser-generated aerosol transport has also been

Table 1.2 Wavelength of the first harmonic (bold type) of various available lasers, and frequency multiplied wavelengths

	IR	VIS	UV
Ti: sapphire, ~30–150 fs	**760–820 nm**		380–410 nm (2nd), 265 nm (3rd)
Nd:YAG, <6 ns	**1064 nm**	532 nm (2nd)	355 nm (3rd), 266 nm (4th), 213 nm (5th)
ArF, 15–20 ns			**193 nm**

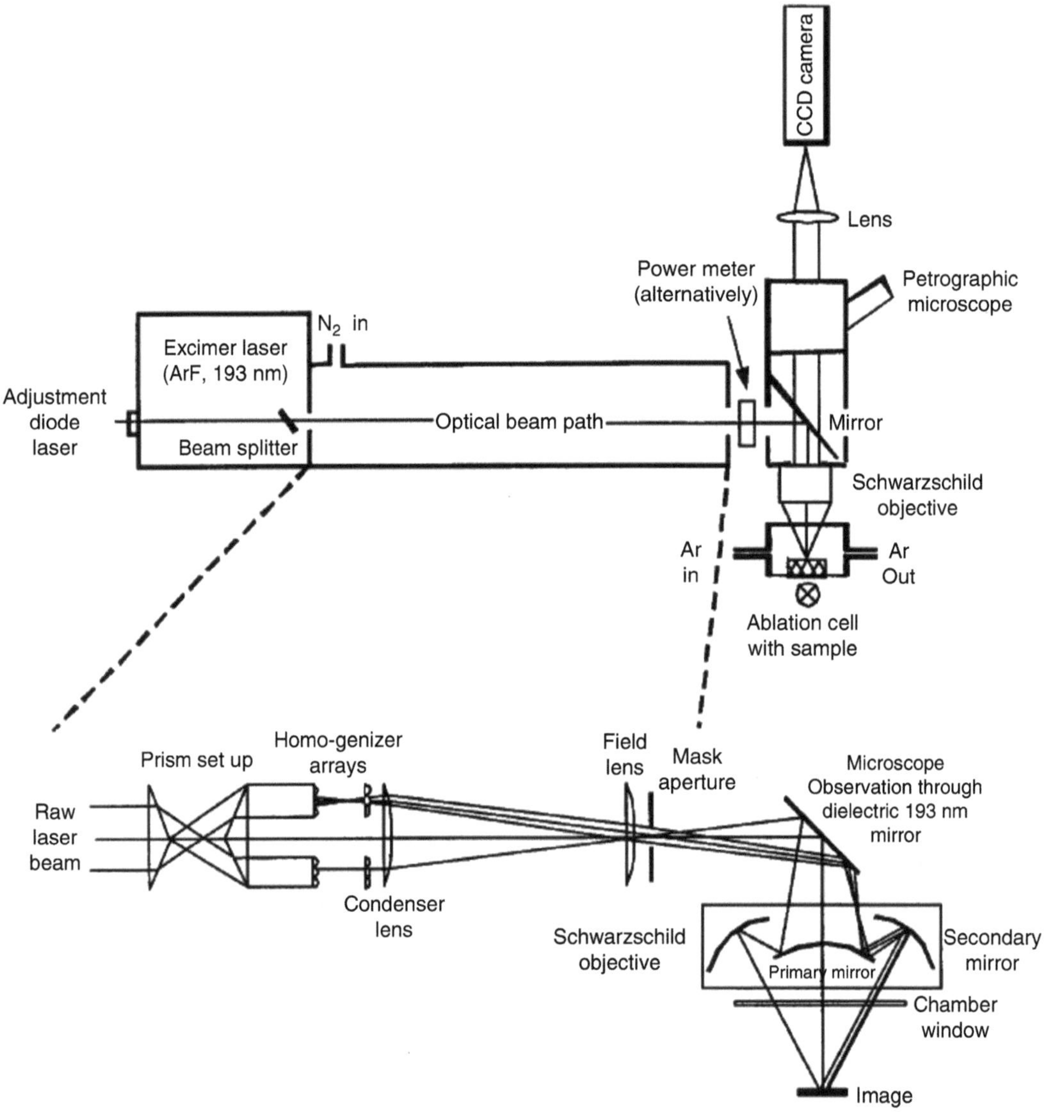

Fig. 1.2 The ArF excimer laser and details of the beam homogenizer (Günther et al. 1997). Reproduced with permission of the Royal Society of Chemistry

studied (Eggins et al. 1998; Günther and Heinrich 1999; Mank and Mason 1999). Eggins et al. (1998) proposed the use of He which allows a more efficient transport of smaller aerosol particles resulting in a three- to fivefold increase in sensitivity and a significant reduction in surface deposition of ablated material around ablation craters. Sensitivity enhancement has also been observed by the addition of small amounts of H_2 to the carrier gas, upstream of the ablation cell (Guillong and Heinrich 2007), which is especially useful for elements with high ionization potentials.

The quantitative analysis of trace concentrations of rare earth elements in a variety of minerals were reported by Jackson et al. (1992), using a 1064 nm Nd:YAG laser. Soon after this study, limitations to the accurate quantitative determination of some analytes were discovered when analyzing a wide variety of elements, including highly volatile ones. This phenomenon was dubbed elemental fractionation

(Fryer et al. 1995), and later shown to originate not only from non-stoichiometric sampling, but also processes taking place during aerosol transport (Koch et al. 2004), and incomplete vaporization in the plasma (Guillong and Günther 2002). Different types of fractionation are mainly caused by the ablation process, and it has been shown that a reduction in laser wavelength towards vacuum ultraviolet (VUV) improves the absorption of laser energy, leading to an aerosol size which can be better transported and vaporized within the ICP. For non-conducting samples, the use of an 193 nm wavelength ArF Excimer laser was shown to reduce elemental fractionation due to improved absorption of the laser by the sample.

This resulted in a more controlled ablation, where controlled means that the rate of ablation was slower and the resulting craters showed reduced amounts of surrounding debris. More homogeneous energy distribution also produced fewer molten particles and flat bottomed craters. Improved ablation behavior was demonstrated on quartz samples when analyzing fluid inclusions in mineral thin sections (Günther et al. 1997, 1998) and many other applications (see Heinrich et al. 2003). Furthermore, the use of He as carrier gas for transporting laser-generated aerosols, first reported by Eggins et al. (1998), led to improved transportation of nm-sized particles and thus to more complete vaporization in the plasma. Other laser systems in the UV range (213 nm and 266 nm, e.g. Jeffries et al. 1998) also became available. The fundamental differences in particle size distribution produced by the three (266, 213, and 193 nm) UV laser wavelengths are reported by Guillong et al. (2003), who showed the advantages of the VUV wavelengths, particularly improved particle size distribution. The importance of particle size distribution has led to studies investigating ICP-induced fractionation on generated aerosol using various laser wavelengths (Guillong and Günther 2002). It has been shown that incomplete vaporization of large particles in the plasma can take place. For example, Guillong et al. (2003) and Kuhn and Günther (2003) demonstrated that larger particle sizes (sizes >150 nm) are not completely vaporized within the ICP. This effect was visualized by Aeschliman et al. (2003), who observed incompletely vaporized particles passing through the plasma. As a result of these issues with fractionation, today most of the laser systems applied for non-conducting samples in geology, medicine, and archaeology are excimer lasers (193 nm) or solid state Nd:YAG lasers operated at 213 nm.

Further developments in laser technology have led to the introduction of femtosecond (fs) lasers as a sampling tool. Studies utilizing fs lasers for LA of metallic samples were first published using Laser-Induced Breakdown Spectroscopy (LIBS) (Margetic et al. 2000). The first application using LA-ICP-MS was Fe isotope ratio determinations performed using a 196 nm fs LA system coupled to a Multi-Collector (MC)-ICP-MS instrument (Horn et al. 2006). The advantages of using fs lasers, in particular when ablating conducting samples, has been demonstrated by analyzing the stoichiometric composition of the aerosols formed during ablation (Garcia et al. 2008, see also a review by Koch and Günther 2011). It has been shown that the elemental composition of the various particles sizes is not equal, which was already known from nanosecond (ns) LA experiments. However, the overall composition of all particle sizes, when combined, represented the stoichiometry of the ablated sample. Given the stoichiometry and knowing that the transport efficiency is comparably high (determined between 78 and 92 %) (Garcia et al. 2008; Wälle et al. 2009) and the fact that fs-LA produces particles small enough to be completely vaporized, it becomes clear that fs-LA is advantageous. Studies on the differences between ns and fs lasers, the expansion of the laser plume and aerosol formation (Koch et al. 2007, 2008a, 2010), transport and transport efficiency of laser-generated aerosols (Garcia et al. 2007, 2008), visualization of the expansion of laser aerosols within the ablation cell (Koch et al. 2008b) and within the transport tubes (Koch et al. 2008a) have been extensively reported. All studies revealed that the expansion of the aerosols is significantly different between ns and fs lasers. However, the original particle size formed during ablation is not so different between ns (193 nm) and fs lasers (265 nm, 780 nm), which indicates that the improved stoichiometry of the generated aerosol must be the major improvement when applying fs lasers for sampling.

In the last 20 years it has been shown that UV-ns laser ablation is not suitable for stoichiometric sampling of conducting samples and/or volatile elements. For example, Chen (1999), Figg et al. (1998), and Outridge et al. (1996) reported temporal changes of element ratios. Fryer et al. (1995) introduced the term *elemental fractionation* (EF) for these temporal changes in elemental ratios. Later, EF indices were revised (Günther et al. 1999) with several other studies investigating fractionation effects using ns and fs LA (Horn et al. 2006; Jackson and Günther 2003; Košler et al. 2005; Kroslakova and Günther 2007; Kuhn and Günther 2003; Longerich et al. 1996a) at different wavelengths (Russo et al. 2000). It has been shown that EF can be significantly reduced for non-conducting samples by using 193 nm lasers when compared to 266 nm Nd:YAG lasers.

The comparatively long pulse width of ns lasers leads to partial heating of conducting samples, which results in non-representative sampling of volatile elements. Therefore, femtosecond lasers with pulse durations smaller than 1 picosecond have been reported to be more suitable for LA-ICP-MS (Bian et al. 2005, 2006; Horn et al. 2006; Koch et al. 2007). Major advantages of fs LA have been reported for the analysis of metallic samples (Mozna et al. 2006; Wiltsche and Günther 2010). This improvement is attributed to the short pulse width of fs lasers, on the order of femtoseconds, being sufficiently short to reduce heat effects in the ablation zone (Koch and Günther 2011; Russo et al. 2002b), thereby decreasing elemental fractionation. However, the use of fs-LA for the analysis of non-conducting materials did not result in superior results as reported by Birbaum (2011) and Glaus et al. (2010). More studies are currently being performed to provide conclusive information.

Further studies have focused on other aspects of laser ablation, for instance the impact of different transport tube materials (Kovacs and Günther 2008) and the use of rectangular rather than circular craters (Eggins et al. 1998). A more recent fundamental study investigated the formation of gas phases during the ablation of carbon containing materials (Frick and Günther 2012), as carbon is one of the most frequently used internal standards for the analysis of tissue, paints, and carbonates (Becker et al. 2008; Feldmann et al. 2002).

1.2.3 Ablation Cells

The ablation cell is amongst the most important parts influencing the performance of LA-ICP-MS and several requirements are indispensable for a routinely used ablation cell. An ablation cell must be airtight, preventing atmospheric gases from entering and extinguishing the ICP, as well as preventing loss of carrier gas and aerosol. Secondly, the carrier gas flowing through the cell must transport as much of the laser-generated aerosol as possible (ideally 100 %) to the ICP. Finally, containment and mounting of at least one sample and one external standard must be possible.

However, generally a number of samples and standards are mounted simultaneously, and it must therefore be possible to position the ablation cell while observing sample positioning. Most commercially available ablations cells meet these requirements, and a wide variety of ablation cells have been developed. However, for more demanding applications than bulk analysis, such as elemental depth profiling (Pisonero et al. 2007; Plotnikov et al. 2001) and/or imaging (Woodhead et al. 2007) or the analysis of fluid inclusions (Heinrich et al. 2003; Pettke et al. 2012), these requirements, particularly for aerosol washout, are more stringent.

The spatial resolution of transient signals generated from line scans is determined by crater diameter, scan speed, and aerosol washout time. For bulk analysis of homogeneous samples the washout time is less critical, as the entire signal of the sample is integrated to calculate elemental concentrations. On the other hand, when performing line scans on heterogeneous samples, the transient signal should reflect the variation of element concentration within the sample. Therefore, a long washout (high dispersion) leads to mixing of aerosol originating from different laser pulses at different locations along the line scan, resulting in distorted signals. As a consequence, spatial information is lost and can only be made available by using complex deconvolution algorithms (Plotnikov et al. 2008). In addition, long washout and mixing of the aerosol may lead to cross contamination, which has a significant influence on the accuracy of the analysis.

Early ablation cells were developed for LA-Graphite Furnace-Atomic Absorption Spectrometry (GF-AAS) (Schrön et al. 1983) and LA-ICP-Atomic Emission Spectrometry (AES) (Carr and Horlick 1982; Ishizuka and Uwamino 1983; Thompson et al. 1981) but are in principle similar to those used for LA-ICP-MS. An early study by Arrowsmith and Hughes (1988) reported a "cell in the cell" approach in order to entrain and transport laser-generated aerosol. This approach was further developed by several groups (Garcia et al. 2007; Müller et al. 2009). Müller and colleagues suggested the use of an aerosol mixing device, placed between the ablation cell and the ICP for smoothing signal. The use of a Volume-Optional Low Memory (VOLM) ablation cell, which uses two cell volumes, has been reported (Liu et al. 2007). These approaches are based on a sample holder and an insert of a small volume cylinder, which is moved together with the laser beam to the location where LA is performed. These cells can host a large number of samples, show less surface contamination between different samples and have reduced signal dispersion.

Fundamental studies on aerosol transport in the ablation cell and transport tubing using powder deposition have been reported (Bleiner and Günther 2001), and reproduced using gas flow modelling studies. The authors report that when using extremely small ablation cells, transport efficiency was limited due to particle-wall interactions. Bleiner and Altorfer (2005) introduced the use of a rotating nozzle reducing the dead volume in a large ablation cell, with the inlet gas sweeping the entire cell volume. A cyclonic flux cell, sweeping the carrier gas from the ablation cell walls to the centre, has been proposed (Monticelli et al. 2009). Tangential or circular gas flows have also been reported by Pisonero et al. (2006) using a High Efficiency Aerosol Dispersion (HEAD) cell. This principle was based on two volumes, in combination with a Venturi-effect extraction. Some approaches to ablation cell design are based on the geometry of an enlarged transport tube resulting in a fast washout for relatively large samples (Gurevich and Hergenröder 2007). Small ablation cells based on the same principle have also been reported, providing shorter washout times (Bi et al. 2000; Horstwood et al. 2003; Wang et al. 2013).

Ultimately, the fast aerosol washout of in-tube ablation can only be made shorter by using an in-torch ablation such as reported by Tanner and Günther (2005, 2006; Tanner 2007), which had severe limitations in sample handling, prohibiting many applications in routine analysis. It must be mentioned that such fast washout cells are only applicable when simultaneous detection of the signals becomes possible, e.g. using an ICP-TOF-MS.

In addition to attempts to reduce washout times using in-torch ablation, further work has focused on ablation cell development for both routine analysis and specialized applications. Analysis of large samples is usually done by cutting the sample, in order to fit it into the small volume ablation cells that are currently available. As the most common applications of LA-ICP-MS have been analyses of geological samples, e.g. minerals in rock thick or thin sections; the volume of ablation cells has not been very critical. These samples are in most cases small enough (a few cm) to be hosted in traditional ablation cells that have a volume of a few hundred cm^3, which includes the sample holder and the effective aerosol transport volume.

However, for many types of samples it is simply impractical or impossible to employ destructive analysis, e.g. precious objects or museum artefacts (see Chap. 4 for instance). Therefore, ablation cell development has focused on various so-called "open configurations." An early approach allowing high flexibility with respect to object size was an open ablation cell with a plasticine seal to prevent carrier gas containing the laser-generated aerosol from being lost (Günther and Gäckle 1988) and applied for LA-ICP-OES. This type of cell has also been applied by Devos et al. (1999) for analysing silver objects. Some limitations of this general idea were later improved upon by Wagner and Jędral (2011), using an ablation cell without a bottom and applying adhesive materials to attach the cell to a sample. Both of these designs are not truly open, as they form an airtight configuration using the sample as part of the cell. The only open and so called "non-contact" ablation cell has been reported by Asogan et al. (2009). This ablation cell uses several concentric gas curtains to exclude atmospheric gases from entering the carrier gas flow. The set-up requires close proximity of the cell to the sample surface and more importantly, it requires a very flat and even sample surface so that the gas curtain seals the inner cell completely. Therefore, a wider application of this cell to different archaeological artifacts is partially restricted to very flat objects.

The literature on ablation cells is extensive, and some very specialized cell designs have been developed—one example is shown in Fig. 1.3 (Fricker 2012), which allows the analysis of laminated sediments of stalagmite samples with reduced sample destruction. Further examples can be found in (Feldmann et al. 2002; Gonzalez et al. 2012; Müller et al. 2009, 2011; Reinhardt et al. 2001, 2003).

The size and shape of archaeological samples would benefit from a sampling approach without any ablation cell. A first possible strategy has been demonstrated using a Gas Exchange Device (GED) (Dorta et al. 2013; Kovacs et al. 2010) and was further optimized by Tabersky et al. (2013). In these devices, the laser aerosol is sampled in air and sucked into a gas exchange device that exchanges atmospheric gases (air, He) for Ar without affecting the laser-generated aerosol. The exchange process allows flow rates of 0.8 L/min, which is a commonly used flow rate in LA-ICP-MS. This approach allows removing the ablation cell completely (see Fig. 1.4). Preliminary studies showed a 20–30 % reduction in sensitivity when using such a device. The use of a plume entrainment device, introduced by Tabersky et al. (2013), significantly improves aerosol sampling in combination with a GED. This open ablation GED configuration is most valuable for archaeological objects, since no sample preparation, cutting or contact with a sample is required. Quantitative results have been reported for various gold objects and steel samples and the results indicated that a closed

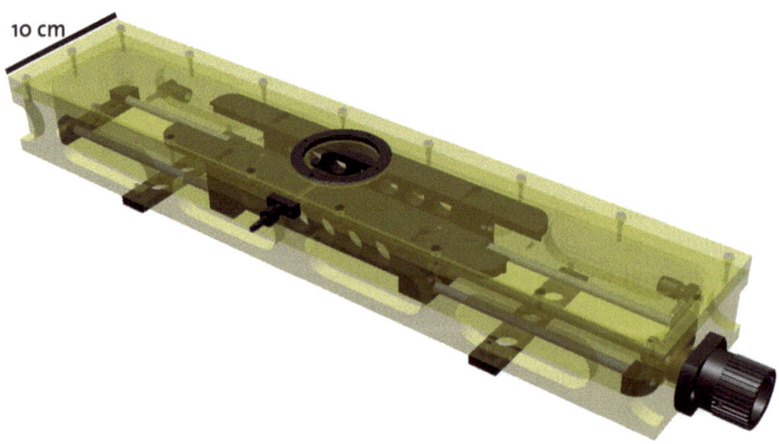

Fig. 1.3 Perspective view of a low dispersion high capacity laser ablation cell as described in Fricker (2012)

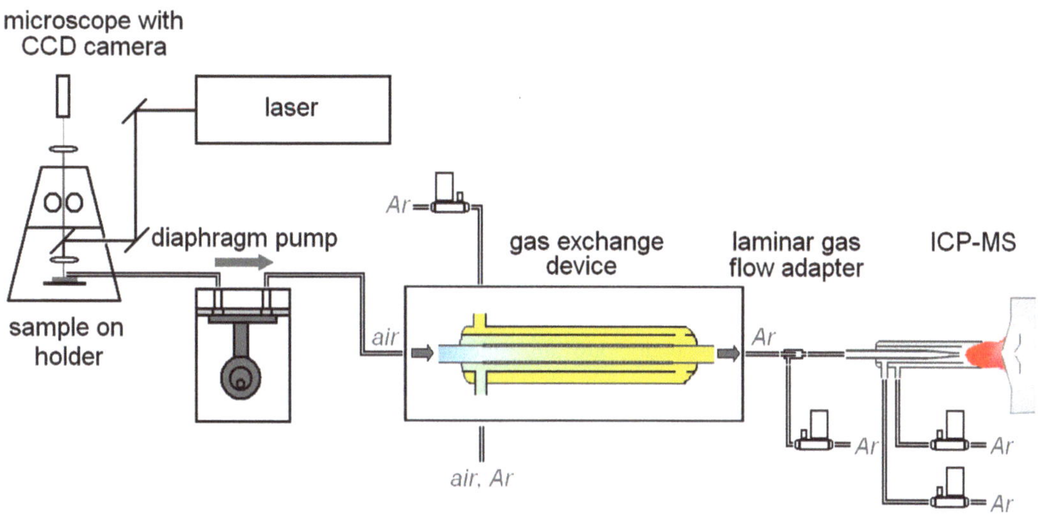

Fig. 1.4 Laser ablation-GED-ICP-MS setup for direct laser sampling in air followed by online ICP-MS measurements of the laser-generated aerosols as described in Kovacs et al. (2010) and further advanced and reported by Tabersky et al. (2013). Reproduced with permission of the Royal Society of Chemistry

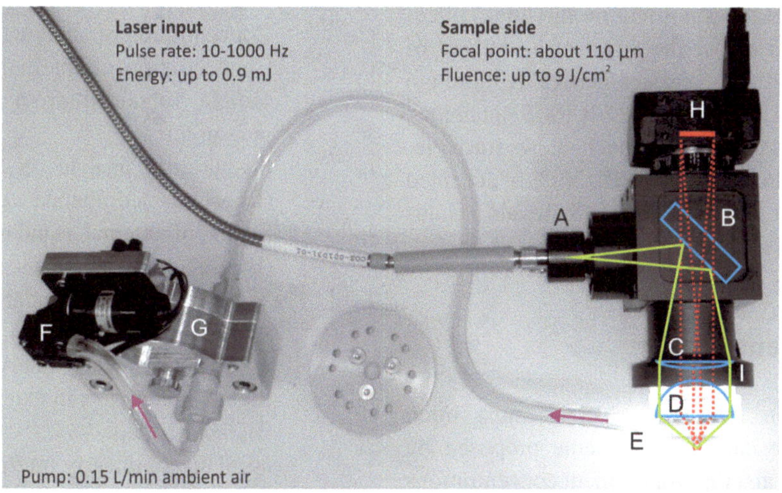

Fig. 1.5 Portable laser ablation system (Glaus et al. 2013). The figure shows the sampling head connected to the diode pumped solid laser (not shown) via fiber optics, which can be placed on any object. The generated aerosols are pumped via a diaphragm pump to a filter collector (Figure is reproduced with permission of the Royal Society of Chemistry)

cell configuration and the "in air" ablation procedure using the GED provide very similar results.

Another more recently introduced approach of interest for archeometrical studies is the concept of a portable laser ablation system (pLA), as introduced by Glaus et al. (2012, 2013). The system weighs 16 kg and can be transported to samples outside the laboratory. The sampling strategy is based on a diode pumped laser ablation system coupled to a sampling device (see Fig. 1.5). The aerosol generated by the laser is sucked from the ablation site via a diaphragm pump onto a filter. A rotating filter plate can host up to 12 filters, which can be stored and finally analyzed in the laboratory. The samples can be processed by complete digestion of the filter and solution nebulization ICP-MS or MC-ICP-MS. Alternatively, the filter can also be analyzed by LA-ICP-MS. This sampling strategy allows access to elemental composition of the ablated sample and/or isotope determinations. One requirement is that filter blanks are also collected. To do so, environmental air at the sampling site is sampled for a similar period as the aerosol collection. This strategy has been tested and validated on pigments and on copper objects. The limits of detection are an order of magnitude above those commonly obtained using a normal LA-ICP-MS setup. However, limits of detection below 1 mg/kg were achieved for most elements, and are therefore still below the detection limits of other direct sampling techniques such as X-ray Fluorescence (XRF) and Laser-Induced Breakdown Spectroscopy (LIBS). This portable instrument configuration can also be used in the lab in combination with the GED, does not require the filter collection step, and allows the direct measurement of samples of all sizes and geometries. A thorough validation of the pLA-GED-ICPMS technique demonstrated that this approach allows trace element determinations within 10–20 % of accepted values when using reference materials (Glaus et al. 2012; Tabersky et al. 2013).

1.3 Quantification

Quantification in LA-ICP-MS is most often based on the calculation scheme proposed by Longerich et al. (1996b). Element concentrations are calculated using external and internal standardization (see Eq. 1.1). To correct for instrumental drift, the external standard signal intensities are interpolated between two external standards analyzed before and after the analyses of the unknown samples. All intensities are background corrected using a "gas background" acquired with the laser not firing on the sample. This "gas background" is commonly acquired for a period of 30 s prior to the acquisition of the analyte signal. Matrix effects and aerosol transport, that is the ratio of the sensitivity (signal per unit concentration) in the external calibration material and the sample is corrected using a "naturally" occurring internal standard (e.g. Ca), which must be known in both the sample of interest and the calibration material. The concentration of the internal standard in the unknown sample can be determined using an alternative analytical technique, most commonly Electron Probe Micro-Analysis (EPMA). In some cases it is also possible to estimate the internal standard concentration from the stoichiometry of the sample. A variant internal standardization procedure uses normalization to 100 % oxide weight. Background-subtracted intensities are used to calculate concentrations using Eq. (1.1).

$$C_X^{SMP} = C_{IS}^{SMP} \times \frac{I_X^{SMP}}{I_{IS}^{SMP}} \times \frac{I_{IS}^{ES}}{I_X^{ES}} \times \frac{C_X^{ES}}{C_{IS}^{ES}} \qquad (1.1)$$

Where:

${C_X}^{SMP}$	Concentration of analyte x in sample
${C_{IS}}^{SMP}$	Concentration of internal standard in sample (obtained from independent measurements)
${I_X}^{SMP}$	Signal of analyte x in sample (BG-subtracted, abundance-normalized)
${I_{IS}}^{SMP}$	Signal of internal standard in sample (BG-subtracted, abundance-normalized)
${I_{IS}}^{ES}$	Signal of internal standard in external standard (BG-subtracted, abundance-normalized, linearly interpolated)
${I_X}^{ES}$	Signal of analyte x in external standard (BG-subtracted, abundance-normalized, linearly interpolated)
${C_X}^{ES}$	Concentration of analyte x in external standard (certificate of ES)
${C_{IS}}^{ES}$	Concentration of internal standard in external standard (certificate of ES).

For bulk quantification the background-subtracted mean signals (SG) are used in Eq. (1.1), whereas for imaging quantification or depth profiling each data point, or integration over a few data values, of the SG interval (background-subtracted) is quantified individually to obtain a concentration profile. In cases

Fig. 1.6 Calculation schemes for the limits of detection estimation. DT refers to dwell time

	BULK	BORE
ND **Normal distribution**	$LOD_x = \frac{3 \cdot \sigma_{BG,x}}{S_x} \sqrt{\frac{1}{m} + \frac{1}{n}}$	$LOD_x = \frac{3 \cdot \sigma_{BG,x}}{S_x}$
PD **Poisson distribution**	$LOD_x = \frac{3.29 \cdot \sqrt{\mu_{counts,BG,x}} \cdot \frac{1}{\sqrt{n}} + 2.71}{S_x \cdot DT_x}$	$LOD_x = \frac{3.29 \cdot \sqrt{\mu_{counts,BG,x}} + 2.71}{S_x \cdot DT_x}$

$\sigma_{BG,x}$	BG standard deviation of analyte x
S_x	Average Sample Sensitivity of analyte x
N	# of data points of BG
M	# of data points of SG
$\mu_{counts,BG,x}$	Mean cps_{BG}*DT_x (after Poisson: $\sigma^2=\mu$)

where the presence of spikes (related to the sample) limits data quality, it is also possible to use the median instead of average signals, which is more robust against outliers than the mean.

Ongoing discussions on estimating limits of detection (LOD) using either the normal distribution (ND) or Poisson distribution (PD) have led to the use of different formulas. The LOD in LA-ICP-MS, and in any elemental analysis technique, is a function of the uncertainty (standard deviation) of the background (blank) measurement, the estimated uncertainty in a theoretical sample with zero concentration of the analyte, and the instrument sensitivity along with its estimated uncertainty. Due to the very low backgrounds (a few counts) for many analytes in ICP-MS, a large portion of the uncertainty in the background is from Poisson counting statistics. Thus in some analytical situations an estimate of background uncertainty derived from Poisson statistics alone can be a good approach. As it is common in LA-ICP-MS to have close to or zero background signal for heavy isotopes, it is important to employ suitable statistics to estimate the standard deviation of the background. This is important for several reasons, including the non-Gaussian distribution obtained when considering a small number of events. For zero or close to zero counts the standard deviation of the background follows, to a large extent, counting statistics, i.e. a Poisson distribution. The two major ablation schemes are considered (bulk and bore). "Bore" considers each data point during drilling into a sample or during scanning on a surface, which means n is equal to one. The calculation schemes for LOD estimation are listed in Fig. 1.6.

The quantification scheme shown above is based on one single external standard and significantly different from the procedure from solution nebulization ICP-MS, where many standard solutions, including matrix matched ones, can be prepared to establish the calibration function. The lack of suitable calibration materials for LA-ICP-MS is one of the major limitations of this technique since the material has to fulfill certain requirements. The ablation behavior of the sample of interest and the calibration material should be similar, which means that similar amounts of aerosol should enter the ICP for complete vaporization (Kroslakova and Günther 2007). Furthermore it is required that the calibration material should be homogeneous on the scale of ablation, which is between 5 and 200 μm. This is rather difficult to achieve by mixing powders or adding solutions to powder samples. Therefore, various approaches to production of in-house prepared calibration materials have been reported based on fusion (Jochum et al. 2011). Due to problems with homogeneity when using alternative preparation techniques, glasses of the NIST series 61× are currently the most frequently applied calibration materials in LA-ICP-MS when analyzing non-conducting samples, due to the large number of elements present in various concentrations and the spatial homogeneity of those elements. Exceptions have been reported by Eggins and Shelley (2002). However, a variety of other glass standards (e.g. MPDING, USGS

GSA-GSE) are also available. For conducting samples more calibration materials (e.g. brass, bronze, copper) are available and need to be more closely matrix matched for quantification, as the ablation process using ns lasers can cause severe fractionation effects hampering quantification.

1.4 Summary and Outlook

Laser ablation-inductively coupled plasma-mass spectrometry (LA-ICP-MS) is a very promising techniques for elemental analysis and has gained a lot of attention during the last decade in a wide variety of applications. Current commercially available laser technology has been tested on a variety of samples and matrix types and it can be concluded that the precision and accuracy of the technique has significantly improved over time. The introduction of VUV ns lasers has reduced particle size distribution and helped to improve the transport efficiency of generated aerosols. Reduction in laser pulse widths with femtosecond lasers has further extended the applications of the technique, especially for conducting samples such as metals. The last decade has shown that a number of applications provide accurate quantitative data using non-matrix matched calibration for quantification.

Major improvements for archaeological samples can be seen in the development of sample size and geometry independent sampling strategies using a newly implemented gas exchange device into LA-ICP-MS. Sampling the aerosol in air and sucking the laser-generated aerosols directly into the ICP-MS will enable analysis of large and valuable samples without any further sample preparation and without direct contact to the sample, which is in most cases a requirement to obtain access to the samples.

The use of small, mobile laser ablation units (diode pumped pulsed laser at 532 nm) with glass-fiber optics is another alternative laser ablation system for archaeological applications. Such a portable laser can be taken into the field for sampling the laser-generated aerosol onto a filter for subsequent analysis in the laboratory. Such a system provides access to samples which are difficult to move (e.g. museums objects) and can be used directly in the field.

Acknowledgements The authors would like to thank Henry Longerich for fruitful discussion and comments on this manuscript.

References

Aeschliman DB, Bajic SJ, Baldwin DP, Houk RS (2003) High-speed digital photographic study of an inductively coupled plasma during laser ablation: comparison of dried solution aerosols from a microconcentric nebulizer and solid particles from laser ablation. J Anal Atom Spectrom 18:1008–1014

Arnold DE, Bohor BF, Neff H, Feinman GM, Williams PR, Dussubieux L, Bishop R (2012) The first direct evidence of pre-columbian sources of palygorskite for Maya Blue. J Archaeol Sci 39:2252–2260

Arrowsmith P (1987) Laser ablation of solids for elemental analysis by inductively coupled plasma mass spectrometry. Anal Chem 59:1437–1444

Arrowsmith P, Hughes SK (1988) Entrainment and transport of laser ablated plumes for subsequent elemental analysis. Appl Specstroc 42:1231–1239

Asogan D, Sharp BL, Connor CJPO, Green DA, Hutchinson RW (2009) An open, non-contact cell for laser ablation-inductively coupled plasma-mass spectrometry. J Anal Atom Spectrom 24:917–923

Barca D, de Francesco AM, Mirocle Crisci G (2007) Application of Laser Ablation ICP-MS for characterization of obsidian fragments from peri-Tyrrhenian area. J Cult Herit 8:141–150

Becker JS, Zoriy MV, Pickhardt C, Palomero-Gallagher-N, Zilles K (2005) Imaging of copper, zinc, and other elements in thin section of human brain samples (hippocampus) by laser ablation inductively coupled plasma mass spectrometry. Anal Chem 77:3208–3216

Becker JS, Dietrich RC, Matusch A, Pozebon D, Dressler VL (2008) Quantitative images of metals in plant tissues measured by laser ablation inductively coupled plasma mass spectrometry. Spectrochim Acta B 63:1248–1252

Bi M, Ruiz AM, Gornushkin I, Smith BW, Winefordner JD (2000) Profiling of patterned metal layers by laser ablation inductively coupled plasma mass spectrometry (LA-ICPMS). Appl Surf Sci 158:197–204

Bian QZ, Koch J, Lindner H, Berndt H, Hergenröder R, Niemax K (2005) Non-matrix matched calibration using near-IR femtosecond laser ablation inductively coupled plasma optical emission spectrometry. J Anal Atom Spectrom 20:736–740

Bian Q, Garcia CC, Koch J, Niemax K (2006) Non-matrix matched calibration of major and minor concentrations of Zn and Cu in brass, aluminium and

silicate glass using NIR femtosecond laser ablation inductively coupled plasma mass spectrometry. J Anal Atom Spectrom 21:187–191
Birbaum K (2011) Capabilities and limitations of inductively coupled plasma mass spectrometry-analyses on engineered and laser-generated nanoparticles. Ph.D. Dissertation, ETH Zurich
Bleiner D, Altorfer H (2005) A novel gas inlet system for improved aerosol entrainment in laser ablation inductively coupled plasma mass spectrometry. J Anal Atom Spectrom 20:754–756
Bleiner D, Günther D (2001) Theoretical description and experimental observation of aerosol transport processes in laser ablation inductively coupled plasma mass spectrometry. J Anal Atom Spectrom 16:449–456
Borovinskaya O, Hattendorf B, Tanner M, Gschwind S, Günther D (2013) A prototype of a new inductively coupled plasma time-of-flight mass spectrometer providing temporally resolved, multi-element detection of short signals generated by single particles and droplets. J Anal Atom Spectrom 28:226–233
Campbell AJ, Humayun M (1999) Trace element microanalysis in iron meteorites by laser ablation ICPMS. Anal Chem 71:939–946
Carr JW, Horlick G (1982) Laser vaporization of solid metal samples into an inductively coupled plasma. Spectrochim Acta B 37:1–15
Chen Z (1999) Inter-element fractionation and correction in laser ablation inductively coupled plasma mass spectrometry. J Anal Atom Spectrom 14:1823–1828
Cottle JM, Horstwood MSA, Parrish RR (2009) A new approach to single shot laser ablation analysis and its application to in situ Pb/U geochronology. J Anal Atom Spectrom 24:1355–1363
Cromwell EF, Arrowsmith P (1996) Novel multichannel plasma-source mass spectrometer. J Am Soc Mass Spectrom 7:458–466
Devos W, Moor C, Lienemann P (1999) Determination of impurities in antique silver objects for authentication by laser ablation inductively coupled plasma mass spectrometry (LA-ICPMS). J Anal Atom Spectrom 14:621–626
Dorta L, Kovacs R, Koch J, Nishiguchi K, Utani K, Günther D (2013) Determining isotope ratios using laser ablation sampling in air with MC-ICPMS. J Anal Atom Spectrom 28:1513–1521
Durrant SF (1999) Laser ablation inductively coupled plasma mass spectrometry: achievements, problems, prospects. J Anal Atom Spectrom 14:1385–1403
Dussubieux L, Golitko M, Williams PR, Speakman RJ (2007) LA-ICP-MS analysis applied to the characterization of Peruvian Wari ceramics. In: Glascock MD, Speakman RJ, Popelka-Filcoff RS (eds) Archaeological chemistry: analytical technique and archaeological interpretation. American Chemical Society, Washington, DC, pp 349–363
Dussubieux L, Deraisme A, Frot G, Stevenson C, Creech A, Bienvenu Y (2008) LA-ICP-MS, SEM-EDS and EPMA analysis of eastern north american copper-based artefacts: impact of corrosion and heterogeneity on the reliability of the LA-ICP-MS compositional results. Archaeometry 50:643–657
Eggins SM, Shelley JMG (2002) Compositional heterogeneity in NIST SRM 610-617 glasses. Geostandard Newslett 26:269–286
Eggins SM, Kinsley LPJ, Shelley JMG (1998) Deposition and element fractionation processes during atmospheric pressure laser sampling for analysis by ICP-MS. Appl Surf Sci 127–129:278–286
Feldmann J, Kindness A, Ek P (2002) Laser ablation of soft tissue using a cryogenically cooled ablation cell. J Anal Atom Spectrom 17:813–818
Felton JA, Schilling GD, Ray SJ, Sperline RP, Denton MB, Barinaga CJ, Koppenaal DW, Hieftje GM (2011) Evaluation of a fourth-generation focal plane camera for use in plasma-source mass spectrometry. J Anal Atom Spectrom 26:300–304
Figg DJ, Cross JB, Brink C (1998) More investigations into elemental fractionation resulting from laser ablation–inductively coupled plasma–mass spectrometry on glass samples. Appl Surf Sci 127–129:287–291
Fontaine G, Hametner K, Peretti A, Günther D (2010) Authenticity and provenance studies of copper-bearing andesines using Cu isotope ratios and element analysis by fs-LA-MC-ICPMS and ns-LA-ICPMS. Anal Bioanal Chem 398(7–8):2915–2928
Frick DA, Günther D (2012) Fundamental studies on the ablation behaviour of carbon in LA-ICPMS with respect to the suitability as internal standard. J Anal Atom Spectrom 27:1294–1303
Fricker MB (2012) Design of ablation cells for LA-ICPMS: from modeling to high spatial resolution analysis applications. Ph.D. Dissertation, ETH Zurich
Fryer BJ, Jackson SE, Longerich HP (1995) The design, operation and role of the laser-ablation microprobe coupled with an inductively coupled plasma; mass spectrometer (LAM-ICPMS) in the earth sciences. Can Mineral 33:303–312
Garcia CC, Lindner H, Niemax K (2007) Transport efficiency in femtosecond laser ablation inductively coupled plasma mass spectrometry applying ablation cells with short and long washout times. Spectrochim Acta B 62:13–19
Garcia CC, Wälle M, Lindner H, Koch J, Niemax K, Günther D (2008) Femtosecond laser ablation inductively coupled plasma mass spectrometry: transport efficiencies of aerosols released under argon atmosphere and the importance of the focus position. Spectrochim Acta B 63:271–276
Geertsen C, Briand A, Chartier F, Lacour J-L, Mauchien P, Sjöström S, Mermet J-M (1994) Comparison between infrared and ultraviolet laser ablation

at atmospheric pressure—implications for solid sampling inductively coupled plasma spectrometry. J Anal Atom Spectrom 9:17–22

Glaus R, Kaegi R, Krumeich F, Günther D (2010) Phenomenological studies on structure and elemental composition of nanosecond and femtosecond laser-generated aerosols with implications on laser ablation inductively coupled plasma mass spectrometry. Spectrochim Acta B 65:812–822

Glaus R, Koch J, Günther D (2012) A portable laser ablation sampling device for elemental fingerprinting of objects outside the laboratory with laser ablation inductively coupled plasma mass spectrometry. Anal Chem 84:5358–5364

Glaus R, Dorta L, Zhang Z, Ma Q, Berke H, Günther D (2013) Isotope ratio determination of objects in the field by portable laser ablation sampling and subsequent multicollector ICPMS. J Anal Atom Spectrom 28:801–809

Gonzalez J, Oropeza DD, Mao X, Longerich HP, Russo RE (2012) Rapid bulk analysis using femtosecond laser ablation inductively coupled plasma Time-of-Flight mass spectrometry. J Anal Atom Spectrom 27:1405–1412

Gratuze B (1999) Obsidian characterization by laser ablation ICP-MS and its application to prehistoric trade in the Mediterranean and the Near East: sources and distribution of Obsidian within the Aegean and Anatolia. J Archaeol Sci 26:869–881

Gray AL (1985) Solid sample introduction by laser ablation for inductively coupled plasma source mass spectrometry. Analyst 110:551–556

Guilhaus M (2000) Essential elements of time-of-flight mass spectrometry in combination with the inductively coupled plasma ion source. Spectrochim Acta B 55:1511–1525

Guillong M, Günther D (2002) Effect of particle size distribution on ICP-induced elemental fractionation in laser ablation-inductively coupled plasma-mass spectrometry. J Anal Atom Spectrom 17:831–837

Guillong M, Heinrich CA (2007) Sensitivity enhancement in laser ablation ICP-MS using small amounts of hydrogen in the carrier gas. J Anal Atom Spectrom 22:1488–1494

Guillong M, Horn I, Günther D (2003) A comparison of 266 nm, 213 nm and 193 nm produced from a single solid state Nd:YAG laser for laser ablation ICP-MS. J Anal Atom Spectrom 18:1224–1230

Günther D, Gäckle M (1988) Laser-Mikro-ICP-Spektrometrie – Integrale Messungen an verschiedenen Matrizes. Z Chem 28:227–228

Günther D, Heinrich CA (1999) Enhanced sensitivity in laser ablation-ICP mass spectrometry using helium-argon mixtures as aerosol carrier. J Anal Atom Spectrom 14:1363–1368

Günther D, Frischknecht R, Heinrich CA, Kahlert HJ (1997) Capabilities of an argon fluoride 193 nm excimer laser for laser ablation inductively coupled plasma mass spectrometry microanalysis of geological materials. J Anal Atom Spectrom 12:939–944

Günther D, Audétat A, Frischknecht R, Heinrich CA (1998) Quantitative analysis of major, minor and trace elements in fluid inclusions using laser ablation–inductively coupled plasma mass spectrometry. J Anal Atom Spectrom 13:263–270

Günther D, Jackson SE, Longerich HP (1999) Laser ablation and arc/spark solid sample introduction into inductively coupled plasma mass spectrometers. Spectrochim Acta B 54:381–409

Gurevich EL, Hergenröder R (2007) A simple laser ICPMS ablation cell with wash-out time less than 100 ms. J Anal Atom Spectrom 22:1043–1050

Halliday AN, Lee D-C, Christensen JN, Rehkämper M, Yi W, Luo X, Hall CM, Ballentine CJ, Pettke T, Stirling C (1998) Applications of multiple collector-ICPMS to cosmochemistry, geochemistry, and paleoceanography. Geochim Cosmochim Acta 62:919–940

Hattendorf B, Günther D (2001) Experimental evidence for the formation of doubly charged polyatomic ions in inductively coupled plasma mass spectrometry. Fresenius J Anal Chem 370(5):483–487

Hattendorf B, Latkoczy C, Günther D (2003) Laser ablation-ICPMS. Anal Chem 75:341A–347A

Heinrich CA, Pettke T, Halter W, Aigner M, Audetat A, Günther D, Hattendorf B, Bleiner D, Guillong M, Horn I (2003) Quantitative multi-element analysis of minerals, fluid and melt inclusions by Laser Ablation Inductively Coupled Plasma Mass Spectrometry. Geochim Cosmochim Acta 67:3473–3497

Horn I, von Blanckenburg F, Schoenberg R, Steinhoefel G, Markl G (2006) In situ iron isotope ratio determination using UV-femtosecond laser ablation with application to hydrothermal ore formation processes. Geochim Cosmochim Acta 70: 3677–3688

Horstwood MSA, Foster GL, Parrish RR, Noble SR, Nowell GM (2003) Common-Pb corrected in situ U–Pb accessory mineral geochronology by LA-MC-ICPMS. J Anal Atom Spectrom 18:837–846

Houk RS, Fassel VA, Flesch GD, Svec HJ, Gray AL, Taylor CE (1980) Inductively coupled argon plasma as an ion source for mass spectrometric determination of trace elements. Anal Chem 52:2283–2289

Ishizuka T, Uwamino Y (1983) Inductively coupled plasma emission spectrometry of solid samples by laser ablation. Spectrochim Acta B 38:519–527

Jackson SE, Günther D (2003) The nature and sources of laser induced isotopic fractionation in laser ablation-multicollector-inductively coupled plasma-mass spectrometry. J Anal Atom Spectrom 18:205–212

Jackson SE, Longerich HP, Dunning GR, Fryer BJ (1992) The application of laser-ablation microprobe; inductively coupled plasma-mass spectrometry (LAM-ICPMS) to in situ trace-element determinations in minerals. Can Mineral 30:1049–1064

Jeffries TE, Perkins WT, Pearce NJG (1995) Comparisons of infrared and ultraviolet laser probe microanalysis inductively coupled plasma mass spectrometry in mineral analysis. Analyst 120:1365–1371

Jeffries TE, Jackson SE, Longerich HP (1998) Application of a frequency quintupled Nd:YAG source (lambda = 213 nm) for laser ablation inductively coupled plasma mass spectrometric analysis of minerals. J Anal Atom Spectrom 13:935–940

Jenner GA, Longerich HP, Jackson SE, Fryer BJ (1990) ICPMS-A powerful tool for high-precision trace-element analysis in Earth sciences: evidence from analysis of selected U.S.G.S. reference samples. Chem Geol 83:133–148

Jochum KP, Weis U, Stoll B, Kuzmin D, Yang Q, Raczek I, Jacob DE, Stracke A, Birbaum K, Frick DA, Günther D, Enzweiler J (2011) Determination of reference values for NIST SRM 610-617 glasses following ISO guidelines. Geostand Geoanal Res 35:397–429

Koch J, Günther D (2011) Review of the state-of-the-art of laser ablation inductively coupled plasma mass spectrometry. Appl Spectrosc 65:155A–162A

Koch J, von Bohlen A, Hergenröder R, Niemax K (2004) Particle size distributions and compositions of aerosols produced by near-IR-femto- and nanosecond laser ablation of brass. J Anal Atom Spectrom 19:267–272

Koch J, Schlamp S, Rösgen T, Fliegel D, Günther D (2007) Visualization of aerosol particles generated by near infrared nano- and femtosecond laser ablation. Spectrochim Acta B 62:20–29

Koch J, Wälle M, Dietiker R, Günther D (2008a) Analysis of laser-produced aerosols by inductively coupled plasma mass spectrometry: transport phenomena and elemental fractionation. Anal Chem 80:915–921

Koch J, Wälle M, Schlamp S, Rösgen T, Günther D (2008b) Expansion phenomena of aerosols generated by laser ablation under helium and argon atmosphere. Spectrochim Acta B 63:37–41

Koch J, Heiroth S, Lippert T, Günther D (2010) Femtosecond laser ablation: visualization of the aerosol formation process by light scattering and shadow graphic imaging. Spectrochim Acta B 65:943–949

Košler J, Wiedenbeck M, Wirth R, Hovorka J, Sylvester P, Míková J (2005) Chemical and phase composition of particles produced by laser ablation of silicate glass and zircon—implications for elemental fractionation during ICPMS analysis. J Anal Atom Spectrom 20:402–409

Kovacs R, Günther D (2008) Influence of transport tube materials on signal response and drift in laser ablation-inductively coupled plasma-mass spectrometry. J Anal Atom Spectrom 23:1247–1252

Kovacs R, Nishiguchi K, Utani K, Günther D (2010) Development of direct atmospheric sampling for laser ablation-inductively coupled plasma-mass spectrometry. J Anal Atom Spectrom 25:142–147

Kroslakova I, Günther D (2007) Elemental fractionation in laser ablation-inductively coupled plasma-mass spectrometry: evidence for mass load induced matrix effects in the ICP during ablation of a silicate glass. J Anal Atom Spectrom 22:51–62

Kuhn H-R, Günther D (2003) Elemental fractionation studies in laser ablation inductively coupled plasma mass spectrometry on laser-induced brass aerosols. Anal Chem 75:747–753

Liu Y, Hu Z, Yuan H, Hu S, Cheng H (2007) Volume-optional and low-memory (VOLM) chamber for laser ablation-ICPMS: application to fiber analyses. J Anal Atom Spectrom 22:582–585

Longerich HP (2012) Inductively coupled plasma-mass spectrometry (ICPMS): a personal Odyssey III. J Anal Atom Spectrom 27:1181–1184

Longerich HP, Günther D, Jackson SE (1996a) Elemental fractionation in laser ablation inductively coupled plasma mass spectrometry. Anal Bioanal Chem 355:538–542

Longerich HP, Jackson SE, Günther D (1996b) Inter-laboratory note. Laser ablation inductively coupled plasma mass spectrometric transient signal data acquisition and analyte concentration calculation. J Anal Atom Spectrom 11:899–904

Mahoney PP, Li G, Hieftje GM (1996) Laser ablation inductively coupled plasma mass spectrometry with a time-of-flight mass analyser. J Anal Atom Spectrom 11:401–405

Mank AJG, Mason PRD (1999) A critical assessment of laser ablation ICPMS as an analytical tool for depth analysis in silica-based glass samples. J Anal Atom Spectrom 14:1143–1153

Margetic V, Pakulev A, Stockhaus A, Bolshov M, Niemax K, Hergenröder R (2000) A comparison of nanosecond and femtosecond laser-induced plasma spectroscopy of brass samples. Spectrochim Acta B 55:1771–1785

Mattauch J, Herzog R (1934) About a new mass spectrograph. Z Phys 89:786–795

Mattey D, Lowry D, Duffet J, Fisher R, Hodge E, Frisia S (2008) A 53 year seasonally resolved oxygen and carbon isotope record from a modern Gibraltar speleothem: Reconstructed drip water and relationship to local precipitation. Earth Planet Sci Lett 269:80–95

Monticelli D, Gurevich EL, Hergenroder R (2009) Design and performances of a cyclonic flux cell for laser ablation. J Anal Atom Spectrom 24:328–335

Mozna V, Pisonero J, Hola M, Kanicky V, Günther D (2006) Quantitative analysis of Fe-based samples using ultraviolet nanosecond and femtosecond laser ablation-ICP-MS. J Anal Atom Spectrom 21:1194–1201

Müller W, Shelley M, Miller P, Broude S (2009) Initial performance metrics of a new custom-designed ArF excimer LA-ICP-MS system coupled to a two-volume laser-ablation cell. J Anal Atom Spectrom 24:209–214

Müller W, Shelley JMG, Rasmussen SO (2011) Direct chemical analysis of frozen ice cores by UV-laser ablation ICPMS. J Anal Atom Spectrom 26:2391–2395

Myers DP, Hieftje GM (1993) Preliminary design considerations and characteristics of an inductively coupled plasma-time-of-flight mass spectrometer. Microchem J 48:259–277

Outridge PM, Doherty W, Gregoire DC (1996) The formation of trace element-enriched particulates during laser ablation of refractory materials. Spectrochim Acta B 51:1451–1462

Pettke T, Oberli F, Audétat A, Guillong M, Simon AC, Hanley JJ, Klemm LM (2012) Recent developments in element concentration and isotope ratio analysis of individual fluid inclusion by laser ablation single and multiple collector ICPMS. Ore Geol Rev 44:10–38

Pisonero J, Fliegel D, Gunther D (2006) High efficiency aerosol dispersion cell for laser ablation-ICPMS. J Anal Atom Spectrom 21:922–931

Pisonero J, Koch J, Wälle M, Hartung W, Spencer ND, Günther D (2007) Capabilities of femtosecond laser ablation inductively coupled plasma mass spectrometry for depth profiling of thin metal coatings. Anal Chem 79:2325–2333

Plotnikov A, Vogt C, Hoffmann V, Taschner C, Wetzig K (2001) Application of laser ablation inductively coupled plasma quadrupole mass spectrometry (LA-ICP-QMS) for depth profile analysis. J Anal Atom Spectrom 16:1290–1295

Plotnikov A, Vogt C, Wetzig K, Kyriakopoulos A (2008) A theoretical approach to the interpretation of the transient data in scanning laser ablation inductively coupled plasma mass spectrometry: consideration of the geometry of the scanning area. Spectrochim Acta B 63:474–483

Raab A, Ploselli B, Munro C, Thomas-Oates J, Feldmann J (2009) Evaluation of gel electrophoresis conditions for the separation of metal-tagged proteins with subsequent laser ablation ICP-MS detection. Electrophoresis 30:303–314

Reinhardt H, Kriews M, Miller H, Schrems O, Lüdke C, Hoffmann E, Skole J (2001) Laser ablation inductively coupled plasma mass spectrometry: a new tool for trace element analysis in ice cores. Fresen J Anal Chem 370:629–636

Reinhardt H, Kriews M, Miller H, Lüdke C, Hoffmann E, Skole J (2003) Application of LA-ICP-MS in polar ice core studies. Anal Bioanal Chem 375:1265–1275

Resano M, McIntosh KS, Vanhaecke F (2012) Laser ablation-inductively coupled plasma-mass spectrometry using a double-focusing sector field mass spectrometer of Mattauch–Herzog geometry and an array detector for the determination of platinum group metals and gold in NiS buttons obtained by fire assay of platiniferous ores. J Anal Atom Spectrom 27:165–173

Russo RE, Mao XL, Borisov OV, Liu H (2000) Influence of wavelength on fractionation in laser ablation ICP-MS. J Anal Atom Spectrom 15:1115–1120

Russo RE, Mao X, Gonzalez JJ, Mao SS (2002a) Femtosecond laser ablation ICP-MS. J Anal Atom Spectrom 7:1072–1075

Russo RE, Mao X, Liu H, Gonzalez J, Mao SS (2002b) Laser ablation in analytical chemistry-a review. Talanta 57:425–451

Schilling GD, Ray SJ, Rubinshtein AA, Felton JA, Sperline RP, Denton MB, Barinaga CJ, Koppenaal DW, Hieftje GM (2009) Evaluation of a 512-channel Faraday-strip array detector coupled to an inductively coupled plasma Mattauch–Herzog mass spectrograph. Anal Chem 81:5467–5473

Schrön W, Bombach G, Beuge P (1983) Schnellverfahren zur flammenlosen AAS-Bestimmung von Spurenelementen in geologischen Proben. Spectrochim Acta B 38:1269–1276

Shuttleworth S (1996) Optimisation of laser wavelength in the ablation sampling of glass materials. Appl Surf Sci 96–98:513–517

Sinclair DJ, Kinsley LPJ, McCulloch MT (1998) High resolution analysis of trace elements in corals by laser ablation ICP-MS. Geochim Cosmochim Acta 62:1889–1901

Tabersky D, Nishiguchi K, Utani K, Ohata M, Dietiker R, Fricker MB, de Maddalena IM, Koch J, Günther D (2013) Aerosol entrainment and a large-capacity gas exchange device (Q-GED) for laser ablation inductively coupled plasma mass spectrometry in atmospheric pressure air. J Anal Atom Spectrom 28: 831–842

Tanner M (2007) Fundamental studies on fast signal generation without an aerosol transport system for Laser Ablation Inductively Coupled Plasma Mass Spectrometry, Ph.D. Dissertation, ETH Zurich

Tanner M, Günther D (2005) In torch laser ablation sampling for inductively coupled plasma mass spectrometry. J Anal Atom Spectrom 20:987–989

Tanner M, Günther D (2006) In torch laser ablation sampling for inductively coupled plasma time of flight mass spectrometry. J Anal Atom Spectrom 21:941–947

Thompson M, Goulter JE, Sieper F (1981) Laser ablation for the introduction of solid samples into an inductively coupled plasma for atomic-emission spectrometry. Analyst 106:32–39

Treble P, Shelley JMG, Chappell J (2003) Comparison of high resolution sub-annual records of trace elements in a modern (1911–1992) speleothem with instrumental climate data from southwest Australia. Earth Planet Sci Lett 216:141–153

Treble PC, Chappell J, Shelley JMG (2005) Complex speleothem growth processes revealed by trace element mapping and scanning electron microscopy of annual layers. Geochim Cosmochim Acta 69:4855–4863

Vaughn KJ, Dussubieux L, Williams RR (2011) A pilot compositional analysis of Nasca ceramics from the Kroeber collection. J Archaeol Sci 38:3560–3567
Wagner B, Jędral W (2011) Open ablation cell for LA-ICP-MS investigations of historic objects. J Anal Atom Spectrom 26:2058–2063
Wälle M, Koch J, Flamigni L, Heirot S, Lippert T, Hartung W, Günther D (2009) Detection efficiencies in nano- and femtosecond laser ablation inductively coupled plasma mass spectrometry. Spectrochim Acta B 64:109–112
Wang HAO, Grolimund D, Giesen C, Borca CN, Shaw-Stewart JRH, Bodenmiller B, Günther D (2013) Fast chemical imaging at high spatial resolution by laser ablation inductively coupled plasma mass spectrometry. Anal Chem 85:10107–10116
Wiltsche H, Günther D (2010) Capabilities of femtosecond laser ablation ICP-MS for the major, minor and trace element analysis of high alloyed steels and super alloys. Anal Bioanal Chem 399:2167–2174
Wirth J, Poletti S, Aeschlimann B, Yakandawala N, Drosse B, Osorio S, Tohge T, Fernie AR, Günther D, Gruissem W, Sautter C (2009) Rice endosperm iron biofortification by targeted and synergistic action of nicotianamine synthase and ferritin. Plant Biotechnol J 7:631–644
Woodhead JD, Hellstrom J, Hergt JM, Greig A, Maas R (2007) Isotopic and elemental imaging of geological materials by laser ablation inductively coupled plasma-mass spectrometry. Geostand Geoanalyt Res 31:331–343

Part I

Sample Introduction

Introduction to Solid Sampling Strategies 2

Mark Golitko

Abstract

This chapter provides a brief introduction to basic concepts of solid sample introduction by laser ablation as used in archaeology. Major methodological advances presented in the chapters in Part I of the volume are briefly highlighted, as are projected future trends in solid sampling by laser ablation.

2.1 Solid Sampling Strategies

Laser Ablation (LA) as a sample introduction method has gained popularity in archaeology for a number of reasons. Besides requiring very limited sample preparation in comparison to liquid introduction, laser ablation allows highly targeted analysis of particular phases or components of a sample, but with lower limits of detection than other micro-sampling techniques like Electron Probe Micro-Analysis (EPMA) or Scanning Electron Microscopy with Energy Dispersive Spectrometry (SEM-EDS), often allowing for measurement of larger numbers of elements. In the case of homogenous materials such as obsidian (see Chap. 10), LA-ICP-MS also allows minimally destructive characterization with precision comparable to true bulk techniques such as Instrumental Neutron Activation Analysis (INAA) or X-ray Fluorescence Spectrometry (XRF). In the case of more heterogeneous sample matrices such as ceramic paste (e.g., Dussubieux et al. 2007; Stoner and Glascock 2012) or fluorite (Goemaere et al. 2013), researchers have utilized LA-ICP-MS to analyze numerous ablation lines or craters and thereby generate an averaged "pseudo-bulk" characterization (see also Chaps. 3, 8, 16, and 17), albeit with slightly more error than might be achievable using true bulk analysis methods such as INAA or ICP-MS with liquid sample introduction.

The chapters in this section highlight attempts to expand the utility of ICP-MS with Laser Ablation sampling, both by redesign of the basic laser unit itself, and by employing the time-resolved analysis capabilities of laser sampling. One limitation of LA-ICP-MS to date has been object size. As Fricker and Günther detail in Chap. 1, cell design has consequently been an area of very active development, and the first paper (Chap. 3) in this section by Golitko and colleagues is a case

M. Golitko (✉)
Department of Anthropology, University of Notre Dame, 621 Flanner Hall, Notre Dame, IN 46556, USA
e-mail: mgolitko@nd.edu

L. Dussubieux et al. (eds.), *Recent Advances in Laser Ablation ICP-MS for Archaeology*, Natural Science in Archaeology, DOI 10.1007/978-3-662-49894-1_2

study highlighting the advantages and limitations of so-called "open-cell" lasers. These systems have been developed in order to allow ablation of very large objects that cannot fit into traditional enclosed cells without destructive sampling. Although LA-ICP-MS is technically speaking a destructive analytical method, in that ablation vaporizes a portion of sample, it has been enthusiastically adopted by museums and in other research programs where minimal destruction (i.e., not visible to the naked eye) and simultaneous measurement of large numbers of elements to very low detection limits is either desirable or essential.

In their case study, Golitko and colleagues analyze ceramics in museum collections, the study of which requires both limited object damage, and measurement of as many elements as possible, often down to <1 ppm. While generating data comparable to that obtained using a commercially available closed-cell laser, their study highlights some of the difficulties that remain in the application of open-cell systems. These include higher limits of detection than when using smaller volume enclosed chamber lasers, difficulties in generating an adequate and reliable seal between the open-cell and object, and much slower sample throughput than when using a closed-cell design. Consequently, the applicability of open-cell laser systems at present seems limited to situations in which one or only a few objects need to be studied. Studies requiring analysis of large numbers of objects will hopefully become feasible in the future as open-cell and other more size-flexible ablation systems are refined and improved.

The open cell systems reported to date (see Chaps. 1 and 3) also lack another principal strength of most LA systems—the ability to focus on and chemically characterize discrete areas or components of archaeological samples using motorized sample stages and high-magnification cameras. For instance, the capability to raster over flat surfaces with minimal depth penetration has been utilized previously for compositional studies of thin pottery slips and glazes (e.g. Oka et al. 2009; Speakman 2005), allowing for distinction between paints, slips, and glazes and underlying ceramic paste. Other papers in this volume highlight the unique ability of LA-ICP-MS to provide spatially resolved chemical information that can push analysis beyond what is possible by bulk techniques or other micro-sampling techniques, for instance the analysis of slag inclusions in iron by L'Héritier and colleagues (Chap. 14) and raster ablation of lead glazes on Panamanian Majolica ceramics by Iñañez and colleagues (Chap. 19). Publications elsewhere have documented the utility of LA-ICP-MS for temper analysis in ceramics, particularly when materials such as obsidian (Palumbi et al. 2014) or volcanic tephra (Neff and Sheets 2005) were incorporated by ancient potters.

These prior studies largely deal with subtle chemical variations across a basically similar sample matrix, a condition not always met with multi-phase objects such as composite glasses. In Chap. 4, van Elteren and colleagues report a novel approach to rastering, applying simulation of ablation conditions to optimize settings prior to actual analysis of the surface of ancient glasses. This approach allows for a minimally destructive yet comprehensive study of differences in composition that can be related back to colorants, opacifiers, and sands used in the production of glass. Doing so allows the analyst to "get it right" the first time, and limits further damage that might occur during re-analysis.

Laser ablation also allows for the opposite approach, namely depth-profiling by drilling craters down into objects. Within limits, this allows for a minimally-destructive analysis of the internal composition of these objects, and the study by Sarah and Gratuze (Chap. 5) tests this method on silver coins. Coins and other metallic objects can be heavily impacted by surface corrosion—in the case of silver, surficial copper enrichment is a particular problem—and obtaining an appropriate sampling of unaltered ore previously required destructive sectioning (e.g., Notis et al. 2007) or removal of corrosion products by chemical means. Because the ablation signal can be time resolved (and thus depth-resolved), it is relatively straight forward to determine when the laser has passed through a corrosion layer into unaltered alloy.

Additionally, elements providing potential provenance information such as Pd, Au, Pt, or Os are detectable by LA-ICP-MS (see Chap. 7), but may be present at concentrations below the detection limits of other methods such as XRF. LA-ICP-MS depth profiling may thus both open up the possibility to analyze unaltered alloy composition in valuable ancient coinage, and provide information of archaeological utility that cannot be obtained with any other technique currently available.

These improvements in cell design and application will no doubt be further augmented by ongoing improvements in laser design itself. The historical development of LA units has witnessed a steady drop in laser wavelength, from early units operated in the 1000 nm range, to modern ArF and excimer lasers operating at 193 nm (see Chap. 1). The majority of studies reported in this volume utilize either 266 or 213 nm frequency-multiplied Nd:YAG lasers. Shorter wavelength lasers produce smaller and more evenly distributed particle sizes upon ablation, which result in less fractionation once ionized in the plasma (Guillong et al. 2003). This is particularly important in allowing use of non-matrix matched standards for quantification. Another area of development is pulse-frequency. Most lasers currently utilized in archaeology are pulsed at a scale of nanoseconds, but some newer models pulse on the order of femtoseconds. In principal, more rapid pulsing allows for reduced fractionation, but to date, the mechanism of ablation with these lasers is not well understood (Diwakar et al. 2014). It is anticipated that issues with short pulse times will be overcome, and that short wavelength femtosecond laser units will eventually become standard equipment for archaeological applications.

References

Diwakar PK, Gonzalez JJ, Harilal SS, Russo RE, Hassanein A (2014) Ultrafast laser ablation ICP-MS: role of spot size, laser fluence, and repetition rate in signal intensity and elemental fractionation. J Anal Atom Spectrom 29:339–346

Dussubieux L, Golitko M, Williams PR, Speakman RJ (2007) LA-ICP-MS analysis applied to the characterization of Peruvian Wari ceramics. In: Glascock MD, Speakman RJ, Popelka-Filcoff RS (eds) Archaeological chemistry: analytical technique and archaeological interpretation. American Chemical Society, Washington, DC, pp 349–363

Goemaere E, Honings J, Jungels C, Golitko M, Degryse P, Schneider J (2013) Discerning geological and geographical sources of Belgian Upper Paleolithic fluorites by rare earth elements and Sr-isotopic geochemistry. J Archaeol Sci 40:2892–2901

Guillong M, Horn I, Günther D (2003) A comparison of 266 nm, 213 nm and 193 nm produced from a single solid state Nd:YAG laser for laser ablation ICP-MS. J Anal Atom Spectrom 18:1224–1230

Neff H, Sheets P (2005) Applications of tephra analysis by LA-ICP-MS. In: Speakman RJ, Neff H (eds) Laser ablation-ICP-MS in archaeological research. University of New Mexico Press, Albuquerque, NM, pp 117–124

Notis M, Shugar A, Herman D, Ariel DT (2007) Chemical composition of the Isfiya and Qumran coin hoards. In: Glascock MD, Speakman RJ, Popelka-Filcoff RS (eds) Archaeological chemistry: analytical technique and archaeological interpretation. American Chemical Society, Washington DC, pp 258–274

Oka RC, Dussubieux L, Kusimba C, Gogte V (2009) The impact of "imitation" industries and Imperial restrictions on Chinese ceramic commercial exports in the Indian Ocean Maritime Exchange, ca. 1200–1700 CE. In: McCarthy B, Salzman Chase E, Cort LA, Douglas JG, Jett P (eds) Scientific research on historic Asian ceramics: Proceedings of the Fourth Forbes Symposium at the Freer Gallery of Art. Archetype Publications, Washington, DC, pp 175–185

Palumbi G, Gratuze B, Harutyunyan A, Chataigner C (2014) Obsidian-tempered pottery in the Southern Caucusus: a new approach to obsidian as a ceramic-temper. J Archaeol Sci 44:43–54

Speakman RJ (2005) Chemical characterization of Mesa Verde and Mancos Black-on-white pottery pigments by LA-ICP-MS. In: Speakman RJ, Neff H (eds) Laser ablation-ICP-MS in archaeological research. University of New Mexico Press, Albuquerque, NM, pp 167–186

Stoner W, Glascock MD (2012) The forest or the trees? Behavioral and methodological considerations for geochemical characterization of heavily-tempered ceramic pastes using NAA and LA-ICP-MS. J Archaeol Sci 39:2668–2683

3 Open-Cell Ablation of Killke and Inka Pottery from the Cuzco Area: Museum Collections as Repositories of Provenience Information

Mark Golitko, Nicola Sharratt, and Patrick Ryan Williams

Abstract

Museum collections are often invaluable repositories of information for archaeologists, as they contain large numbers of intact, well provenanced objects. However, the utility of such collections for chemical analysis is often limited by the inability to perform destructive analysis on them. We report here on the analysis of Killke (AD 1000–1400) and Inka (AD 1400–1532) ceramics from the Cuzco area of Peru housed in the Field Museum of Natural History South American collections using a specially modified open-cell laser to generate compositional data from complete intact vessels in a minimally destructive manner. These data are compared to measurements performed using a conventional laser cell on Killke and Inka sherds from the same collections and regional raw material samples. We discuss both the utility and shortcomings of the open-cell system as a means of ceramic analysis, as well as implications for archaeological understanding of prehistoric ceramic production and economy in the Cuzco area.

3.1 Introduction

Chemical source studies of archaeological ceramics rely on several lines of evidence to establish place of production. Ideally, one can measure the chemical composition of ceramics known to have been produced in a particular location, then compare archaeological specimens of unknown production location to these ceramics to establish with great certainty where the archaeological ceramics were produced (e.g., Joyce et al. 2006). In the case where a survey of raw materials has been conducted, the

M. Golitko (✉)
Department of Anthropology, University of Notre Dame, 621 Flanner Hall, Notre Dame, IN 46556, USA
e-mail: mgolitko@nd.edu

N. Sharratt
Department of Anthropology, Georgia State University, 33 Gilmer Street, Suite 335, Atlanta, GA 30303, USA
e-mail: nsharratt@gsu.edu

P.R. Williams
Integrative Research Center, Social Sciences, Field Museum of Natural History, Chicago, IL 60605, USA
e-mail: rwilliams@fieldmuseum.org

L. Dussubieux et al. (eds.), *Recent Advances in Laser Ablation ICP-MS for Archaeology*, Natural Science in Archaeology, DOI 10.1007/978-3-662-49894-1_3

composition of clays potentially used in the past can provide another valuable line of evidence for establishing production provenance (e.g., Golitko 2011; Golitko and Bosquet 2011; Neff and Bove 1999; Sharratt et al. 2009; Vaughn and Neff 2004). Often, archaeologists are forced to rely on the so-called "criterion of abundance" (Rice 1987: 177)—the assumption that the most frequently represented compositional profile at a particular archaeological site or set of proximate sites is associated with local production, and that less well represented chemical profiles represent imports from more distant places of production.

Clearly, the last solution is far from optimal, as ethnographic and historical accounts of high volume ceramic producers and exporters abound, suggesting that in many prehistoric cases, the majority of ceramics at any given site may have been imported from more distant places of production. Museum collections—which often contain ethnographically collected, well provenanced ceramics—provide a valuable repository of information against which to compare prehistoric sherds. However, it is desirable to limit any damage to objects in museum collections, making the application of techniques such as Instrumental Neutron Activation Analysis (INAA), X-ray Fluorescence Analysis (XRF), and petrography to such ceramics problematic. While these techniques are in principal non-destructive—in other words, no material is physically removed, atomized, or otherwise lost during actual measurement—most require highly destructive sample preparation via powdering, polishing, or other physical destruction or alteration of the sample. Recent non-destructive applications of portable X-ray Fluorescence (PXRF) to ceramic analysis have shown some promise (Forster et al. 2011; Frankel and Webb 2012; Goren et al. 2011), but are limited both by calibration issues (related to density variability in un-powdered ceramics that make comparison with standards difficult) and high detection limits for most elements that allow for the measurement of 10–20 elements at most (Speakman et al. 2011). Unlike analysis of other kinds of materials such as metal alloys, obsidian, or glass (see other chapters in this volume), there is no *a priori* way to know which elements in clays and ceramics will prove effective for establishing provenance, and as such, ceramic sourcing requires application of a technique that can measure a large number of major, minor, and trace elements (Harbottle 1976).

Many researchers, both at museums and elsewhere, have therefore turned to Inductively Coupled Plasma-Mass Spectrometry with Laser Ablation (LA-ICP-MS) as a minimally-destructive alternative to other methods of chemical characterization (e.g., Beck and Neff 2007; Eckert and James 2011; Fitzpatrick et al. 2006; Golitko 2011; Golitko and Terrell 2012; Niziolek 2013; Vaughn et al. 2011). LA-ICP-MS, while technically a destructive technique, in practice removes such a small amount of material from a sample as to leave no visible damage. Advantages of LA-ICP-MS include sensitivity, ability to analyze a wide array of elements, and rapidity of analysis. While ceramics are relatively heterogeneous, and LA-ICP-MS measures only a small amount of material, several studies have shown that it can produce reliable characterization of the paste fraction of ceramic matrix (e.g., Dussubieux et al. 2007; Stoner and Glascock 2012).

However, most laser ablation systems require that the sample to be analyzed fit within a relatively small chamber (typically ~6–10 cm maximal diameter), sometimes requiring destructive sub-sampling of any objects to be measured. As a result, a number of research groups have developed open-cell laser systems during the last three decades (Arrowsmith and Hughes 1988; Devos et al. 1999; Glaus et al. 2012; Wagner and Jędral 2011). Rather than enclosing the sample to be analyzed in a chamber, these systems typically function by sealing a chamber with one open side onto the object surface or in some cases, allow ablation without any direct contact at all between the ablation cell and sample surface (Asogan et al. 2009; Kovacs et al. 2010).

We report on the results of analyzing Killke and Inka style ceramic vessels from the Cuzco region of Peru—the heartland of the Inka empire—housed in the Field Museum's South American collections using a specially modified

NewWave Micro 266 open-cell laser that allows for the analysis of large vessels with damage no greater than that produced by a conventional laser ablation unit. These analyses are compared to data generated on sherds from the same collections using a NewWave UP213 closed-cell system. We address both the utility and drawbacks of the open-cell laser, and assess its performance relative to the closed-cell laser. Finally, we briefly review some of the results of our study as they pertain to the provenience potential of the studied ceramics by comparing them to a large number of clay samples collected in the Cuzco region.

3.2 Open-Cell Ablation

Typical laser ablation units produced commercially by manufacturers such as NewWave, Cetac, and others, utilize relatively small enclosed chambers in which samples are placed for ablation. The laser passes through a thin window and is focused at the sample surface. A gasket or seal of some sort prevents external atmosphere from entering, and carrier gas (usually Ar or He) is introduced into the chamber to sweep ablated material towards the ICP-MS. Establishing a reliable seal is critical, as atmospheric gases can interfere with ablation and introduce background noise, but also promote fractionation effects. Both the shape of the chamber (e.g., square, circular, bottle-shaped, etc...) and the flow dynamics of the carrier gas can impact the stability and magnitude of measured signal intensity resulting from ablation (Bleiner and Bogaerts 2007). Commercially available ablation chambers are typically relatively small, however, and as a result, it is often necessary to destructively sample objects to fit them into the ablation chamber. Consequently, a number of laboratories have developed larger cells for analysis of a variety of archaeological and historical materials—see Sarah and Gratuze (Chap. 5) and Baron and Gratuze (Chap. 17) in this volume for descriptions of these chambers.

Another approach that has been utilized to eliminate the need for destructive sampling of large objects for ablation is the development of open-cell designs. Some such systems utilize a small mobile cell with an ablation window on one side and an open aperture on the other that can be sealed to the surface of the object to be analyzed (Devos et al. 1999; Wagner and Jędral 2011). Several other approaches have been attempted to date as well. Asogan et al. (2009) and Arrowsmith and Hughes (1988) report cells that utilize an outer annular gas flow ring (typically Ar) to create a "gas-sleeve" that prevents ambient atmosphere from reaching the ablation surface of the object. An inner flow of He or Ar serves as carrier gas. Kovacs et al. (2010) remove even this gas-sleeve and ablate directly at atmospheric pressure, subsequently replacing oxygen with argon via a gas-exchange device before introducing the sample aerosol into the ICP-MS. Glaus et al. (2012) have utilized a different approach, using a portable laser to ablate material onto filter papers first. These filter papers are then themselves ablated in a laboratory and fed into an ICP-MS for elemental measurements. This approach eliminates the need to bring the object to the lab, but does require the application of ablated standards onto filters using the same method as for sample ablation. All three of these approaches eliminate the need to have any direct contact between an ablation cell and the object surface.

The unit housed at the Field Museum of Natural History Elemental Analysis Facility (EAF) is of the first type, utilizing a small cell with an open aperture on one side. Ablation is achieved using a NewWave Micro 266 laser ablation unit modified by Richard Cox (Université du Québec à Chicoutimi) to an open-cell configuration (Fig. 3.1). The laser unit is supported on an adjustable hydraulic jack that allows it to be raised or lowered. Analyzed objects are similarly supported on a mechanical jack system that

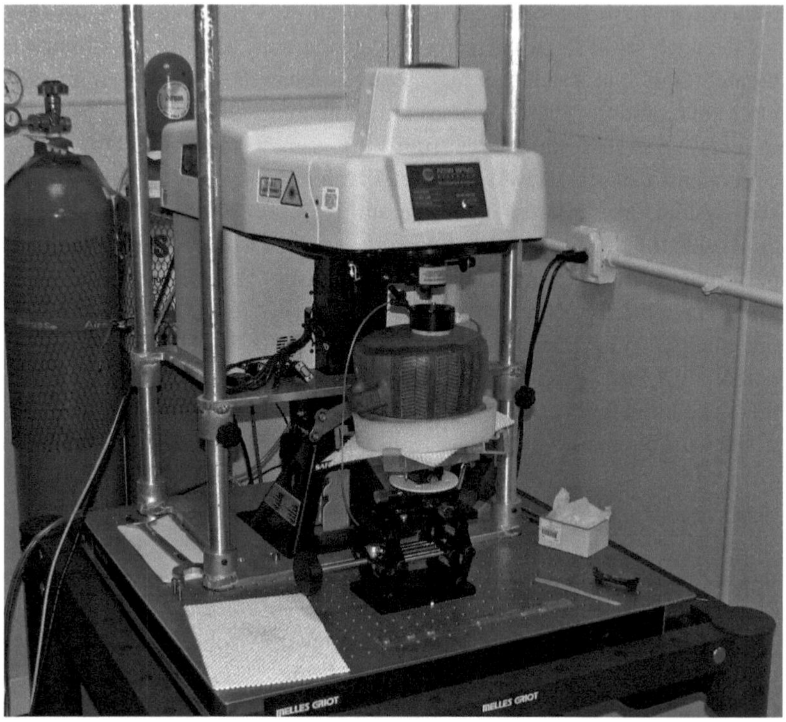

Fig. 3.1 The Field Museum's open-cell laser, a New Wave Micro 266 laser modified to seal to large objects. An Inka style ceramic vessel (FM2765) is positioned for analysis

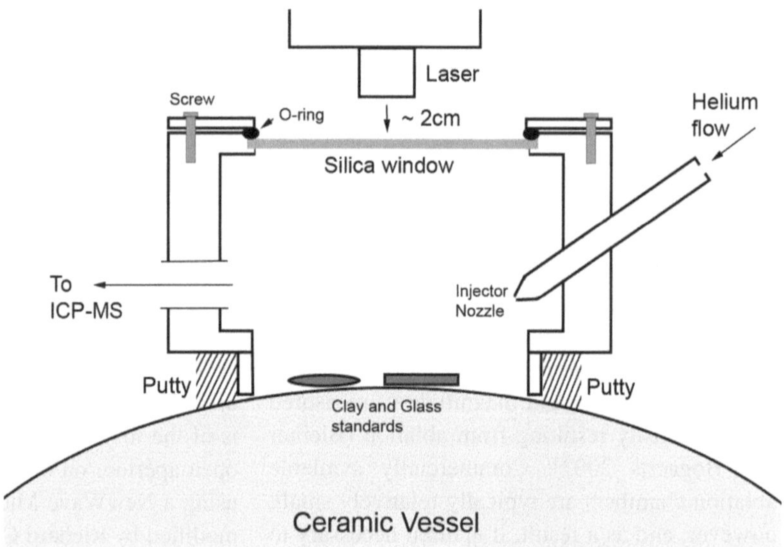

Fig. 3.2 Schematic cross-section of the EAF open-cell chamber shown positioned for analysis on a ceramic vessel

allows the object to be raised and lowered, moved laterally in both X- and Y-directions, and to be slightly tilted to level the chamber window relative to the laser aperture. The cell itself is a 4 cm diameter plastic cylinder with a window sealed on top and a flange on bottom. Helium is introduced into the cell via an injector nozzle on one side and collected through an aperture and connected hose on the opposite side of the cell (Fig. 3.2).

3.3 Samples

3.3.1 Analyzed Ceramics

Our analysis was undertaken on ceramics contained in the Field Museum collections from the Cuzco area (Fig. 3.3) of the south-central Peruvian Highlands as part of a larger scale analysis of ceramic production and distribution in the heartland of the Inka Empire both during the Late Intermediate period (LIP—AD 1000–1400) and the Classic Inka phases (AD 1400–1532) (Bauer 2004). The analyzed Killke (LIP) and Inka ceramics were drawn from two Field Museum of Natural History collections, the Montez and Collier collections (Bauer 1999). The Montez collection contains 1258 individual catalog numbers, the majority recorded as from Cuzco. Many are Inka in style, but the collection also contains Killke style vessels. Some of the Killke ceramics in the Montez collection are attributed to specific sites in the Cuzco area, and three of the vessels have handwritten tags that were likely with the objects when they arrived in Chicago (Bauer and Stanish 1990), and indicate that the vessels came from three sites in the Urubamba Valley; Ollantaytambo, San Salvador and Pisac. Ollantaytambo and San Salvador vessels were not sampled as part of this study.

The Collier collection also includes both Killke and Inka ceramics (Bauer 1999). It was acquired by the Field Museum following an archaeological expedition to Peru in 1946 by curator Donald Collier. The principal focus of the expedition was an intensive study of the Virú Valley, 20 miles south of Trujillo, and most of the 186 catalogued materials are from there. However, Collier also made a brief visit to Cuzco, where he collected ceramics at sites including Sacsahuaman and Qoripata. In total, we selected 42 separate vessels for analysis (Table 3.1)—32 of these were broken in transport and ablated as small fragments using a closed-cell laser, while ten of these were intact vessels

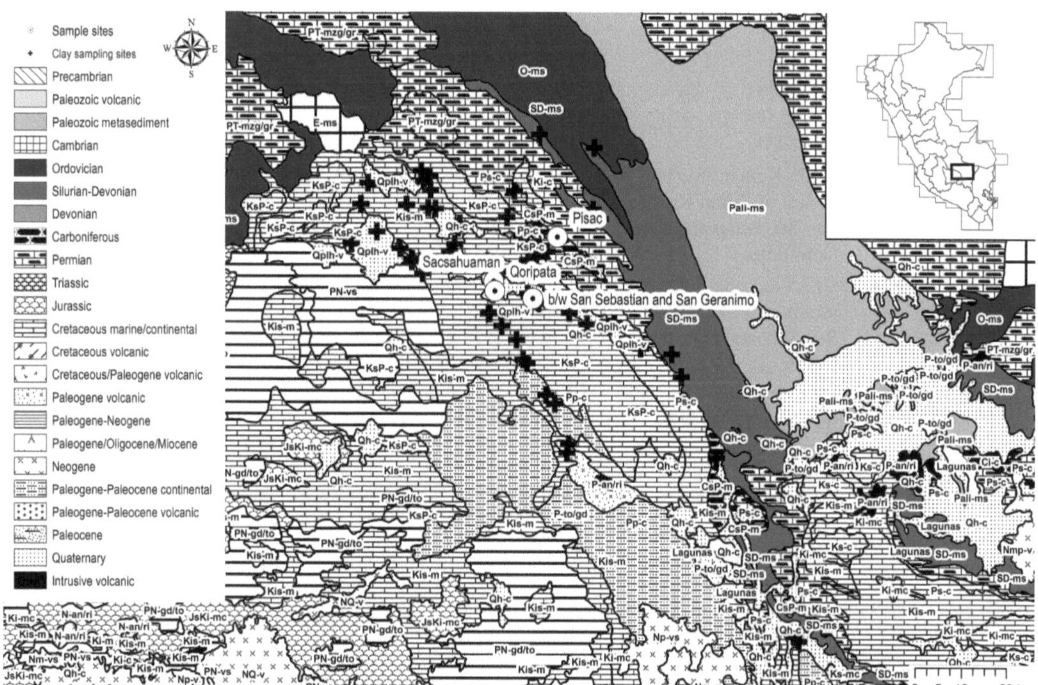

Fig. 3.3 Geological map of the Cusco region with ceramic collection locales and clay sampling points shown

Table 3.1 Inca and Killke style sherds and vessels analyzed by both closed-cell and open-cell LA-ICP-MS

ANID	FMNH#	Site	Style	Form	Laser	Chemical group
CSC001	241810	Sacsahuaman (SW1 C2-5)	Kilke Series	Painted body sherd	Closed-cell	Group 1
CSC002	241810	Sacsahuaman (SW1 C2-5)	Kilke Series	Painted rim	Closed-cell	Group 1
CSC003	241810	Sacsahuaman (SW1 C2-5)	Kilke Series	White painted rim	Closed-cell	Group 1
CSC004	241810	Sacsahuaman (SW1 C2-5)	Kilke Series	Painted body sherd	Closed-cell	Group 1
CSC005	241810	Sacsahuaman (SW1 C2-5)	Kilke Series	Painted rim	Closed-cell	Group 1
CSC006	241810	Sacsahuaman (SW1 C2-5)	Kilke Series	Plain body sherd with lug	Closed-cell	Group 1
CSC007	241810	Sacsahuaman (SW1 C2-5)	Kilke Series	Painted body sherd	Closed-cell	Group 1
CSC008	241810	Sacsahuaman (SW1 C2-5)	Kilke Series	Painted body sherd	Closed-cell	Group 1
CSC019	241814	Qoripata	Inca	Decorated body sherd	Closed-cell	Group 1
CSC020	241814	Qoripata	Inca	Decorated body sherd	Closed-cell	Group 1
CSC023	241814	Qoripata	Inca	Decorated neck	Closed-cell	Group 1
CSC025	241814	Qoripata	Inca	Decorated body sherd	Closed-cell	Group 1
CSC027	241812	Qoripata	Inca	Decorated body sherd, flower motif	Closed-cell	Group 1
CSC028	241812	Qoripata	Inca	Decorated plate	Closed-cell	Group 1
CSC029	241812	Qoripata	Inca	Decorated plate	Closed-cell	Group 1
CSC030	241812	Qoripata	Inca	Decorated shallow bowl	Closed-cell	Group 1
CSC031	241813	SW 1934	Killke Series?	Decorated base/body	Closed-cell	Group 1
CSC032	241813	SW 1934	Inca	Decorated plate	Closed-cell	Group 1
CSC035	241811	CZ35 Kencha-Kencha at mouth of Pillan River b/w San Sebastian and San Geronimo	Killke Series	Decorated body sherd	Closed-cell	Group 1
CSC037	241811	CZ35 Kencha-Kencha at mouth of Pillan River b/w San Sebastian and San Geronimo	Killke Series	Decorated body sherd	Closed-cell	Group 1
CSC038	241811	CZ35 Kencha-Kencha at mouth of Pillan River b/w San Sebastian and San Geronimo	Killke Series	Decorated body sherd (bowl?)	Closed-cell	Group 1
CSC039	241811	CZ35 Kencha-Kencha at mouth of Pillan River b/w San Sebastian and San Geronimo	Killke Series	Decorated base	Closed-cell	Group 1
CSC040	241811	CZ35 Kencha-Kencha at mouth of Pillan River b/w San Sebastian and San Geronimo	Killke Series	Decorated body sherd	Closed-cell	Group 1

(continued)

Table 3.1 (continued)

ANID	FMNH#	Site	Style	Form	Laser	Chemical group
CSC041	241811	CZ35 Kencha-Kencha at mouth of Pillan River b/w San Sebastian and San Geronimo	Killke Series	Decorated body sherd	Closed-cell	Group 1
CSC042	241811	CZ35 Kencha-Kencha at mouth of Pillan River b/w San Sebastian and San Geronimo	Killke Series	Decorated body sherd	Closed-cell	Group 1
CSC026	241814	Qoripata	Inca	Decorated base	Closed-cell	Group 2
CSC033	241813	SW?	Inca	Decorated rim	Closed-cell	Group 2
CSC034	241813	SW?	Inca	Decorated body sherd	Closed-cell	Group 2
CSC036	241811	CZ35 Kencha-Kencha at mouth of Pillan River b/w San Sebastian and San Geronimo	Killke Series	Decorated rim	Closed-cell	Group 2
CSC021	241814	Qoripata	Inca	Decorated body sherd	Closed-cell	Group 3
CSC024	241814	Qoripata	Inca	Decorated rim	Closed-cell	Group 3
CSC022	241814	Qoripata	Inca	Decorated body sherd	Closed-cell	Unassigned
CSC009	2637	Cusco area	Kilke Series	Decorated vase	Open-cell	Group 1
CSC010	2633	Cusco area	Kilke Series	Decorated pitcher	Open-cell	Group 1
CSC011	2809	Cusco area	Kilke Series	Decorated pitcher	Open-cell	Group 1
CSC012	2789	Cusco area	Kilke Series	Decorated pitcher	Open-cell	Group 1
CSC013	2805	Cusco area	Kilke Series	Decorated pitcher	Open-cell	Unassigned
CSC014	2763	Pisac	Kilke Series	Decorated pitcher	Open-cell	Group 1
CSC015	3139	Cusco area	Inca		Open-cell	Group 2
CSC016	2765	Cusco area	Inca	Decorated bowl	Open-cell	Unassigned
CSC017	2618	Cusco area	Inca Colonial	Decorated arriballoid	Open-cell	Group 1
CSC018	2622	Cusco area	Inca	Decorated pitcher (large)	Open-cell	Group 2

that could only be sampled using an open-cell system. While both Inka and Killke style ceramics are typically painted with black, tan, and white pigments, we focused on ablating either broken cross-sections (closed-cell), or else unpainted bases (open-cell) to ensure that only paste was sampled. Several of the vessels and sherds we analyzed had previously been analyzed by Bray et al. (2005) using INAA and assigned to local Cuzco area production—as this included both broken sherds and whole vessels, we have no particular reason to suspect that the ten samples analyzed using the open-cell should differ significantly compositionally from the 32 analyzed using the closed-cell laser, suggesting that the same compositional profiles should be identifiable in both datasets.

3.3.2 Cuzco Area Clay Survey

Ethnohistoric data indicate that imperial Inka ceramic production occurred at workshops in the area between San Sebastian and San

Jeronimo, today modern suburbs of the city of Cuzco (Bauer 1999). Consequently, our analysis also draws on a geological survey of the broader Cuzco region undertaken by two of the authors during 2007. The survey spanned approximately 300 km^2 and encompassed the Vilcanota/Urubamba River Valley (including the Sacred Valley), the Cuzco Basin, the Oropesa Basin, the Lucre Basin, the Maras plain, and the Province of Paruro (Fig. 3.3). Our survey utilized a combination of culturally biased and systematic sampling, as adopted in other clay surveys in the Andes (Sharratt et al. 2009; Vaughn and Neff 2004), and included large clay sources that are used today in industrial production of bricks and tiles. Local potters also indicated the location of smaller deposits. However, we sought to document the range of clays available to potters during the LIP and Late Horizon, not just ascertain sources in use today. We therefore made extensive use of geological maps to identify other potential locales of clay in the region and to ensure that we sampled all of the geological formations present in the area.

The Cusco area encompasses four major geographical and topographic areas, the Cordillera Oriental, Altiplano-Cordillera intermediate zone, the Altiplano, and the northeastern edge of the Cordillera Occidental. The archaeological sites included in this study are located either on the Altiplano or the Altplano-Cardillera Oriental intermediate area, and those in the immediate vicinity of Cusco are located on Quaternary geology (Qplh-v) surrounded by areas of Cretaceous geology (Kis-m and KsP-c units on Fig. 3.3). These geographical regions are comprised of a series of synclines and anticlines formed during the Andean orogeny (beginning during the Cretaceous) uplifting and folding numerous geographical units, primarily shales, sandstones, gypsum, and limestones dating between the Siluro-Devonian and the Miocene. Igneous rocks are present as intrusives, including the massive plutonic rocks of the Andahuaylas-Yauri Batholith (Pp-c on Fig. 3.3). In some places, basalts and dacites belonging to the Mitu Group outcrop (Ps-c, present around Pisac), as well as andesitic volcanic rocks of the Quaternary Rumicolca formation (included under Qplh-v on Fig. 3.3), on which a number of the study sites are located. These later rocks were used as building material in the Cusco area in Prehispanic times (Caillaux et al. 2011).

Fifty-six clay samples were collected and analyzed as fired briquettes. These briquettes were produced by mixing homogenized and powdered clay with ultra-pure deionized water. After a period of drying, the briquettes were fired to 900 °C for 1 h to completely solidify them. Repeated studies have demonstrated that within this temperature range, there is no significant loss of elemental content via volatilization from clays (Cogswell et al. 1996; Kilikoglou et al. 1988; Schwedt and Mommsen 2007).

3.4 Method

Methodological details for both closed-cell and open-cell ablations can be found in Table 3.2. To compare the performance of the two chambers, we repeatedly analyzed a quality assurance standard, New Ohio Red Clay (NORC) (Glascock 1992; Kuleff and Djingova 1998) with both lasers during analysis of all samples. A number of comparison samples that had previously been measured using the closed-cell UP213 laser were also ablated using the open-cell Micro 266 unit—these included five samples from the Cuzco ceramic collections, several ceramic samples and clay briquettes previously analyzed at the EAF, and a commercially purchased green glass wine bottle. The later pieces were included to span a wider range of concentrations than were present in the Cuzco ceramics, and to examine the impact of material type on instrument response in the case of the glass bottle.

3.4.1 Closed-Cell Ablation

Closed-cell ablation of clay samples and ceramics was performed at the EAF using an Analytik Jena (formerly Varian) quadrupole ICP-MS (Elliot et al. 2004) coupled to a NewWave UP213 laser ablation system. Analysis parameters were identical to those utilized in previous published studies of

Table 3.2 Details of method employed for both closed- and open-cell ablation

	Closed-cell	Open-cell
LA system	NewWave UP213	NewWave UP266
Type	ND:YAG (5th harmonic)	ND:YAG (4th harmonic)
Wavelength	213 nm	266 nm
Carrier gas	He	He
Flow rate	0.5 l/min	~2 l/min
Energy	0.2 mJ	~1.3 mJ
Fluence	~2.5 J/cm^2	~17 J/cm^2
Pulse frequency	10 Hz	10 Hz
Spot size	100 μm	100 μm
# Spots	5 (standards), 10 (samples)	5 (standards), 10 (samples)
Cell dimension	Maximal 52 mm × 52 mm	3.6 cm × 2 cm
ICP-MS	Analytik Jena (Varian) Quadrupole	Analytik Jena (Varian) Quadrupole
Plasma gas	Ar	Ar
Mode	Peak jumping	Peak jumping
# Isotopes measured[a]	57	57
Calibration		
External standards	NIST610, NIST612, NIST679	NIST610, NIST679
Calibration strategy	Oxide-summed internal standardized	Oxide-summed internal standardized
Internal standard	^{29}Si	^{29}Si
Calibration curve	Linear	Linear
Quality assurance standard	New Ohio Red clay	New Ohio Red clay

[a]See Table 3.3 for a complete listing

Andean ceramics generated at the same laboratory (Dussubieux et al. 2007; Sharratt et al. 2009; Vaughn et al. 2011). The laser was run at an energy output of 0.2 mJ and a pulse frequency of 15 Hz. Ten 100 μm spots were ablated per specimen and resulting signals were averaged after blank subtraction. A helium carrier gas at a flow rate of 0.50 l/min was used to transport ablated material to the argon plasma. The ICP-MS was set to peak-jumping mode with three scans of the measured mass range per replicate and nine replicates per analysis—the first three of these replicates were omitted to account for surface contamination and allow the signal to stabilize. Isotopes of 57 major, minor, and trace elements were measured (see Table 3.3 for a full listing), although P and Cl were omitted from consideration as they ionize inefficiently and measure with poor precision. These elements are measured to assess qualitatively whether these anions are present at high concentrations or not, which might impact the summation approach to calibration utilized to calculate final concentrations (see below).

Three standard reference materials were run before and after each batch of 4–5 samples—NIST610 and 612 glass standards, and NIST standard clay 679 (Brick Clay). Concentrations for NIST679 were taken from the certification sheet provided, while values for NIST glasses published in Pearce et al. (1997) were assumed. While updated concentrations have since been published for the NIST glass standards (Jochum et al. 2011), the EAF has continued to use the 1997 published values to retain consistency with earlier analyses. Concentrations were calculated by comparison of silica normalized signal counts to a linear regression line of silica-normalized signal for NIST610, 612, and 679 collected on the same day against certified or reported concentrations for those three Standard Reference Materials (SRMs). These silica normalized concentrations were then converted to parts per million concentrations by summing all measured elements as oxides to 100 % (see Gratuze et al. 2001; Speakman and Neff 2005).

Table 3.3 Comparison of measurements of NORC (ppm) with published values and comparison measurements on 5 Cusco ceramic samples using both the closed and open-cell lasers, and correlation coefficients derived from measurement of 12 comparison samples

	NORC			CSC024		CSC025		CSC027		CSC031		CSC033		
	Closed	Open	Kuleff and Djingova	Closed	Open	Closed	Open	Closed	Open	Closed	Open	Closed	Open	r^2
^{7}Li	148 ± 3	140 ± 12	139	29	24	51	47	56	67	55	60	65	64	0.97
^{9}Be	3.87 ± 0.15	3.68 ± 0.7	–	3.39	3.33	2.99	1.99	3.02	2.08	2.53	2.18	2.58	2.55	0.98
^{11}B	115.8 ± 11.5	121.5 ± 16.2	–	30.3	30.6	44.8	43.7	37.1	44.9	55.8	36.8	107.9	73.9	0.98
^{23}Na	1257 ± 299	1637 ± 285	1400 ± 70	2701	9372	4520	17,015	8171	17,645	7642	11,393	7359	13,343	0.98
^{24}Mg	9628 ± 773	10,469 ± 771	7300	5401	5919	23,368	23,422	25,888	24,266	23,308	26,784	19,139	19,530	0.97
^{27}Al	102,681 ± 4175	105,368 ± 7124	94,500 ± 3800	132,864	131,581	100,372	101,620	102,925	98,236	90,040	98,751	94,953	98,597	0.91
^{29}Si	312,150 ± 7445	305,269 ± 13,199	309,000	300,545	289,815	293,928	275,533	288,704	289,442	300,879	286,886	306,685	289,290	0.93
^{39}K	36,095 ± 2122	37,549 ± 4049	35,500 ± 2000	20,885	24,028	25,681	29,843	27,260	26,526	33,542	32,789	35,229	35,770	0.95
^{44}Ca	1422 ± 7070	1523 ± 349	4500 ± 700	6598	10,713	20,693	29,477	20,321	25,356	16,283	24,086	6975	14,697	0.99
^{45}Sc	19.5 ± 1.4	21.5 ± 2.5	18.95 ± 1.17	14.9	14.7	18.0	18.6	21.6	16.8	18.1	17.2	18.5	17.1	0.73
^{49}Ti	5213 ± 587	6255 ± 936	5940 ± 630	3415	3880	3298	3914	4496	3960	4465	5078	4886	5936	0.96
^{51}V	230.4 ± 17.2	225 ± 22.4	211 ± 26	102.9	85.0	197.4	160.9	156.8	147.3	143.5	163.5	147.4	148.3	0.87
^{53}Cr	95.2 ± 5.1	90 ± 7.1	90.7 ± 7.4	22.3	19.7	41.6	35.0	36.9	28.6	43.2	25.3	71.8	72.0	0.98
^{55}Mn	312 ± 43	337 ± 58	253 ± 10	637	596	871	826	1254	985	1048	923	825	811	0.94
^{57}Fe	42,774 ± 2951	45,597 ± 7299	53,300 ± 2000	31,732	35,090	44,633	46,051	40,183	34,851	396,91	34,270	38,329	43,720	0.87
^{59}Co	24.4 ± 2.4	26.8 ± 5.3	20.2 ± 0.6	9.8	9.7	19.2	16.3	19.7	17.7	19.5	17.7	20.2	19.1	0.91
^{60}Ni	80.1 ± 6.2	87.5 ± 9.5	75.9	14.1	13.7	23.9	19.6	22.5	18.7	21.9	17.6	44.2	42.7	0.98
^{65}Cu	24.2 ± 11.2	27.5 ± 5	36	24.7	29.5	31.3	35.6	44.6	49.6	42.2	39.5	24.1	29.7	0.84
^{66}Zn	121.9 ± 44.4	126.8 ± 23.6	101 ± 6	77.5	84.1	108.1	119.5	139.1	178.5	172.9	183.1	227.4	235.6	0.95
*^{75}As	11.4 ± 2.9	18.2 ± 8.8	14.1 ± 0.9	6.3	3.9	6.0	3.1	12.4	13.4	17.1	13.7	38.5	25.5	0.72
^{85}Rb	194.5 ± 11.1	205.1 ± 16.8	184 ± 16	100.5	115.2	129.0	116.2	123.8	123.9	128.8	154.6	130.4	137.7	0.90
^{88}Sr	71.3 ± 6.7	70.8 ± 9.6	66 ± 14	324.0	368.2	395.5	323.0	531.4	509.3	435.5	514.1	434.8	592.0	0.99
^{89}Y	28 ± 6.5	32.3 ± 8.5	–	20.3	25.5	19.8	21.2	31.8	25.4	21.4	27.6	24.6	27.1	0.99
*^{90}Zr	133 ± 34	210 ± 188	170	77	73	111	96	148	119	126	176	156	188	0.63
^{93}Nb	21.74 ± 1.79	23.42 ± 3.7	–	11.91	8.94	11.86	10.57	16.29	11.71	13.96	15.58	18.84	18.09	0.86
^{98}Mo	1.07 ± 0.11	1.43 ± 0.38	2.2	0.43	0.43	0.99	0.74	0.75	0.97	1.19	1.31	1.27	1.44	0.86
*^{107}Ag	0.12 ± 0.1	0.64 ± 0.66	–	0.20	0.15	0.15	0.13	0.07	0.32	1.26	0.85	0.05	0.21	0.002
*^{111}Cd	0.36 ± 0.11	1.04 ± 1.05	–	0.53	0.60	0.51	0.29	0.97	0.47	1.00	0.34	0.59	0.50	0.05
*^{115}In	0.1 ± 0.04	0.11 ± 0.07	–	0.08	0.04	0.07	0.06	0.09	0.06	0.08	0.04	0.06	0.15	0.01
*^{118}Sn	3.7 ± 0.7	4.9 ± 0.6	–	1.7	1.7	2.3	2.3	1.7	2.4	2.3	2.2	1.9	2.7	0.88
*^{121}Sb	1.05 ± 0.31	1.23 ± 0.3	1.39 ± 0.09	0.48	0.43	0.84	0.61	0.70	1.14	1.15	0.69	1.09	1.50	0.72
^{133}Cs	10.5 ± 0.8	11.8 ± 1.4	10.6 ± 0.7	7.0	7.5	9.2	8.5	10.9	11.3	8.7	10.9	19.9	20.3	0.89
^{137}Ba	626 ± 43	634 ± 59	668 ± 43	1654	1518	812	679	755	692	733	1148	742	851	0.96
^{139}La	39.68 ± 7.89	43.96 ± 13.34	50.9 ± 3.1	32.56	37.77	28.21	27.43	32.33	26.94	37.69	54.81	52.59	64.67	1.00
^{140}Ce	95.6 ± 16.7	96.5 ± 27.4	109.7 ± 6.5	48.9	51.5	64.5	52.1	62.3	55.0	80.9	108.9	114.0	128.8	1.00

^{141}Pr	10.3 ± 2.3	10.2 ± 3.3	–	7.3	9.0	6.9	6.3	8.0	6.9	9.7	12.8	13.1	15.3	1.00
^{146}Nd	37.8 ± 8.2	37 ± 9.9	42.3 ± 2.8	29.0	32.0	28.6	23.5	32.1	25.2	35.5	45.6	46.7	52.8	1.00
^{147}Sm	7.3 ± 1.5	7.2 ± 1.8	8 ± 0.6	5.7	6.5	5.8	5.1	7.1	5.6	7.5	8.8	9.9	9.3	0.99
^{153}Eu	1.68 ± 0.33	1.75 ± 0.48	1.54 ± 0.14	1.76	1.89	1.57	1.60	1.69	1.71	1.57	2.35	2.10	2.69	1.00
^{157}Gd	5.47 ± 1.27	6.77 ± 1.75	–	4.43	4.90	4.09	4.43	4.61	4.82	4.32	6.44	5.67	6.84	1.00
^{159}Tb	0.81 ± 0.18	1.23 ± 0.34	1.06 ± 0.04	0.68	0.88	0.65	0.79	0.81	0.95	0.61	1.06	0.78	1.19	0.99
^{163}Dy	4.68 ± 1.39	5.87 ± 1.69	7.4 ± 1.4	3.55	4.25	3.75	4.18	5.34	4.48	4.21	5.40	4.85	5.29	1.00
^{165}Ho	0.97 ± 0.26	1.33 ± 0.44	–	0.72	1.02	0.75	0.96	1.11	1.10	0.79	1.21	0.96	1.18	1.00
^{166}Er	2.91 ± 0.71	3.48 ± 0.73	–	1.89	2.37	2.07	2.50	3.04	2.50	2.47	2.87	3.00	2.75	1.00
^{169}Tm	0.41 ± 0.14	0.57 ± 0.13	–	0.25	0.46	0.28	0.45	0.42	0.50	0.36	0.48	0.45	0.51	1.00
^{172}Yb	3.1 ± 0.58	3.75 ± 0.83	4.27 ± 0.16	1.78	2.59	2.20	2.67	3.49	2.72	2.97	2.94	3.21	2.64	1.00
^{175}Lu	0.48 ± 0.12	0.68 ± 0.23	0.64 ± 0.09	0.28	0.42	0.32	0.47	0.54	0.57	0.48	0.53	0.50	0.52	0.99
*^{178}Hf	4.63 ± 0.58	12.33 ± 14.24	7.16 ± 0.54	2.32	11.10	2.92	3.82	4.95	3.26	4.45	1.53	6.46	6.29	0.36
^{181}Ta	2.65 ± 0.92	2.34 ± 0.91	1.65 ± 0.21	1.64	0.64	0.63	0.70	0.79	0.89	0.98	1.02	1.87	1.18	0.73
^{182}W	3.16 ± 0.29	3.64 ± 0.66	5.7	0.75	0.72	1.34	0.96	1.08	1.60	1.40	1.65	1.27	1.57	0.83
*^{197}Au	0.12 ± 0.06	0.15 ± 0.1	0.013	0.10	0.54	0.07	0.61	0.21	0.06	0.12	0.04	0.18	0.03	0.08
*^{206,207,208}Pb	15.3 ± 4.1	18.9 ± 5.1	–	18.4	17.6	22.7	21.2	24.4	28.3	35.7	34.6	41.2	36.1	0.95
*^{209}Bi	0.26 ± 0.37	0.88 ± 0.69	–	0.20	0.13	0.05	0.30	0.00	0.12	0.02	0.07	0.01	0.17	0.21
^{232}Th	13.4 ± 2.2	14.4 ± 3.3	15.1 ± 0.8	7.4	8.9	7.7	8.9	11.3	9.8	11.5	17.8	15.6	17.1	0.78
^{238}U	3.1 ± 0.5	3.7 ± 0.8	2.96 ± 0.34	1.9	1.9	2.6	2.7	2.9	3.3	2.7	4.1	4.1	4.2	0.98

Elements marked with an asterisk were removed from further analysis due to poor comparability between ablation cells, low precision, or high fractionation

3.4.2 Open-Cell Ablation

Whole vessels were ablated using the EAF open-cell laser. The laser was run at 100 % energy output, but settings were otherwise identical to those used with the 213 nm laser. He carrier gas flow was adjusted for each sample to achieve maximal signal strength. The chamber was positioned against the bottom of the vessels, which are both unpainted and relatively flat in the case of the Inka and Killke style ceramics analyzed. The chamber flange was then sealed to the vessel surface using Duck© brand poster-putty. While this material adheres reasonably well to both the plastic flange and the ceramic vessel surface, it does tend to leave a slight discoloration on the surface of the ceramic that disappears after several months. Additionally, it often proved difficult to establish a reliable seal with the sample, particularly for vessels with dusty or friable surfaces. It was sometimes necessary to make several attempts to produce a reliable seal, and even then at times the integrity of the seal progressively degraded such that the magnitude of background signal increased and stability decreased over the course of an analysis, introducing a source of uncontrolled error. Comparison samples analyzed using both chambers were affixed to a flat plastic sheet, to which the open-cell was then sealed—this resulted in a far more robust seal that did not degrade during the course of ablation.

Standards (NIST610, NIST679, and NORC) were placed directly on the sample surface within the chamber and analyzed (five ablation spots) before and after ten ablations of the object surface. Calibration was performed by the same method as utilized for data generated with the closed-cell. The EAF open-cell design produces an initial sharp peak in signal followed by a continuous attenuation as ablation proceeds, and as such, we initially included data from the first three measured replicates, but as this proved to introduce large amounts of error, we subsequently processed data in an identical manner as for closed-cell ablation, removing the first three replicates from consideration. This did however result in relatively low signal to background ratios for many elements, introducing another potential source of error. We additionally observed that movement across the sample surface sometimes resulted in highly variable signal strength, an effect which has been noted in other open-cell ablation systems using an injector nozzle when ablation is performed outside of the immediate area of gas flow against the sample surface (Bleiner and Bogaerts 2007).

Unlike most laser ablation units, which utilize pure quartz windows, the window provided with the EAF unit is coated with a polarizing film. After test measurements revealed unusually high concentrations of Zr, Sn, and Hf in samples measured with the open-cell laser, we performed a test ablation on the window using the closed-cell 213 nm laser. Signal counts per second measured on the window compared to a blank (Fig. 3.4) suggest that a number of elements are present at measurable concentrations in this film or within the window material itself. While in principal the laser is focused beyond the window, the presence of a polarizing film suggests that the laser should interact with the surface and therefore the potential exists for material present on the window surface to enter the cell atmosphere. How much this impacts measurement is also a function of how concentrated a particular element is in any sample being measured—comparative overlays of signal counts for ablation of NIST610 glass and NIST679 ceramic, for instance, suggest that while NIST610 (trace elements at ~500 ppm) measures well above both the blank and concentrations in the cell window, for some elements, elemental concentrations in NIST679 and other ceramics might be overwhelmed by background noise. While the actual amount of any given element that might enter the open cell during ablation is unknown and may differ from that actually present in the cell window, relatively high concentrations of Zr, As, Se, Cd, Ag-Sb, Au, Hf and Pb may impact measurements on ceramic samples using the open-cell.

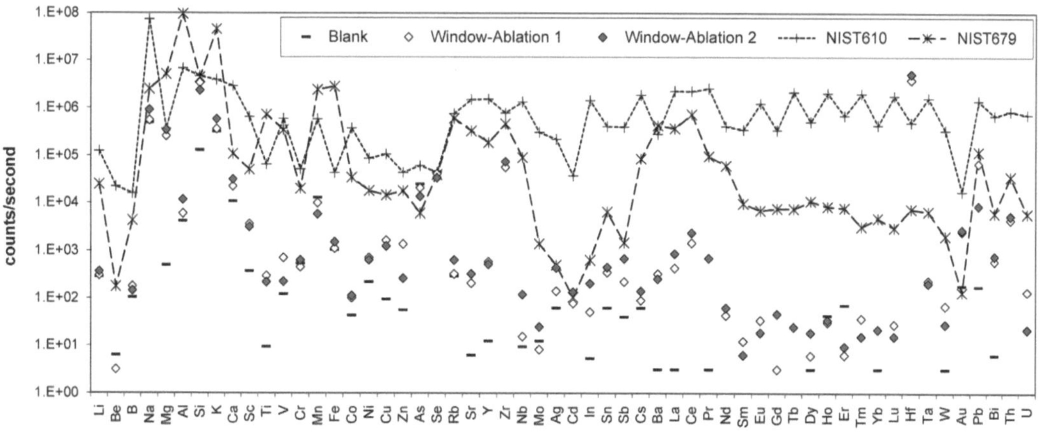

Fig. 3.4 Results of two ablations of the open-cell quartz window compared to a blank measurement showing the presence of significant concentrations of many elements in the coating applied to the window. Signal counts for ablations of NIST610 and NIST679 are shown for comparison. Counts per second are displayed on a logarithmic scale

3.5 Comparison of Chamber Performance

3.5.1 Limits of Detection

Limits of detection (LODs) for the regular and open-cells were compared by measuring ten blanks in a chamber containing only NIST610 glass. NIST610 was then ablated, and the major element concentrations were summed to 100 % and used to calculate minor and trace concentrations for measured blank replicates. Limits of detection were taken as three times the standard deviation for these ten blank measurements. In order to test the impact of seal quality on background, the performance of the open-cell was assessed in two configurations—with the chamber sealed to a flat plastic surface, and with the chamber sealed onto a ceramic vessel.

In both configurations, limits of detection (Fig. 3.5) were generally higher for the open-cell than for the closed-cell for most elements, with the exception of Cu, Sn, and Pb. The lower LODs for these three elements may reflect the shorter wash-out time characteristic of large volume open-cell chambers (Wagner and Jędral 2011). LODs for the open-cell either sealed to a ceramic vessel or to a flat plastic backing were generally comparable, although interestingly, were in some cases lower when sealed to a ceramic vessel. Whether this results from differences in gas flow and wash-out because of slightly different chamber geometry, or from differences in the quality of seal produced is unknown.

3.5.2 Precision

Figure 3.6 shows estimated precision—percent relative standard deviation (1-σ) between repeated measurements of NORC and NIST610 using both the open- and closed-cell lasers, arranged from highest to lowest precision for closed-cell measurements of NORC. NIST610 glass is taken to represent the base level of precision achievable on a highly homogenous sample matrix with elemental concentrations generally well above detection limits. Percent relative standard deviations between ten measurements of NIST610 are generally below 4 % for the majority of elements (and below 1 % for some) using both chambers, although values measured on the open-cell are slightly higher in general than those obtained using the closed-

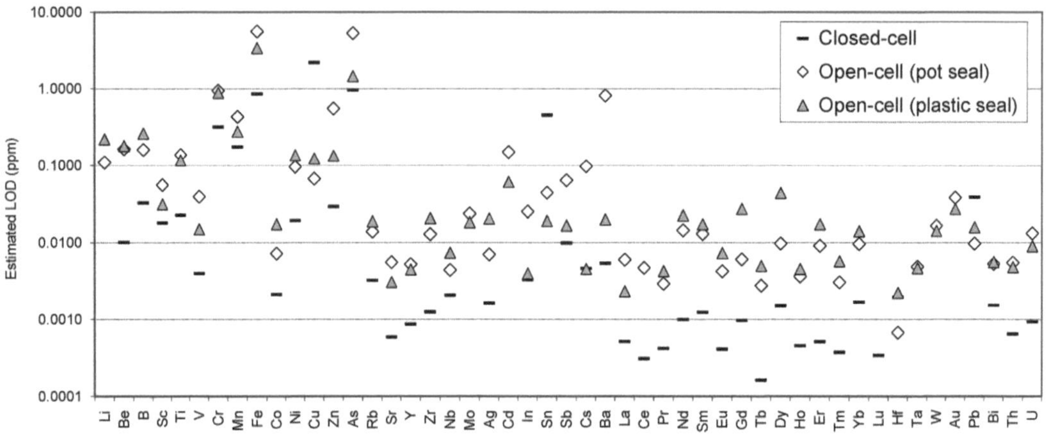

Fig. 3.5 Limits of detection measured for minor and trace elements for closed-cell and open-cell laser chambers, with the later sealed either to a flat plastic backing or directly to the surface of a ceramic vessel. Concentrations are displayed on a logarithmic scale

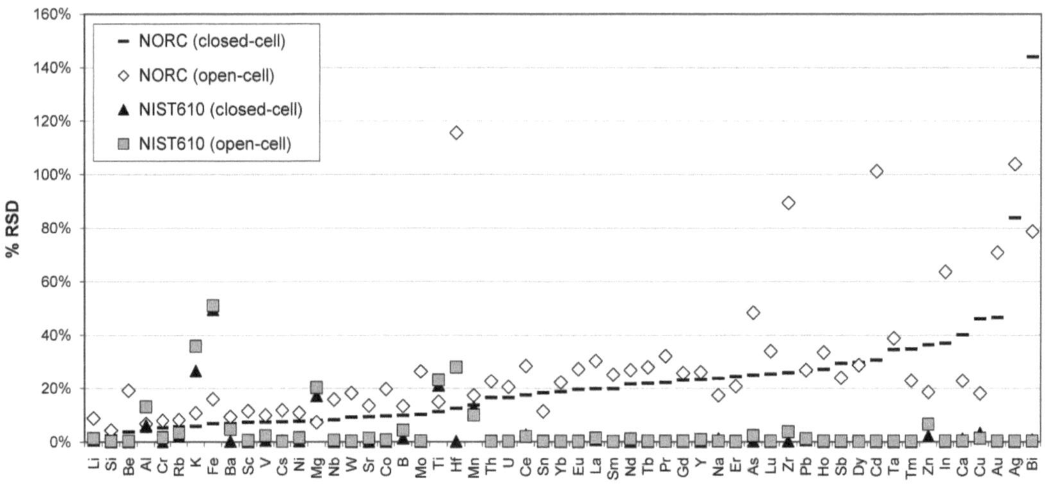

Fig. 3.6 Precision estimated from repeated measurements of New Ohio Red Clay (NORC) and NIST610 standard glass using both closed-cell and open-cell lasers

cell. The only exceptions are elements that are near detection limits (Mg, K, Ti, Fe), and Hf, which is present at high concentration in the cell window as noted.

Repeated measurements of NORC are presented as an estimate of error when measuring ceramic samples. In general, deviations between measurements of NORC are marginally greater using the closed-cell and were actually lower for some elements such as Sn and Cu, perhaps reflecting the faster wash-out rate achieved with the open cell. However, for other elements, the open-cell system performed substantially worse than the closed-cell, particularly for Zr, As, Cd, Hf, Ag, Au, and Bi, elements that are present at very low concentrations in ceramic samples and standards, and at relatively high concentrations in the cell window as noted. For elements which do not present these problems, precision appears generally comparable between the two chambers.

3.5.3 Fractionation

Fractionation—differential elemental removal and entrapment in the ablation plume—results from a variety of factors including differences in ablation efficiency, transport efficiency, and ionization efficiency (Chen 1999; Eggins et al. 1998; Jackson 2001; Speakman and Neff 2005). The amount of fractionation can vary dramatically between different material types, and is dependent on laser wavelength, gas flow, ablation time, and other specifics of chamber and laser specification.

Previous studies have indicated that ablation using shorter wavelength lasers generally results in reduced fractionation. Longer wavelength laser such as the 266 nm unit utilized in this study produce ablated particle size distributions weighted towards heavier particles relative to shorter wavelength (193 or 213 nm) lasers. These heavier, larger particles tend to more readily settle back to the sample surface as they interact with the carrier gas in the chamber, tend to be transported less efficiently to the plasma, and are less efficiently ionized in the plasma (Alexander et al. 1998; Arrowsmith and Hughes 1988; Gonzalez et al. 2002; Guillong and Günther 2002). Additionally, these heavier particles may subsequently interact with the laser after ablation (plasma ablation), preferentially leading to deposition of more refractory elements on the sample surface, while less refractory elements are preferentially transported to the plasma (Eggins et al. 1998). Furthermore, differences in sample opacity can significantly impact ablation efficiency (Gaboardi and Humayun 2009) and induce fractionation, an effect which is also impacted by laser wavelength. In particular, for more opaque materials such as NIST610 glass, Guillong et al. (2003) found that 213 nm lasers produce far less fractionation than 266 nm lasers.

Additionally, differences in chamber design can contribute to fractionation (e.g., Eggins et al. 1998). The closed-cell UP213 laser used in the present study introduces carrier gas in a laminar fashion across the sample surface, while the open-cell uses an injector nozzle to introduce a jet of gas incident upon the sample surface at approximately a 45° angle. The later arrangement might exacerbate plasma ablation as heavier particles are driven back down more rapidly towards the ablation crater rather than being swept towards the plasma. Chamber volume and shape can also result in different sized particles taking very different trajectories from the time of ablation and entrapment until they reach the exit opening and are transported to the ICP-MS for ionization (Bleiner and Bogaerts 2007).

To assess the degree to which fractionation occurs using the two lasers, we calculate the Fractionation Index (FI—Fryer et al. 1995; Guillong et al. 2003) for samples and corresponding standards for both chambers. This index is defined as the ratio of the average of the second half of the ablation signal normalized to an internal standard element divided by the internal standardized average signal for the first half of the ablation. While prior applications of this index have utilized Ca as an internal standard, we here utilize Si, as we use this element as our internal standard for purposes of calibration. An FI of close to one indicates that little fractionation occurs—any signal variability or attenuation occurs at the same rate in both a given element and the internal standard element. Conversely, values less than or greater than one indicate that fractionation occurs during ablation. Large differences in the degree of fractionation between samples and standards suggest that close matrix-matching is necessary to generate a reliable calibration. The FI was calculated by taking the ratio of the average Si-normalized signal for the third-fifth measured replicates to the Si-normalized signal for the sixth-ninth replicates, i.e., for ~20–40 seconds and 40–-60 seconds of ablation. The FI was calculated for all measured standards and a Cusco ceramic sample (CSC033) that was measured using both the open-cell and closed-cell.

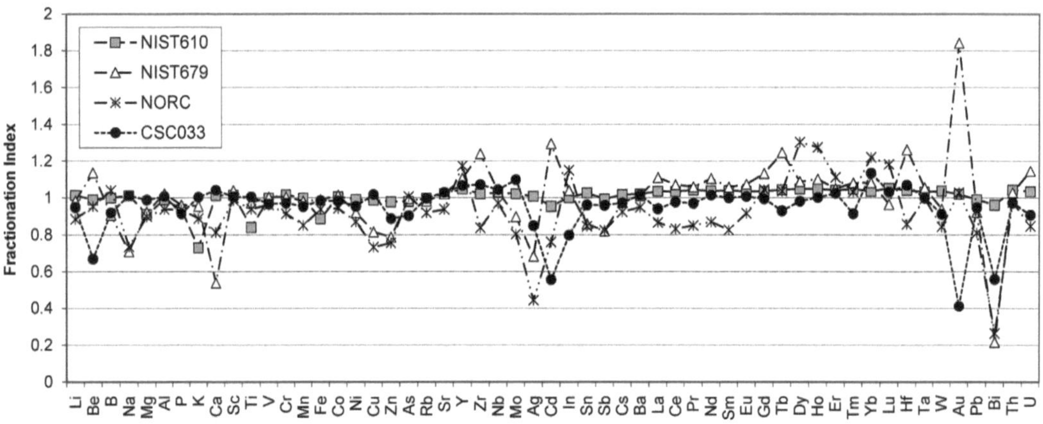

Fig. 3.7 Fractionation Index (relative to Si) calculated for NIST standards 610 (glass) and 679 (ceramic), New Ohio Red Clay (NORC), and pottery sample CSC033 from measurements obtained using the closed-cell 213 nm laser

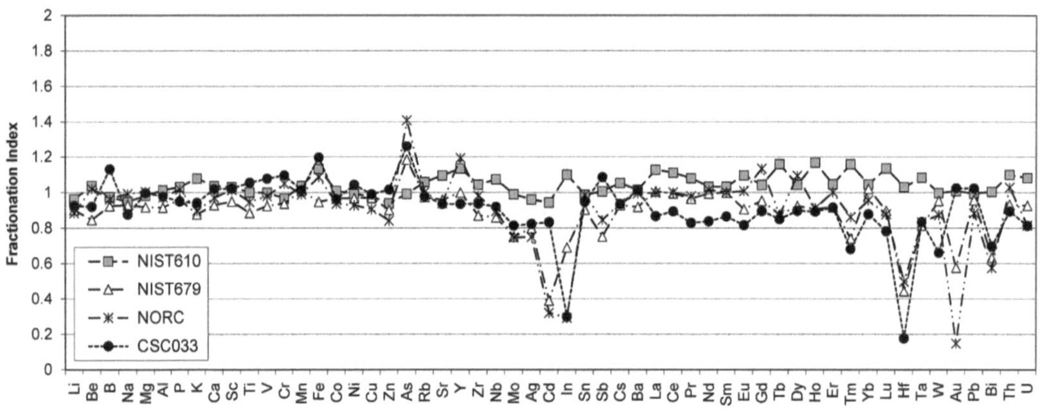

Fig. 3.8 Fractionation Index (relative to Si) calculated for NIST standards 610 (glass) and 679 (ceramic), New Ohio Red Clay (NORC), and pottery sample CSC033 from measurements obtained using the open-cell 266 nm laser

FI results for the closed-cell UP213 laser (Fig. 3.7) suggest that during ablation of NIST679 and NORC, some fractionation does occur, particularly for Na, Zr, Ag, Au, and Bi. For NIST610 standard glass, little fractionation is evident, with only K, Ti, and Fe—all of which are present in very low concentrations and measure with poor precision—showing any significant deviation from unity. CSC033 likewise demonstrates little evidence for fractionation for the majority of elements. FIs deviate more substantially from unity for analysis performed using the open-cell Micro 266 laser (Fig. 3.8), even for NIST610 standard glass. In part, the elements that appear most impacted may reflect some degree of input from the cell window (As, Cd, In, Hf, Au, and Bi), but fractionation appears to be greater for most measured elements than for ablation performed using the 213 nm laser.

3.5.4 Accuracy and Comparability of Open- and Closed-Cell Ablation Results

We report accuracy measured on Ohio Red clay in comparison to values published in Kuleff and Djingova (1998), summarizing repeat

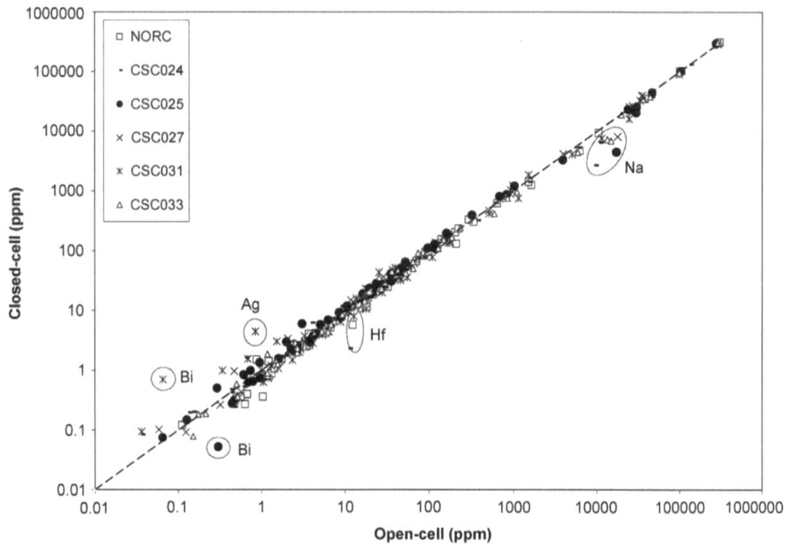

Fig. 3.9 Comparison of concentrations for all measured elements in New Ohio Red Clay (NORC) and five Cusco area ceramic vessels using the open-cell (X-axis) and closed-cell (Y-axis) lasers. The *dashed line* represents a 1:1 ratio between measurements for the two chambers. Elements that diverge significantly between the two chambers in some or all comparison samples are labeled. Concentrations on both axes are displayed on a logarithmic scale

measurements from ten laboratories using INAA and ICP-AES (Table 3.3). It should be noted that the methods used to analyze Ohio Red clay at other laboratories reported in the literature are all bulk methods, rather than the spot analysis employed at the EAF. NORC contains inclusions visible under magnification in the laser chamber, which are avoided during analysis, and consequently, our values and reported values by bulk analysis may not always match for every element. For instance, published values for Ca, Zr, and Hf (comparing only the closed-cell for Hf) are particularly divergent, possibly because earlier studies sampled larger proportions of large zircon grains and calcium-rich particles in the NORC matrix. For other measured elements, both our closed-cell and open-cell results are within methodological error ranges of published values.

Despite issues with fractionation, signal to background ratios, and precision, for most elements, the open-cell Micro 266 laser appears to produce comparable results to those obtained using the closed-cell UP213 system. Comparison of measurements for Cuzco ceramics and NORC indicates no significant deviations between measurements for most elements (Fig. 3.9, Table 3.3). Exceptions include Na, Hf, Bi, and Ag, all elements for which a substantial degree of fractionation, poor precision and potential impact from the cell window were noted.

Direct comparison on an element by element basis for all comparison samples similarly demonstrates a high level of linear correlation for most elements, including Na (Fig. 3.10, r^2 values reported in Table 3.3). The measured piece of glass did not deviate in any significant way from lines established using ceramics and clays, suggesting that for most elements, fractionation on the basis of material type or opacity does not prohibit good calibration of results. Most elements for which problems have already been noted correlated poorly between the two laser chambers, and were omitted from further analysis, as was Pb—although Pb values correlated reasonably well between the two lasers, anomalously high values were measured in some Cuzco area samples using the open-cell.

While Pb is sometimes used in white paints, and can leach into the underlying ceramic matrix,

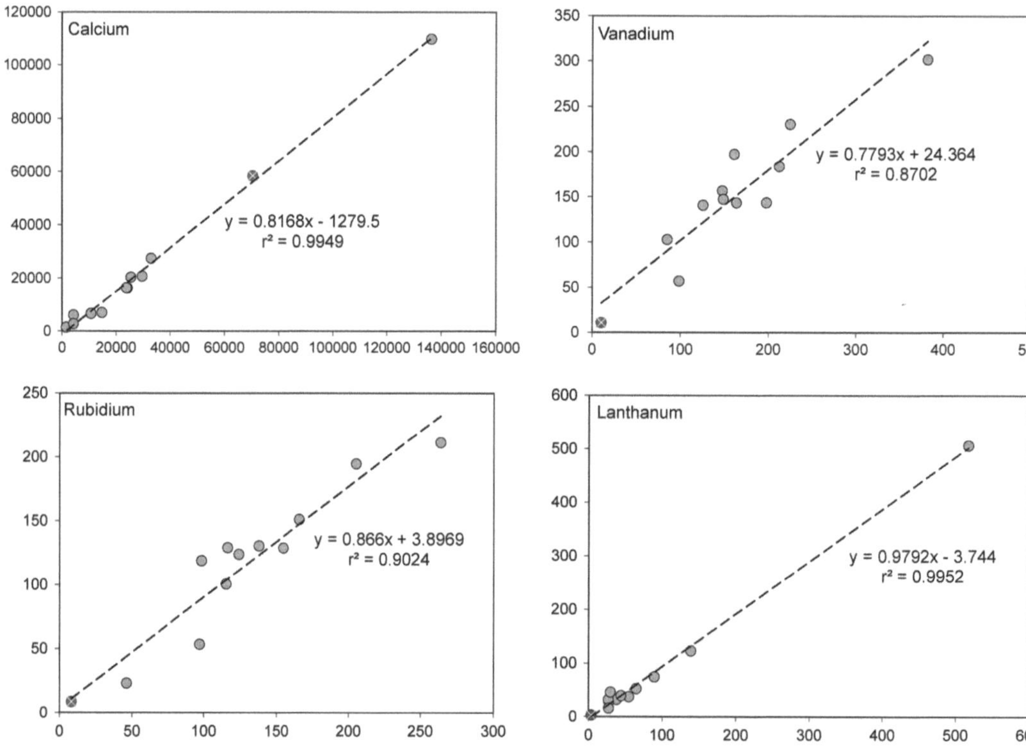

Fig. 3.10 Results of comparison measurements for selected elements with open-cell results on the X-axis and closed-cell values on the Y-axis, with linear least-squares regression lines displayed. *Circles* represent ceramic samples, and *crosses* in circles glass. All concentrations are in parts per million

there is little textual or archaeological evidence that lead-based paints were used on Inka or Killke ceramics (P. VanValkenberg, personal communication), lead glazing was introduced only during the colonial period (see Chap. 19), and only one of the vessels sampled in the current study had white paint applied. That particular sherd (CSC003) was not one of the samples run by both methods, and measurements on its cross-section using the closed-cell do not exhibit elevated Pb concentrations. While studies of Aguada tri-color ceramics (fourth to twelfth centuries AD) from the Catamarca region of Argentina have identified use of lead-based white paints there (Bertolino et al. 2008), studies of earlier ceramics from elsewhere in the Andes (Nasca, northern Peru, first to seventh centuries AD, and Diaguita and Aconagua cultures, central Chile, ~AD 1200) identified Ca and Ti as the primary constituents of white paints applied to their surfaces (Dinator and Morales 1990; Vaughn et al. 2005). Most of the vessels included in our study were decorated with red, brown, or black paints likely produced using Fe and Mn. Pb is however present at relatively high levels in the open-cell window (Fig. 3.4), likely explaining high Pb concentrations in some open-cell measurements, and consequently, only 44 elements were retained for statistical analysis (see Tables 3.3 and 3.4).

3.6 Preliminary Results of Analysis of Cuzco Area Ceramics and Clays

3.6.1 Ceramics Analyzed by Closed-Cell Ablation

All chemical measurements were logged (base 10) to normalize their distribution and minimize scaling differences between high- and

Table 3.4 Mean concentration values (ppm) and standard deviations for identified chemical groups measured by both closed-cell ablation and open-cell ablation

	Group 1 (closed), n = 25	Group 1 (open), n = 6	Group 2 (closed), n = 4	Group 2 (open), n = 2	Group 3 (closed), n = 2	CSC022 (closed)	CSC016 (open)
^{7}Li	76.4 ± 24.1	76.1 ± 44.9	79.9 ± 32.3	50.0 ± 14.5	27.9 ± 1.8	54.1	71.7
^{9}Be	2.86 ± 0.64	4.30 ± 1.17	2.91 ± 0.37	3.86 ± 0.46	3.35 ± 0.06	2.24	4.55
^{11}B	60.1 ± 22.4	67.6 ± 9.3	76.3 ± 32.4	57.6 ± 9.9	26.9 ± 4.8	21.8	85.9
^{23}Na	6750 ± 2375	8017 ± 2835	6142 ± 1543	4222 ± 1027	3816 ± 1577	5986	8226
^{24}Mg	36,175 ± 15,663	22,478 ± 5749	33,910 ± 17,925	17,938 ± 5749	6223 ± 1162	8071	25,108
^{27}Al	97,961 ± 4966	100,671 ± 12,118	97,941 ± 6864	89,853 ± 12,209	136,953 ± 5782	77,689	102,967
^{29}Si	288,117 ± 13,235	317,389 ± 24,017	289,715 ± 16,584	326,691 ± 15,002	298,447 ± 2968	333,062	278,043
^{39}K	26,300 ± 5409	26,983 ± 5148	32,872 ± 2474	36,106 ± 9198	17,762 ± 4417	37,845	68,253
^{44}Ca	19,049 ± 9121	13,235 ± 11,986	12,449 ± 6203	13,527 ± 2166	5745 ± 1207	10,404	23,040
^{45}Sc	18.0 ± 2.7	16.6 ± 4.9	17.6 ± 1.8	18.5 ± 4.8	16.6 ± 2.3	9.9	20.1
^{49}Ti	3921 ± 787	3624 ± 996	4370 ± 616	3969 ± 673	3465 ± 72	3245	4444
^{51}V	138.4 ± 24.7	112.3 ± 14.5	140.7 ± 12.0	139.2 ± 24.8	84.2 ± 26.4	121.5	161.0
^{53}Cr	52.5 ± 13.0	33.5 ± 15.7	72.6 ± 26.1	38.2 ± 12.5	20.5 ± 2.6	6.1	39.0
^{55}Mn	991 ± 351	1274 ± 626	774 ± 199	2534 ± 2734	604 ± 46	631	848
^{57}Fe	38,714 ± 4220	29,692 ± 2341	39,322 ± 1972	34,412 ± 7504	31,585 ± 208	30,870	33,488
^{59}Co	20.1 ± 3.0	16.6 ± 1.7	22.6 ± 3.1	30.6 ± 23.5	10.5 ± 1.0	5.6	18.1
^{60}Ni	28.3 ± 7.5	26.5 ± 4.4	46.5 ± 12.2	31.6 ± 8.9	13.5 ± 0.9	5.7	28.8
^{65}Cu	34.6 ± 10.5	63.1 ± 21.8	39.4 ± 18.2	53.6 ± 1.9	17.8 ± 9.7	16.5	54.9
^{66}Zn	151.8 ± 66.0	250.3 ± 200.3	219.7 ± 103.9	177.7 ± 19.8	76.5 ± 1.4	108.8	117.2
*^{75}As	9.4 ± 5.0	34.4 ± 18.8	15.2 ± 15.7	20.3 ± 1.9	6.4 ± 0.2	9.5	24.0
^{85}Rb	105.7 ± 26.9	92.4 ± 20.8	127.0 ± 15.4	95.4 ± 32.6	90.9 ± 13.5	145.9	111.9
^{88}Sr	362.1 ± 129.1	369.7 ± 150.8	977.3 ± 725.5	584.5 ± 275.9	370.1 ± 65.2	988.3	1687.9
^{89}Y	25.5 ± 5.0	19.1 ± 6.0	25.6 ± 3.5	21.3 ± 8.7	23.8 ± 4.9	24.8	21.7
*^{90}Zr	119.9 ± 27.0	104.8 ± 27.8	175.2 ± 29.2	155.4 ± 12.0	77.9 ± 1.4	253.1	197.6
^{93}Nb	14.5 ± 3.4	13.6 ± 2.1	19.2 ± 0.3	16.2 ± 4.5	11.0 ± 1.3	22.4	22.2
^{98}Mo	0.80 ± 0.31	1.38 ± 0.26	1.22 ± 0.24	2.17 ± 0.30	0.35 ± 0.11	1.96	1.79
*^{107}Ag	0.13 ± 0.24	3.93 ± 2.81	0.13 ± 0.09	2.12 ± 0.09	0.12 ± 0.12	0.04	3.43
*^{111}Cd	0.53 ± 0.26	1.93 ± 0.75	0.59 ± 0.07	1.38 ± 0.32	0.43 ± 0.14	0.42	1.67
*^{115}In	0.07 ± 0.04	0.13 ± 0.07	0.07 ± 0.01	0.09 ± 0.01	0.07 ± 0.02	0.03	0.10
*^{118}Sn	2.3 ± 1.6	8.3 ± 3.9	2.3 ± 0.6	4.2 ± 3.0	1.4 ± 0.4	1.0	3.2
*^{121}Sb	0.76 ± 0.45	2.09 ± 0.91	0.79 ± 0.25	1.62 ± 0.59	0.39 ± 0.14	0.20	1.10
^{133}Cs	7.5 ± 2.3	10.4 ± 6.1	11.8 ± 5.6	6.0 ± 4.6	7.0 ± 0.1	3.0	3.5

(continued)

Table 3.4 (continued)

	Group 1 (closed), n = 25	Group 1 (open), n = 6	Group 2 (closed), n = 4	Group 2 (open), n = 2	Group 3 (closed), n = 2	CSC022 (closed)	CSC016 (open)
^{137}Ba	707 ± 283	764 ± 292	1504 ± 1157	1512 ± 386	1198 ± 645	1475	2469
^{139}La	31.7 ± 7.3	27.5 ± 8.7	70.9 ± 13.3	53.6 ± 2.8	35.3 ± 3.9	111.5	133.3
^{140}Ce	64.6 ± 15.5	53.0 ± 16.5	143.6 ± 20.7	107.9 ± 2.4	54.9 ± 8.6	216.3	226.8
^{141}Pr	8.2 ± 1.8	6.0 ± 2.0	16.5 ± 2.5	10.9 ± 0.2	8.3 ± 1.4	26.1	23.3
^{146}Nd	31.3 ± 6.0	26.9 ± 8.0	60.6 ± 9.5	46.9 ± 1.6	32.2 ± 4.5	95.9	90.4
^{147}Sm	6.4 ± 1.2	5.7 ± 2.1	10.7 ± 1.4	8.8 ± 0.4	6.3 ± 0.9	14.9	15.6
^{153}Eu	1.6 ± 0.3	1.5 ± 0.8	2.5 ± 0.3	1.7 ± 0.3	1.9 ± 0.2	2.8	3.9
^{157}Gd	4.7 ± 0.9	5.2 ± 2.1	6.8 ± 1.1	6.9 ± 1.6	4.6 ± 0.2	7.3	14.2
^{159}Tb	0.71 ± 0.12	0.86 ± 0.63	0.91 ± 0.11	0.92 ± 0.27	0.69 ± 0.01	0.86	1.60
^{163}Dy	4.6 ± 0.8	4.0 ± 1.1	5.3 ± 1.0	4.5 ± 0.9	4.0 ± 0.7	4.6	4.9
^{165}Ho	0.94 ± 0.18	0.73 ± 0.28	1.01 ± 0.14	0.87 ± 0.20	0.82 ± 0.15	0.85	1.05
^{166}Er	2.7 ± 0.4	2.5 ± 1.3	3.0 ± 0.6	2.7 ± 1.0	2.2 ± 0.4	2.3	3.2
^{169}Tm	0.42 ± 0.12	0.27 ± 0.12	0.48 ± 0.12	0.31 ± 0.00	0.25 ± 0.00	0.17	0.30
^{172}Yb	2.91 ± 0.51	2.37 ± 0.29	3.24 ± 0.72	2.37 ± 0.66	2.25 ± 0.65	2.25	2.24
^{175}Lu	0.44 ± 0.09	0.36 ± 0.09	0.48 ± 0.11	0.42 ± 0.03	0.32 ± 0.06	0.30	0.41
$*^{178}Hf$	3.9 ± 0.8	6.2 ± 5.9	6.4 ± 1.2	4.0 ± 5.2	2.6 ± 0.3	6.7	8.3
^{181}Ta	0.87 ± 0.19	1.45 ± 0.31	1.57 ± 0.37	1.44 ± 0.16	1.44 ± 0.28	1.54	1.48
^{182}W	1.18 ± 0.28	1.69 ± 0.53	1.32 ± 0.21	1.72 ± 0.17	0.65 ± 0.13	1.52	1.62
$*^{197}Au$	0.29 ± 0.83	0.31 ± 0.20	0.12 ± 0.05	0.24 ± 0.07	64.51 ± 91.09	14.24	0.00
$*^{206,207,208}Pb$	21.2 ± 9.5	164.8 ± 223.9	34.5 ± 5.7	110.1 ± 87.0	15.7 ± 3.8	14.2	27.0
$*^{209}Bi$	0.04 ± 0.08	0.25 ± 0.26	0.05 ± 0.08	0.04 ± 0.03	0.10 ± 0.14	0.00	0.07
^{232}Th	10.6 ± 2.6	7.9 ± 2.9	17.8 ± 2.8	13.7 ± 0.3	8.7 ± 1.9	31.4	21.4
^{238}U	2.8 ± 0.5	2.4 ± 0.7	4.0 ± 0.3	4.4 ± 1.0	1.7 ± 0.3	5.8	5.2

Elements marked with an asterisk were removed from further analysis due to poor comparability between ablation cells, low precision, or high fractionation

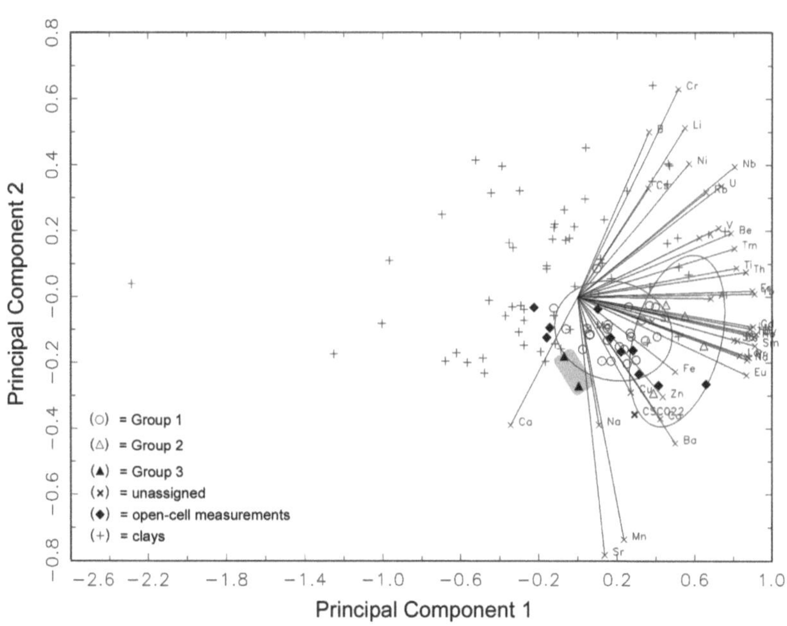

Fig. 3.11 R-Q mode biplot of principal components 1 (48 % of total data variance) and 2 (9.8 % of total data variance) showing the three group structure identified among the sherds analyzed using closed-cell ablation, with clay samples and pots measured with the open-cell Micro 266 laser projected against the 90 % confidence *ellipses* for chemical Groups 1 and 2

low-concentration elements. An R-Q mode (Baxter 1992; Neff 1994, 2002) principal components analysis (PCA) based on the correlation matrix for the combined dataset of ceramics (run with both chambers) and clays was then performed to allow for identification and display of patterning in the dataset. R-Q mode analysis allows simultaneous display of patterning on both matrix rows (samples) and columns (elements), allowing for concise display of relationships between samples and how elemental patterning relates to sample patterning, as well as correlations between elements. Elemental correlations can be particularly important for hypothesizing about what mineralogical and geological factors may account for differences between identified distinct chemical groupings. This type of graph is known as a biplot, and is commonly utilized in archaeological ceramic analysis as both a means of pattern recognition and data display (e.g., Arnold et al. 2000; Glascock et al. 2004).

Both hierarchical cluster analysis and examination of PCA biplots suggests the presence of a three group structure in the sherds analyzed by closed-cell ablation. The bulk of the analyzed ceramic samples fall into a single chemically homogenous group termed Group 1, while Groups 2 and 3 are much smaller, containing four and two sherds respectively (see Table 3.4 for summary statistics). CSC022 is classified as a chemical outlier, and may represent a fourth chemical type present among the analyzed sherds and vessels. Display of the results on the first, second, and third principal component axes (Figs. 3.11 and 3.12) shows much of the patterning present in the data. As can be noted, Group 2 is distinguished from sherds in Group 1 largely due to higher concentrations of Rare Earth Elements, but also some transition metals (Fe, Zn) and actinoids (Th, U). Group 3 sherds are distinguished by lower Li, Mg, Co, and Ca concentrations, and elevated Na concentrations. All three chemical groups appear to overlap with locally available clays, suggesting that all recognized chemical types could have been produced from raw materials available in the broader Cuzco area.

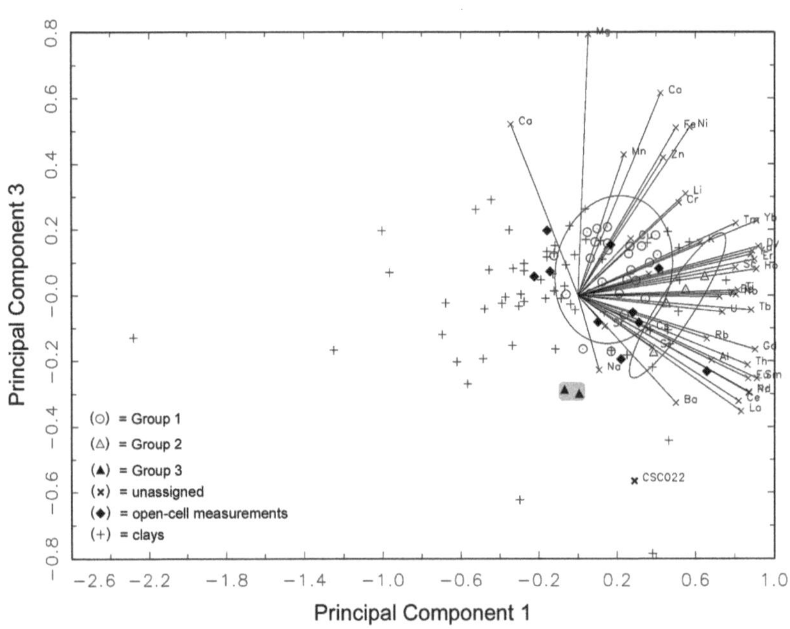

Fig. 3.12 R-Q mode biplot of principal components 1 (48 % of total data variance) and 3 (8.4 % of total data variance) showing the three group structure identified among the sherds analyzed using closed-cell ablation, with clay samples and pots measured with the open-cell Micro 266 laser projected against the 90 % confidence *ellipses* for chemical Groups 1 and 2

Chemical Groups 1 and 2 identified in the current study may be identical to the two Cuzco linked compositional profiles identified by Bray et al. (2005) in their INAA study of ceramics drawn from the same Field Museum collection. The two chemical signatures identified in that study were principally differentiated by REE concentrations as well as some transition metals (principally Cr), possibly reflecting the addition of basaltic tempers to some vessels and andesitic tempers to others (Ixer and Lunt 1991). While our analysis in principal avoided larger temper grains, it is possible that smaller grains were ablated and may account for the identification of two distinct compositional profiles. Alternatively, some of the chemical differences noted by Bray and colleagues may also be reflected in the fine-fraction of the Cuzco area ceramics. Our identification of a third and possibly fourth chemical profile among the Collier and Montez collection ceramics likely represents new previously unidentified production regimes in the broader Cuzco region.

3.6.2 Open-Cell Analyses

The ten vessels analyzed using our open-cell system appear to match up compositionally to those analyzed by closed-cell ablation, and overlap with both Groups 1 and 2 on both principal components plots. This overlap is also evident on bivariate plots, for instance La and U concentrations (Fig. 3.13). However, the degree of dispersion evident for the samples analyzed using the open-cell laser is greater than for those analyzed using the closed-cell laser (Table 3.3), suggesting that the reduced precision noted for some elements may have an impact. Even so, most vessels measured using the open-cell are assignable to Groups 1 (6) and 2 (2), while vessel CSC016 consistently plots with unassigned sherd CSC022, suggesting that they represent the same chemical paste type (see Table 3.4). One other vessel measured with the open-cell, CSC013, remains unassigned due to anomalously high Sr content (4764 ppm).

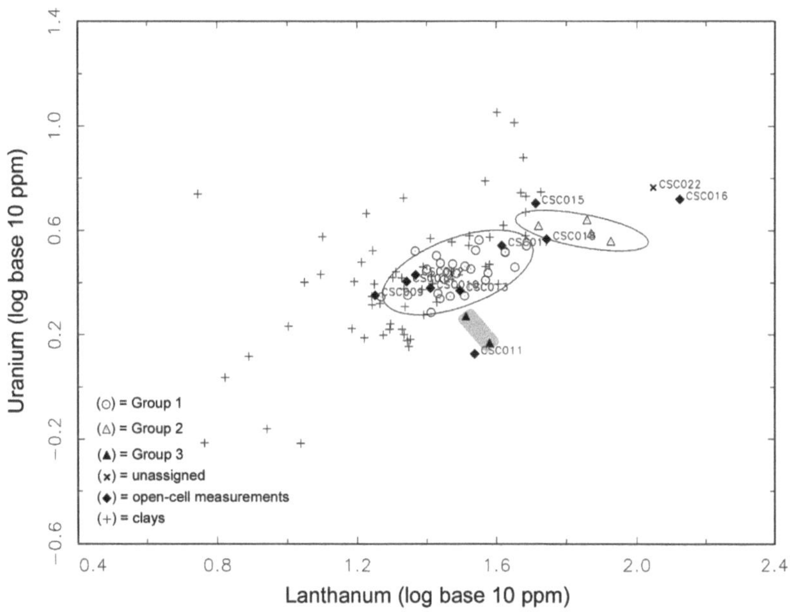

Fig. 3.13 Bivariate plot of logged (base 10) La and U concentrations showing the three group structure identified among the sherds analyzed using closed-cell ablation, with clay samples and pots measured with the open-cell Micro 266 laser projected against the 90 % confidence *ellipses* for chemical Groups 1 and 2

3.6.3 Comparison to Cuzco Area Clays

Detailed comparison of clays to ceramics is possible only for Group 1 ceramics at present. Mahalanobis distance based probabilities of group membership were calculated for all clay samples relative to Group 1 ceramic samples using the first 12 principal components (~91 % of total data variance). In general, membership probabilities were relatively low, with no clay sample exceeding a 7 % membership probability. In part, this may reflect changes in composition in prepared pastes relative to raw clays due to tempering, levigation, or other clay processing (e.g., Kilikoglou et al. 1988) that was not undertaken for the geological samples included in this analysis, but may have been by ancient potters. However, comparison of group membership probabilities for clays relative to Group 1 suggests a reasonable similarity to clays available within and along the edges of the Cusco basin. In particular, most clays with high membership probabilities in Group 1 were collected from areas associated with the Cretaceous fluvial, lacustrine, and eolian Huancané, Lucre, and Puraro Formations (Fig. 3.3—collectively grouped as KsP-c), suggesting derivation of the clays used to produce them from those units. This similarity suggests that Group 1 ceramics, whether Killke or Inka stylistically, were produced from clays collected very near to the archaeological contexts from which they were recovered.

Unfortunately, the small number of samples included in Groups 2 and 3 prohibit such robust comparison to analyzed clays at present. All such ceramics identified to date were collected at the sites of Qoripata and San Sebastian/San Geranimo, and may have been produced in that area of the southwestern Cuzco Basin (Table 3.1). However, further analysis of larger numbers of archaeological specimens will be required to identify potential extraction locations for the clays used to produce Group 2 and 3 vessels.

3.7 Conclusion

Analysis of Cuzco-area Killke and Inka style ceramics drawn from the Field Museum Montez and Collier collections by both conventional closed-cell ablation and using a modified open-cell laser chamber suggests that the later method holds great promise for incorporating museum collections into archaeological studies of ancient ceramic production and distribution. However, there are some problems that were identified with the Field Museum open-cell system that should be corrected by future modifications. Issues related to the coating applied to the window of the EAF open-cell are easily fixed. However, we also experienced difficulties producing a reliable seal between the open-cell and some vessel surfaces, or else to a progressive degradation of seal integrity over time, leading to leaching of atmospheric gases into the chamber. This resulted in substantial increases in analysis time, making it unfeasible at present to run more than three or four vessels per day. It is likely that some of the reduced precision we observed in open-cell measurements as compared to those performed using a closed-cell result from inconsistent chamber atmosphere, particularly for elements with low signal to background ratios. Additionally, introduction of oxygen into the ablation atmosphere can increase fractionation and further impact measurement using the open-cell (Košler et al. 2002).

We are currently testing several solutions to this issue, including different types of sealants, and the containment of samples within large vacuum bags which are then pumped down to low pressure. As previously discussed, the recent development of open-cells that are not physically sealed to the sample surface (Asogan et al. 2009; Kovacs et al. 2010) represents another potential means of analyzing materials such as ceramic vessels that often have irregular or dusty/friable surfaces.

Additionally, we observed inconsistent signal strength as the chamber was moved relative to the inflow of He from the injector nozzle, resulting in poor signal to background ratios for some measurements. The introduction of a micro jet-pump for gas-extraction from the chamber such as that utilized by Asogan et al. (2009) would allow for more uniform extraction and mixing of ablated material from the chamber and the introduction of a steady gas flow to the plasma. Alternatively, it might be necessary to alter the shape of the chamber to an elongated or bottle shaped design, both of which have been shown to improve performance (Bleiner and Bogaerts 2007).

Despite these issues, the data generated using the open-cell laser compared favorably to those generated using the closed-cell, suggesting that with methodological improvements, the EAF open-cell laser system will prove a viable means of incorporating collections objects into archaeological ceramic sourcing studies in the future. In the present preliminary study of ceramic production in the Cuzco area during both the Late Intermediate and Inka periods, we were able to link samples analyzed using the open-cell laser to chemical profiles identified among the samples analyzed using the closed-cell, and to link both to potential clay extraction locations along the edges of the Cuzco Basin.

Acknowledgments The ICP-MS lab at the Field Museum Elemental Analysis Facility was constructed with funding from the National Science Foundation (BCS-0320903), the Museum's Anthropology Alliance, and an anonymous donation. The open-cell laser used in this study (projects EAF047 and EAF049) was modified by Richard Cox (Université du Québec à Chicoutimi) and purchased with funding from the Grainger Foundation. Analysis was funded by NSF Archaeometry Grant BCS-0726651. Many thanks to Laure Dussubieux and Brian Bauer for their assistance, and to two anonymous reviewers for an abundance of useful comments on an earlier draft. All remaining errors are the sole responsibility of the authors.

References

Alexander ML, Smith MR, Hartman JS, Mendoza A, Koppenaal DW (1998) Laser ablation inductively coupled plasma mass spectrometry. Appl Surf Sci 127–129:255–261

Arnold DE, Neff H, Glascock MD (2000) Testing assumptions of Neutron Activation Analysis: communities, workshops and paste preparation in Yucatán, Mexico. Archaeometry 42:301–316

Arrowsmith P, Hughes SK (1988) Entrainment and transport of laser ablated plumes for subsequent elemental analysis. Appl Spectrosc 42:1231–1239

Asogan D, Sharp BL, O'Connor CP, Green DA, Hutchinson RW (2009) An open, non-contact cell for laser ablation-inductively coupled plasma-mass spectrometry. J Anal Atom Spectrom 24:917–923

Bauer BS (1999) The early ceramics of the Inca Heartland, vol 31, Fieldiana Anthropology New Series. Field Museum of Natural History, Chicago, IL

Bauer BS (2004) Ancient Cuzco: heartland of the Inca. University of Texas Press, Austin, TX

Bauer BS, Stanish C (1990) Killke and Killke-related pottery from Cuzco, Peru, in the Field Museum of Natural History, vol 15, Fieldiana Anthropology New Series. Field Museum of Natural History, Chicago, IL

Baxter MJ (1992) Archaeological uses of the biplot—a neglected technique? In: Lock G, Moffett J (eds) Computer applications and quantitative methods in archaeology. BAR International Series S577. Tempvs Reparatvm, Archaeological and Historical Associates, Oxford, pp 141–148

Beck ME, Neff H (2007) Hohokam and Patayan interaction in southwestern Arizona: evidence from ceramic compositional analyses. J Archaeol Sci 34:289–300

Bertolino SR, Josa VG, Carreras AC, Laguens A, de la Fuente G, Riveros JA (2008) X-ray techniques applied to surface paintings of ceramic pottery pieces from Aguada Culture (Catamarca, Argentina). X-Ray Spectrom 38:95–102

Bleiner D, Bogaerts A (2007) Computer simulations of sample chambers for laser ablation-inductively coupled plasma spectrometry. Spectrochim Acta B 62:155–168

Bray TL, Minc LD, Constanza Ceruti M, Chávez JA, Perea R, Reinhard J (2005) A compositional analysis of pottery vessels associated with the Inca ritual of capacocha. J Anthropol Archaeol 24:82–100

Caillaux VC, Cárdenas Roque J, Carlier G (2011) Geología del Cuadrángulo de Cusco. Hoja 28-s Boletín no. 138 Serie A Carta Geológica Nacional Escala 1:50,000. INGEMMET, Lima

Chen Z (1999) Inter-element fractionation and correction in laser ablation inductively coupled plasma mass spectrometry. J Anal Atom Spectrom 14:1823–1828

Cogswell JW, Neff H, Glascock MD (1996) The effect of firing temperature on the elemental characterization of pottery. J Archaeol Sci 23:283–287

Devos W, Moor C, Lienemann P (1999) Determination of impurities in antique silver objects for authentication by laser ablation inductively coupled plasma mass spectrometry (LA-ICP-MS). J Anal Atom Spectrom 14:621–626

Dinator MI, Morales JR (1990) Characterization of colour pigments in pre-Columbian Chilean potteries by PIXE elemental analysis. J Radioanal Nucl Chem 140:133–139

Dussubieux L, Golitko M, Williams PR, Speakman RJ (2007) LA-ICP-MS analysis applied to the characterization of Peruvian Wari ceramics. In: Glascock MD, Speakman RJ, Popelka-Filcoff RS (eds) Archaeological chemistry: analytical technique and archaeological interpretation. American Chemical Society, Washington, DC, pp 349–363

Eckert SL, James WD (2011) Investigating the production and distribution of plain ware pottery in the Samoan archipelago with laser ablation-inductively coupled plasma-mass spectrometry (LA-ICP-MS). J Archaeol Sci 38:2155–2170

Eggins SM, Kinsley LPJ, Shelley JMG (1998) Deposition and element fractionation processes during atmospheric pressure laser sampling for analysis by ICP-MS. Appl Surf Sci 127–129:278–286

Elliot S, Knowles M, Kalinitchenko I (2004) A new direction in ICP-MS. Spectroscopy 19:30–38

Fitzpatrick SM, Takamiya H, Neff H, Dickinson WR (2006) Compositional analysis of Yayoi-Heian period ceramics from Okinawa: examining the potential for provenance study. Geoarchaeology 21:803–822

Forster N, Grave P, Vickery N, Kealhofer L (2011) Non-destructive analysis using PXRF: methodology and application to archaeological ceramics. X-Ray Spectrom 40:389–398

Frankel D, Webb JM (2012) Pottery production and distribution in prehistoric Bronze Age Cyprus. An application of pXRF analysis. J Archaeol Sci 39:1380–1387

Fryer BJ, Jackson SE, Longerich HP (1995) The design, operation and role of the laser-ablation microprobe coupled with an inductively coupled plasma-mass spectrometer (LAM-ICP-MS) in the earth sciences. Can Mineral 33:303–312

Gaboardi M, Humayun M (2009) Elemental fractionation during LA-ICP-MS analysis of silicate glasses: implications for matrix-independent standardization. J Anal Atom Spectrom 24:1188–1197

Glascock MD (1992) Neutron activation analysis. In: Neff H (ed) Chemical characterization of ceramic pastes in archaeology. Prehistory Press, Madison, WI, pp 11–26

Glascock MD, Neff H, Vaughn KJ (2004) Instrumental neutron activation analysis and multivariate statistics for pottery provenance. Hyperfine Interact 154:95–105

Glaus R, Koch J, Günther D (2012) A portable laser ablation sampling device for elemental fingerprinting of objects outside the laboratory with laser ablation inductively coupled plasma mass spectrometry. Anal Chem 84(12):5358–5364

Golitko M (2011) Provenience investigations of ceramic and obsidian samples using laser ablation-inductively coupled plasma-mass spectrometry (LA-ICP-MS) and portable X-ray fluorescence (p-XRF). In: Terrell JE, Schechter EM (eds) Exploring prehistory on the Sepik Coast of Papua New Guinea, vol 42, Fieldiana Anthropology New Series. Field Museum of Natural History, Chicago, IL, pp 251–287

Golitko M, Bosquet D (2011) Implications des Analyses de la Composition de la Céramique pour la transition Mesolithique-Neolithique dans L'Europe du Nord-Ouest. In: Hauzeur A, Jadin I, Jungels C (eds) 5000 ans avant J.-C.: la grande migration? Le Néolithique ancien dans la collection Louis Éloy, vol 3, Collections du Patrimoine culturel no. Édition du Service Patrimoine culturel de la Fédération Wallonie-Bruxelles, Brussels, pp 86–93

Golitko M, Terrell JE (2012) Mapping prehistoric social fields on the Sepik coast of Papua New Guinea: ceramic compositional analysis using laser ablation-

inductively coupled plasma-mass spectrometry. J Archaeol Sci 39:3568–3580
Gonzalez J, Mao XL, Roy J, Mao SS, Russo RE (2002) Comparison of 193, 213 and 266 nm laser ablation ICP-MS. J Anal Atom Spectrom 17:1108–1113
Goren Y, Mommsen H, Klinger J (2011) Non-destructive provenance study of cuneiform tablets using portable X-ray fluorescence. J Archaeol Sci 38:684–696
Gratuze B, Blet-Lemarquand M, Barrandon J-N (2001) Mass spectrometry with laser sampling: a new tool to characterize archaeological materials. J Radioanal Nucl Chem 247:645–656
Guillong M, Günther D (2002) Effect of particle size distribution on ICP-induced elemental fractionation in laser ablation-inductively coupled plasma-mass spectrometry. J Anal Atom Spectrom 17: 831–837
Guillong M, Horn I, Günther D (2003) A comparison of 266 nm, 213 nm and 193 nm produced from a single solid state Nd:YAG laser for laser ablation ICP-MS. J Anal Atom Spectrom 18:1224–1230
Harbottle G (1976) Activation analysis in archaeology. In: Newton GWA (ed) Radiochemistry, vol 3. The Chemical Society, London, pp 33–72
Ixer RA, Lunt S (1991) Petrography of certain pre-Spanish pottery of Peru. In: Middleton A, Freestone I (eds) Recent developments in ceramic petrology. British Museum Occasional Paper no. 81, London, pp 137–164
Jackson S (2001) The application of Nd:YAG lasers in LA-ICP-MS. In: Sylvester P (ed) Laser-ablation ICPMS in the earth sciences: principles and applications, vol 29, Short Course Series. Mineralogical Association of Canada, St. Johns, pp 29–46
Jochum KP, Weis U, Stoll B, Kuzmin D, Yang Q, Raczek I, Jacob DE, Stracke A, Birbaum K, Frick DA, Günther D, Enzweiler J (2011) Determination of reference values for NIST SRM 620-617 glasses following ISO guidelines. Geostand Geoanal Res 35:397–429
Joyce AA, Neff H, Thieme MS, Winter M, Elam JM, Workinger A (2006) Ceramic production and exchange in late/terminal formative period Oaxaca. Lat Am Antiq 17:579–594
Kilikoglou V, Maniatis Y, Grimanis AP (1988) The effect of purification and firing of clays on trace element provenance studies. Archaeometry 30:37–46
Košler J, Longerich HP, Tubrett MN (2002) Effect of oxygen on laser-induced elemental fractionation in LA-ICP-MS analysis. Anal Bioanal Chem 374:251–254
Kovacs R, Nishiguchi K, Utani K, Günther D (2010) Development of direct atmospheric sampling for laser ablation-inductively coupled plasma-mass spectrometry. J Anal Atom Spectrom 25:142–147
Kuleff I, Djingova R (1998) Mean concentrations of elements determined in Ohio Red Clay. J Radioanal Nucl Chem 237:3–6
Neff H (1994) R-Q mode principal components analysis of ceramic compositional data. Archaeometry 36:115–130
Neff H (2002) Quantitative techniques for analyzing ceramic compositional data. In: Glowacki DM, Neff H (eds) Ceramic production and circulation in the Greater Southwest: source determination by INAA and Complementary Mineralogical Investigations. Monograph 44. The Cotsen Institute of Archaeology, University of California, Los Angeles, CA
Neff H, Bove FJ (1999) Mapping ceramic compositional variation and prehistoric interaction in Pacific Coastal Guatemala. J Archaeol Sci 26:1037–1051
Niziolek LC (2013) Earthenware production and distribution in the Prehispanic Philippine polity of Tanjay: results from laser ablation-inductively coupled plasma-mass spectrometry (LA-ICP-MS). J Archaeol Sci 40:2824–2839
Pearce NJG, Perkins WT, Westgate JA, Gorton MT, Jackson SE, Neal CR, Chenery SP (1997) A compilation of new and published major and trace element data for NIST SRM610 and SRM612 glass reference materials. Geostandard Newslett 21:114–115
Rice PM (1987) Pottery analysis: a sourcebook. University of Chicago Press, Chicago, IL
Schwedt A, Mommsen H (2007) On the influence of drying and firing of clay on the formation of trace element concentration profiles within pottery. Archaeometry 49:495–509
Sharratt N, Golitko M, Williams PR, Dussubieux L (2009) Ceramic production during the Middle Horizon: Wari and Tiwanaku clay procurement in the Moquegua Valley, Peru. Geoarchaeology 24:792–820
Speakman RJ, Neff H (2005) The application of laser ablation ICP-MS to the study of archaeological materials—an introduction. In: Speakman RJ, Neff H (eds) Laser ablation ICP-MS in archaeological research. University of New Mexico Press, Albuquerque, NM, pp 1–16
Speakman RJ, Little NC, Creel D, Miller MR, Iñañez JG (2011) Sourcing ceramics with portable XRF spectrometers? A comparison with INAA using Mimbres pottery from the American Southwest. J Archaeol Sci 38:3483–3496
Stoner WD, Glascock MD (2012) The forest or the trees? Behavioral and methodological considerations for geochemical characterization of heavily-tempered ceramic pastes using NAA and LA-ICP-MS. J Archaeol Sci 39:2668–2683
Vaughn KJ, Neff H (2004) Tracing the clay source of Nasca polychrome pottery: results from a preliminary raw material survey. J Archaeol Sci 31:1577–1586
Vaughn KJ, Conlee CA, Neff H, Schreiber KJ (2005) A compositional analysis of Nasca pigments: implications for craft production on the pre-hispanic South Coast of Peru. In: Speakman RJ, Neff H (eds) Laser ablation ICP-MS in archaeological research. University of New Mexico Press, Albuquerque, NM, pp 139–153
Vaughn KJ, Dussubieux L, Williams PR (2011) A pilot compositional analysis of Nasca ceramics from the Kroeber collection. J Archaeol Sci 38:3560–3567
Wagner B, Jędral W (2011) Open ablation cell for LA-ICP-MS investigations of historic objects. J Anal Atom Spectrom 26:2058–2206

Optimization of 2D LA-ICP-MS Mapping of Glass with Decorative Colored Features: Application to Analysis of a Polychrome Vessel Fragment from the Iron Age

4

Johannes T. van Elteren, Serena Panighello, Vid S. Šelih, and Emilio F. Orsega

Abstract

2D elemental mapping of glass surfaces by LA-ICP-MS is an interesting technique to elucidate past technologies, establish provenance or understand deterioration processes of ancient, polychrome glass by visualization of the elemental distribution of the glass surface. However, selection of the appropriate LA-ICP-MS conditions for generation of high-quality elemental maps with the highest spatial resolution, lowest signal-to-noise ratio and shortest analysis time is normally a trial-and-error process. In this chapter a computational-experimental strategy is described to optimize the LA-ICP-MS conditions for 2D elemental mapping of polychrome glass by finding the best balance between fluence, beam diameter, repetition rate, scanning speed, gas flow rate and acquisition time. To aid in the initial selection of the optimal LA-ICP-MS conditions for spatial resolution and analysis time, a digital image of the glass was subjected to virtual 2D mapping, using existing software which simulates the actual LA-ICP-MS mapping process. To verify whether these initial conditions would result in an acceptable signal-to-noise ratio during the actual LA-ICP-MS mapping process, they were used to experimentally determine the detection limits for each element via a simple line scan on a "blank"

J.T. van Elteren (✉) • V.S. Šelih
Laboratory for Analytical Chemistry, National Institute of Chemistry, Hajdrihova 19, 1000 Ljubljana, Slovenia
e-mail: elteren@ki.si; vid.selih@ki.si

S. Panighello
Laboratory for Analytical Chemistry, National Institute of Chemistry, Hajdrihova 19, 1000 Ljubljana, Slovenia

Department of Molecular Sciences and Nanosystems, University Ca'Foscari of Venice, Calle Larga S. Marta 2137, 30123 Venice, Italy
e-mail: serena.panighello@unive.it

E.F. Orsega
Department of Molecular Sciences and Nanosystems, University Ca'Foscari of Venice, Calle Larga S. Marta 2137, 30123 Venice, Italy
e-mail: ors-ef@unive.it

L. Dussubieux et al. (eds.), *Recent Advances in Laser Ablation ICP-MS for Archaeology*, Natural Science in Archaeology, DOI 10.1007/978-3-662-49894-1_4

glass, and consequently predict the noise floor in the maps. This strategy was successfully validated (using a modern *murrina*) and applied to a polychrome glass from the Iron Age yielding more insight into its elemental composition and the mineral sources involved.

4.1 Introduction

The raw materials for manufacturing of ancient glass were siliceous sand with sodium or potassium compounds used as fluxes. In the Egyptian area of Wadi-El-Natrun, an evaporitic source of sodium (natron) was used as flux since the early fourth millennium BC (Turner 1956; Shortland 2004; Shortland et al. 2006). In the Bronze Age the most used sources of sodium were plant ashes. From the early first millennium BC, to the end of the first millennium AD, almost all glass was natron-based. During the ninth century AD natron flux was progressively replaced by soda or potash plant ashes (Henderson 1985, 2000; Freestone et al. 2002; Tite et al. 2006; Freestone 2006; Barkoudah and Henderson 2006; Aerts et al. 2003; Henderson et al. 2004; Mirti et al. 2009). Other minerals from different sources were added to the glass melting batch to serve as chromophores and opacifiers. Their main occurrence is shown in the timetable in Fig. 4.1 and their effect on the colouring of glass is shown in Table 4.1. For several centuries manganese and antimony compounds were employed alternately or together to decolourize transparent glasses (Baxter et al. 2005; Jackson 2005; Sayre 1963; Silvestri et al. 2008).

Elemental analysis is the primary tool to characterize ancient glass samples concerning their raw materials, decolourizers, chromophores and opacifiers, and to confirm hypotheses about the mineral sources of the added compounds. The data obtained can then be related to other groups of glasses in order to resolve assumptions about their primary and secondary provenance and

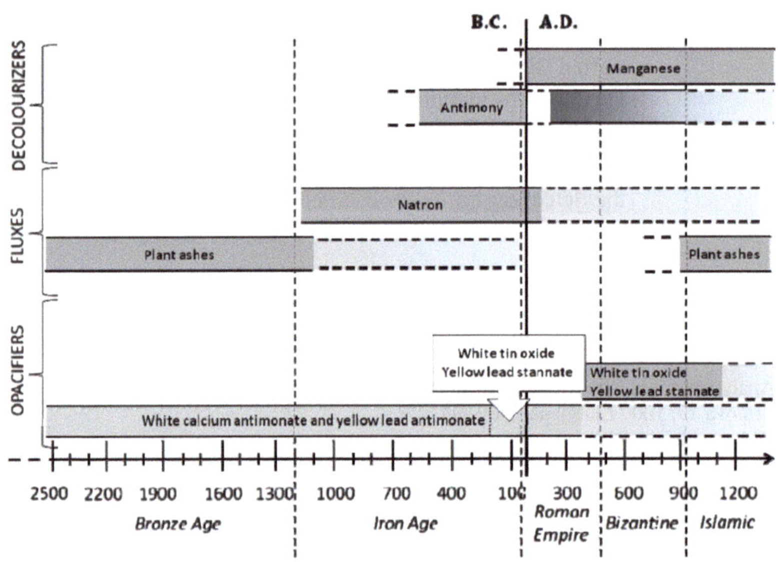

Fig. 4.1 Timeline of the use of opacifiers, fluxes and decolourizers in the production of glass from the Bronze Age to the Islamic Age (Turner and Rooksby 1961; Sayre and Smith 1967; Henderson and Warren 1983; Henderson 1985; Nicholson and Henderson 2000; Mass et al. 2002; Shortland 2002; Tite et al. 2008; Lahlil et al. 2010a)

Table 4.1 Principal ancient chromophoric elements and their relative colouring effects based on oxidizing or reducing melting conditions (Henderson 2000; Shortland 2002; Verità et al. 2002)

	Colour	
Element/compound	Oxidizing conditions	Reducing conditions
Cobalt	Blue (Co^{2+})	
Copper	Aquamarine (Cu^{2+})	Uncoloured (Cu^{+}) or Red (Cu+/Cu°)
Manganese	Violet (Mn^{3+})	
Manganese (high concentration)	Black (Mn^{3+})	
Iron	Yellow (Fe^{3+})	Light blue (Fe^{2+})
Iron (high concentration)	Very dark green or Black (Fe^{3+} + Fe^{2+})	
Lead antimonate	Opaque yellow	
Calcium antimonate	Opaque white	
Lead stannate	Opaque yellow	
Tin oxide	Opaque white	
Sulphur-iron		Amber-yellow (Fe^{3+}–S^{2-})
Sulphur-iron (high concentration)		Brown or Black (Fe^{3+}–S^{2-})

production period. A large variety of multi-elemental techniques exist for the microanalysis of glass and glazes: SEM-EDXS (e.g., Lahlil et al. 2010b), PIXE-PIGE (e.g., Šmit et al. 2012), EPMA (e.g., Arletti et al. 2010; Alberta et al. 2011), TEM (e.g., Mata et al. 2002; Lombardo et al. 2013), micro-XRF (e.g., Naes et al. 2008), SIMS (e.g., Rutten et al. 2009) and LA-ICP-MS (e.g., Duwe and Neff 2007; Pérez-Arantegui et al. 2008; Robertshaw et al. 2010). Often a combination of techniques is used to investigate elemental composition or distribution (e.g., Sokaras et al. 2009; Carmona et al. 2010; Šmit et al. 2013). They all have specific advantages and limitations such as spatial resolution, range of detectable elements, the feasibility of quantitative analysis, detection limits, dynamic range, required sample preparation, cost and accessibility (Behrends and Kleingeld 2009). In spite of the fact that LA-ICP-MS has a comparatively low native lateral resolution, and leads to sometimes serious damage to the sample, the technique is very versatile with a very large dynamic range (analysis of concentrations from ng g^{-1} to % *w*/*w*), and the potential for imaging on scales from several μm to tens of mm (Woodhead et al. 2007).

Compared to the commonly available contenders such as SEM-EDXS and micro-XRF, LA-ICP-MS has a lower lateral resolution than SEM-EDXS but is far superior with regards to obtainable detection limits, whereas LA-ICP-MS is relatively comparable to micro-XRF but with better accuracy, precision, sensitivity and analysis speed (Naes et al. 2008). With recent technical and instrumental improvements and improved 2D mapping protocols and data deconvolution approaches lateral resolutions $\leq$1 μm have been demonstrated for LA-ICP-MS (Van Malderen et al. 2015; Wang et al. 2013).

Although 2D mapping by LA-ICP-MS is destructive to the sample, often many fragments of sufficiently small dimensions are available from museums or other sources on which this invasive sampling is allowed. Moreover, the mapping may be very informative in the study of mechanisms and effects of weathering and corrosion on ancient (and modern) glass, glaze and ceramic surfaces. Several papers have reported on the use of LA-ICP-MS for visualization of the elemental distribution of glass surfaces using depth profiling (Panighello et al. 2015), 2D mapping (Rusk et al. 2011; Šelih and Van Elteren 2011; Wagner and Jedral 2011) or 3D mapping (Van Elteren et al. 2013). One of the difficulties of high-quality 2D elemental mapping by LA-ICP-MS is the selection of appropriate mapping parameters; this chapter will look into ways to improve upon this situation as will be explained in the background section below.

4.2 Background

Selection of the appropriate LA-ICP-MS conditions (fluence, beam diameter, repetition rate, scanning speed, gas flow rate and acquisition time) for generation of high-quality 2D elemental image maps is complex due to the mutual effects of the LA-ICP-MS conditions on the quality of the maps in terms of spatial resolution and signal-to-noise ratio within a given time frame for analysis (Fig. 4.2). While it is possible to increase the resolution by decreasing the beam diameter, this invariably increases the analysis time and potentially the signal-to-noise ratio when trace level concentrations are to be measured.

Since decorative coloured features in glass are determined by the presence of chromophoric elements (especially heavy metals in the form of ions or compounds, see Table 4.1), a digital image of the glass gives a rough representation of the chromophoric elemental distribution, and as such may be used for virtual mapping to simulate the 2D LA-ICP-MS elemental mapping process. Virtual mapping software has been previously developed (Triglav et al. 2010); the conceptual model behind the software can be explained in a simple way by illustrating what happens from ablation to detection during a single line scan on a sample with two rectangular objects (Fig. 4.3). When a laser beam with diameter D (μm) traverses the surface of the sample with a scanning speed S (μm s^{-1}), and the laser firing at an optimal fluence and a repetition rate R (s^{-1}), the amount of material sampled per pulse is directly proportional to the cross-sectional area A (μm^2) of the laser beam and the object. Since the ablation cell volume V (l) is relatively large and the gas flow rate F (l s^{-1}) has an upper limit, the washout of aerosol particles from the ablation cell is controlled by their residence time σ ($= V/F$) in the cell, resulting in an exponential washout of the particles from the cell expressed by $A \times \exp(-t/\sigma)$ for each pulse, assuming aerosol ablation under turbulent conditions. Thus, the aerosol concentration leaving the ablation cell is represented, as a function of time, by superimposition of tailing peaks as a result of exponential washout of subsequent pulses. When this continuous time-related concentration signal is presented to the ICP-MS instrument, discrete measurement takes place as a function of the acquisition time T (s). The software translates this 1D model into a virtual 2D mapping tool, with the understanding that maps consist of many parallel lines.

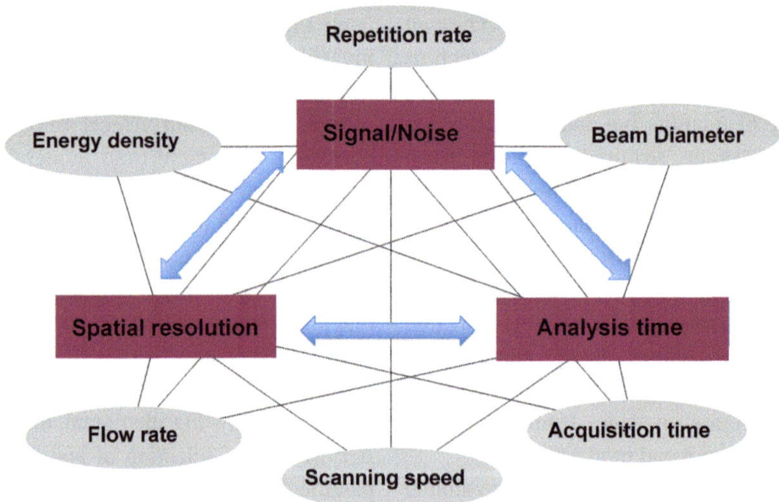

Fig. 4.2 The influence of LA-ICP-MS parameters (*ovals*) on the delicately balanced quality criteria (*rectangles*) for 2D elemental mapping by LA-ICP-MS

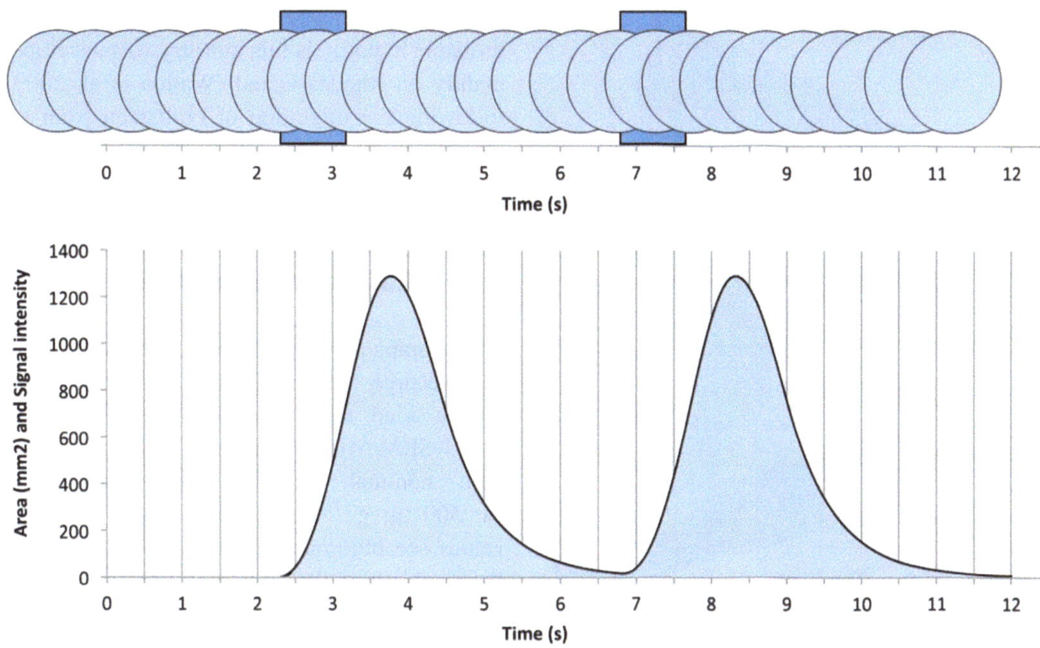

Fig. 4.3 Illustration of the LA-ICP-MS mapping process in 1D upon virtual ablation of two *rectangular objects* assuming a homogeneous density distribution; we assume that the background has a density distribution of zero

With this 2D mapping tool the quality of the maps can be accurately predicted by virtual mapping of a digital image of the polychrome glass sample as a function of the LA-ICP-MS conditions (see Fig. 4.2). However, the predicted quality can only be assessed in terms of 2D spatial resolution and analysis time as signal-to-noise ratios are based on actual elemental concentrations and can only be investigated experimentally. To get a handle on these signal-to-noise ratios we performed a simple line scan on a "blank" glass to retrieve the detection limits for each element as they are a direct indicator for the expected noise floor in the elemental maps. In an iterative process the software-optimized LA-ICP-MS conditions for spatial resolution and analysis time can be experimentally verified and fine-tuned with regards to acceptable noise levels. This is undertaken prior to actual elemental image mapping of the glass, thereby preventing unnecessary damage to the glass and yielding the best possible maps in the shortest amount of time. The predictive capabilities of this computational-experimental strategy were demonstrated by actual LA-ICP-MS elemental image mapping of a modern *murrina*—a slice of a group of parallel glass canes melted together, whose cross section reveals multiple layers of colours and shapes. Application of this approach served to generate 2D elemental image maps with a high spatial resolution and a low noise floor for a polychrome glass vessel fragment from the Iron Age. In all cases 54 elements were quantified in the image maps using a so-called sum normalization calibration procedure by summating the 54 elements in each pixel as their oxides to 100 % *w*/*w* (Šelih and Van Elteren 2011).

4.3 Materials and Methods

4.3.1 Archaeological Glass Sample

The example presented in Fig. 4.4 is an opaque polychrome glass vessel fragment from the Iron Age (fifth century BC) from the museum of Adria (Northern Italy). More information on the provenance of the sample can be found in Panighello et al. (2012). It presents a blue bulk with opaque yellow and turquoise decorations. The body part is translucent and shows the presence of bubbles

Fig. 4.4 Digital image of the polychrome glass vessel fragment; the *rectangle* indicates the area (5 mm × 5 mm) which was mapped

and devitrification crystals. This vessel was made using the "core-forming technique." Before the invention of glass-blowing around the first century AD, this was the most common method for the manufacture of glass vessels. This technique involved "the shaping of a clay core, which was then dipped into molten glass. Decoration was then trailed onto the surface of the vessel, often in contrasting colours. These trails of glass could then be combed into close-set zigzag patterns and either left in relief or, more commonly, flattened by marvering. Finally, a rim-disc, foot-ring or handles could be added. On cooling, the core was carefully scraped out" (Harden and Tatton-Brown 1981). Stylistically, this vessel falls within Harden's Mediterranean Group I, *aryballos* form 2 (Harden and Tatton-Brown 1981). This group denotes the first of three successive industries which manufactured large amounts of core-formed vessels traded in the Mediterranean context. According to the most probable hypothesis this industry was developed mainly on Rhodes Island (Walton et al. 2009), following the migration of craftsmen from the Near East.

4.3.2 Sample Handling and Glass Calibration Standards

For mapping purposes the archaeological artifact was securely mounted on a glass slide with double-sided tape, next to a reference glass (NIST SRM 610) serving as an external standard with nominal elemental concentrations of ca. 500 μg g^{-1} (for published and consensus values see http://georem.mpch-mainz.gwdg.de). For establishing the elemental detection limits a synthetic "blank" glass (DLH6, P&H Developments Ltd.), prepared according to the coprecipitated gel technique (Hamilton and Hopkins 1995), was used with a base composition of 72 % *w*/*w* SiO_2, 2 % *w*/*w* Al_2O_3, 12 % *w*/*w* CaO and 14 % *w*/*w* Na_2O. To validate the computational-experimental strategy for 2D LA-ICP-MS mapping a modern *murrina* from Murano, Venice (Italy) was used, both for LA-ICP-MS elemental oxide mapping and mapping simulation based on a digital image of the object. The *murrina* was embedded in epoxy resin in which carbon powder was mixed in for better visibility of the glass structures. Hence, in the polished block (Fig. 4.5a) the translucent parts of the glass are black and the dark blue colour of the outer rim of the *murrina* is obscured by the black background colour.

4.3.3 LA-ICP-MS Instrumentation and Mapping Conditions

A quadrupole ICP-MS (Agilent 7500ce, Palo Alto, USA) interfaced with a LA system (New Wave Research UP 213, Fremont, USA) was used to quantify elements present in the glass by rastering with the laser beam over the surface. The laser ablation device contained a frequency-quintupled Nd:YAG laser (wavelength, 213 nm;

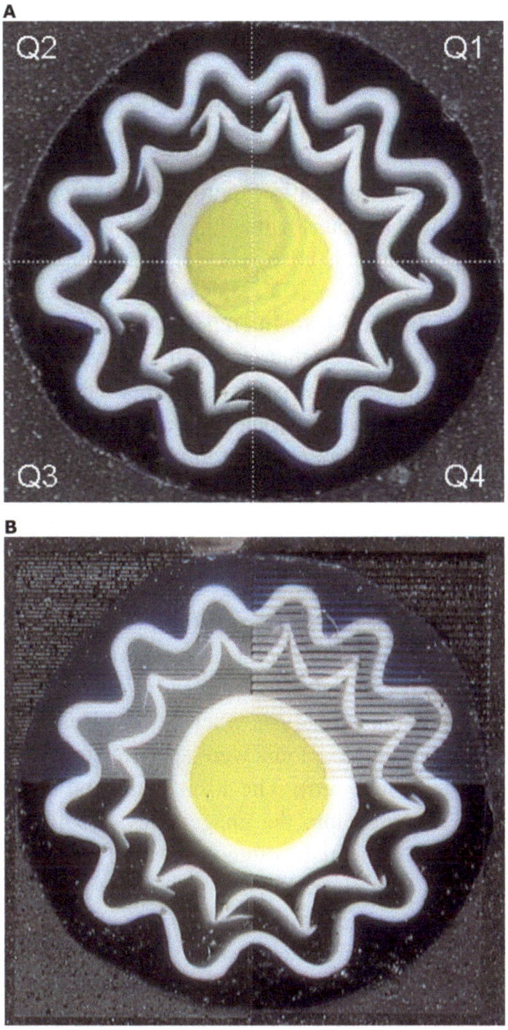

Fig. 4.5 Digital image of the modern *murrina* used for virtual and experimental LA-ICP-MS image mapping (**a**) and damage induced by LA scanning of the quadrants Q1–Q3 of the glass using three different sets (Table 4.2) of LA-ICP-MS conditions (**b**); quadrant Q4 served as a visual reference in the elemental image maps

pulse width, 4 ns) with a motorized stage and a laser ablation chamber. Laser ablation took place in a so-called SuperCell™ (New Wave Research). Helium was used to transport the ablated material from the ablation chamber to the inductively coupled plasma (ICP); argon was added as a make-up gas before the torch of the ICP. The mass spectrometer was set up in time-resolved analysis mode, measuring one point per mass (TRA(1)), and acquiring 54 masses. Measurement of the background

Table 4.2 Operating conditions of the LA-ICP-MS instrument for 2D mapping purposes

Laser ablation	New Wave Research UP 213
Wavelength	213 nm
Pulse length	4 ns
Spot size	100 (Q1), 55 (Q2), 25 (Q3) μm
Fluence	ca. 7 J cm^{-2}
Repetition rate	10 Hz
Scanning speed	9 (Q1), 16 (Q2), 36 (Q3) μm s^{-1}
Grid spacing	100 (Q1), 55 (Q2), 25 (Q3) μm
He flow rate	0.95 l min^{-1}
Make-up Ar flow rate	0.75 l min^{-1}
ICP-MS	Agilent 7500ce
Rf power	1500 W
Sampling depth	6.5 mm
Isotopes measured (54 masses in total)	^{7}Li, ^{9}Be, ^{11}B, ^{23}Na, ^{24}Mg, ^{27}Al, ^{29}Si, ^{31}P, ^{39}K, ^{43}Ca, ^{45}Sc, ^{47}Ti, ^{51}V, ^{53}Cr, ^{55}Mn, ^{57}Fe, ^{59}Co, ^{60}Ni, ^{63}Cu, ^{66}Zn, ^{69}Ga, ^{75}As, ^{82}Se, ^{85}Rb, ^{88}Sr, ^{89}Y, ^{90}Zr, ^{93}Nb, ^{95}Mo, ^{107}Ag, ^{111}Cd, ^{115}In, ^{118}Sn, ^{121}Sb, ^{137}Ba, ^{139}La, ^{140}Ce, ^{141}Pr, ^{146}Nd, ^{147}Sm, ^{153}Eu, ^{157}Gd, ^{159}Tb, ^{163}Dy, ^{165}Ho, ^{166}Er, ^{169}Tm, ^{172}Yb, ^{175}Lu, ^{197}Au, ^{208}Pb, ^{209}Bi, ^{232}Th, ^{238}U
Acquisition time/ pixel	11.11 (Q1), 3.44 (Q2), 0.69 (Q3) s
Measurement mode	Time-resolved, TRA(1)
Plasma gas flow rate	15 l min^{-1}
Auxillary gas flow rate	1 l min^{-1}
Total analysis time/ map	2.8 h

Reference to conditions for mapping of the quadrants (Q1–Q3) of the modern *murrina* (Fig. 4.3a)

gases (He/Ar mixture) served to establish a gas blank signal for all masses.

Each map consisted of many parallel line scans, with no space between adjacent scans, which were acquired and saved individually by a remotely triggered ICP-MS line scan sequence routine. The individually collected line scans with time-resolved data for each element (in cps) were saved and later converted into elemental oxide concentrations (in % *w*/*w*) (see Sect. 4.3.4). To test the LA-ICP-MS conditions for mapping of the *murrina* (Fig. 4.5a), three sets of LA-ICP-MS conditions were selected (Table 4.2) and the imaginary quadrants (Q1–Q3) mapped; quadrant Q4 served as a visual reference. As can be seen from Fig. 4.5b the damage to the surface of the *murrina* is very dependent on the LA-ICP-MS conditions selected (fluence and repetition rate were kept constant for mapping of the quadrants), resulting in minimal damage to Q3 (25 μm/36 μm s^{-1}) and destructive damage to Q1 (100 μm/9 μm s^{-1}) and Q2 (55 μm/16 μm s^{-1}); ablation of Q3 can be regarded as a kind of dulling of the surface.

4.3.4 Quantification and Image Analysis

The individual line scans making up the elemental maps were sequentially stitched together into a 2D CSV file using an in-house macro. This macro converts the response of the line scans into 2D matrices with pixels conveying their content (cps/element). Since the content of often tens of thousands of pixels needs to be converted into % *w*/*w* for each element, a custom quantification routine (in Microsoft® Office Excel 2003, Microsoft Corporation, Redmond, WA, USA) was used based on summating 54 elements as their oxides to 100 % *w*/*w*. Similar approaches have been successfully applied before, using an array of glass standards or a single standard (Gratuze et al. 2001; van Elteren et al. 2009; Liu et al. 2008; Šelih and van Elteren 2011). Since the use of a single standard greatly reduces the analysis time and simplifies processing of the pixels, NIST SRM 610 was used for calibration purposes to generate quantitative data, similar to earlier published work (Šelih and van Elteren 2011). In short, the sum normalization calibration approach converts the raw signal intensity I_i (in cps) of the elemental oxide i (1 to n) in each pixel into an unnormalized concentration $c_i(unnorm)$ (in % *w*/*w*):

$$c_i(unorm) = \frac{I_i}{I_{i,N610}/c_{i,N610}} \tag{4.1}$$

with $I_{i,N610}$ and $c_{i,N610}$ the raw signal intensity (in cps) and the elemental oxide concentration i (in % *w*/*w*) in NIST SRM 610, respectively. From Eq. (4.1) it is clear that after cumulation and normalization to 100 % *w*/*w* the normalized elemental oxide concentrations $c_i(norm)$ (in % *w*/*w*) can be retrieved:

$$c_i(norm) = 100 \times \frac{c_i(unnorm)}{\sum_{i=1}^{n} c_i(unnorm)} \tag{4.2}$$

This sum normalization approach not only yields accurate data, but for mapping of large surfaces the procedure is also not very susceptible to defocusing issues as a result of surface unevenness (Šelih and Van Elteren 2011). However, it should be stressed that for small beam diameters it is difficult to keep the laser focused (in the order of the beam diameter) for extended periods of time (many hours); we found out that a beam diameter ≤12 μm yielded elemental image maps of varying quality. Processing of the elemental oxide concentrations (in % *w*/*w*) into elemental maps (in.JPEG format) was achieved by Origin 7.5 SR4 (OriginLab Corporation, Northampton, MA, USA) and Sigmaplot (Systat Software Inc.); in general 256 shades of gray (8-bit image) were used for image analysis whereas for visual presentation pseudocolours were applied (from dark blue to bright red). The 8-bit elemental maps were imported into the image analysis software ImageJ (National Institute of Health, USA) for detailed digital analysis of the glass artefacts, using a simple calibration which links the shades of gray (256 brightness levels from 0 [black] to 255 [white]) to the oxide concentrations. With a tracing tool areas on the glass artefact were selected for localized elemental analysis.

4.3.5 Software for Virtual LA-ICP-MS Mapping

To aid in the selection of appropriate LA-ICP-MS mapping conditions, finding the best balance between spatial resolution, analysis time and signal-to-noise ratio, a previously developed LA-ICP-MS mapping software package (Triglav et al. 2010) was used (upon request available from one of the authors [elteren@ki.si]). After uploading a digital image in the software (the pre-Roman glass shown in Fig. 4.4) it is subjected to permutations simulating the LA-

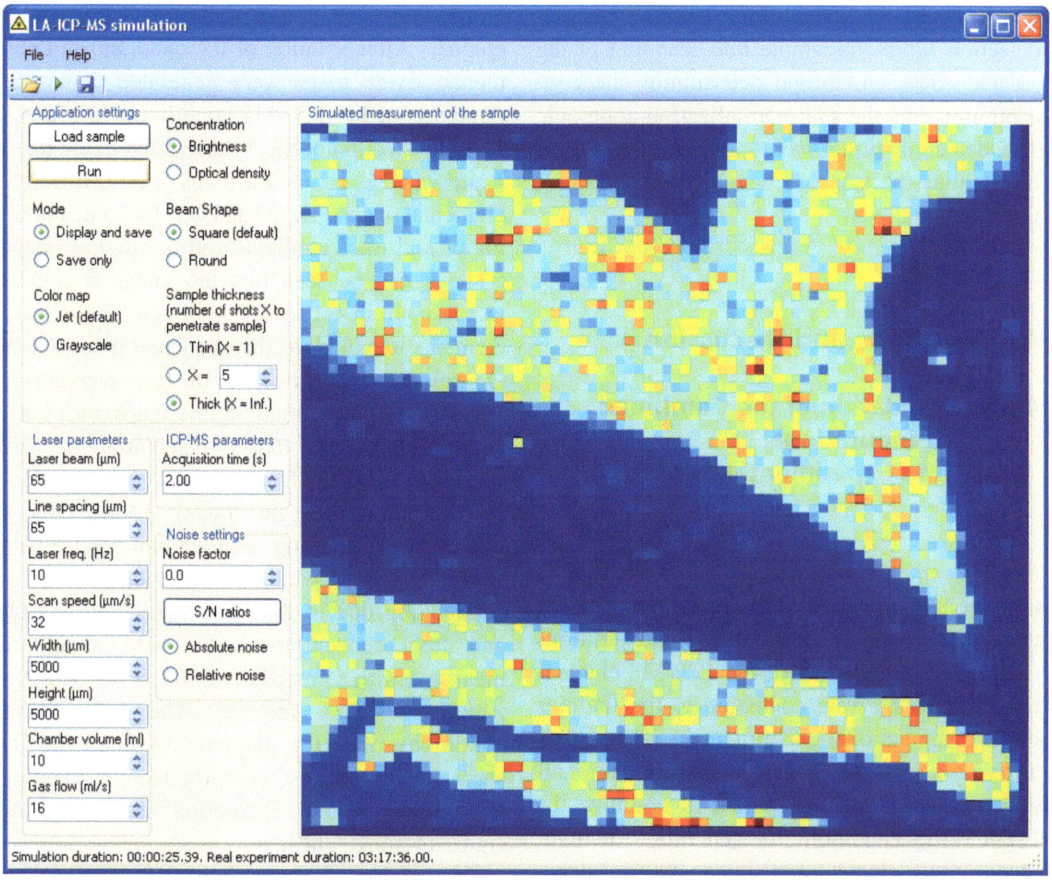

Fig. 4.6 Screenshot of the LA-ICP-MS software for virtual 2D mapping; in this instance a digital image of the pre-Roman glass shown in Fig. 4.4 has been uploaded in the software and the resulting virtual image map for the optimized LA-ICP-MS conditions is shown

ICP-MS mapping action (see the screenshot in Fig. 4.6). The virtual LA-ICP-MS conditions (beam diameter, repetition rate, scanning speed, flow rate and acquisition time) can be freely selected; laser fluence cannot be selected as optimum coupling to the target material occurs in a fairly narrow fluence range (Phipps et al. 2006; Xu 2002). The virtual map generated is based on a computational approach comprising the following steps:

1) scaling of the digital image to its actual size (5 mm $\times$ 5 mm in this case),
2) creating of the LA-ICP-MS pixel grid and averaging of the RGB 0÷255 content of the photographic pixels,
3) determining the distortion of each pulse caused by exponential washout from the ablation cell,
4) gathering of the appropriate number of pixels ($=R \times T$) to generate the final virtual map in pseudocolours.

Since the computational approach does not take into account the elemental concentrations it is essential to investigate how the noise levels in the maps are related to the trace elemental concentrations for selected sets of LA-ICP-MS conditions. Although the software is kitted out with a noise generator, in this work on glass mapping it was found more convenient to investigate the anticipated noise in the elemental image maps by establishing multi-elemental detection limits via simple line scanning of the "blank" glass DLH6 and determining the concentrations via the sum normalization approach. The detection limits were calculated based on three times the standard deviation of the noise in this "blank" glass.

4.4 Results and Discussion

4.4.1 Validation of the Combined Computational and Experimental Strategy

Experimental Murrina *2D Mapping Details* A modern *murrina* (Fig. 4.5a) was subjected to three sets of LA-ICP-MS mapping conditions (Table 4.2), and the actual LA-ICP-MS elemental image maps were compared to the findings of the computational-experimental strategy. This allowed us to prove that the computational-experimental strategy is suited to optimize the LA-ICP-MS conditions for generation of high-quality multi-elemental 2D image maps of decorative features in glass artefacts. The sample was divided into four imaginary quadrants (each of 3.2 mm × 3.0 mm) of which three were virtually and experimentally mapped, each with a different set of LA-ICP-MS conditions but equal analysis times (2.8 h). Square pixels were considered, implying that D (beam diameter) = S (scanning speed) × T (acquisition time). Based on the LA-ICP-MS conditions in Table 4.2 the relative theoretical elemental sensitivity values (in signal [cps] per concentration [% *w*/*w*]) are 1:0.30:0.06 for Q1:Q2:Q3. The relative experimental elemental sensitivity values (averaged over 54 elements upon ablation of NIST SRM 610) are 1:0.38 ± 0.05:0.12 ± 0.04, which is relatively close to the theoretical values. Deviations are possible due to the fact that scanning speeds are different, resulting in different degrees of collateral damage to the glass surface and hence slight sensitivity variations.

Evaluation of Experimental 2D Map Quality After ablation of the quadrants, 54 elemental oxide maps were generated under the three sets of LA-ICP-MS conditions. In Fig. 4.7 a small selection of the 54 elemental oxide maps is presented as composite maps of three different elemental maps (Q1, Q2 and Q3) for some major, minor and trace elements and the digital image (Q4). It can be seen that the image sharpness increases from Q1 to Q3, although for the trace elements it is obvious that the small pixel size in Q3 causes the noise floor for the trace elements to rise so much that unintelligible maps with inaccurate elemental oxide concentrations are obtained (visible from variations in colour patterns in the three quadrants). Additionally, it is clear that retrieving accurate and precise data from the narrow, white, wavy patterns for Q1 is challenging due to the large pixelation; As_2O_3 and PbO concentrations in these areas are measured with a precision of ~60 % for Q1 compared to ~20 % for Q2. The LA-ICP-MS conditions used for mapping of quadrant Q2 seem to be the best compromise for yielding appropriate spatial resolution and adequate signal-to-noise ratios.

Technological Fabrication Information Extracted from the Maps The compositional data extracted from all 54 maps (based on the measurement of Q2) reveal the following (concentrations ± standard errors): (1) the opaque, white, wavy patterns are associated with lead arsenate (10.8 ± 0.8 % *w*/*w* As_2O_5 and 30.6 ± 2.8 % *w*/*w* PbO), which is a weak (*opaline*) opacifier; (2) the yellow spot in the *murrina* centre is produced by colloidal dispersion of cadmium sulphide (measured as CdO [0.31 ± 0.16 % *w*/*w*]), with a minor part of selenium-cadmium sulphide, under reducing conditions; the reducing agent is zinc metal powder that oxidizes to zinc oxide (9.1 ± 2.1 % *w*/*w*); the latter acts also as a stabilizer in this glass poor in lime (1.3 ± 0.3 % *w*/*w*); (3) the dark blue outer edge is due to the presence of cobalt oxide (0.31 ± 0.08 % *w*/*w*), with a minor contribution of copper oxide (0.0020 ± 0.0005 % *w*/*w*);

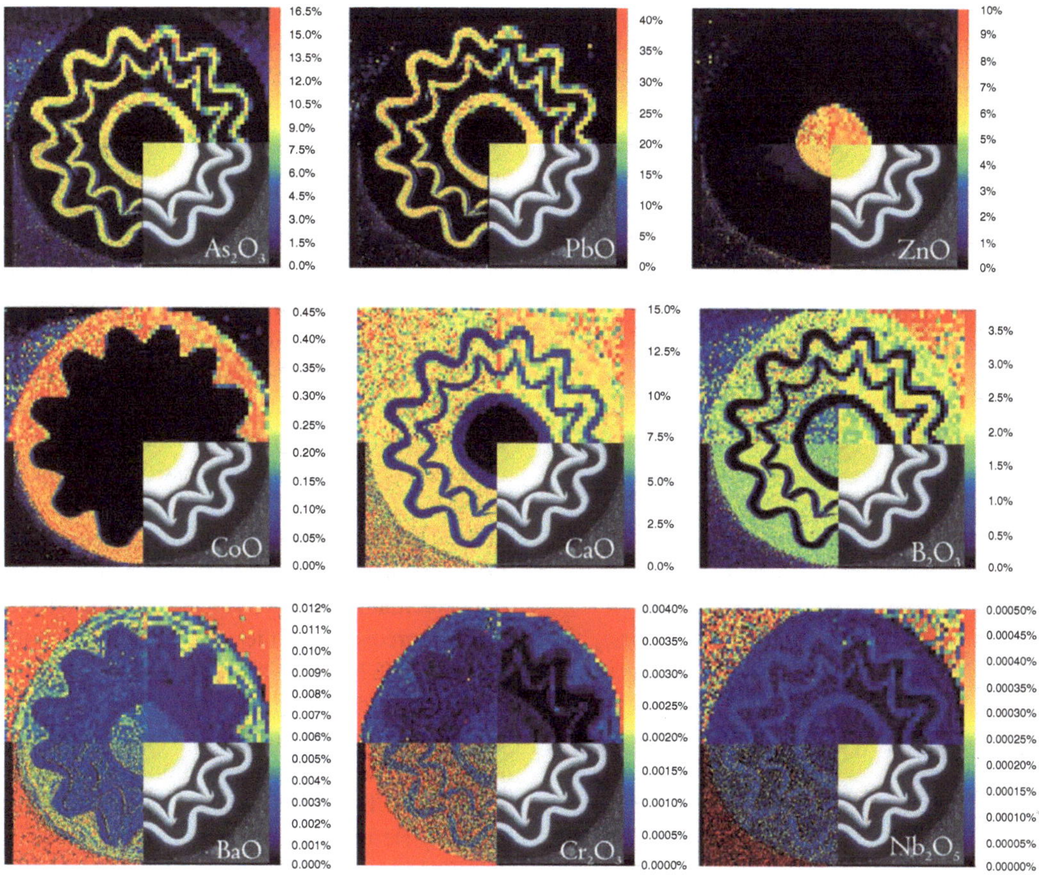

Fig. 4.7 Selected composite elemental oxide image maps after actual 2D LA-ICP-MS mapping of three quadrants of the *murrina* (Fig. 4.5a) using three different conditions (Table 4.2); the visual image in the fourth quadrant serves as a reference. Lower image densities are mapped onto "cool" colours and higher densities onto "hot" colours. Low density images are shown in *blue* or *violet*, and high density images appear as *orange* or *red*

(4) the translucent bulk glass matrix consists of SiO_2 (66.1 ± 4.1 % *w/w*), Na_2O (21.3 ± 3.6 % *w/w*), CaO (7.9 ± 1.8 % *w/w*), B_2O_3 (1.3 ± 0.3 % *w/w*), K_2O (1.6 ± 0.2 % *w/w*) and Al_2O_3 (0.81 ± 0.12 % *w/w*). These data indicate that the *murrina* is a silica-soda-lime glass where boron has been added as a fluidifying agent.

4.4.2 Computational and Experimental Strategy

Virtual Murrina *Mapping Details* To investigate whether the best LA-ICP-MS conditions can be singled out by a computational–experimental strategy as well, we uploaded not only the digital image of the *murrina* (Fig. 4.5a) in the software but also its negative image to emphasize both high and low concentrations in similar parts of the glass. E.g., the translucent part of the *murrina* contains high main elemental concentrations but most other elements are expected to be present at trace levels; thus, uploading the (positive) digital image of the *murrina* will yield an image map with a low contribution of the main elements since the black in the image is associated with a zero RGB brightness value. However, the negative of the digital image will appropriately address this issue. In Fig. 4.8, the results of the virtual 2D maps of the *murrina* (both of the positive and negative digital image) for the three

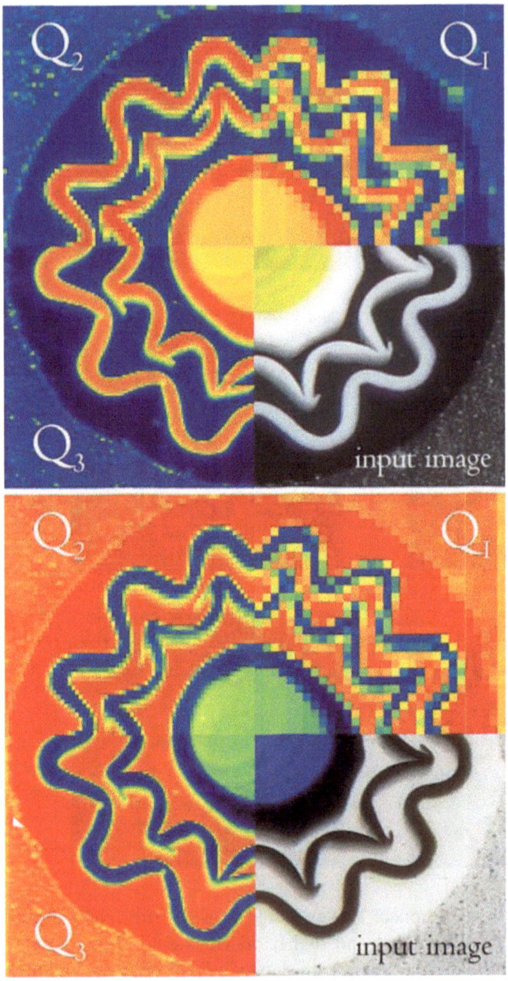

Fig. 4.8 Composite 2D image maps after virtual LA-ICP-MS mapping of the three quadrants of the *murrina* (Fig. 4.5a) using three different conditions (Table 4.2) for the positive (*top*) and negative (*bottom*) digital image; the visual image in the fourth quadrant serves as a reference. For the explanation of colours see Fig. 4.7

different quadrants are shown. As expected, the virtual 2D maps increase in quality from quadrant Q1 to Q3, with quadrant Q1 also showing too large pixelation for accurate and precise measurement on the white, wavy patterns, in agreement with the actual LA-ICP-MS mapping results. Since the LA-ICP-MS conditions used for virtual mapping of quadrant Q3 yield the best virtual map, they seem the most obvious choice for experimental mapping. However, due to introduction of instrumental noise associated with the small pixel size (25 μm × 25 μm), especially for trace level concentrations, these conditions might not be indicative for the best possible elemental image maps.

Evaluation of Virtual Map Quality To envisage how these conditions might influence the noise floor of the 2D maps, the "blank" glass DLH6 was measured in single line scan mode to retrieve the concentrations of 54 elemental oxides according to the sum normalization procedure. In Fig. 4.9 the detection limits for 50 elemental oxides (except the base compositional elemental oxides SiO_2, Al_2O_3, CaO and Na_2O) are presented as three times the standard deviation of the noise in the "blank" glass (in % *w/w*) for the three different LA-ICP-MS conditions. As a general trend the detection limits increase from Q1 to Q3; averaged over all the elemental oxides the detection limit ratios are 1:3.3:40 for Q1:Q2:Q3. For prediction of the anticipated noise in the actual LA-ICP-MS elemental image maps, using Fig. 4.9, we can assume that the main elements (Ca, B, As, Pb, Zn and Co) have concentration levels well above their detection limits for all three LA-ICP-MS conditions. This suggests that all these elements can be measured satisfactorily in all three quadrants, as is confirmed by the actual elemental oxide image maps in Fig. 4.7. For the trace elements Ba, Cr and Nb, the LA-ICP-MS conditions for quadrant Q3 have elemental oxide detection limits of 3.59×10^{-3}, 4.50×10^{-3} and 2.47×10^{-3} % *w/w*, respectively. Knowing that the actual concentrations of BaO, Cr_2O_3 and Nb_2O_5 in the bulk of the *murrina* (measured on the outer rim of quadrant Q2) are $5 \times 10^{-3} \pm 2 \times 10^{-3}$, $9 \times 10^{-4} \pm 3 \times 10^{-4}$ and $9 \times 10^{-5} \pm 3 \times 10^{-5}$ % *w/w*, respectively, it is clear that the LA-ICP-MS conditions for Q3 cannot produce reliable 2D maps, as proven in Fig. 4.7. The LA-ICP-MS conditions for Q2 fare much better, with elemental oxide detection limits of 1.26×10^{-4} % *w/w* BaO, 3.62×10^{-4} % *w/w* Cr_2O_3 and 6.30×10^{-5} % *w/w* Nb_2O_5, which are all below the actual measured concentrations in the *murrina*. Thus, our computational-experimental strategy has faithfully predicted the LA-ICP-MS conditions for Q2 to yield

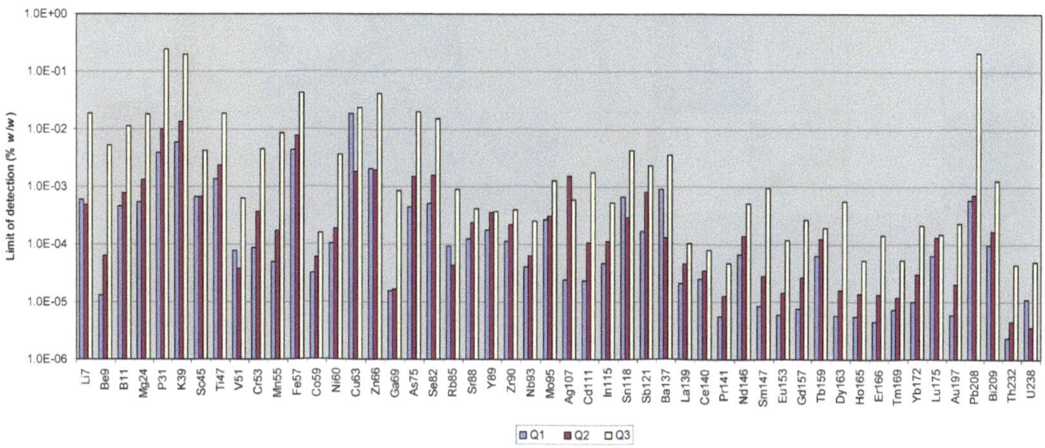

Fig. 4.9 Detection limits for the elemental oxides based on three times the standard deviation of the noise in the "blank" glass DLH6 (400 pixels) upon LA-ICP-MS single line scanning under three different conditions (Table 4.2)

the best quality elemental image maps. Although pre-knowledge on elemental oxide concentrations has been used in the prediction, in general we may assume that concentrations of ca. 10^{-4} % *w*/*w* are borderline cases and LA-ICP-MS conditions yielding detection limits below 10^{-4} % *w*/*w* are better excluded. Something that cannot be predicted is the elemental heterogeneity, although this will only give rise to more noise in the maps and not necessarily to unintelligible maps.

4.4.3 Application to 2D Mapping of a Pre-Roman Glass

The computational-experimental strategy described above was used to predict the best possible LA-ICP-MS conditions for experimental 2D mapping of the pre-Roman glass shown in Fig. 4.4. These optimal conditions were as follows: laser beam, 65 μm (square); grid spacing, 65 μm; repetition rate, 10 Hz; scanning speed, 32 μm s^{-1}; acquisition time/pixel, 2 s (the remaining standard conditions can be found in Table 4.2). The resulting 2D map upon virtual ablation of the pre-Roman glass is shown in Fig. 4.6 and illustrates that a high-quality map can be obtained in a relatively short analysis time (3.3 h).

LA-ICP-MS Mapping of the Iron Age Polychrome Glass Preliminary observations of the glass sample by optical and electron microscopy detected a highly inhomogeneous texture, both on the external surface and on the broken section indicated in Fig. 4.4. The glass bulk showed a very rough general texture and a wide-scale distribution of crystalline aggregates and bubbles in the glass. These features can considerably affect the precision of the elemental quantification by a microanalytical technique such as LA-ICP-MS. From the elemental mapping results (Fig. 4.10) this glass can be classified as a soda-lime-silica glass, based on its contents (concentrations ± standard errors): silica (70.0 ± 6.8 % *w*/*w*), soda (17.0 ± 3.9 % *w*/*w*) and lime (7.0 ± 1.4 % *w*/*w*), and low potash (0.50 ± 0.02 % *w*/*w*) and magnesia content (0.54 ± 0.01 % *w*/*w*). Moreover, it presents a very low level of heterogeneously distributed phosphorous oxide (0.041 ± 0.004 % *w*/*w*). Low K_2O, MgO and P_2O_5 contents correspond to the use of natron as the fluxing agent (Verità et al. 2002; Shortland et al. 2006; Reade et al. 2009; Shortland and Schroeder 2009). Particular elemental maps from Fig. 4.10 were compiled (see Fig. 4.11) to highlight correlations between elements and their association with the features of the artefact concerning chromophores, opacifiers and their mineral sources.

Fig. 4.10 LA-ICP-MS elemental oxide image maps of the Iron Age glass. Since the elemental oxide concentrations in all maps are scaled from 10^{-6} to 10^{-2} % *w/w*, a logarithmic scale was applied for visualization of all elemental oxides simultaneously. For the explanation of colours see Fig. 4.7

Analysis of Chromophores and Opacifiers The Co and Cu maps (Fig. 4.11a) show that the deep blue features can be attributed to the presence of CoO (0.088 ± 0.021 % *w/w*) and CuO (0.55 ± 0.22 % *w/w*). The application of elemental mapping proves to be particularly suitable for the analysis of opaque glasses. The opacity of glass is due to the presence of a dispersion of crystals in the glass matrix (Henderson 2000). The 2D maps in Fig. 4.11b indicate that the opaque turquoise and yellow decorations contain opacifiers based on antimony. Antimony-based opacifiers were used from the beginning of glass production in the Near East and Egypt around 1500 BC (Tite et al. 2008). The opacity and colour of the yellow glass can be attributed to the presence of lead antimonate ($Pb_2Sb_2O_7$; measured as Sb_2O_5 [3.2 ± 1.3 % *w/w*] and PbO [30.0 ± 6.0 % *w/w*]). White calcium antimonate opacifier ($Ca_2Sb_2O_7$ or $Ca_2Sb_2O_6$; measured as CaO [6.8 ± 0.4 % *w/w*] and Sb_2O_5 [3.2 ± 1.2 % *w/w*]) and aquamarine-blue copper oxide (2.5 ± 0.6 % *w/w*) are responsible for the turquoise hue. Calcium was not intentionally added to the glass batch since lime was a natural component of the sand. Lead is present in strong excess in the yellow glass and in a more limited amount in the turquoise glass (2.3 ± 1.1 % *w/w*). This element, with its low concentration and heterogeneous distribution in the blue glass (0.8 ± 0.5 % *w/w*), found also in some other Iron Age glasses (Arletti et al. 2010), probably entered the batch with the cobalt mineral. The presence of lead in the turquoise glass may have led to the formation of a small amount of lead antimonate, or more probably it completely dissolved in the batch, without contributing to its opacity (Arletti et al. 2008). Figure 4.11c also shows the 2D maps rendered as 3D maps for Sb and Pb with linear scaling, convincingly demonstrating their distribution in the glass and the potential of LA-ICP-MS for elemental mapping of glass.

Mineral Sources Elemental image maps also allow us to speculate on the composition of the ores used as the sources of chromophores and opacifiers based on the correlation between the

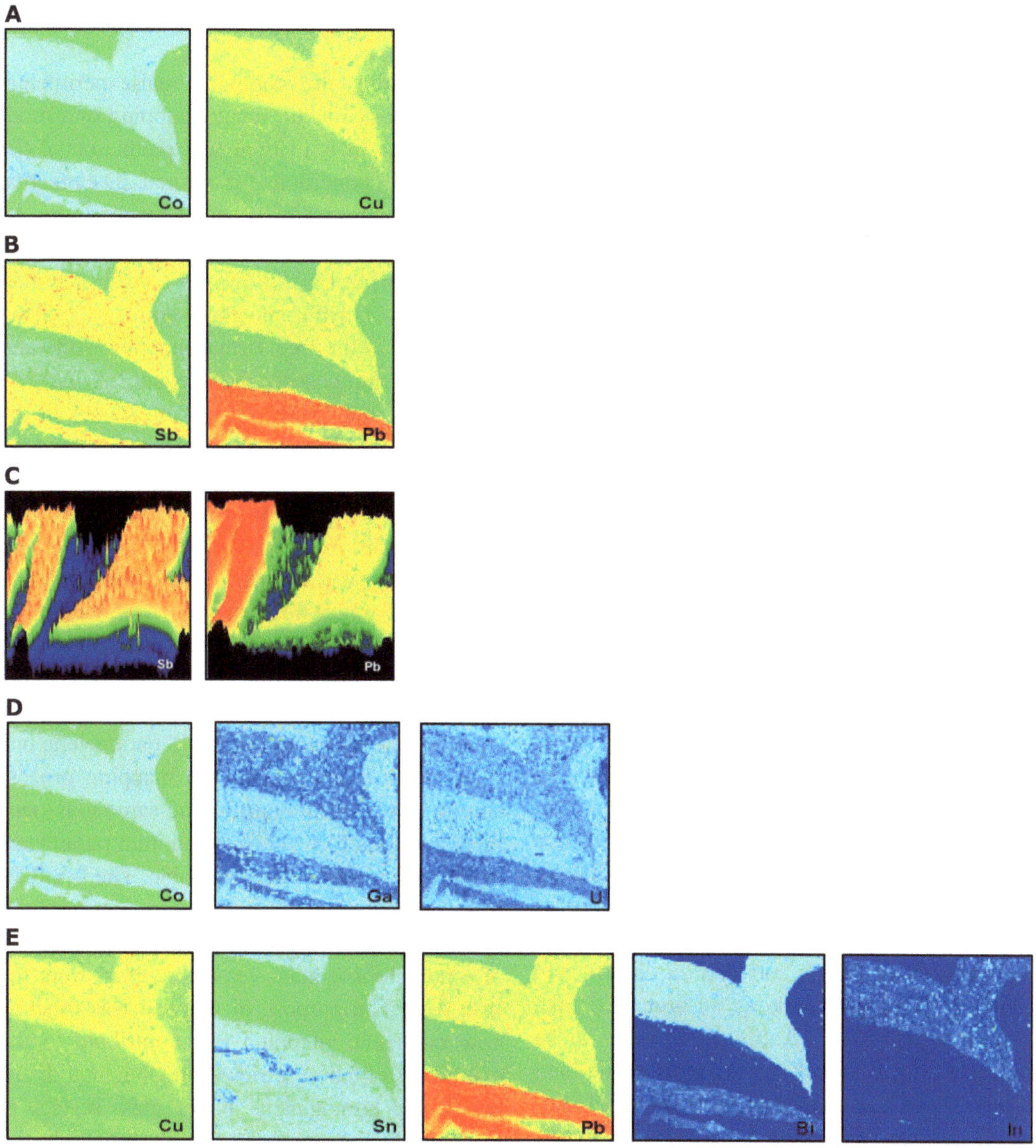

Fig. 4.11 Selected elemental oxide image maps of the Iron Age glass highlighting: (1) the correlations with the *blue* colourations (**a**) and the *opaque* and *yellow* decorations in 2D (**b**) and 2D rendered in 3D (**c**); (2) the provenance of some elements (**d** and **e**). For the explanation of colours see Fig. 4.7. All maps (**a**)–(**e**) had logarithmic scaling as in Fig. 4.10, except the 2D maps rendered in 3D in (**c**) which had linear scaling

elements. The study of the mineral sources for chromophores and opacifiers is rather complex and it is often difficult to state useful associations between elements. Several studies have investigated the nature and origin of the cobalt ores used for blue coloured glass (Henderson 1985; Gratuze and Picon 2006; Gratuze et al. 1992, 1996; Gratuze 1997; Shortland et al. 2006; Towle et al. 2001). They were based on the analysis of the elements associated with cobalt, such as aluminium, copper, manganese, iron, nickel, zinc, arsenic, antimony and lead.

From the elemental oxide maps in Fig. 4.11d it seems that in the dark blue glass cobalt is associated with copper and some impurities such as gallium and uranium. Cobalt may occur in combination with copper in heterogenites (mixtures of minerals as mindigite $2Co_2O_3{\cdot}CuO{\cdot}3H_2O$ and trieuite $2Co_2O{\cdot}CuO{\cdot}6H_2O$) (Brady et al. 2002), so that the presence of copper could be related to the addition of this kind of ore as a colouring agent. However, we cannot exclude that copper has been added to the glass batch independently to modify the cobalt blue hue, probably in the form of brass, as suggested by a % *w/w* Cu/Zn ratio of about 7, the same as one would expect in brass. This could also explain the correlation of cobalt and copper with Ga_2O_3 (0.0023 ± 0.0011 % *w/w*), which is usually present in trace amounts in zinc. Uranium, as UO_2 (0.0019 ± 0.0011 % *w/w*), might originate from the ores of copper and/or cobalt. The elemental oxide maps of the turquoise decoration in Fig. 4.11e show that copper is associated with tin, lead and traces of indium and bismuth, although the occurrence of ^{115}In is most likely an interference caused by ^{115}Sn (natural abundance, 0.34 %) upon measuring higher levels (ca. 0.01 % *w/w* in this case) of ^{118}Sn (natural abundance, 24.44 %). The presence of copper, lead and tin suggests that two distinct possible materials could be identified as sources of copper: (1) an ore containing all these three elements (Santopadre and Verità 2000) and (2) slags from the working of bronze (Peake and Freestone 2012). Antimony appears to be associated with traces of silver, arsenic and nickel, while lead correlates with traces of cadmium, arsenic and molybdenum. This suggests that the possible sources of lead and antimony could be sulphide ores containing galena (PbS) and stibnite (Sb_2S_3), respectively. The correlation of silver with antimony might be related to the use of cupellation litharge as a source for lead antimonate in the decorations (Mass et al. 2002). This could also explain the presence of lead in the turquoise decoration. However, according to Rehren and Shortland (Rehren and Shortland 2003; Shortland et al. 2000), litharge as the source for antimony would be rather unlikely.

4.5 Conclusion

By evaluating the computational-experimental strategy for elemental image mapping of a modern *murrina*, it is evident that prediction of the LA-ICP-MS conditions for optimal 2D elemental image mapping prior to actual mapping can be carried out efficiently and effectively. The steps to be taken are as follows: (1) make a digital image of the polychrome glass sample; (2) upload this image in the free virtual mapping software; (3) plug in numbers for the relevant LA-ICP-MS parameters and "play" with them to obtain the best image with the shortest analysis time and the highest spatial resolution; (4) use these virtually optimized parameters to make a line scan on a reference glass (NIST610) and a "blank" glass; (5) calculate the elemental detection limits to predict the noise floor in the elemental maps. When set up properly this process should take no longer than ca. 1 h but might save many hours of analysis time or even prevent one from choosing the wrong LA-ICP-MS parameters; one should also be aware that the mapping process cannot be repeated again on the same glass surface, and therefore choosing optimal parameters in advance is particularly valuable. The LA-ICP-MS optimization strategy was successfully applied for quantitative multi-elemental mapping of a polychrome glass vessel fragment from the Iron Age. The purpose of this test was to show how elemental mapping can contribute to the overall information retrieval on elemental distribution and homogeneity in the glass surface to gain insight into the presence of chromophores, opacifiers and their associated ores in an easy way, i.e. by visual inspection of the maps.

References

Aerts A, Velde B, Janssens K, Dijkman W (2003) Change in silica source in Roman and post-Roman glass. Spectrochim Acta B 58:659–667

Alberta S, Gianmario M, Valentina P (2011) The stained glass window of the southern transept of St. Anthony's

Basilica (Padova, Italy): study of glasses and grisaille paint layers. Spectrochim Acta B 66:81–87

Arletti R, Vezzalini G, Quartieri S, Ferrari D, Merlini M, Cotte M (2008) Polychrome glass from Etruscan sites: first non-destructive characterization with synchrotron μ-XRF, μ-XANES and XRPD. Appl Phys A 92:127–135

Arletti R, Maiorano C, Ferrari D, Vezzalini G, Quartieri S (2010) The first archaeometric data on polychrome Iron Age glass from sites located in Northern Italy. J Archaeol Sci 37:703–712

Barkoudah Y, Henderson J (2006) Plant ashes from Syria and the manufacture of ancient glass: ethnographic and scientific aspects. J Glass Stud 48:297–321

Baxter MJ, Cool HEM, Jackson CM (2005) Further studies in the compositional variability of colourless Romano-British vessel glass. Archaeometry 47:47–68

Behrends T, Kleingeld P (2009) Bench-top micro-XRF – a useful apparatus for geochemists? Geochemical News 138

Brady GS, Clauser HR, Vaccari JA (2002) Materials handbook, 15th edn. McGraw-Hill, New York, NY

Carmona N, Ortega-Feliu I, Gomez-Tubio B, Villegas MA (2010) Advantages and disadvantages of PIXE/PIGE, XRF and EDX spectrometries applied to archaeometric characterisation of glasses. Mater Charact 61:257–267

Duwe S, Neff H (2007) Glaze and slip pigment analyses of Pueblo IV period ceramics from east-central Arizona using time of flight-laser ablation-inductively coupled plasma-mass spectrometry (TOF-LA-ICP-MS). J Archaeol Sci 34:403–414

Freestone IC (2006) Glass production in Late Antiquity and the Early Islamic period: a geochemical perspective. In: Maggetti M, Messiga B (eds) Geomaterials in cultural heritage. Geological Society of London, London, pp 201–216

Freestone IC, Greenwood R, Gorin-Rosen Y (2002) Byzantine and early Islamic glassmaking in the eastern Mediterranean: production and distribution of primary glass. In: Kordas G (ed) Hyalos-Vitrum-Glass. History, technology and conservation of glass and vitreous materials in the Hellenic world. 1st International Conference, Rhodes, Greece. Glasnet, Athens, p 167–174

Gratuze B (1997) L'apport des analyses de verres archéologiques: etudes de cas. In: Techne, Verres–Emaux–Glaçures. Paris Laboritoire de recherche des musées de France n. 6, p 8–18

Gratuze B, Picon M (2006) Utilisation par l'industrie verrière des sels d'aluns des oasis égyptiennes au début du premier millénaire avant notre ère. In: Borgard P, Brun J-P, Picon M (eds) L'Alun de Mediterranie. Naples/Aix-en-Provence, Centres Jean Bérard et Camille Jullian, Collection du Centre Jean Bérard 23, p 269–276

Gratuze B, Soulier I, Barrandon J-N, Foy D (1992) De l'origine du cobalt dans les verres. Revue d'Archéométrie 16:97–108

Gratuze B, Soulier I, Blet M, Vallari L (1996) De l'origine du cobalt: du verre à la céramique. Revue d'Archéométrie 20:77–94

Gratuze B, Blet-Lemarquand M, Barradon J-N (2001) Mass spectrometry with laser sampling: a new tool to characterize archaeological materials. J Radioanal Nucl Chem 247:645–656

Hamilton DL, Hopkins TC (1995) Preparation of glasses for use as chemical standards involving the coprecipitated gel technique. Analyst 120:1373–1377

Harden DB, Tatton-Brown VA (1981) Catalogue of Greek and Roman glass in the British Museum, vol 1, Core and rod-formed vessels and pendants and Mycenaean cast objects. British Museum Publications, London

Henderson J (1985) The raw materials of early glass production. Oxford J Archaeol 4:267–291

Henderson J (2000) The science and archaeology of materials: an investigation of inorganic materials. Routledge, New York, NY

Henderson J, Warren SE (1983) Analysis of prehistoric lead glass. In: Aspinall A, Warren SE (eds) Proceedings of 22nd International Symposium on Archaeometry. Schools of Physics and Archaeological Sciences, Bradford, pp 168–180

Henderson J, McLoughlin SD, McPhail DS (2004) Radical change in Islamic glass technology: evidence for conservatism and experimentation with new glass recipes from early and middle Islamic Raqqa, Syria. Archaeometry 46:439–468

Jackson CM (2005) Making colourless glass in the Roman period. Archaeometry 47:763–780

Lahlil S, Biron I, Cotte M, Susini J, Menguy N (2010a) Synthesis of calcium antimonate nano-crystals by the 18th Dynasty Egyptian glassmaker. Appl Phys A 98:1–8

Lahlil S, Biron I, Cotte M, Susini J, Menguy N (2010b) New insight on the in situ crystallization of calcium antimonate opacified glass during the Roman period. Appl Phys A 100:683–692

Liu Y, Hu Z, Gao S, Gunther D, Xu J, Gao C, Haihong C (2008) In situ analysis of major and trace elements of anhydrous minerals by LA-ICP-MS without applying an internal standard. Chem Geol 257:34–43

Lombardo T, Gentaz L, Verney-Carron A, Chabas A, Loiseb C, Neff D, Leroy E (2013) Characterization of complex alteration layers in medieval glasses. Corros Sci 72:10–19

Mass JL, Wypyski MT, Stone RE (2002) Malkata and lisht glassmaking technologies: towards a specific link between second millennium B.C. metallurgist and glassmakers. Archaeometry 44:67–82

Mata MP, Peacor DR, Gallart-Marti MD (2002) Transmission electron microscopy (TEM) applied to ancient pottery. Archaeometry 44:155–176

Mirti P, Pace M, Malandrino M, Negro Ponzi M (2009) Sasanian glass from Veh Ardašīr: new evidences by ICP-MS analysis. J Archaeol Sci 36:1061–1069

Naes BE, Umpierrez S, Ryland S, Barnett C, Almirall JR (2008) A comparison of laser ablation inductively

coupled plasma mass spectrometry, micro X-ray fluorescence spectroscopy, and laser induced breakdown spectroscopy for the discrimination of automotive glass. Spectrochim Acta B 63:1145–1150
Nicholson PT, Henderson J (eds) (2000) Glass in ancient Egyptian materials and technology. Cambridge University Press, Cambridge
Panighello S, Orsega EF, van Elteren JT, Šelih VS (2012) Analysis of polychrome Iron Age glass vessels from Mediterranean I, II and III groups by LA-ICP-MS. J Archaeol Sci 39(9):2945–2955
Panighello S, Van Elteren JT, Orsega EF, Moretto LM (2015) Laser ablation-ICP-MS depth profiling to study ancient glass surface degradation. Anal Bioanal Chem. doi:10.1007/s00216-015-8568-7
Peake RN, Freestone IC (2012) Cross-craft interactions between metal and glass working: slag additions to early Anglo-Saxon red glass. In: Meulebroeck W, Nys K, Vanclooster D (eds) Integrated approaches to the study of historical glass – IAS12. Proceedings of SPIE 8422, 842204, pp 1–12
Pérez-Arantegui J, Resano M, García-Ruiz E, Vanhaecke F, Roldán C, Ferrero J, Coll J (2008) Characterization of cobalt pigments found in traditional Valencian ceramics by means of laser ablation-inductively coupled plasma mass spectrometry and portable X-ray fluorescence spectrometry. Talanta 74:1271–1280
Phipps C, Luke J, Funk D, Moore D, Glownia J, Lippert T (2006) Laser impulse coupling at 130 fs. Appl Surf Sci 252:4838–4844
Reade W, Freestone IC, Bourke S (2009) Innovation and continuity in Bronze and Iron Age glass from Pella in Jordan. In: Janssens K, Degryse P, Caen J, Van't dack L (eds) Annales du 17 Congrès de l'Association Internationale pour l'Histoire du Verre, Anvers, 2006. AIHV and University Press Antwerp, Brussels, p 47–54
Rehren T, Shortland AJ (2003) Comments on J.L. Mass, M.T. Wypyski and R.E. Stone, Malkata and Lisht glassmaking technologies: towards a specific link between second millennium B.C. metallurgists and glassmakers and reply. Archaeometry 44:67–82
Robertshaw P, Benco N, Wood M, Dussubieux L, Melchiorre E, Ettahiri A (2010) Chemical analysis of glass beads from medieval Al-Basra (Morocco). Archaeometry 52:355–379
Rusk B, Koenig A, Lowers H (2011) Visualizing trace element distribution in quartz using cathodoluminescence, electron microprobe, and laser ablation-inductively coupled plasma-mass spectrometry. Am Mineral 96:703–708
Rutten FJM, Briggs D, Henderson J, Roe MJ (2009) The application of time-of-flight secondary ion mass spectrometry (ToF-SIMS) to the characterization of opaque ancient glasses. Archaeometry 51:966–986
Santopadre P, Verità M (2000) Analyses of the production technologies of Italian vitreous materials of the Bronze Age. J Glass Stud 42:25–40
Sayre EV (1963) The intentional use of antimony and manganese in ancient glasses. In: Matson FR, Rindone GE (eds) Advances in glass technology, Part 2. Plenum Press, New York, NY, pp 263–282
Sayre EV, Smith RW (1967) Some materials of glass manufactory in antiquity. In: Levey M (ed) Archeological chemistry: 3rd symposium on archeological chemistry, Atlantic City, New Jersey. University of Pennsylvania Press, Philadelphia, PA, pp 279–312
Šelih VS, Van Elteren JT (2011) Quantitative multi-element mapping of ancient glass using a simple and robust LA-ICP-MS rastering procedure in combination with image analysis. Anal Bioanal Chem 401:745–755
Shortland AJ (2002) The use and origin of antimonate colorants in early Egyptian glass. Archaeometry 44:517–530
Shortland AJ (2004) Evaporites of the Wadi Natrun: seasonal and annual variation and its implication for ancient exploitation. Archaeometry 46:497–516
Shortland AJ, Schroeder H (2009) Analysis of first millennium BC glass vessels and beads from the Pichvnari Necropolis, Georgia. Archaeometry 51:947–965
Shortland AJ, Nicholson PT, Jackson CM (2000) Lead isotopic analysis of Eighteenth-Dynasty Egyptian eyepaints and lead antimonate colorants. Archaeometry 42:153–157
Shortland AJ, Schachner L, Freestone I, Tite M (2006) Natron as a flux in the early vitreous materials industry: sources, beginnings and reason for decline. J Archaeol Sci 33:521–530
Silvestri A, Molin G, Salviulo G (2008) The colourless glass of Iulia Felix. J Archaeol Sci 35:331–341
Šmit Ž, Knific T, Jezeršek D, Istenič J (2012) Analysis of early medieval glass beads – glass in the transition period. Nucl Instr Meth B 278:8–14
Šmit Ž, Milavec T, Fajfar H, Rehren T, Lankton JW, Gratuze B (2013) Analysis of glass from the post-Roman settlement Tonovcov grad (Slovenia) by PIXE-PIGE and LA-ICP-MS. Nucl Instr Meth Phys Res B 311:53–59
Sokaras D, Karydas AG, Oikonomou A, Zacharias N, Beltsios K, Kantarelou V (2009) Combined elemental analysis of ancient glass beads by means of ion beam, portable XRF, and EPMA techniques. Anal Bioanal Chem 395:2199–2209
Tite MS, Shortland A, Maniatis Y, Kavoussanaki D, Harris SA (2006) The composition of the soda-rich and mixed alkali plant ashes used in the production of glass. J Archaeol Sci 33:1284–1292
Tite MS, Pradell T, Shortland AJ (2008) Discovery, production and use of Tin-based opacifiers in glasses, enamels and glazes from the Late Iron Age onwards: a reassessment. Archaeometry 50:67–84
Towle A, Henderson J, Bellintani P, Gambacurta G (2001) Frattesina and Adria: report of scientific analyses of early glass from the Veneto. PADUSA, Bollettino del Centro Polesano di studi storici, archeologici ed etnografici; anno XXXVII, Rovigo, p 7–68
Triglav J, Van Elteren JT, Šelih VS (2010) Basic modeling approach to optimize imaging by laser ablation ICPMS. Anal Chem 82:8153–8160

Turner WES (1956) Studies in ancient glasses and glassmaking processes. Part V. Raw materials and melting processes. J Soc Glass Technol 40:277–300

Turner WES, Rooksby HP (1961) Further historical studies based on X-ray diffraction methods of the reagents employed in making opal and opaque glasses. Jahrbuch des Römisch – Germanischen Zentral – Museums 8:1–16

van Elteren JT, Tennent NH, Šelih VS (2009) Multi-element quantification of ancient/historic glasses by laser ablation inductively coupled plasma mass spectrometry using sum normalization calibration. Anal Chim Acta 644:1–9

van Elteren JT, Izmer A, Šala M, Orsega EF, Šelih VS, Panighello S, Vanhaecke F (2013) 3D laser ablation-ICP-mass spectrometry mapping for the study of surface layer phenomena – a case study for weathered glass. J Anal Atom Spectrom 28:994–1004

Van Malderen SJM, van Elteren JT, Vanhaecke F (2015) Development of a fast laser ablation-inductively coupled plasma-mass spectrometry cell for sub-μm scanning of layer materials. J Anal Atom Spectrom 30:119–125

Verità M, Renier A, Zecchin S (2002) Chemical analyses of ancient glass findings excavated in the Venetian lagoon. J Cult Herit 3:261–271

Wagner B, Jedral W (2011) Open ablation cell for LA-ICP-MS investigations of historic objects. J Anal Atom Spectrom 26:2058–2063

Walton MS, Shortland AJ, Kirk S, Degryse P (2009) Evidence for the trade of Mesopotamian and Egyptian glass to Mycenaean Greece. J Archeolog Sci 36:1496–1503

Wang HAO, Grolimund D, Giesen C, Borca CN, Shaw-Stewart JRH, Bodenmiller B, Günther D (2013) Fast chemical imaging at high spatial resolution by laser ablation inductively coupled plasma mass spectrometry. Anal Chem 85:10107–10116

Woodhead JD, Hellstrom J, Hergt JM, Greig A, Maas R (2007) Isotopic and elemental imaging of geological materials by laser ablation inductively coupled plasma-mass spectrometry. Geostand Geoanal Res 31:331–343

Xu X (2002) Phase explosion and its time lag in nanosecond laser ablation. Appl Surf Sci 197–198:61–66

LA-ICP-MS Analysis of Ancient Silver Coins Using Concentration Profiles

5

Guillaume Sarah and Bernard Gratuze

Abstract

This chapter presents the main features of depth profile analysis for silver coins using LA-ICP-MS. The first section contains general information about ancient silver coins and their characteristics. Next, we discuss the implications of these characteristics for the analysis of these objects, followed by a presentation of the possibilities and limitations of the analytical methods used to date. The next section deals with previous attempts to analyze depth-heterogeneous materials using depth profile LA-ICP-MS analysis (DP-LA-ICP-MS). Finally, we present some aspects of the development and validation of LA-ICP-MS for the analysis of ancient silver coins.

5.1 Introduction

As of 2005, although Laser Ablation-Inductively Coupled Plasma-Mass Spectrometry (LA-ICP-MS) had already been applied successfully to various archaeological samples, the use of this technique had not yet been attempted for the characterization of ancient silver coins. It was then discovered that due to the presence of a surface layer on silver coins whose composition is not representative of the entire alloy, typical LA-ICP-MS analysis with an accumulation of signal during a determined duration cannot be applied. The only way to provide reliable results for the analyst and the historian is to use the progressive penetration of the laser by coupling it with a time-resolved measurement of the signal, in order to represent the evolution of the composition of the alloy dependent on ablation time–a concentration profile, an approach which has since been experimentally validated. It is now considered the most relevant approach for analysis of ancient silver coins, as it yields accurate and reliable results. Experiments have been carried out both on ancient coins and modern alloys by LA-ICP-MS and other techniques to study the composition and the structure of ancient silver-based alloys and to validate LA-ICP-MS protocol. Furthermore, the concentration profile approach for the analysis of ancient

G. Sarah (✉) • B. Gratuze
Institut de Recherche sur les Archéomatériaux, Centre Ernest Babelon, CNRS/Université d'Orléans, 3D rue de la Férollerie, 45071 Orléans Cedex 2, France
e-mail: guillaume.sarah@cnrs-orleans.fr; gratuze@cnrs-orleans.fr

L. Dussubieux et al. (eds.), *Recent Advances in Laser Ablation ICP-MS for Archaeology*, Natural Science in Archaeology, DOI 10.1007/978-3-662-49894-1_5

objects has since been applied to other materials such as gold-based coins, which in some cases revealed the unexpected presence of a similar surface-enrichment layer.

5.2 Background

5.2.1 Ancient Silver Coins

The analysis of archaeological artifacts, and especially of precious objects such as ancient coins, requires both a good understanding of their inner structure and the use of a minimally or non-destructive method of characterization. Unlike modern samples whose manufacturing process is known precisely, there is no certainty as to the metallurgical, mechanical and chemical processes involved in the production of ancient coins. It is also noteworthy that the sum of these processes, combined with potential subsequent alteration of the coin, be it during circulation or burial, entails changes in the composition and structure of the alloy, especially on the surface. These preliminary observations are the basis for development of any methodology for the analysis of ancient coins.

The term "silver" is conventionally used by numismatists for very diverse alloy compositions. It can sometimes refer to very pure silver with trace elements at very low concentrations. However, in many cases another metal, usually copper, is alloyed to the silver. Copper concentration can range from a few percent to tens of percent; in some cases copper can become the main element of the alloy, and the coin can still look exactly like pure silver visually (Arles and Téreygeol 2011). "Billon" refers to debased silver coins whose fineness ranges from about 50 % to as low as 25 % silver content. The structure of the alloy and its alteration depends on the nature and quantity of the metal added to the silver. It is therefore nearly impossible to predict the inner structure of a silver coin prior to its analysis. Along with the major elements of the alloy, ancient silver coins also contain minor and trace elements at very low concentrations. These components are usually not a deliberate addition by the ancient metallurgist or moneyer, but were part of the silver or other metal of the alloy. Determining the concentration of these minor and trace elements can help distinguish coins containing metal of different origins, and can sometimes be used to link the metal with its extraction source. An effective method for analysis of silver coins should therefore not only enable the precise determination of the fineness of the coin, but also the concentration of other elemental components of the alloy, some of them at very low levels.

Ancient silver coins are especially likely to present variations in their composition depending on depth. Complex physical and chemical phenomena such as the oxidation of copper in the alloy entail changes in both structure and composition. This is known as "silver surface enrichment" or "surface enrichment".[1] It results from the combination of phenomena occurring during the production, circulation, and finally deposition of the coin. Our understanding of this phenomenon is however not yet sufficient to allow its estimation without studying a cross-section, resulting in the destruction of the coin. A silver-enriched layer can be observed in Fig. 5.1, a Scanning Electron Microscopy (Back-Scattered Electron mode) image of the cross-section of a silver coin. The surface layer, whose composition has changed over time, presents cavities from which oxidized copper was leached. The elongated shape of these cavities, as well as the eutectic and copper-rich phases under this layer, result from the hammering of the blank during the production of the coin.

Silver surface enrichment is a major impediment to the analysis of silver coins, as it can cause a significant overestimation of the silver content if not taken into account. The depth of this layer can range from a few micrometers to hundreds of micrometers. Although debased coins are more likely to present a thick surface-

[1] For a review of the causes of silver surface enrichment, see Arles (2009), Beck et al. (2004), Carter (1964a, b, 1977), and Condamin and Picon (1972).

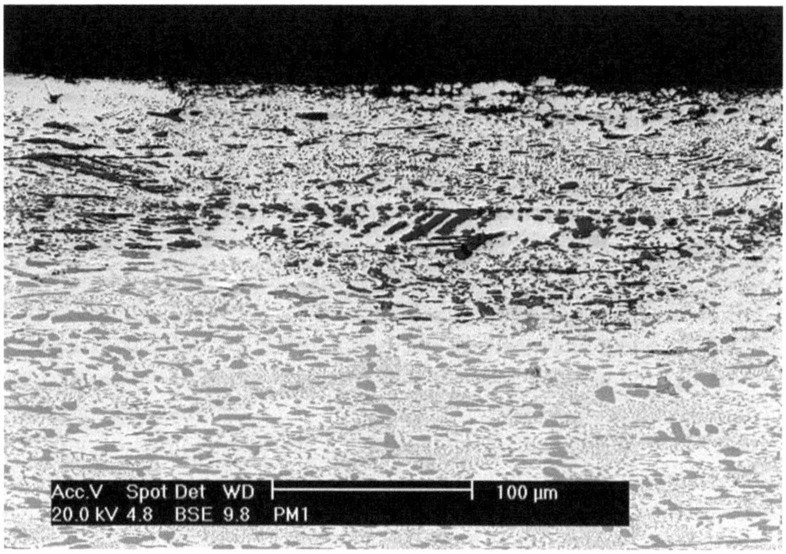

Fig. 5.1 Scanning Electron Microscopy image (Back Scattered Electron mode) of the cross-section of a silver coin presenting silver surface enrichment (CNRS–IRAMAT)

enriched layer, this is by no means systematic, and coins from the same hoard that have been subjected to the same burial conditions can possess very different internal structures. The main problem for the analyst aiming to characterize the alloy in its original state–that is, when the coin was struck–is to get rid of this surface layer, whose composition differs from that of the interior, without causing visible damage.

5.2.2 Analysis of Ancient Silver Coins: Different Approaches and Their Limitations

The first analyses of silver coins were destructive chemical analyses. They entailed the destruction of the whole object or of a substantial sample for analytical purposes. This is obviously unsatisfactory for precious ancient objects such as coins, so this type of analysis has been abandoned. Later, and particularly from the 1960s onwards, the development of spectrometric and nuclear methods gave rise to non-destructive analytical techniques applicable to ancient coins. Experiments were then carried out both to understand the silver surface enrichment phenomenon and to improve these non-destructive techniques for analysing silver coins (Condamin and Picon 1964, 1972; Hall 1960, 1961). Two main approaches have been widely implemented—surface methods and bulk methods.

The presence of an altered superficial layer can cause a significant overestimation of the silver content if a method which only takes the surface of the coin into consideration is used. This was noted as early as the 1960s by researchers using surface analytical methods such as X-Ray Fluorescence/milliprobe analysis (XRF), and later Particle Induced X-Ray Emission (PIXE) (Cowell 1998; Guerra and Barrandon 1998; Hall 1960; Klockenkämper et al. 1999; Linke et al. 2003). Researchers first attempted to solve the surface enrichment problem by "cleaning" a small area of the sample, usually on the face or the edge of the coin, and performing repeated analyses in order to remove the silver-enriched surface layer and to reach the unaffected core alloy (Cowell 1998; Carter 1964a, 1977; Schweizer 1972). This preparation of the samples obviously yielded better results, but the repeated process of cleaning and analysis was time consuming. Furthermore, it soon became apparent that "analysis on an abraded section on the edge of the coin should in theory discount surface effects, such as corrosion or surface enrichment, and should measure the unaltered alloy; in practice, however, it is a delicate and sometimes perhaps an impossible task, within the limitation of acceptable aesthetic damage, to obtain even an approximately correct result from these tiny coins, particularly when they are debased" (Metcalf 1983; see also Linke and Schreiner 2000). Although they are to this day the most widespread analytical tools available for coin analysis, surface methods such as XRF should therefore only be used to distinguish coins of very different compositions, or to identify the major components of the alloy, but not to determine their precise and complete composition.

The second of the main approach that has been used to analyze silver coins is nuclear bulk analysis (Barrandon 1980; Barrandon et al. 1988; Gordus 1967; Meyers 1969). Such analyses provide an average content of the coin as a whole for all detectable elements. Although the bulk analysis of silver coins is more reliable than surface analysis, the presence of a modified superficial layer can still cause a significant difference between the results and the actual content of the original unaffected alloy (Guey and Condamin 1961). In this case, the geometry of the coin is crucial: the thickness of the silver-enriched layer on each face of the coin and the overlapping of these layers at the edge can have a significant impact on measured composition (Condamin and Picon 1964). Moreover, bulk analysis only provides a single value for each element and gives no indication as to the thickness and contribution of the silver surface enriched layer to the measured bulk composition.

In addition to the major limitations caused by the silver surface enrichment phenomenon, both surface and bulk analytical methods, when applied to ancient silver coins, only enable identification of a limited number of trace elements, as their detection limits are often too high for many elements of interest. Because of these limitations of traditional methods of analysis, we endeavoured to test another approach, seeking to get rid of the silver surface enrichment problem and to measure the minor and trace elements that are potentially useful for provenance studies. The successful use of LA-ICP-MS for the characterization of archaeological objects and modern multi-layer samples strongly recommended it for the task of analysing ancient silver coins.

5.3 LA-ICP-MS for the Analysis of Depth-Heterogeneous Materials

LA-ICP-MS is a particularly suitable method for the analysis of a wide range of solid samples. It requires no sample preparation and allows for the determination of major, minor and trace elements with good accuracy and low detection limits. This technique has been applied in archaeometry beginning in the early 1990s to analyse vitreous materials, stone, and ceramics (Gratuze 1999; Gratuze et al. 1993; Neff et al. 2003; Speakman et al. 2002), and more recently to gold coins, especially for provenance purposes (Dussubieux and van Zelst 2004; Gratuze et al. 2004). The micro-samples resulting from laser ablation are small enough to be invisible to the naked eye (Fig. 5.2), but can also be large enough to be representative of the entire object. Later analysis of depth-heterogeneous materials such as ancient silver coins was to yield even more interesting results, as it was discovered that the progressive penetration of the laser into the sample could be coupled with time-resolved acquisition of the signal—concentration profiles representing the evolution of the composition of the coin depending on the duration of the analysis, corresponding to the progression of the laser into the sample, could be obtained using this technique. This allows for identification of the surface enrichment layer, and for only the part of the concentration curve corresponding to the unaffected alloy to be used for historical interpretation.

Development of the analytical parameters for depth-profile analysis of ancient silver coins was based on approaches found in various publications, including studies on the characterization of archaeological metals and depth-profile analysis by LA-ICP-MS. In the case of ferrous metals, LA-ICP-MS is used only to determine the fingerprint of the vitreous inclusions embedded

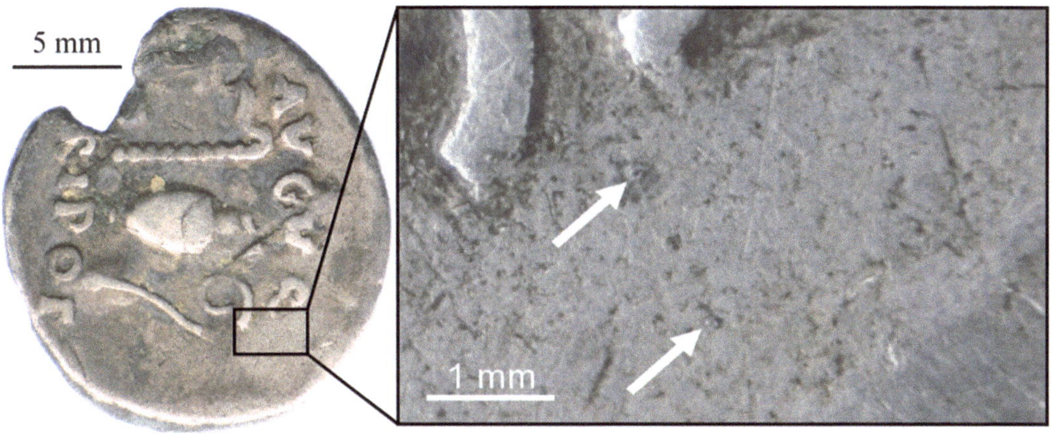

Fig. 5.2 Two laser ablation spots at the surface of a silver coin (CNRS–IRAMAT)

in the matrix.[2] For non-ferrous alloys, the aforementioned studies regarding gold coins, and others on some of the wide range of possible copper-based alloys, have recently proved successful (Dussubieux et al. 2008), although LA-ICP-MS is as of yet an inadequate method of analysis in other cases. In the case of silver, experiments aiming to determine the level of impurities have been undertaken for the authentication of works of art (Devos et al. 1999). Qualitative LA-ICP-MS analysis has also been used to reveal the presence of mercury deposits on the surface of silver coins, allowing identification of instances of silvering using the amalgam technique (Vlachou-Mogire et al. 2007).

Outside of archaeometry, LA-ICP-MS is not a very widespread method for the characterization of metallic samples, as its main advantage from an archaeological point of view, namely that it is relatively non-destructive, is irrelevant to modern samples, for which destructive methods have therefore been generally preferred. However, some publications have been dedicated to the study of isotopic fractionation of copper during laser ablation in modern copper-based alloys (Kuhn et al. 2006, 2007). Other researchers have studied the improvements obtained by the use of a femtosecond laser for the sampling of brass alloys (Bian et al. 2006; Liu et al. 2004).

[2] See the contribution of L'Héritier et al. (Chap. 14).

Experiments have also sought to determine the concentration of Pd, Pt and Rh in lead samples obtained from the cupellation of platinum-rich ores (Vanhaecke et al. 2004). Trace-element contents can also be studied to determine the provenance of modern gold and silver (Kogan et al. 1994). None of this research however has had any direct influence on the development of the analytical protocol for the characterization of ancient silver coins, as it does not seek to determine the composition of the major elements or to remove surface enrichment.

If we consider the results of past analyses of multi-layered samples by the means of LA-ICP-MS, it should first be noted that LA-ICP-MS had never before been used to study the evolution of the composition of such samples according to depth. Previous studies including depth-profile analysis by LA-ICP-MS have focused developing depth resolution for the characterization of multi-layered samples, usually by means of producing a thin layer of metal deposited on another material (Coedo et al. 2005; Kanický et al. 2004; Margetic et al. 2001; Mason and Mank 2001; Plotnikov et al. 2001). Similar experiments have been carried out in archaeometry to study gold or silver deposited on the surface of coins using the mercury amalgam technique (Vlachou-Mogire et al. 2007). However, no attempts were made in these studies to determine the bulk composition of alloy under a surface layer whose composition has been modified over time. Every experimental aspect of the LA-ICP-MS depth-profile analytical protocol for ancient silver coins therefore had to be first studied in order to verify the suitability of this approach to analysis of silver coins, and to determine the best combination of parameters for analysis.

5.4 Protocol

5.4.1 Instrumentation

Our LA-ICP-MS protocol for analysis ancient silver coins was developed as part of PhD research carried out between 2005 and 2008 at the CNRS IRAMAT–Centre Ernest-Babelon (Sarah 2008). This laboratory was at the time equipped with a Fisons VG Plasma Quad PQXS quadrupole ICP-MS. Since 2009, analyses have been carried out using a Sector Field Thermo Fisher Element XR, which allows for significant improvements both in sensitivity and mass resolution. The laser utilized in our studies is generated by a Nd:YAG pulsed laser whose frequency is quadrupled to operate in the ultraviolet range at 266 nm.

The ablation cells used at Centre Ernest-Babelon for the analyses of ancient coins have been specially designed for this purpose. One of them, used mostly for silver coins, is 8 cm in diameter and 3 cm in height. This ablation cell allows room for 6–20 coins at a time, depending on their diameter and on the number of reference materials required for calibration. Additional ablation cells of various dimensions are available to adapt analysis to any coinage. Analysis is carried out on the face of the coins rather than on the edge, as the surface enrichment layer can be very thick on that part of the coin due to the overlapping of the layers of each coin face.

DP-LA-ICP-MS analysis is a representation of the evolution of the concentration of each element detected depending on the duration of the ablation. This requires that a composition be calculated for each time interval for which a measurement is obtained. The blank measured before turning on the laser is deduced from the signal measured for each isotope and each time slice, and a calculation procedure based on the determination of response coefficient factors theoretically identical for the reference materials and samples is applied (Gratuze 1999; Chap. 12). The internal standard used in the case of silver coins is ^{107}Ag.

5.4.2 Development of the Protocol

The DP-LA-ICP-MS analytical protocol for ancient silver coins was developed and validated through the study of more than 1000 objects. This corpus consisted mostly of medieval Western coins covering a wide range of compositions, containing from a few to tens of percent of copper, as well as minor and trace elements at significant levels. It also included very pure silver coins, containing trace elements at only very low concentrations, as well as billon coins whose major component is copper. The protocol as described in the following paragraphs should be regarded only as a guideline for the LA-ICP-MS analysis of silver coins—the recommendations given here must be adapted to the specificities of a particular coinage when necessary.

The size of the laser ablation spot should be chosen in order to provide representative analysis of a silver-copper two-phase alloy. Scanning Electron Microscopy observations and LA-ICP-MS experiments revealed that a laser diameter of 80 μm is wide enough, yet invisible to the naked eye (Fig. 5.3). Three profiles are usually generated on the surface of each coin in order to verify the reproducibility and the representativeness of the analysis. The most suitable repetition rate for the laser is 6 Hz. This is the best compromise between speed of penetration of the laser, the instantaneous amount of matter ablated and sent to the detector, and the in-depth resolution required for the characterization of depth-heterogeneous materials. This repetition rate also prevents significant mass fractionation from occurring.

At a rate of 6 Hz, and considering the usual possible thickness of the silver surface-enriched layer, a duration of 300 s of ablation is sufficient

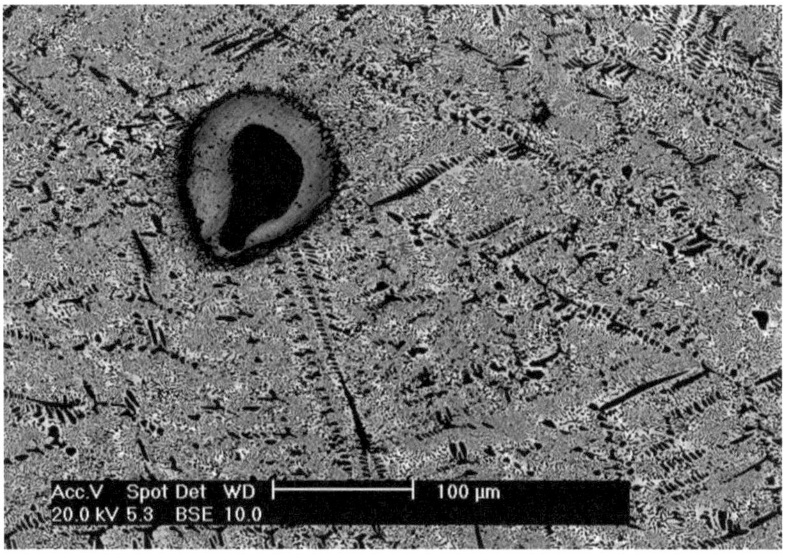

Fig. 5.3 Scanning Electron Microscopy image (Back-Scattered Electron mode) of a two-phase silver-copper alloy and comparison with the size of a laser ablation crater (CNRS–IRAMAT)

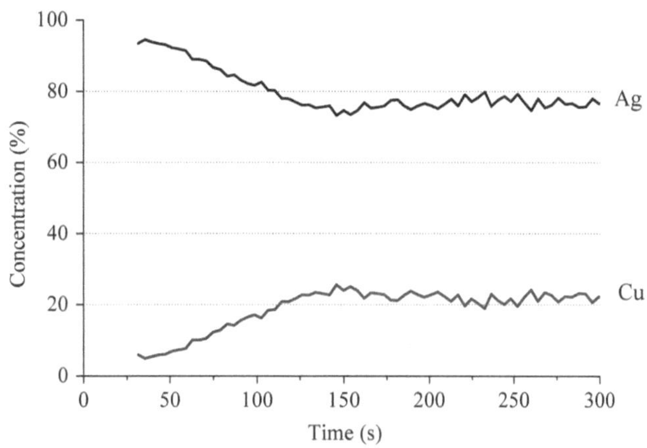

Fig. 5.4 Example of concentration profiles of silver and copper for an ancient coin presenting significant silver surface enrichment

in most cases to penetrate through a region of decreasing silver content corresponding to the altered alloy, and to obtain a stable plateau for signal strength. This first part of the curve (corresponding to the surface enrichment layer) is not used for the calculation of the coin's composition; an average value is determined for each element only for the stable signal corresponding to the unaffected alloy beneath the surface layer. This can be observed in Fig. 5.4, which shows the silver and copper concentration profiles for a medieval coin with significant silver surface enrichment. However, visualization of the evolution of composition depending on the duration of ablation is a valuable source of information, as it allows the analyst to ensure that the unaltered core has been reached; it also confirms the reliability of the results by demonstrating that a plateau of stability is obtained for each of the three profiles measured, and that the values are comparable for each element across each of the three profiles.

Experiments carried out on cross-sections of ancient coins revealed that 300 s of laser ablation at a rate of 6 Hz provides information for a superficial layer of about 300 µm on average for silver-copper alloys. In cases when the alloy is very pure, the silver surface enrichment phenomenon is not likely to be very pronounced, and ablation can be shortened to 200 s for such coins. In any study, preliminary experiments must be carried out in order to determine the most suitable analytical parameters for a given corpus of coins if they have never been studied previously by LA-ICP-MS.

The example in Fig. 5.4 shows how the profiles obtained by DP-LA-ICP-MS for silver and copper appear to be an effective approach for representing the evolution of the composition from the surface to the interior of ancient coins. The profile shows a significant silver-enriched surface layer: the concentration of silver progressively decreases during 150 s of ablation beginning at more than 95 % near the surface to about 75 % in the unaffected part of the coin. This representation makes it possible to distinguish the part of the curve corresponding to the unaffected alloy, so that silver surface enrichment does not affect the accuracy of the results—the composition provided to the historian in this case would be Ag-Cu 75 %–25 %, corresponding to that of the unaffected core of the coin. The DP-LA-ICP-MS concentration profiles also provide confidence in the results, as it can be determined that a plateau of stability is reached at some point, and that consequently, the entire coin is not corroded. In other words, depth-profiling demonstrates that original alloy is present in the coin (or that it is not in some cases), which cannot be determined using superficial or bulk methods.

The Thermo Fisher Element XR can be used in Low Resolution, Medium Resolution and High Resolution modes, depending on the mass selectivity necessary for the separation of the peaks corresponding to the ions detected. The analysis in concentration profiles as used for analysis of silver coins requires a time-resolved acquisition of signal, instead of the usual accumulation of signal. For determination of trace elements, it is necessary to select the resolution mode with which the most important elemental signal values can be obtained. This means that detection of isotopes corresponding to major, minor and trace elements can take place in the same menu in low resolution mode. This makes it possible to minimize the total duration of the analysis of a sample and to obtain better accuracy for trace element content, although some elements cannot be satisfactorily detected due to mass interferences when present at very low concentrations. Most isotopes impacted by interferences have proven of limited use in provenance studies, however, so use of the low resolution mode to speed analysis results in no significant loss of historically useful information.

In most cases, ancient silver coins are either silver-copper alloys with minor and trace elements at various levels, or pure silver containing trace elements only, but there is no general rule concerning the concentration of minor and trace elements. Previous analyses have revealed that the concentration of these elements, although sometimes present at only trace levels, can in other cases reach 1 % or more. These "main minor elements" are Zn, Sn, Au, Pb and Bi. Other elements are present at very low concentrations, except for As and Sb, concentrations of which can reach thousands of ppm. These "main minor" elements, as well as As and Sb, can be used for provenance purposes and historical interpretation to separate or group coins with similar elemental patterns. These observations served as the basis for selection of the alloys used as reference materials for calibration.

These reference materials are analysed regularly during each analytical session, and used in the calculation procedure to determine the concentrations in the samples (see Gratuze 1999). Four of them are pure silver-copper alloys, whose silver content ranges from about 60 % to 97 % (RAgA1, RAgA3, RAgA4 and RAgA5, purchased from Rand Refinery, South Africa). Four others are various silver-copper alloys containing the five main minor elements (Zn, Sn, Au, Pb and Bi) as well as As and Sb at high levels (EV1, EV2, EV3 and EV4). They were prepared at our request by the Centre d'Études de Chimie Métallurgique (CNRS, Vitry-sur-Seine–France). The last three reference materials utilized, SF1A, RAgP3 and RAgP7, were purchased from Rand Refinery, South Africa, and Gliwice, Poland; they are pure silver containing a wide range of trace elements. Compositions for these 11 reference samples are given in Table 5.1.

Table 5.1 Composition of reference materials used for calibration of silver coins analyses

Element	RAgA1	RAgA3	RAgA4	RAgA5	EV1	EV2	EV3	EV4	RAgP3	RAgP7	SF1A
Ag	96.94 %	91.50 %	79.94 %	59.89 %	75.0 %	45.0 %	80.0 %	65.0 %	99.97 %	99.77 %	99.86 %
Cu	3.03 %	8.47 %	20.03 %	40.10 %	15.0 %	45.0 %	18.2 %	32.8 %	29.9	209	857
Zn					20,000	20,000	1000	2000	28.7	209	79.1
Sn					20,000	20,000	1000	2000	27.2	192	
Au					20,000	20,000	2000	5000	39.4	216	21.9
Pb					20,000	20,000	10,000	10,000	26.7	194	21.0
Bi					20,000	20,000	1000	2000	24.5	231	53.7
As							2000	1000			41.6
Sb							1000	500	26.7	185	58.1
Fe									17.4	88	42.5
Ni									24.4	181	58.1
Se											45.0
Pd									25.3	193	
Te									26.1	181	43.9
Pt									26.7	187	20.9

Unless otherwise indicated, all values are expressed in ppm

The list of detected elements has been determined in accordance with past analyses. Silver, the major constituent of the coins, and copper, which is most often alloyed to the silver, are the two main elements to be determined. They are usually detected using the Faraday mode, which is well adapted to the detection of the elements composing the matrix of the sample. In cases when the silver is very pure, silver is the only element whose ions need to be detected in Faraday mode. There is no significant interference for either of the two isotopes of silver (^{107}Ag and ^{109}Ag) or for those of copper (^{63}Cu and ^{65}Cu).

The concentration of the five main minor elements Zn, Sn, Au, Pb and Bi can vary from a few ppm to one or several percent. These elements must be detected in Triple Mode, which enables adaptation of the detection mode depending on the quantity of ions measured during the first part of acquisition. Au and Bi are monoisotopic, so there is only one choice of measurement, ^{197}Au and ^{209}Bi. In contrast, Zn, Sn and Pb have several stable isotopes. Usually, the most abundant is preferable in order to obtain the highest signal and lowest detection limit. Consequently, we measure ^{64}Zn and ^{118}Sn. As silver usually derives from argentiferous lead, lead concentration is often relatively high—consequently, we measure ^{204}Pb: although it is the Pb isotope whose isotopic abundance is the lowest, its signal is sufficient in most cases to be detected in Faraday mode, which allows detection in the same mode for all samples. Another Pb isotope can be chosen if concentration is too low.

For other minor and trace elements, the most abundant isotope is measured, unless an interference causes a significant error in measurement of the resulting signal. The elements and isotopes typically measured during analysis of silver coins are listed in Table 5.2, along with their isotopic abundance, detection mode, and possible interferences.

Some of the elements listed in Table 5.2 are absent from the reference alloys used in our calibration procedure—specifically, Ru, Rh, Cd, In, Os, Ir and Tl. In these cases, the K_y response coefficient factor for another element of similar mass, and whose instrument response should be similar, is used for calibration purposes. The signal measured for isotopes for which an

Table 5.2 List of isotopes measured with their isotopic abundance, detection mode, possible interferences and detection limits

Element	Isotope	Isotopic abundance (%)	Detection mode	Possible interference(s)	Detection limit (ppm)
Ni	^{60}Ni	26.1	Triple		4.2
Cu	^{63}Cu	69.2	Triple/Faraday		5.0
Zn	^{64}Zn	48.6	Triple		0.61
As	^{75}As	100	Triple		0.61
Ru	^{101}Ru	17.0	Triple		0.03
Rh	^{103}Rh	100	Triple	^{40}Ar-^{63}Cu	0.08
Pd	^{104}Pd	11.1	Triple		0.62
	^{105}Pd	22.3	Triple	^{40}Ar-^{65}Cu	
Ag	^{107}Ag	51.8	Faraday		
Cd	^{110}Cd	12.5	Triple	^{110}Pd	0.11
In	^{115}In	95.7	Triple	^{115}Sn	0.12
Sn	^{118}Sn	24.2	Triple		4.5
Sb	^{121}Sb	57.3	Triple		0.15
Te	^{128}Te	31.7	Triple		0.87
Os	^{189}Os	16.1	Triple		0.04
Ir	^{193}Ir	62.7	Triple		0.02
Pt	^{195}Pt	33.8	Triple		0.04
Au	^{197}Au	100	Triple		0.12
Tl	^{205}Tl	70.5	Triple		0.01
Pb	^{204}Pb	1.4	Triple		7.0
Bi	^{209}Bi	100	Triple		0.15

isobaric interference can occur can be corrected by taking into account the signal of the interfering isotope. Detection limits are also presented in Table 5.2. Detection limits were calculated as a signal value corresponding to three standard deviations on an average of 20 blank measurements. Most detection limits, especially for trace elements, are near 1 ppm or less.

5.5 Validation of the Protocol

Several approaches have been utilized to validate our analytical protocol by DP-LA-ICP-MS for ancient silver coins. For the purpose of brevity, we present two approaches for validating our analytical protocol here. Firstly, reproducibility and the variability of results for both modern alloys and ancient coins have been calculated. Secondly, cross-sections of coins have been studied by several other methods and compared with the LA-ICP-MS concentration profile results.

5.5.1 Reproducibility of Analysis of Silver Coins by LA-ICP-MS

Firstly, we carried out experiments studying the reproducibility of the micro-analysis on a single object in order to validate our DP-LA-ICP-MS analytical protocol for ancient silver coins. The concentration profiles obtained for a single coin have been superimposed for comparison purposes, and the relative standard deviation determined from the averaged value for each profile was calculated. The results for silver are presented in Fig. 5.5. The superposition of three profiles for silver obtained from laser ablation carried out at different points on individual coins is shown for 11 examples with varying alloy compositions. Whatever the fineness of the coin, the three micro-analyses generate similar compositions. The relative standard deviation for silver measured in each of the three profiles on a given coin is in every case less than 5 %, and as low as 1 % for coins whose fineness is greater than the silver-copper eutectic composition (72 %). These results show that analysis by LA-ICP-MS is a suitable approach not only for modern alloys, but also for ancient coins, and that billon coins, although copper is their main component, can also be characterised by this method. Similar results were observed for the main minor elements, for which the relative standard deviation across three ablations is generally below 5 %. Even for trace elements, standard deviations were below 10 % across three depth-profiles.

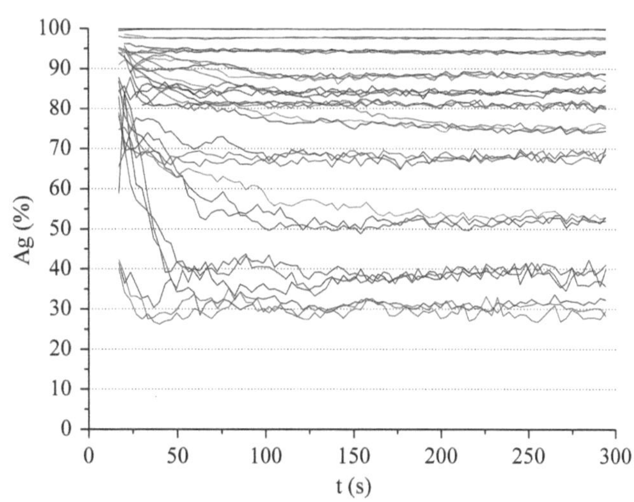

Fig. 5.5 Superposition of three concentration profiles of silver obtained from laser ablation at different spots on the same coin for various alloy compositions

5.5.2 Analysis of Cross-Sections

Although damaging a large number of coins for analytical purposes is unacceptable practice, it was deemed necessary in this case to study the cross-sections of a few coins in order to compare the results obtained via DP-LA-ICP-MS and the actual evolution of internal composition in the same coins. A few coins were sectioned for this purpose, and analysed by DP-LA-ICP-MS, Fast Neutron Activation Analysis (a bulk method), and XRF and SEM-EDS at the surface and on the cross-section. These experiments demonstrate that LA-ICP-MS concentration profiles correspond to the actual evolution of the composition depending on depth. Figure 5.6. shows the comparison between LA-ICP-MS concentration profiles and standard analyses carried out on the surface of the coin, as well as repeated average single spot analyses by LA-ICP-MS performed on the cross-section of the coin from the surface to a depth of 500 μm. The resulting ablation craters are shown in Fig. 5.7, as well as the silver surface enrichment layer at two different locations on the same coin. These two figures show the concurrence between concentration profiles generated by our protocol and the actual internal structure of these coins. LA-ICP-MS is the only analytical approach that enables measurement of the depth evolution of silver-based alloy composition without any visible damage to the coins.

5.5.3 Limitations

Though in most cases LA-ICP-MS provides reproducible analyses for multiple concentration profiles, a few coins could not be satisfactorily characterised by spot analysis using this method.

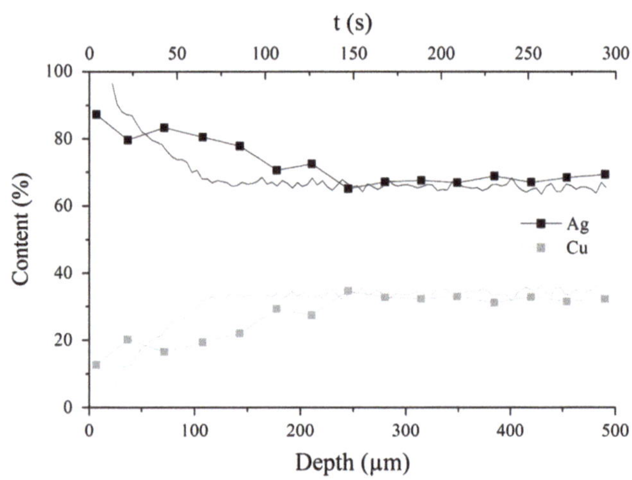

Fig. 5.6 Comparison of LA-ICP-MS concentration profiles of silver and copper with spot analysis on the cross-section of the same coin

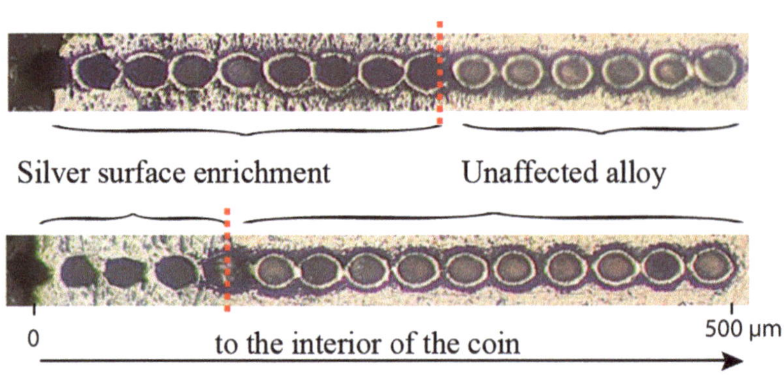

Fig. 5.7 Cross-section images of a coin showing craters generated during repeated LA-ICP-MS analysis performed from the surface to the interior (CNRS–IRAMAT)

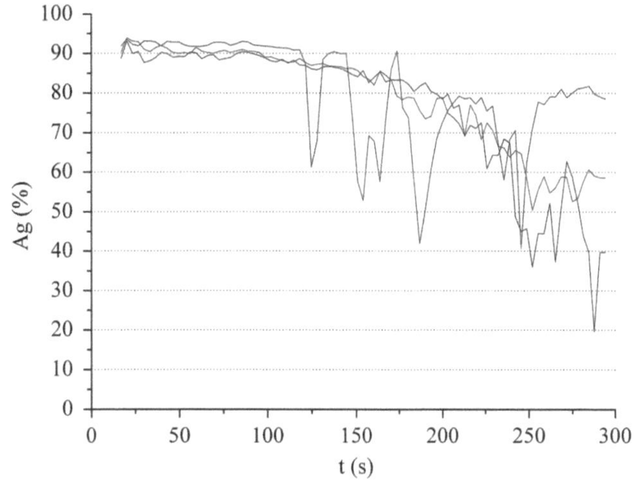

Fig. 5.8 Superposition of three LA-ICP-MS concentration profiles for silver on a coin producing non-reproducible analyses

In some cases, the resulting three profiles are not superimposable, or no signal plateau can be reached. The worst case scenario is presented in Fig. 5.8—none of the three depth profiles shown can be used to determine the original silver content of the coin. This may be caused by the preferential oxidation of copper after burial. In such a case, a spot analysis approach by LA-ICP-MS is irrelevant for characterization of the alloy; however, the resulting concentration profiles still provide information about the inner structure of the alloy. Although other analyses can be performed on such coins to determine their bulk composition, erratic LA-ICP-MS profiles should inspire caution in interpretation of such results.

5.6 Conclusion

Only the main features of DP-LA-ICP-MS analysis for silver coins are presented here. The results nonetheless reveal that spot depth analysis provides representative results for the entire coin, even in the case of two-phase silver-copper alloys. Concentration profile analysis has also been shown to be a suitable approach for removing the silver surface enrichment layer and characterizing the underlying unaffected alloy. This is a significant improvement for the analysis of ancient silver coins, and LA-ICP-MS is the only method allowing reliable measurement of compositional evolution with depth without causing visible damage to the analysed coins. The analytical parameters presented here should be regarded as a guideline, and must be adapted to the characteristics of the coins to be analysed and to the goals of the analyses. This versatility is another major advantage of the application of LA-ICP-MS to the study of ancient silver coinage. Concentration profiles generated by LA-ICP-MS constitute a significant step in the analysis of archaeomaterials in general, and not only of silver coinage. There is no doubt that this approach will be applied to other samples presenting depth-heterogeneous structure or multiple layers with different compositions in the future. This type of analysis allows for improved understanding of the inner structure of the samples and better confidence in the results provided by the analyst to the historian.

References

Arles A (2009) Entre monnayage official et faux-monnayage. La fabrication de la monnaie au marteau en France (XIIIe-XVIIe siècles). Approche physico-chimique, expérimentale et historique. Ph.D Dissertation, Université d'Orléans

Arles A, Téreygeol F (2011) Le procédé de blanchiment dans les ateliers monétaires français au XVIème siècle: approche archéométrique et expérimentale. Anuario de Estudios Medievals 41(2):699–721

Barrandon JN (1980) Analyse par activation et numismatique. J Radioanal Nucl Chem 55:317–327

Barrandon JN, Guerra MF, Magro FA (1988) Chemical compositions of Portuguese dinheros. In: Marques MG, Metcalf DM (eds) Problems of Medieval coinage in the Iberian Area: a symposium held by the Sociedad Numismática Scalabitana and the Instituto de Sintra on 4–8 October, 1988. Sociedad Numismática Scalabitana and the Instituto de Sintra, Santarém, pp 343–372

Beck L, Bosonnet S, Réveillon S, Eliot D, Pilon F (2004) Silver surface enrichment of silver-copper alloys: a limitation for analysis of ancient silver coins by surface techniques. Nucl Instrum Meth B 226:153–162

Bian Q, Garcia CC, Koch J, Niemax K (2006) Non-matrix matched calibration of major and minor concentrations of Zn and Cu in brass, aluminium and silicate glass using NIR femtosecond laser ablation inductively coupled plasma mass spectrometry. J Anal Atom Spectrom 21:187–191

Carter GF (1964a) Preparation of ancient coins for accurate X-ray fluorescence analysis. Archaeometry 7 (1):106–113

Carter GF (1964b) X-ray fluorescence analysis of Roman coins. Anal Chem 36(7):1264–1268

Carter GF (1977) Reproducibility of X-ray fluorescence analyses of Septimius Severus denarii. Archaeometry 19:67–73

Coedo AG, Dorado T, Padilla I, Fariñas JC (2005) Depth profile analysis of copper coating on steel using laser ablation inductively coupled plasma mass spectrometry. J Anal Atom Spectrom 20:612–620

Condamin J, Picon M (1964) The influence of corrosion and diffusion on the percentage of silver in Roman Denarii. Archaeometry 7:98–105

Condamin J, Picon M (1972) Changes suffered by coins in the course of time and the influence of these on the results of different methods of analysis. In: Hall ET, Metcalf DM (eds) Methods of chemical and metallurgical investigation of ancient coinage. Royal Numismatic Society, London, pp 49–66

Cowell M (1998) Coin analysis by energy dispersive X-ray fluorescence spectrometry. In: Oddy WA, Cowell MR (eds) Metallurgy in numismatics v 4. Royal Numismatic Society, London, pp 448–460

Devos W, Moor C, Lienemann P (1999) Determination of impurities in antique silver objects for authentication by laser ablation inductively coupled plasma mass spectrometry (LA-ICP-MS). J Anal Atom Spectrom 14:621–626

Dussubieux L, van Zelst L (2004) LA-ICP-MS analysis of platinum-group elements and other elements of interest in ancient gold. Appl Phys A-Mater 79:353–356

Dussubieux L, Deraisme A, Frot G, Stevenson C, Creech AMY, Bienvenu Y (2008) LA-ICP-MS, SEM-EDS and EPMA analysis of Eastern North American copper-based artefacts: impact of corrosion and heterogeneity on the reliability of the LA-ICP-MS compositional results. Archaeometry 50:643–657

Gordus AA (1967) Quantitative non-destructive neutron activation analysis of silver in coins. Archaeometry 10:78–86

Gratuze B (1999) Obsidian characterization by laser ablation ICP-MS and its application to prehistoric trade in the Mediterranean and the near East: sources and distribution of Obsidian within the Aegean and Anatolia. J Archaeol Sci 26:869–881

Gratuze B, Barrandon JN, Al Isa K, Cauvin MC (1993) Non-destructive analysis of obsidian artefacts using nuclear techniques: investigation of provenance of Near Eastern artefacts. Archaeometry 35(1):11–21

Gratuze B, Blet-Lemarquand M, Barrandon J-N (2004) Caractérisation des alliages monétaires à base d'or par LA-ICP-MS. Bulletin de la Société Française de Numismatique 59:163–169

Guerra MF, Barrandon JN (1998) Ion beam activation analysis with a cyclotron. In: Oddy WA, Cowell MR (eds) Metallurgy in Numismatics, vol 4. Royal Numismatic Society, London, pp 15–34

Guey J, Condamin J (1961) Exploration spectrographique d'un antoninianus de Caracalla. Revue Numismatique 6:51–73

Hall ET (1960) X-ray fluorescent analysis applied to archaeology. Archaeometry 3(1):29–35

Hall ET (1961) Surface-enrichment of buried metals. Archaeometry 4:29–35

Kanický V, Kuhn HR, Günther D (2004) Depth profile studies of ZrTiN coatings by laser ablation inductively coupled plasma mass spectrometry. Anal Bioanal Chem 380:218–226

Klockenkämper R, Bubert H, Hasler K (1999) Detection of near-surface silver enrichment on Roman imperial silver coins by X-ray spectral analysis. Archaeometry 41:311–320

Kogan VV, Hinds MW, Ramendik GI (1994) The direct determination of trace metals in gold and silver materials by laser ablation inductively coupled plasma mass spectrometry without matrix matched standards. Spectrochim Acta B 49:333–343

Kuhn HR, Pearson NJ, Jackson SE (2006) Laser ablation ICP-MS: particle size-dependent isotopic fractionation of copper in laser-generated aerosols. Geochim Cosmochim Acta 70:A337

Kuhn HR, Pearson NJ, Jackson SE (2007) The influence of the laser ablation process on isotopic fractionation of copper in LA-MC-ICP-MS. J Anal Atom Spectrom 22:547–552

Linke R, Schreiner M (2000) Energy dispersive X-ray fluorescence analysis and X-ray microanalysis of Medieval silver coins. Microchim Acta 133:165–170

Linke R, Schreiner M, Demortier G, Alram M (2003) Determination of the provenance of medieval silver coins: potential and limitations of x-ray analysis using photons, electrons or protons. X-Ray Spectrometry 32 (5):373–380

Liu C, Mao XL, Mao SS, Zeng X, Greif R, Russo RE (2004) Nanosecond and femtosecond laser ablation of brass: particulate and ICP-MS measurements. Anal Chem 76:379–383

Margetic V, Bolshov M, Stockhaus A, Niemax K, Hergenröder R (2001) Depth profiling of multi-layer samples using femtosecond laser ablation. J Anal Atom Spectrom 16:616–621

Mason PRD, Mank AJG (2001) Depth-resolved analysis in multi-layered glass and metal materials using laser ablation inductively coupled plasma mass spectrometry (LA-ICP-MS). J Anal Atom Spectrom 16:1381–1388

Metcalf DM (1983) Interpreting the alloy of the Merovingian silver coinage. In: Brooke CNL, Stewart BHIH, Pollard JG, Volk TR (eds) Studies in numismatic method presented to Philip Grierson. Cambridge University Press, Cambridge, pp 113–126

Meyers P (1969) Non-destructive activation analysis of ancient coins using charged particles and fast neutrons. Archaeometry 11:67–84

Neff H, Cogswell JW, Ross LM Jr (2003) Supplementing bulk chemistry in archaeological ceramic provenance investigations. In: van Zelst L (ed) Patterns and process: a Festschrift in honor of Dr Edward V. Sayre. Smithsonian Center for Materials Research and Education, Uitland, pp 201–224

Plotnikov A, Vogt C, Hoffmann V, Täschner C, Wetzig K (2001) Application of laser ablation inductively coupled plasma quadrupole mass spectrometry (LA-ICP-QMS) for depth profile analysis. J Anal Atom Spectrom 16:1290–1295

Sarah G (2008) Caractérisation de la composition et de la structure des alliages argent-cuivre par ICP-MS avec prélèvement par ablation laser. Application au monnayage carolingien. Ph.D Dissertation, Université d'Orléans

Schweizer F (1972) Analysis of ancient coins using a point source linear x-ray spectrometer: a critical review. In: Hall ET, Metcalf DM (eds) Methods of chemical and metallurgical investigation of ancient coinage. Royal Numismatic Society, London, pp 153–169

Speakman RJ, Neff H, Glascock MD, Higgins B (2002) Characterization of archaeological materials by LA-ICP-MS. In: Jakes K (ed) Archaeological chemistry: materials, methods, and meaning. American Chemical Society, Washington, pp 48–63

Vanhaecke F, Resano M, Garcia-Ruiz E, Balcaen L, Koch KR, McIntosh K (2004) Laser ablation-inductively coupled plasma-dynamic reaction cell-mass spectrometry (LA-ICP-DRC-MS) for the determination of Pt, Pd and Rh in Pb buttons obtained by fire assay of platiniferous ores. J Anal Atom Spectrom 19:632–638

Vlachou-Mogire C, Stern B, McDonnell JG (2007) The application of LA-ICP-MS in the examination of the thin plating layers found in late Roman coins. Nucl Instrum Meth B 265:558–568

Part II

Application to Non-siliceous Materials

Analysis of Non-siliceous Archaeological Materials by LA-ICP-MS

6

Laure Dussubieux

Abstract

This chapter introduces some of advantages of using LA-ICP-MS to analyze non-siliceous materials (metals), highlighting the approaches utilized in the studies included in Part II of this volume.

6.1 Analysis of Non-siliceous Materials

A significant proportion of the projects involving LA-ICP-MS in archaeology deal with metals. Among them, gold, silver, copper and any of their alloys are the most commonly studied materials, due to their economical and practical value in the daily life of ancient communities. The goal of such studies is generally to understand the technology used by ancient craftsman, the provenance of the raw materials utilized, and the circulation of the resulting finished objects by obtaining major, minor and trace element concentrations in their constituent metals.

The study of metals is generally difficult due to the different processes applied to ores to extract and concentrate the metal of interest as well as the metallurgical operations necessary to transform this metal into an object, including alloying and recycling, which can cause dramatic changes in the metal composition. This makes the provenance study of metals particularly difficult and generally requires the complimentary use of other analytical techniques such as isotope analysis (e.g. lead) to obtain more reliable and accurate information about metal provenience.

There are a number of challenges, however, that are more specific to the use of LA-ICP-MS on metals. First, metals are often subject to corrosion that modifies their surface composition to a depth that is difficult to assess non-destructively. For this reason, it is generally recommended to use single spot analysis to reach deep into the artifacts being analyzed. For instance, Dussubieux et al. (2008) analyzed copper and brass artifacts covered with corrosion layers up to 80 μm thick. Two ablations where performed at the same location, the first one to remove the corroded material, and the second, using a smaller laser beam diameter to avoid any contamination from the walls of the initial crater, to measure the newly exposed intact metal. Another way to proceed consists in monitoring the composition of the ablated material in real

L. Dussubieux (✉)
Integrative Research Center, Elemental Analysis Facility, Field Museum of Natural History, 1400 S. Lakeshore Drive, Chicago, IL 60605, USA
e-mail: ldussubieux@fieldmuseum.org

L. Dussubieux et al. (eds.), *Recent Advances in Laser Ablation ICP-MS for Archaeology*, Natural Science in Archaeology, DOI 10.1007/978-3-662-49894-1_6

time to make sure intact metal is reached. Sarah and Gratuze (Chap. 5) and Sarah et al. (2007) use depth profile analysis to measure the compositions of silver coins alloyed with copper starting at the surface, which is generally depleted in copper, until the signal produced for both metals reaches a plateau indicating that intact material deeper in the coin is being sampled.

Metal alloys are also intrinsically heterogeneous in some instances. Gold in particular often contains platinum group elements (Pt, Pd, Rh, Ir, Os, Ru) that co-precipitate, forming inclusions within the gold matrix (Ogden 1977; Meeks and Tite 1980). If this problem cannot be properly avoided by modifying the LA-ICP-MS analytical protocol, its impact can be minimized during data interpretation by considering ratios of platinum group elements that are generally constant in a given source of gold rather than absolute concentrations that can vary tremendously from one ablation to another (e.g. Jett et al. 2007). Lead is also a very problematic element as it is immiscible in copper even at low concentrations (Massalski et al. 1986: 946, 1861). This causes some areas of an object to have very high lead concentrations compared to the surrounding matrix (Dussubieux et al. 2008). In this case, bulk composition obtained with another analytical technique will be the only way to accurately measure the lead concentration in a copper based metal.

It is often difficult to obtain ideal standard reference materials with a composition as close as possible to that of samples to avoid matrix effects. Modern metals are much purer than ancient metals, and generally do not include toxic elements such as lead or arsenic that were often incorporated in very large quantities in ancient metal alloys. Several avenues have been explored to overcome these problems with standardization. Firstly, it is possible to combine several standards, including in-house ones, to cover a wide range of concentrations and elements as proposed in Leusch et al. (Chap. 7) for gold and Peterson et al. (Chap. 8) for copper. The use of standard solutions with elements not available in solid standards is also possible, although this approach induces a higher background signal and consequently results in higher detection limits (Chap. 7; Schlosser et al. 2009). In order to simultaneously introduce liquid and solid into the torch, a T-shaped connection is used to introduce 2 % nitric solution at the same time as the material ablated from a solid sample to maintain wet plasma condition during ablation.

Finally, there are problems with fractionation that are particular to metals. This is especially so when low melting point metals are present in large quantities in an alloy (e.g. lead, zinc). This may cause a preferential volatilization of some elements during ablation, jeopardizing any attempt at accurate measurement. Shorter laser wavelengths for nanosecond lasers greatly reduce fractionation, while femtosecond lasers used with optimized parameters can eliminate such fractionation altogether, providing the possibility of developing a non-matrix matched calibration system for metal (Russo et al. 2002; Shaheen et al. 2015). However, the high cost attached to buying and operating a femtosecond laser may limit its wide application in the field of archaeology in the near future.

Despite these difficulties, the use of LA-ICP-MS for metal characterization is becoming more frequent. This is particularly the case in the field of numismatics, where visible damage to artifacts is unacceptable, and the study of a large corpus of coins is typically necessary. Major and minor elements indicate debasement and trace element concentrations can show changes in metal sources or re-melting of metal stocks (e.g. Blet-Lemarquand et al. 2015; Geneviève and Sarah 2010; Lehmann 2011). Other more general studies involving LA-ICP-MS include those on gold (Brostoff et al. 2009; Nocete et al. 2014; Schlosser et al. 2012) or copper (e.g. Hill 2012; Dussubieux and Williams 2009). The direct analysis of iron by LA-ICP-MS is feasible (Devos et al. 2000); however, the study of slag inclusion has proved more effective for identifying different batches or provenances (Dillmann and L'Héritier 2007; Chap. 14).

References

Blet-Lemarquand M, Suspène A, Amandry M (2015) Augustus' gold coinage: investigating mints and provenance through trace element concentrations. In: Hauptmann A, Modarressi-Tehrani D (eds) Archaeometallurgy in Europe III, Proceedings of the 3rd international conference, Deutsches Bergbau – Museum Bochum, 29 June–1 July 2011, p 107–113

Brostoff LB, González JJ, Jett P, Russo RE (2009) Trace element fingerprinting of ancient Chinese gold with femtosecond laser ablation – inductively coupled plasma mass spectrometry. J Archaeol Sci 36:461–466

Devos W, Senn-Luder M, Moor C, Salter C (2000) Laser ablation inductively coupled plasma mass spectrometry (LA-ICP-MS) for spatial resolved trace analysis of early-medieval archaeological iron finds. Fresen J Anal Chem 366:873–880

Dillmann P, L'Héritier M (2007) Slag inclusion analyses for studying ferrous alloys employed in French medieval buildings: supply of materials and diffusion of smelting processes. J Archaeol Sci 34:1810–1823

Dussubieux L, Williams PR (2009) Elemental analysis of Peruvian copper-based artifacts using LA-ICP-MS. In: 2nd international conference archaeometallurgy in Europe 2007, Aquileia, Italy, 17–21 June 2007. Associazione Italina de Metalulurgica, Milan, p 489–497

Dussubieux L, Deraisme A, Frot G, Stevenson C, Creech A, Bienvenu Y (2008) LA-ICP-MS, SEM-EDS and EPMA analysis of Northeastern American copper-based artefacts: impact of corrosion and heterogeneity on the reliability of LA-ICP-MS compositional results. Archaeometry 50(4):643–657

Geneviève V, Sarah G (2010) Le trésor de deniers mérovingiens de Rodez (Aveyron). Circulation et diffusion des monnayages d'argent dans le Sud de la France au milieu du VIIIe siècle. La Revue Numismatique 166:477–507

Hill M (2012) Tracing social interaction: perspectives on archaic copper exchange from the Upper Great Lakes. Am Antiquity 77(2):279–292

Jett P, Brostoff L, Dussubieux L (2007) Technical study and elemental analysis of Chinese gold from the Late Eastern Zhou period. In: Douglas J, Jett P, Winter J (eds) Scientific research on the sculptural art of Asia, Proceedings of the third forbes symposium at the Freer Gallery of Art. Freer Gallery of Art, Smithsonian Institution, Washington, DC, pp 53–62

Lehmann R (2011) Archäometallurgie von mittelalterlichen deutschen Silberbarren und Münzen. Unpublished Ph.D. Dissertation, Leibniz Universität, Hannover

Massalski TB, Murray JL, Bennett LH, Baker H (eds) (1986) Binary alloy phase diagrams, vols 1 and 2. American Society for Metals, Materials Park, OH

Meeks ND, Tite MS (1980) The analysis of platinum-group element inclusions in gold antiquities. J Archaeol Sci 7(3):265–275

Nocete F, Sáez R, Bayona MR, Nieto JM, Peramo A, López P, Gil-Ibarguchi JI, Inácio N, García S, Rodríguez J (2014) Gold in Southwest of the Iberian Peninsula during the 3rd millennium BC. J Archaeol Sci 41:691–704

Ogden JM (1977) Platinum Group inclusions in ancient gold artifacts. Hist Metallurgy 11(2):53–72

Russo RE, Mao X, Gonzalez JJ, Mao SS (2002) Femtosecond laser ablation ICP-MS. J Anal Atom Spectrom 17:1072–1075

Sarah G, Gratuze B, Barrandon J-N (2007) Application of laser ablation inductively coupled plasma mass spectrometry (LA-ICP-MS) for the investigation of ancient silver coins. J Anal Atom Spectrom 22:1163–1167

Schlosser S, Kovacs R, Pernicka E, Günther D, Tellenbach M (2009) Fingerprints in gold. In: Reindel M, Wagner GA (eds) New technologies for archaeology, multidisciplinary investigations in Palpa and Nasca, Peru. Springer, Berlin, pp 409–436

Schlosser S, Reinecke A, Schwab R, Pernicka E, Sonetra S, Laychour V (2012) Early Cambodian gold and silver from Prohear: composition, trace elements and gilding. J Archaeol Sci 39:2877–2887

Shaheen ME, Gagnon J, Fryer BJ (2015) Elemental fractionation in 785 nm picosecond and femtosecond laser ablation inductively coupled plasma mass spectrometry. Spectrochim Acta B 107:97–109

7 Precise and Accurate Analysis of Gold Alloys: Varna, the Earliest Gold of Mankind—A Case Study

Verena Leusch, Michael Brauns, and Ernst Pernicka

Abstract

During the last decade an analytical routine was established that allows determining the composition of gold artifacts for major and trace elements by LA-ICP-MS. This micro-invasive method has proven highly suitable for detailed characterization of prehistoric gold in order to investigate its *chaîne opératoire*. At the CEZA laboratory (Mannheim, Germany) two different analytical set-ups were developed that yielded accurate and precise results for determining the whole range of matrix components within the gold. They are based upon different calibration strategies—liquid calibration with the ICP-MS operating under wet plasma conditions, and external calibration by solid standard reference materials with the ICP-MS operating under dry plasma conditions—that enable flexible adjustment according to different sample properties. They are described and compared within this paper. An archaeological case study of gold finds from the chalcolithic cemetery of Varna (Bulgaria) demonstrates the applicability of the analytical methods for archaeometallurgical studies. These analyses yielded valuable information for defining and comparing groups of gold artifacts that allude to their distribution within the burial site. However, the comparison of artifact gold with geological gold samples proved problematic, and clear relations between artefacts and specific gold occurrences are difficult to demonstrate.

7.1 Introduction

It has always been of major importance in archaeological science to establish methods of *fingerprinting* materials with regard to their classification and provenance. The idea that the chemical composition of an object may be used to determine the provenance of raw materials

V. Leusch (✉) • M. Brauns • E. Pernicka
Curt-Engelhorn-Zentrum Archäometrie, D6 3, 68159 Mannheim, Germany
e-mail: verena.leusch@cez-archaeometrie.de; michael.brauns@cez-archaeometrie.de; ernst.pernicka@cez-archaeometrie.de

L. Dussubieux et al. (eds.), *Recent Advances in Laser Ablation ICP-MS for Archaeology*, Natural Science in Archaeology, DOI 10.1007/978-3-662-49894-1_7

arose in the nineteenth century with an emphasis on metal objects, thereby providing the impulse for the integration of scientific methods in archaeological research. By chemical analysis of the objects, it was possible to shed light on economic networks and hence to gain a more detailed knowledge about archaeological cultures and their ways of interaction. After the success of trace element and lead isotope analyses in reconstructing metallurgical chains of prehistoric copper the desire to develop effective methods of analysing other materials of interest grew continuously and became an important tool for archaeology (Pernicka 1986, 2014a; Pernicka et al. 1997; Gratuze 1999; Gale et al. 2003; Guerra and Calligaro 2003).

This paper will outline the use of Laser Ablation–Inductively Coupled Plasma–Quadrupole–Mass Spectrometry (LA-ICP-Q-MS) in archaeological science with a special focus on analysing prehistoric gold (see also: Bendall 2003; Dussubieux and Van Zelst 2004; Guerra et al. 1999; Kovacs et al. 2009; Leusch et al. 2014; Leusch et al. 2015; Schlosser et al. 2009; Schlosser et al. 2012). The general aims are to investigate (1) the geological provenance of this raw material, (2) the manufacturing processes employed and (3) to identify forgeries.

At the Curt-Engelhorn-Center for Archaeo metry (CEZA, Mannheim, Germany), LA-ICP-Q-MS has routinely been used for the analysis of archaeological artefacts since 2006. Our specific method for gold was developed in co-operation with the research group of Detlef Günther at the ETH Zürich, Switzerland. During the last 6 years, the method has been improved within projects dealing with the analysis of gold from a variety of regions and chronological contexts.[1] After describing the method itself, this paper will briefly present a study of prehistoric gold conducted at the CEZA: the so-called "earliest gold of mankind" from the Chalcolithic cemetery at Varna (Bulgaria), dated to the fifth millennium BC. The analytical objective is the definition of gold types regarding their specific chemical composition and their interpretation concerning provenance and processing of these early gold finds.

7.2 General Background

LA-ICP-MS as a method for fingerprinting gold artifacts was first established in pursuit of economic interests. The identification of the provenance of modern gold ingots from illegal circulation was and still is a major problem as one tries to trace back the gold to large production companies (Grigorova et al. 1998a, b). Soon afterwards, the possibilities for applications in archaeology were recognized (Guerra et al. 1999), putting the emphasis on the identification of prehistoric gold sources. The development of mass spectrometry provided for the first time the possibility to analyse a wide range of major and trace elements in archaeological gold artifacts and thus a new analytical basis for research on provenance of archaeological gold.

LA-ICP-MS has become increasingly important as an analytical method for metallurgical questions.[2] The general advantages of laser ablation as a sampling method for ICP-MS have already been addressed, e.g. in Durrant (1999). Concerning the analyses of prehistoric gold finds, LA-ICP-MS facilitates the measurement of the whole gold matrix in a micro-invasive way. The required sample size is very low. One single shaving of usually less than 1 mm is sufficient for analysis. It is also possible to analyse complete objects up to certain dimensions by LA-ICP-MS, depending on the size of the ablation cell. New developments have shown that it is in principle possible to analyse objects of any size (Kovacs et al. 2010; Glaus et al. 2013).

[1] These projects examined Early Bronze Age gold on the Nebra Sky Disc (Germany), Peruvian gold, and gold items from the Copper Age cemetery at Varna (Bulgaria), gold from Cambodia, and recently, early Iron Age gold from Germany and France (Schlosser et al. 2009, 2012; Ehser et al. 2011; Pernicka 2014b).

[2] LA-ICP-MS is also being used to study the production and provenance of prehistoric iron (Brauns et al. 2013; Leroy et al. 2011).

Table 7.1 Instrumentation and experimental setup at the CEZA (Mannheim, Germany)

Instrument	Technical details	Operating parameters
Q-ICP-MS	Thermo Fisher Elemental XSeries^{II} Quadrupole ICP-MS (Q-ICP-MS) with Collision Cell Technology (CCT)	Argon plasma; possible collision gases (H_2 and NH_4)
Laser	**Microprobe II with LUV213, 213 nm**	**Variable spot size and energy can be adapted to the sample's properties; ablation gas = He (+N)**
Laser	Resonetics Resolution M-50-E ArF Excimer Laser (193 nm)	Variable spot size and energy can be adapted to the sample's properties; ablation gas = He (+N)
Nebulizer	Standard-Nebulizer Thermo	0.6–0.97 bar operating pressure, max. 0.9 l/min flow rate
Cetac ASX 260	**Auto sampler**	**Connected to the ICP-Computer**
Particle filter	**Tygon hose; inner diameter 1.5 mm; seven coils**	**Replaced continuously after 10 h of ablation**

Printed in bold are the components which are used for the "wet plasma" setup and the 213 nm laser that recently has been replaced

However, for documentation it is certainly better to remove a small quantity of material for analysis so that the sample can be stored and used again for later analytical purposes. When analysing complete objects, the LA patterns are hardly visible to the naked eye. This is of great importance from a conservation point of view.

7.3 Methods

Presently, two setups for LA-ICP-MS are used at the CEZA. They are based on different calibration strategies and plasma conditions:

1) Liquid calibration using three different standard solutions with the ICP-MS operating under wet plasma conditions (Kovacs et al. 2009).
2) External calibration by solid standard reference materials (SRM), e.g. FAU 7 (NIST 8053), FAU 10 (NIST 8062) and our in-house standards NA1 and NA2 (see Appendix Table 7.7) with the ICP-MS operating under dry plasma conditions.

Both setups have different advantages. Liquid calibration facilitates the quantification of elements which are absent in the SRMs. The composition of the standard solutions can be chosen according to elements of interest. This calibration procedure has been used since the method was established at the CEZA in 2006. Using the external calibration strategy, the quantifiable elements are limited to those present in the SRMs. In comparison to the first setup a much quicker measurement procedure is possible, however.

7.3.1 Instrumentation

In the CEZA laboratory a Resonetics Resolution M50-E ArF excimer laser operating at a wavelength of 193 nm, a repetition rate of 8 Hz, a spot size of 44/58 μm and an energy of 4–8 J/cm^2 is used.[3] Data are collected using a Thermo Fisher Elemental XSeries^{II} Q-ICP-MS in time resolved mode. Table 7.1 shows the different instruments used for the experimental setups at the CEZA. Before starting measurement, the ICP-MS is tuned to optimise it to high sensitivity, low background count rates and low oxide rates. The operating conditions of the ICP-MS are summarised in Table 7.2. Data acquisition begins after c. 20 s measuring the pure gas background. The LA signal is acquired for c. 30–40 s, followed again by a washout of c. 20–30 s. The analytical procedure begins with measurement of the SRMs, which are ablated under the same

[3] These settings were empirically determined to be most suitable for the precise, accurate and reproducible measurement of gold. Depending on the sample's properties (e.g. size and thickness) these settings must be adjusted. Other materials require different settings.

Table 7.2 Instrumental settings of the ICP-MS and the excimer laser for dry plasma conditions (and wet plasma conditions with CCT)

ICP-MS settings	**Forward power**	**Nebulizer gas (Ar)**	**Cooling gas (Ar)**	**Auxiliary gas (Ar)**	**CCT (H2)**	**Detector mode**
	1400 W	0.8–0.91 (0.8) l/min	13 l/min	0.7 l/min	0 (6 l/min)	dual

Laser settings	**Energy**	**Spot size PA**	**Spot size A**	**Rep. rate PA**	**Rep. rate A**	**Scan speed**	**He gas flow**	
	8 mJ	58/73 μm	44/58 μm	5 Hz (2 passes; 30% overlap)	8 Hz	10 μm/s	600 ml/min	3 measurements/object

The instrument settings are optimized before each measurement sequence. This affects other settings such as torch position, focus, and lens settings for which no mean parameters can consequently be given here

conditions as the samples. Helium is used as scavenging gas. For statistical evaluation, usually up to three analyses are performed per sample and SRM. To cover the dynamic range of the whole gold matrix, a dual mode detector is used, operating either in a pulse counting mode (high resolution for elements present in high concentrations) or in an analogue counting mode (standard resolution for elements present in low concentrations). To normalise both concentration ranges cross calibration is necessary. By this approach a linear dynamic range of up to 8–9 orders of magnitude is possible.

The laser ablation cell has the dimensions of 50 × 50 mm with a height of approximately 25 mm. Randomly shaped samples (as small objects) can also be adjusted within the cell. It is possible to analyse individual features of a sample by choosing special laser patterns (spot, line or bulk ablation) and thus profile specific areas. Especially when dealing with inclusions, this feature is valuable because different parts of the signal can be evaluated separately. A sample aerosol is formed by adding an ablation/scavenging gas (in our case He) and transported by the gas flow towards the Ar plasma torch (laser settings are listed in Table 7.2).

Generally, the laser beam causes different effects when hitting the sample's surface. The energy of the laser beam is absorbed by the sample which leads to the ablation of material and the formation of the sample aerosol. The particle size of the sample aerosol turned out to have a huge effect on the "vaporization and ionisation efficiency [. . .] in the inductively coupled plasma (ICP)" (Guillong et al. 2003: 211) and is causally linked to three main factors (Guillong et al. 2003):

1) Wavelength, energy density and pulse duration of the operating laser
2) Ablation gas
3) Sample properties (chemical and morphological)

The thermal impact on the sample surface causes different effects within the matrix.[4] Fractionation is a problem, especially when dealing with sample aerosols consisting of large particles (Fig. 7.1). Within the Ar plasma particles up to 300 nm can be ionised completely (Guillong et al. 2003). Comparative studies revealed that "the elemental composition of different particle size fractions shows more significant compositional differences when ablating with a 266 nm Nd:YAG laser compared with a 193 nm ArF excimer laser" (Halicz and Günther 2004: 1540). As can be seen in Table 7.1, two laser systems were used in the experimental setup. From 2006 to 2011 a Nd:YAG laser with a wavelength of 213 nm was used. In 2011 a 193 nm ArF excimer laser was installed. The effect of the particle size on the ionisation and smoothness of the signal using the different laser systems is

[4] The interactions between laser and sample are complex and still a matter of investigation (Durrant 1999; Cromwell and Arrowsmith 1995).

Fig. 7.1 Sample aerosol particles generated using a 213 nm Nd:YAG laser (Schmiderer 2008: Fig. 41)

illustrated in Günther and Heinrich (1999) and also could be observed when changing the laser systems at the CEZA (see below).

7.3.2 Liquid Calibration (Wet Plasma Conditions)

A general lack of solid reference materials made calibration difficult for LA-ICP-MS. Following the approach of Halicz and Günther (2004), a combination of liquid calibration and the measurement of SRMs and solid samples was established at the CEZA (Schmiderer 2008: 55–68) to enable standardised and matrix matched measurements of solid gold samples. This measuring procedure is well described in Schmiderer (2008), Kovacs et al. (2009) and Schlosser et al. (2009).

Liquid calibration is performed using three synthetic solutions (Appendix Table 7.8) to obtain a calibration curve. The solutions were prepared with regard to the chemical compatibility of the respective elements. In Kovacs et al. (2009) the calibration is performed by using Cu as an internal standard, however, this approach caused problems in the quantification procedure. By changing the internal standard to ^{169}Tm a significant improvement was achieved. The SRMs for matrix matched standardisation and quantification are the in-house standards NA1 and NA2. Additionally FAU 7 (NIST 8053) and FAU 10 (NIST 8062) were used (Appendix Table 7.7). Before entering the Ar plasma, a nebuliser gas and a 2 % HNO_3 blank is added via a spray chamber to the ablated sample aerosol to maintain stable wet plasma conditions. The ICP-MS operates in collision cell mode (CCT) to reduce polyatomic interferences. Time resolved analysis (TRA) is performed. The standard solutions are re-measured at the end of each sample sequence, which facilitates a drift correction performed by the Plasma Lab (Thermo) software. By this setup an accuracy and precision of 1–15 % and 5–10 % respectively for most of the determined elements were achieved (Kovacs et al. 2009). Its applicability was confirmed by the measurement of early Peruvian gold artefacts (Schlosser et al. 2009).

Quantification is carried out as follows. In the Plasma Lab software of the ICP-MS time slices of the TRA for background and signal can be chosen according to their stability. Their counts per second (cps) and concentrations are recorded and calculated by the software. For this purpose the sensitivity of each element is calculated according to their concentration in the liquid standards. The raw data (concentrations) are exported via a.csv file into an Excel sheet and data reduction is performed by normalising the whole matrix of each SRM and sample to 100 %. Correction factors are calculated according to the certified values of the SRMs. This procedure is explained in detail in Longerich et al. (1996). As mentioned above, this measurement procedure facilitates a flexible adaption to the sample's composition by adding the elements of interest to the liquid standards. This is a valuable feature when dealing with archaeological samples.

7.3.3 External Calibration by SRMs (Dry Plasma Conditions)

Measurement under "dry plasma" conditions was performed using the 193 nm ArF excimer laser connected to the quadrupole ICP-MS. Before entering the Ar plasma a nebulizer gas (Ar) is added to the sample aerosol (Fig. 7.2). Because of instrumental drift that can be observed by the repeated measurement of the SRMs, quantification has to be carried out by bracketing the analytical units, measuring (1) the SRMs, followed by (2) 3–4 samples and again (3) by SRMs. Quantification is carried out directly using the SRMs. The raw data of the cps are exported via a.csv file into Excel and data reduction is carried out over several steps following the approach of Gäbler et al. (2011). It is an external, matrix matched calibration based on the normalisation to 100 % of the matrix.

The exported.csv file contains the cps for each analysed element. The SRMs are evaluated according to their specific sensitivities. Dividing the average cps by the certified value (C_{SRM}) yields the sensitvity ($S_{element}$) of each element

$$S_{element} = cps_{SRM}/C_{SRM} \quad (7.1)$$

The mean values of $S_{element}$ of the two standards are used to calculate the concentration of each element. First a normalisation factor (N) according to the Cu, Ag and Au ratios (main matrix components) is calculated.

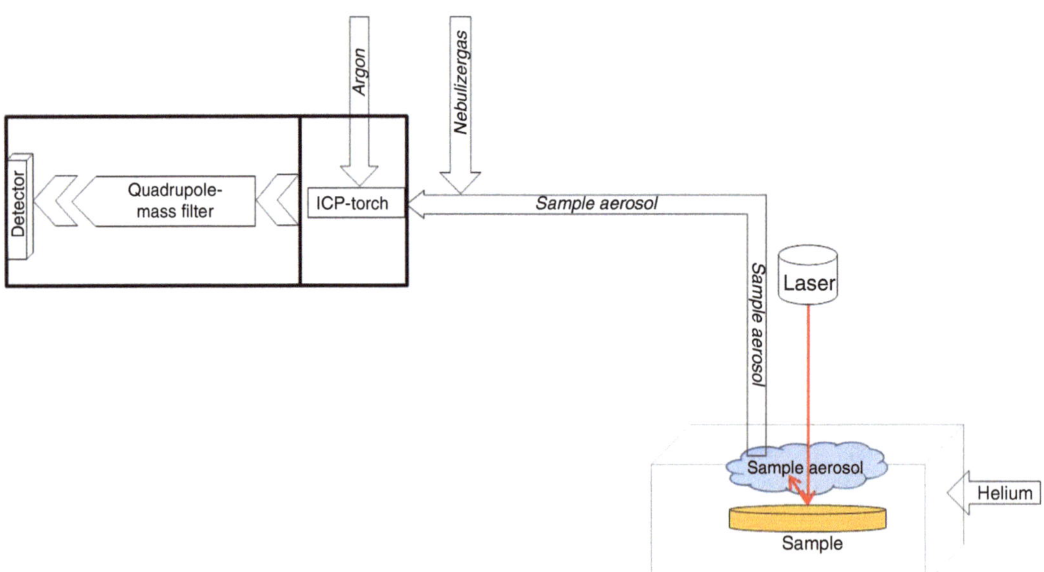

Fig. 7.2 Diagram of setup with the ICP-MS operating under dry plasma conditions. (diagram: V. Leusch)

$$N = 1{,}000{,}000 / \left(\frac{cps(Cu)_{(sample)}}{S_{Cu}} + \frac{cps(Ag)_{(sample)}}{S_{Ag}} + \frac{cps(Au)_{(sample)}}{S_{Au}} \right) \quad (7.2)$$

The concentration (C) is then calculated by multiplying (N) with the quotient of the respective cps of each element in the sample (cps_{sample}) and its sensitivity ($S_{element}$).

$$C = N \times cps_{sample} / S_{element} \quad (7.3)$$

In comparison to liquid calibration, this setup facilitates a much faster measuring process and an improvement of accuracy and precision of 1–7 % and <7 % respectively. A limitation is the suite of the quantifiable elements that are restricted to the certified ones contained within the SRMs.

7.4 Results

The measurement routines developed at the CEZA are based on different studies on the influences of laser systems (different wavelengths, carrier gases, particle filters, etc...), calibration strategies, and optimising procedures concerning the ICP-MS. As already mentioned, the calibration of the LA-ICP-MS was one of the major problems of the method due to the lack of solid standard reference materials. According to Pickhardt et al. (2000) and Halicz and Günther (2004), an experimental setup was chosen that combined liquid calibration with laser ablation under wet plasma conditions. It is described in detail in Kovacs et al. (2009).

The measurement of the goldmatrix of the SRMs showed several problems that are interpreted to be systematic errors. The elements Rh, Pd, Cd and Te for instance are strongly affected by matrix effects and the addition of liquid reagents in wet plasma conditions (Table 7.3). Other elements like Pb exhibit somewhat erratic behaviour. Most likely this can be attributed to inhomogeneous distribution due the immiscibility of lead in gold in the solid state (similar to the behaviour of lead in copper).

Table 7.3 Isotopic interferences and/or tailing effects of Pd, Cd and Te within the gold matrix (information about interferences: ICP-MS Plasma Lab Software)

Element	Isotope	Natural abundance [%]	Important interferences or tailing effects
Rh	103	100	$^{63}Cu^{40}Ar$
Pd	105	22.3	$^{65}Cu^{40}Ar$
Cd	111	12.8	$^{109}Ag+H_2$. (A high concentration of Ag can cause a tailing effect)
Te	125	7.14	$^{109}Ag^{16}O$

The interferences with copper argides (especially with Rh and Pd) depend on the concentration of copper in the measured sample. For natural gold they can usually be neglected due to low copper content

An improvement was achieved after the installation of the 193 nm ArF excimer laser in December 2011, which produced much smoother signals, better signal stability over time, and better sensitivity. The latter is a result of using the 193 nm laser in combination with the use of He as carrier gas, which leads to a two- to threefold enhancement of signal intensity (Günther and Heinrich 1999). Higher precision and accuracy were achieved for Cu and Ag, values which can probably be ascribed to reduced fractionation due to the different wavelengths of the laser systems (smaller mean particle size resulting in better ionisation). By changing the setup to dry plasma conditions, an improvement could be achieved in the quantification of Cd and Te. This can be ascribed to absence of HNO_3 within this setup, which contributed to the formation of polyatomic species of Ag, causing interferences.

In Fig. 7.3, the transient signals of different laser setups are shown. The diagrams show cps plotted against data acquisition time (the cps-rates on the Y-axis are linearly scaled). We decided to use line ablation where possible, because of the resulting stability of the signal intensity over time (Fig. 7.3a, b). Using the 193 nm ArF eximer laser, significantly smaller fluctuations in the signal could be achieved (Fig. 7.3b). As small shavings of the objects are

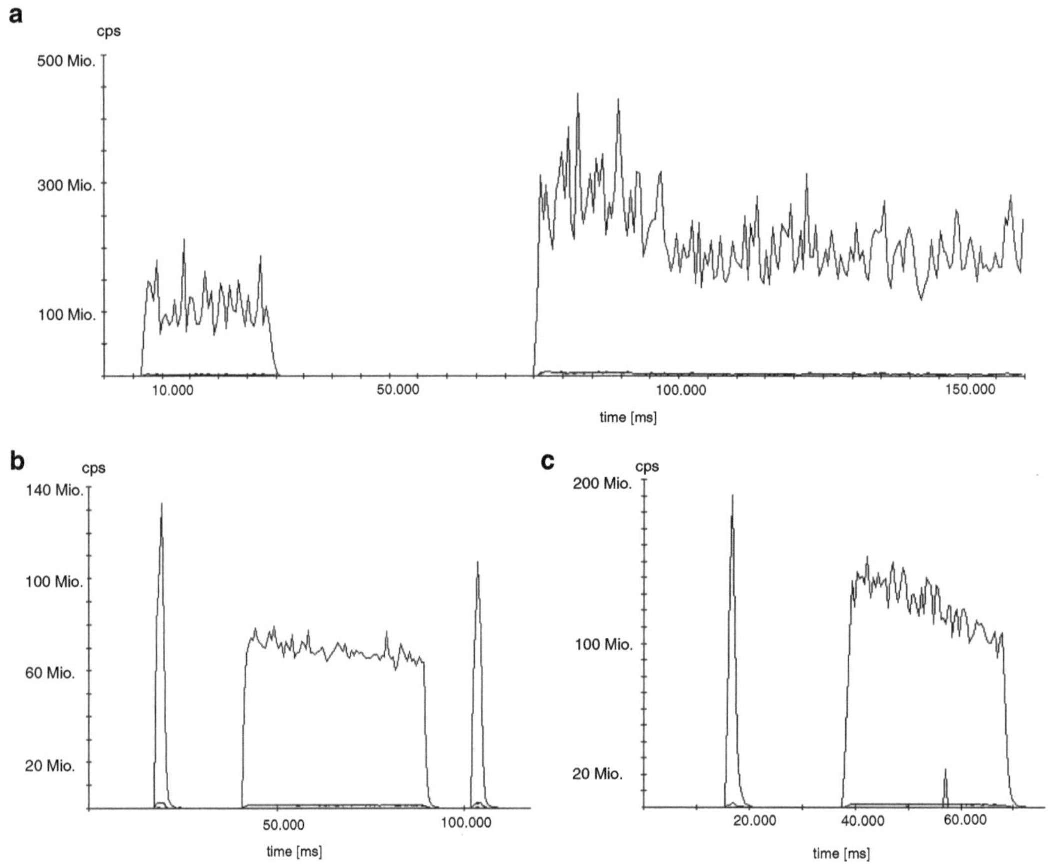

Fig. 7.3 Different laser signals (**a**) line ablation using the 213 nm Nd:YAG laser (very erratic signal, Au deviation c. 15–20 % at spot size of 50 μm; 10 Hz, 4 μm/s, 10–12 J/cm^2), (**b**) line ablation using the 193 nm laser (stable signal, Au deviation c. 4–5 %, spot size 58 μm; 8 Hz; 10 μm/s, 6 mJ) and (**c**) spot ablation using the 193 nm laser (time-attenuation evident, spot size 44 μm; 3 Hz, 90 pulses, 8 mJ)

usually analysed, it is possible to prepare them such that the freshly cut surface is hit by the laser beam. As such it is not a surface analysis and depletion effects can be neglected. Before starting the analysis, potential surface contamination is removed by a pre-ablation. Figure 7.3c shows the signal of a spot ablation. Since we measured a SRM (NA1) a homogeneous distribution of the elements can be assumed. A slight decrease in signal intensity can be observed. Irradiance probably decreases "as the crater deepens due to changes of effective area exposed to the laser beam" (Russo et al. 2002: 440). All of the signals in Fig. 7.3 were acquired under wet plasma conditions.

Dry plasma conditions without liquid calibration using just our SRMs as external standards have been applied since April 2012. These changes led to several improvements:

1) Significant reduction of the background signal, hence better LODs (Appendix Table 7.9).
2) Isobaric interferences by polyatomic species were significantly reduced (especially for Cd and Te).
3) Better reproducibility of the data (precision).
4) Better accuracy.

These improvements can be ascribed to better matrix matched measurement, but also to the

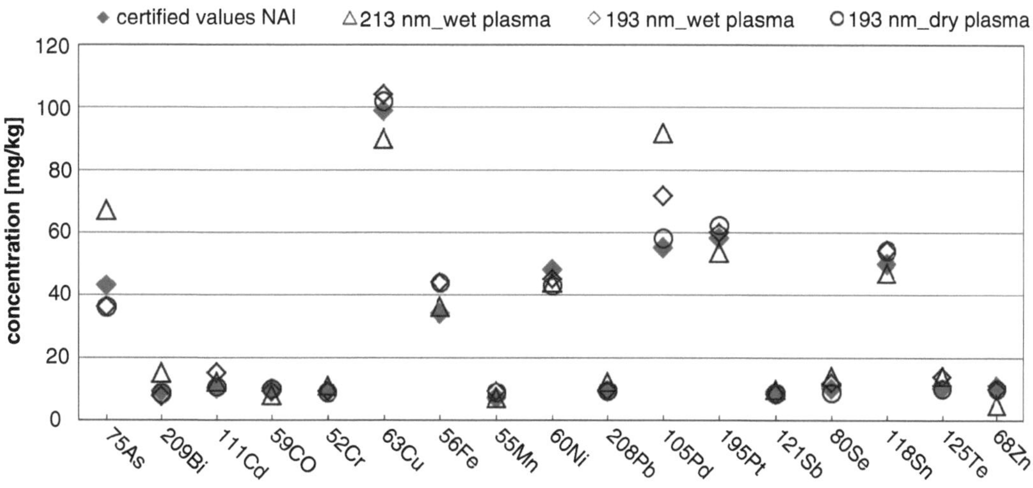

Fig. 7.4 Comparison of the certified values of SRM NA1 and the measured values under dry and wet plasma conditions with the 193 nm ArF excimer laser. The measurements were accurate under dry plasma conditions and generally in better agreement with the certified values than under wet plasma conditions. NA2 was used as an external standard. (diagram: V. Leusch)

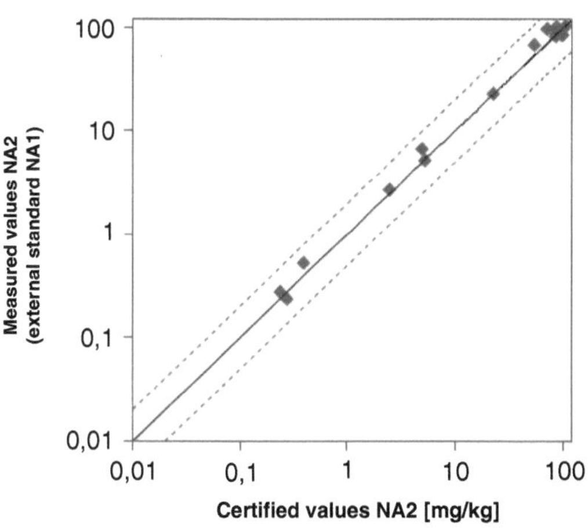

Fig. 7.5 Accuracy of measurements under dry plasma conditions. Trace elements up to 100 mg/kg are plotted (Ti, Cr, Mn, Co, Zn, As, Ru, Rh, Cd, Sb, Ir, Pb, Bi). The measured values are in good agreement with certified values (all within a range of max. ±30 % relative deviation [*dashed lines*]). (diagram: V. Leusch)

observation that the "vaporization, atomization and ionization process of particles within the ICP must be significantly different for liquid droplets and solid particles" (Halicz and Günther 2004: 1540). As is shown in Figs. 7.4 and 7.5, the results measured under dry plasma conditions are in very good agreement with the certified values for SRMs. Figure 7.4 shows the accuracy of the different measuring setups and Fig. 7.5 separately the accuracy for dry plasma conditions. In Table 7.4 the certified values of SRM NA1 are listed together with the measured values of the three different experimental setups. The reproducibility (precision) of the values is described by the standard deviations. To calculate the limits of detection (LOD) listed (see Appendix), the approach explained in Longerich et al. (1996) was used, based on the sensitivity and the triple standard deviation of the averaged background signal of each element ($n > 5$). The comparability of the data acquired either by wet or dry plasma conditions is shown in Fig. 7.6 (as examples the Pt/Pd ratio and the Sn values are compared). For archaeometric purposes, samples analysed either by the wet or dry plasma setup can be evaluated in combination so that the two setups can be arranged flexibly.

7.5 Varna: An Archaeometric Case Study

By presenting a case study, the applicability of the above mentioned methods for archaeological research shall be discussed. The analysed gold artifacts are from the Copper Age cemetery of

Table 7.4 SRM NA1 measured using the three different experimental setups

Element	Certified value [mg/kg]	Values published in (Kovacs et al. 2009)[a]	213 nm wet plasma		193 nm wet plasma		193 nm dry plasma	
			Measured [mg/kg]	SD %	Measured [mg/kg]	SD %	Measured [mg/kg]	SD %
As	43	45.7	67	7	36	8	36	8
Bi	9	11.1	15	6	7.5	9	8.6	7
Cd	10	11.8	12	7	15	8	10.5	6
Co	10	10.7	8	9	9.1	6	10.0	5
Cr	9	10.5	11	5	9	8	8.9	6
Cu	99	99	90	6	104	7	108	6
Fe	34	43.4	36	5	44	9	44	7
Mn	7	7.73	7.1	6	9.1	8	8.7	7
Ni	48	53.0	44	7	45	8	43	5
Pb	9	11.5	12	12	9.4	13	9.3	10
Pd	55	58.7	91.8	5	71.5	7	57.9	6
Pt	58	68.0	53.6	5	60	6	62.3	4
Sb	9	12.1	9.6	8	8.2	7	8.5	6
Se	10	10.2	14	13	11.7	30	8.9	20
Sn	50	53.2	47	7	54	7	54	8
Te	10	11.1	14	27	14	20	9.9	6
Zn	11	12.4	5	13	9.6	21	10	9
Ag	10,000	0.9 [%]	9700	5	11,700	4	10,140	5

The values that were acquired under wet plasma conditions and with liquid calibration are uncorrected. Using dry plasma conditions, the sensitivities were immediately calculated over the SRMs repetitively. NA2 was used as external standard. Concentrations are given in [mg/kg]. Precision is given as relative standard deviation (SD %) to the measured values

[a]Measured concentrations of NA1 using Cu as internal standard and wet plasma conditions

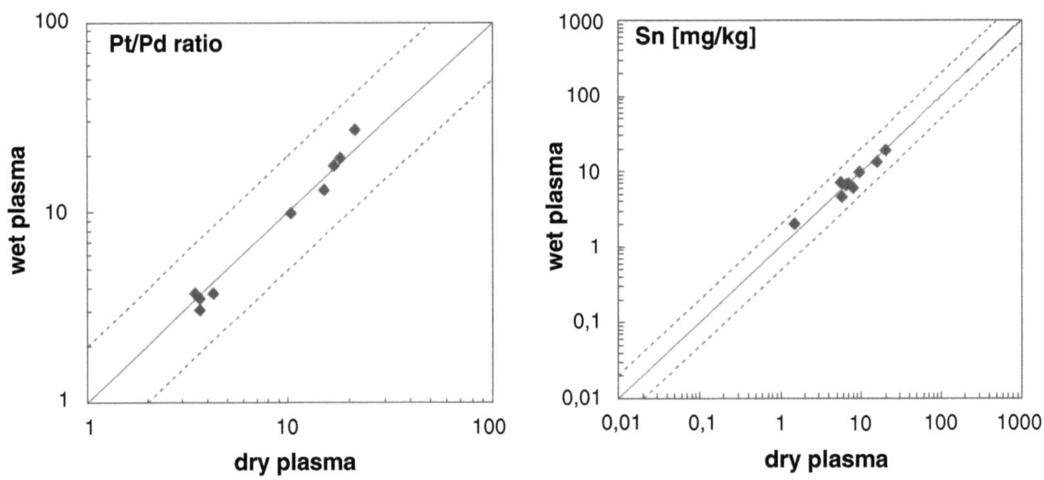

Fig. 7.6 Comparison of data acquired using wet and dry ablation. Both data sets are in good agreement (all within a range of max. ±30 % relative deviation [*dashed lines*]) and comparable. This is demonstrated by the Pt/Pd ratio and Sn concentrations of a selection of samples. (diagrams: V. Leusch)

Fig. 7.7 A selection of Chalcolithic gold finds from the burial site of Varna, Bulgaria. The beads, the bow, and sceptre decoration are from grave 43. The horn-shaped and hemispherical appliqués are from complex 36 (Photos: B. Armburster; objects are part of the Prehistoric collection in the Archaeological Museum Varna (Bulgaria))

Varna (Bulgaria, Fig. 7.7).[5] Presently, Varna is the earliest site where an elaborate gold and copper metallurgy are evident, dating to the second half of the fifth millennium BC. This makes it most suitable for studying the emergence of metallurgy and its social prerequisites and impact. The analytical task is to distinguish gold with different major and trace elemental patterns that may lead the way to determining the provenance and workmanship of the gold, and thus to gain insight into networks of supply and workshop organisation, i.e. *chaîne opératoire*.[6] Thus

[5] The analyses were performed within a bilateral project between German and Bulgarian research institutions and universities. The project focuses on the culture historic investigation of the Varna cemetery and is financed by the German Research Foundation (DFG). For further information about the site see: Fol and Lichardus (1988); Echt et al. (1991); Hansen (2009); Krauß (2010); Todorova (1991); Lichardus (1991).

[6] Small shavings from a representative sample of objects were taken at the Museum of National History in Sofia and the Historical Regional Museum in Varna by Kalin Dimitrov (Archaeological Institute of the Bulgarian Academy of Sciences).

Table 7.5 Element classification according to Pernicka (1999: 170) and Schmiderer (2008: 109, Table 13)

Elements sensitive to melting conditions	"Robust" elements
Cr, Mn, Fe, Co, Zn, As, Se, Cd, In, Sn, Sb, Te, Pb, Bi, Hg	Ag, Cu, Rh, Pd, Pt, Ir, Ge, Ni, Os, Ru

The elements are categorised according to their behaviour during melting. Many geologically indicative elements (like Fe, Se, Sn, Te, Hg) are affected by the heat treatment of gold in an oxidising atmosphere

the aim is the reconstruction of this workflow and furthermore of the underlying social prerequisites that are necessary to sustain this economic niche, obviously exclusively serving the need for social/religious representation.

Since the gold from the Varna cemetery presented in this case study does not seem to be intentionally alloyed,[7] many disturbing influences caused by the addition of other metals can be neglected.[8] The basic working hypothesis for provenance studies is that batches of gold from different sources were not mixed and the trace elemental pattern is consequently interpreted to be indicative of provenance. However, one has to be well aware of the general problem of lacking information concerning the initial stages of the *chaîne opératoire*, i.e. the localisation of the exploited deposits and the reconstruction of the trade or exchange of the raw material. It is for instance conceivable that for trading purposes the raw gold was exchanged as ingots which already had undergone melting processes influencing the elemental pattern of the worked gold. Regarding possible chemical alterations connected to heat treatment, the accompanying elements can be divided into two groups (Table 7.5). The robust elements are of special interest for the comparison of geological gold with artifact gold.

Before addressing the problem of provenancing prehistoric gold finds some geological background information is necessary. The geochemical fingerprint of placer gold is affected by various factors that are not totally traceable. Expressed in simplified terms, the formation of gold deposits is linked to so-called hydrothermal solutions enriched in gold that transport the metal to the earth's crust (Boyle 1987; Mcdonald 2007; Morteani 1995):

> "Hydrothermal solutions leach other elements as well as gold from the rocks through which the solutions pass. Some of these elements are present in trace quantities only; the proportions of others such as silver and tellurium may be significant and materially effect fineness. Gold forms natural alloys with silver, copper, mercury and tellurium; [. . .] Varieties in primary ores include cuproaurite (copper gold), porpezite (palladium gold) and bismuthaurite (bismuth gold). Whilst these minerals are seldom found in alluvial detritus, their presence in a weathering zone may help unravel the geological history of an area under review" (Mcdonald 2007: 10).

Hence, when we try to discriminate different gold occurrences we are confronted with many uncertainties and inhomegeneities due to geological factors. In particular, the high variability of element concentrations within single gold deposits makes their geochemical fingerprinting difficult. Moreover, it has been shown that many trace elements which are detected in artifact gold are absent in gold nuggets and may rather be the result of accessory heavy minerals that occur in placers (e.g. Hauptmann et al. 2010: 150, Fig. 7).[9] Such accessory minerals may

[7] This assumption is based on the analytical observations of low copper and trace element concentrations that are in agreement with what we know about the composition of natural gold. There is for the most part no clear analytical indication for intentional alloying with copper and/or silver. However, the mixing of gold from different origins cannot be excluded.

[8] In this context it is noteworthy that all analysed objects can be confirmed as having been cast. In the literature about early gold working it is often assumed that "in the very beginning of handling gold it was most likely completely or partially sintered and never really melted" (Raub 1995: 243) and that "[a]lthough copper was frequently cast, this technique was hardly ever used for gold." (Eluère 1989: 37). Based on our new analyses (that also comprise technological examinations) these assumptions can be dismissed.

[9] Recent investigations of placer gold conducted at the CEZA demonstrate this inhomogeneous character very clearly.

include cassiterite, cuprite, pyrite, and PGE that apparently enter the artifact gold by subsequent chemical homogenisation during the melting process.

However, much effort has been made to solve this problem during recent years. The basic assumption underlying provenance studies is that "inter-source variation must be greater than intra-source variation" (Wilson and Pollard 2001: 508) which for (placer) gold deposits was confirmed by the studies of Schmiderer (2008) in an attempt to identify the source of the gold inlayed on the Sky Disc of Nebra (Germany).[10] For his comparison with artifact gold, the statistical evaluation of geological samples is based on median values, which avoids the overinterpretation of outliers. In a continuation of the Nebra project, Ehser et al. (2011) found Co, Ni, Cu, Ru, Pd, Ag, Sn, Sb, Ir and Pt to be best suited for provenance studies of gold.[11]

Despite these promising results, there is still a need for a more widespread prospection and analyses of gold occurrences and deposits in many areas of archaeological interest. The geochemical characterisation of (prehistoric) gold often is based on the analyses of archaeological artifacts alone, as the possibilities for direct comparison between artifact gold and geological gold are scarce and not as simple as they may seem. Beside the above mentioned high geochemical variability within single gold occurrences, this is due to another crucial problem that is connected to the lack of archaeological evidence from prehistorically exploited placer occurrences.[12] The precise location of these cultural sites generally remains unknown. Additionally, panning for gold nowadays produces only small amounts of gold in the form of small nuggets, upon which geochemical characterisation is typically based. Taking into account that ancient gold is the result of melting, the current sampled geochemical variabilty of these small nuggets[13] may not be representative of the bulk composition of artifacts produced with much larger amounts of gold.

Within the Varna-project both perspectives (archaeological and geological) are being pursued by different research groups. Intensive geological prospection in Bulgaria has begun and has already shed light on possible prehistoric gold sources not far from Varna that were previously unknown (Yovchev 2014).[14] In co-operation with the Geological Institute of the Sofia University, it was possible to locate and analyse placers from gold bearing rivers in south-eastern Bulgaria.[15] The case study presented here, however, focusses on the analyses of the artifacts from Varna. Small shavings (usually one per object) of maximum 1 mm length were removed from each sampled artifact. This sampling is barely visible to the naked eye and is justified by providing valuable information about prehistoric gold metallurgy. The analytical results typically show a homogeneous composition of the gold. This is demonstrated in Table 7.6 by the mean values and standard deviations (%) of a random choice of six objects. Each sample was analysed three times using line ablation (Table 7.2), which turned out to be a suitable analytical procedure for our purpose.

At present, about 300 objects from the Varna cemetery (Varna I) have been analysed by LA-ICP-MS. They represent a choice of objects

[10] A. Schmiderer was able to integrate about 150 gold occurrences from the alpine region, the Carpathians, the Czech Republic, and the German regions of Thuringia and Saxony in his PhD study of the possible gold sources used to produce the Nebra Disc.

[11] A. Ehser was able to prospect occurrences in Romania and the Balkans as well as in southwestern Europe (Spain, Portugal and the UK).

[12] Except for piles of river sediment, which are usually difficult to date, there are hardly any archaeological traces of this activity.

[13] One has to reckon as well with changes in the geochemical structure over time as placers are "dynamic" systems that are impacted by numerous environmental influences.

[14] Personal information by Danail Yovchev. Until quite recently information about Bulgarian gold deposits was difficult to access. "Due to the restrictive information policy of the Bulgarian government [...] only limited information *was* [adapted by the author] available on Bulgarian gold deposits." (Lehrberger 1995: 137).

[15] This is the topic of a PhD thesis by Danail Yovchev at the Geological Department of Sofia University (supervisor: Prof. Veselin Kovachev).

Table 7.6 Mean values of the analytical results of six gold artefacts based on 2–3 line ablations on each sample (elements below LOD are not given here)

Sample	Object	Ag	Au	Cu	Mn	Fe	Ni	Zn	Sn	Sb	Pb	Bi	Pd	Ir	Pt
LOD [mg/kg]		15	100	2	1	10	2	2	0.5	0.5	0.5	0.5	0.2	0.1	0.1
121750	Hornshaped	8.7	91	3000	1.1	67	<2	9.0	7.0	1.2	6.6	3.9	20	0.23	60
SD %	appliqué	1	0.1	4	6	5		1	5	7	1	1	16	2	1
121751	Hemispherical	9.9	90	5100	1.5	78	55	8.3	4.9	0.45	3.3	1.5	3.5	0.11	57
SD %	appliqué	5	0.7	15	12	9	11	6	16	18	10	14	6	17	6
121752	Hemispherical	11	89	4200	3.9	150	21	15	7.5	1.3	14	6.8	1.6	0.13	20
SD %	appliqué	4	0.6	2	14	15	34	25	10	11	12	16	6	39	20
121753	Hemispherical	9.3	90	3700	1.4	95	2.2	14	4.7	1.6	11	7.5	1.0	0.065	14
SD %	appliqué	2	0.2	2	16	20	15	6	47	18	26	10	17	2	3
121754	Bead	7.4	92	2500	7.2	280	4.3	14	2.0	1.7	11	3.7	1.3	0.026	15
SD %		1	0.1	6	6	6	4	4	2	5	4	6	4	37	7
121755	Hemispherical	9.4	90	3600	3.5	210	19	11	3.2	2.4	17	6.0	2.7	0.20	54
SD %	appliqué	3	0.2	6	7	3	2	13	11	3	2	12	4	27	6

The relative standard deviations show reproducible measurements (relative SD of most values below 15 %). Concentrations are given as follows: Au and Ag in [%], other elements in [mg/kg]. (table: V. Leusch)

from four major contexts (complex 36 and graves 3, 4 and 43) and include different artifact types (e.g. beads, hemispherical appliqués, ring idols, piercings, etc...). Trace element concentrations in the artifact gold are usually very low. Pd and Pt were detected in all samples and indicate the use of placer gold as raw material.[16]

Prior studies have revealed that the Pt/Pd ratio can be used to discriminate between gold groups (Schlosser et al. 2012). This is also possible at Varna, where four major gold groups can be distinguished within the available data set. In Fig. 7.8 these gold groups are visualised by the different slopes of regression lines along which the data plot. Furthermore, the diagram reflects an interrelation between the different gold groups and the archaeological contexts they derive from. By trend, the gold groups from complex 36 differ from those in grave 43. It seems that access to

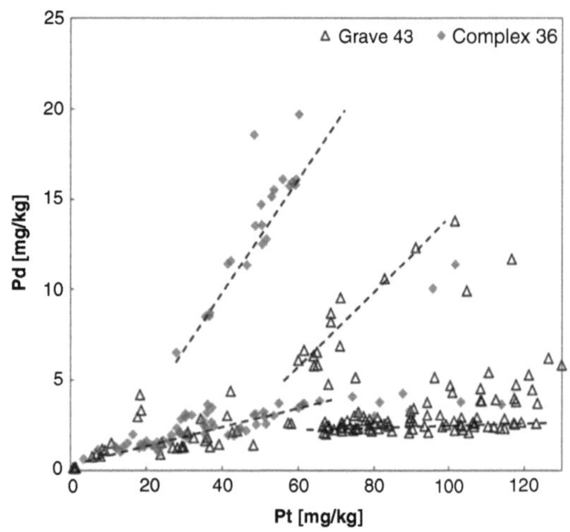

Fig. 7.8 Pt and Pd concentrations measured in the analysed objects. Four major gold groups can be defined according to regression lines (*dashed lines*) along which the data plot. A correlation between these gold groups and different archaeological contexts is evident. Here complex 36 and beads from grave 43 are chosen as examples. (diagram: V. Leusch)

certain gold groups or products of certain workshops was restricted to specific burials or deposition context. This can be interpreted as being socially patterned (e.g. workshops or suppliers work selectively for specific "clients") or as reflecting time-variant changes within the *chaîne opératoire*. Furthermore, the analyses

[16] "Geologically, economic concentrations of gold and platinum-group metals do not occur in the same primary deposit types but, rather, are found independently from each other in different environments. While platinum-group deposits are restricted to magmatic processes (e.g., layered intrusions), gold deposits are formed by hydrothermal processes [...]. Therefore, platinum or platinum-group metals in gold objects are interpreted as indicators for placer gold deposits in which the tributaries collected gold and platinum from both mineralization styles." (Junk and Pernicka 2003: 314).

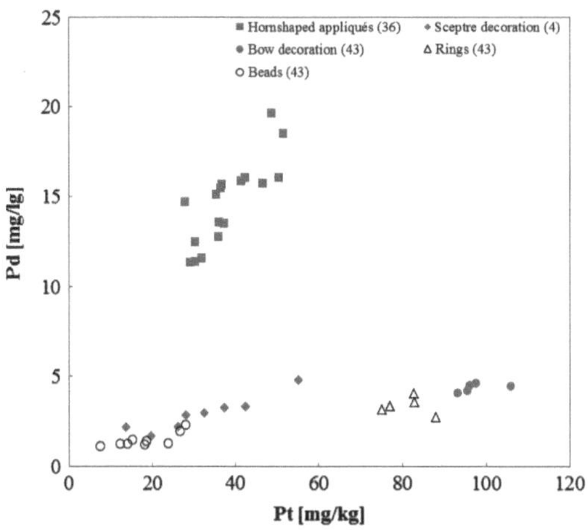

Fig. 7.9 Pt and Pd concentrations displayed by object type (archaeological context is displayed in parentheses). Chemical groups coincide with typological groups, reflecting a targeted series production of certain objects. (diagram: V. Leusch)

reveal a strong correlation between chemical and typological groups (Fig. 7.9) indicating series production suggestive of specialised workshop activities.

Due to the above mentioned new geological findings from eastern Bulgaria, the generally established opinion that the Varna gold was imported from distant regions like the southern Caucasus (as assumed by Hartmann 1982) must be reconsidered, and evidently should be abandoned in favour of a regional supply. This matches well the geographic frame of the contemporaneous cultural and economic sphere that is apparent when considering other commodities (e.g. copper, *Spondylus* shell, flint, etc...) represented in the graves at Varna (Krauß 2010). Our case study of the Chalcolithic gold from Varna attests to exploitation of placer gold in the Copper Age and preliminarily suggests a selective distribution of gold from certain origins among the burials and/or its use within the funeral practices carried out at Varna. Various gold sources (whether they are indicative of discrete geological occurrences, suppliers or workshops remains to be seen) reflected by different Pt/Pd ratios become apparent. Moreover, the results indicate a well organised and elaborate workmanship that may be deduced from the quality and diversity of the gold objects (Fig. 7.7) in addition to the chemical analyses.

7.6 Discussion and Conclusion

Since establishing the method of LA-ICP-MS for gold analyses at the CEZA in 2006, several improvements of the measuring setup employed there have been implemented. The analysis of gold can now be performed under wet plasma conditions with solution calibration as well as under dry plasma conditions with external calibration by SRMs. The advantages of each of the two approaches are summarised below:

Liquid calibration (wet plasma conditions) allows us to quantify elements that are not certified in available SRMs (e.g. Os). This enables a flexible adjustment of the standard solutions to the composition of the analyte and also facilitates, for example, the pre-evaluation of samples of unknown composition.

External calibration by SRMs (dry plasma conditions) permits much faster and more efficient analysis. Lower background levels and higher sensitivities, higher precision and accuracy and a simple quantification procedure are the major advantages of this analytical setup. The quantifiable elements are limited to those certified in the SRMs, which sometimes do not cover all elements of interest. There still is a need for more matrix-matched SRMs in different concentration ranges of major, minor and trace elements.

A great improvement within our measuring setup was the installation of a 193 nm ArF excimer laser. It led to smaller fluctuations of the signal and thus to better analytical statistics. Data that were acquired by both setups turned out to be comparable (Fig. 7.6). Hence two stable and compatible methods are available and can be adjusted to the analytical task.

As a case study, gold samples from the Chalcolithic cemetery Varna I (Bulgaria) were analysed to study the *chaîne opératoire* of these earliest gold objects. The objective is to set the analytical data into their cultural context. They have to be looked upon as "products" of different networks of supply and/or workshops that were the basis for the high abundance of gold objects recovered at the site. The material classification of the gold finds is aimed at the identification of possible changes related to chronology and/or grave groups, and at the evaluation of the social meaning of the metal work. Hence the analytical data are treated as indicators of social activities. Several methodological problems occur when dealing with these tasks that shall be outlined briefly:

Geological samples from Bulgarian rivers are still scarce and there hardly exist sufficient data for the geochemical characterisation of each prospected occurrence yet. Generally, the geochemical fingerprinting of placer gold deposits is problematic (see above), due to the high variations of major, minor and trace elements within single occurrences, which has been widely discussed (e.g. Boyle 1987; Mcdonald 2007; Schmiderer 2008). Concerning questions about provenance we further have to face the general lack of archaeological evidence for placer gold exploitation. Therefore the prehistoric situation and structures of early gold-winning activities remain difficult to reconstruct. Beside these uncertainties in localising prehistorically exploited placers, chemical alterations of the raw gold by melting (and possible re-melting) must be considered when comparing artifact gold with geological gold. Hence there are limitations as to the elements that are suited for provenance studies (Table 7.5). Taking into account that we are still dealing with a quite new field in archaeological science, hopefully these methodological problems will decrease with a growing database, both archaeological and geological.

The archaeological objects from Varna that were analysed so far show discrete trace elemental patterns that can be divided into four main gold groups. The discriminating elements Pt and Pd indicate provenance specific patterning. Even if it is not yet possible to link these features to precise occurrences, we are able to draw several conclusions. Firstly, by the detection of Pt and Pd within the gold artifacts the use of placer gold could be substantiated empirically. The geological record provides evidence for numerous placer gold occurrences in eastern Bulgaria. The closest is situated approximately 40 km south of Varna (personal information Danail Yovchev, Geological Institute, Sofia University) and can be considered as one possible source for the Varna gold. Secondly, the four distinguishable gold groups represent the products of a series of activities within the *chaîne opératoire*, starting with the prospection of the gold deposits and their exploitation, followed by the trade/exchange of the raw gold until its transformation by the goldsmith into an adornment and its final use. These activities were carried out by different persons and were embedded in a social framework. As such, the objects together with the chemical analyses provide valuable information about the different stages of the metallurgical process and its social prerequisites and impact. At Varna, an unequal distribution of gold, and a selective circulation of gold groups become apparent (Figs. 7.7, 7.8 and 7.9) (Leusch et al. 2014, 2015).

In summary, the new analyses by LA-ICP-MS provide important information for the reconstruction of prehistoric gold metallurgy and the prevailing social and economic circumstances in which it first occurred. Nevertheless, numerous methodological problems have to be considered when addressing the question of provenancing gold and a comprehensive approach, combining different archaeological and geological information (e.g. about other commodities) must be pursued.

Acknowledgements The studies of the gold finds from Varna were funded by the German Research Foundation (DFG, Pe 405-25) and are the subject of a PhD thesis (Verena Leusch). For their co-operation and help in sampling we sincerely thank Ivelin Kuleff (Chemical Institute, Sofia University), Raiko Krauß (University Tübingen), Vladimir Slavchev, Olga Pelevina (Historical Regional Museum in Varna), Kalin Dimitrov (Archaeological Institute of the Bulgarian Academy of Sciences) and Svetla Tsaneva (Museum of National History in Sofia). Special thanks go to Barbara Armbruster (CNRS, Toulouse), who provided us with valuable photos and information about the production technology of prehistoric gold. The integration of the latest geological information was only possible due to the dedicated research of Danail Yovchev and Veselin Kovachev (Geological Institute, Sofia University). Furthermore, the authors want to thank Nicole Lockhoff, Ursula Rothe and René Kunze for improving the language of this paper, and for helpful discussions. Finally, we would like to thank two anonymous reviewers for their helpful comments.

Appendix

Table 7.7 Composition of the in-house SRMs NA1 and NA2, the NIST standards FAU 7 and FAU 10

Element	NA1	NA2	FAU7	FAU10
As	43	112	10.1	29.4
Bi	9	100	24	53.9
Cd	10	82		
Co	10	124		
Cr	9	25	32.6	4.9
Cu	99	1062	98.1	9.8
Fe	34	806	11.6	90.4
Mn	7	62	58.9	64.3
Ni	48	1092	32.5	14.6
Pd	55	1112	43.1	80
Pt	58	1152	87.1	5.1
Sb	102	102		0.1
Se	114	114		
Sn	50	773	33.8	33
Te	10	112		
Ti			12.7	2.6
Zn	11	114	45.6	20.9
Pb	9	90	21.9	49.7
Ag [%]	1.0	5.45	20.3	36

Elemental concentrations in [mg/kg] except for Ag in NA1 and NA2

Table 7.8 Standard solutions

Element	Solution 1	Solution 2	Solution 3
Ti		39.3	
Cr	38.7	40.1	
Mn	38.7	40.5	
Fe	38.3	40.1	
Co	38.3	39.7	
Ni	38.7	39.7	
Cu	38.3	40.9	30.9
Zn	38.3	40.1	
As	38.3	40.1	
Se	37.9	40.1	
Rh	22.1	25.1	
Pd		25.8	
Cd	37.9	39.7	
Ag	2013		
Sn			31.2
Sb		39.0	
Te		69.2	
Tm	17.2	17.4	12.7
Ir			32.3
Pt			33.3
Au			1796
Tl	38.7	40.5	
Pb	38.7	40.1	14.6
Bi	38.7		

(continued)

Table 7.8 (continued)

Concentrations are given in [μg/kg]. Nitric acid is the solvent of solutions 1 and 2. *Aqua regia* is the solvent of stock solution 3. As internal standard ^{169}Tm is used

Table 7.9 List of measured elements (isotopes), dwell time, resolution (settings of the ICP-MS) and mean LOD (calculated for dry plasma conditions; estimated values italicised)

Element (Isotope)	Dwell (ms)	Resolution	LOD [mg/kg]
^{48}Ti	10	Standard	1
^{52}Cr	10	Standard	1
^{55}Mn	10	Standard	1
^{56}Fe	10	High	10
^{59}Co	10	Standard	0.2
^{60}Ni	10	Standard	2
^{63}Cu	10	High	2
^{68}Zn	10	Standard	2
^{75}As	10	Standard	5
^{82}Se	10	Standard	40
^{101}Ru	10	Standard	*0.2*
^{103}Rh	10	Standard	*0.1*
^{105}Pd	10	Standard	0.2
^{107}Ag	10	High	15
^{111}Cd	10	Standard	1
^{118}Sn	10	Standard	0.5
^{121}Sb	10	Standard	0.5
^{125}Te	10	Standard	1
^{189}Os	10	Standard	*0.2*
^{193}Ir	10	Standard	0.1
^{195}Pt	10	Standard	0.1
^{197}Au	10	High	100
^{208}Pb	10	Standard	0.5
^{209}Bi	10	Standard	0.5

References

Bendall C (2003) The application of trace element and isotopic analyses to the study of celtic gold coins and their metal sources. Ph.D Dissertation, University of Frankfurt

Boyle RW (1987) Gold – history and genesis of deposits. Springer, New York, NY

Brauns M, Schwab R, Gassmann G, Wieland G, Pernicka E (2013) Provenance of Iron Age iron in Southern Germany: a new approach. J Archaeol Sci 40 (2):841–849

Cromwell EF, Arrowsmith P (1995) Semiquantitative analysis with laser ablation inductively coupled plasma mass spectrometry. Anal Chem 67:131–138

Durrant SF (1999) Laser ablation inductively coupled plasma mass spectrometry: achievements, problems, prospects. J Anal Atom Spectrom 14:1385–1403

Dussubieux L, Van Zelst L (2004) LA-ICP-MS analysis of platinum-group elements and other elements of interest in ancient gold. Appl Phys A: Mater Sci Process 79:353–356

Echt R, Thiele W-R, Ivanov IS (1991) Varna – Untersuchungen zur kupferzeitlichen Goldverarbeitung. In: Lichardus J (ed) Die Kupferzeit als historische Epoche – Symposium Saarbrücken und Otzenhausen 1988. Saarbrücker Beiträge zur Altertumskunde 55, Rudolf Habelt GmbH, Bonn, pp 633–691

Ehser A, Borg G, Pernicka E (2011) Provenance of the gold of the Early Bronze Age Nebra Sky Disk, central Germany: geochemical characterization of natural gold from Cornwall. Eur J Mineral 23:895–910

Eluère C (1989) Secrets of Ancient Gold. Trio, Guin-Düdingen

Fol A, Lichardus J (eds) (1988) Macht, Herrschaft und Gold: das Gräberfeld von Varna (Bulgarien) und die Anfänge einer neuen europäischen Zivilisation. Stiftung Saarländ, Kulturbesitz, Saarbrücken

Gäbler H-E, Melcher F, Graupner T, Bahr A, Sitnikova MA, Henjes-Kunst F, Oberthür T, Brätz H, Gerdes A (2011) Speeding up the analytical workflow for coltan fingerprinting by an integrated mineral liberation analysis/LA-ICP-MS approach. Geostand Geoanal Res 35 (4):431–448

Gale N, Stos-Gale Z, Raduncheva A, Panayotov I, Ivanov I, Lilov P, Todorov T (2003) Early metallurgy in Bulgaria. In: Craddock P, Lang J (eds) Mining and metal production through the Ages. British Museum Press, London, pp 122–173

Glaus R, Dorta L, Zhang Z, Ma Q, Berke H, Günther D (2013) Isotope ratio determination of objects in the field by portable laser ablation sampling and subsequent multicollector ICPMS. J Anal Atom Spectrom 28:801–809

Gratuze B (1999) Obsidian characterization by laser ablation ICP-MS and its application to prehistoric trade in the Mediterranean and the Near East: sources and distribution of Obsidian within the Aegean and Anatolia. J Archaeol Sci 26:869–881

Grigorova B, Anderson S, de Bruyn J, Smith W, Stülpner K, Barer A (1998a) The AARL gold fingerprinting technology. Gold Bull 31(1):26–29

Grigorova B, Smith W, Stülpner K, Tumilty JA, Miller D (1998b) Fingerprinting of Gold Artefacts from Mapungubwe, Bosutswe and Thulamela. Gold Bull 31(3):99–102

Guerra MF, Calligaro T (2003) Gold cultural heritage objects: a review of studies of provenance and manufacturing technologies. Meas Sci Technol 14:1527–1537

Guerra M, Sarthre C-O, Gondonneau A, Barrandon J-N (1999) Precious metals and provenance enquiries using LA-ICP-MS. J Archaeol Sci 26:1101–1110

Guillong M, Kuhn H-R, Günther D (2003) Application of a particle separation device to reduce inductively coupled plasma-enhanced elemental fractionation in laser ablation-inductively coupled plasma-mass spectromentry. Spectrochim Acta B 58:211–220

Günther D, Heinrich CA (1999) Comparison of the ablation behaviour of 266 nm Nd:YAG an 193 nm ArF excimer lasers for LA-ICP-MS analysis. J Anal Atom Spectrom 14:1369–1374

Halicz L, Günther D (2004) Quantitative analysis of silicates using LA-ICP-MS with liquid calibration. J Anal Atom Spectrom 19:1539–1545

Hansen S (2009) Kupfer, Gold und Silber im Schwarzmeerraum währen des 5. und 4. Jahrtausends v. Chr.. In: Apakidze J, Govekarica B, Hänsel B (eds) Der Schwarzmeerraum vom Äneolithikum bis in die Früheisenzeit (5000-500 v.Chr.). Kommunikationsebenen zwischen Kaukusus und Karpaten. Internationale Fachtagung von Humboldtianern für Humboldtianer im Humboldt-Kolleg in Tiflis/ Georgien. Prähistorische Archäologie in Südosteuropa, Bd. 25. Verlag Marie Leidorf GmbH, Rahden/Westfalen, p 11–50

Hartmann A (1982) Prähistorische Goldfunde aus Europa II - Spektralanalytische Untersuchungen und deren Auswertung. In: Bittel K, Junghans S, Otto H, Sangmeister E, Schröder M (eds) Studien zu den Anfängen der Metallurgie 5. Gebr Mann Verlag, Berlin

Hauptmann A, Bendall C, Brey G, Japardize I, Gambaschidze I, Klein S, Prange M, Stöllner T (2010) Gold in Georgien. Analytische Untersuchungen an Goldartefakten und an Naturgold aus dem Kaukasus und dem Transkaukasus. In: Hansen S, Hauptmann A, Motzenbäcker I, Pernicka E (eds) Von Majkop bis Trialeti – Gewinnung und Verarbeitung von Metallen und Obsidian in Kaukasien im 4.-2. Jt. v. Chr. Kolloquien zur Vor- und Frühgeschichte, Bd. 13, Bonn, p 139–160.

Junk S, Pernicka E (2003) An assessment of osmium isotope ratios as a new tool to determine the provenance of gold with platinum-group metal inclusions. Archaeometry 45(2):313–331

Kovacs R, Schlosser S, Staub SP, Schmiderer A, Pernicka E, Günther D (2009) Characterisation of calibration materials for trace element analysis and fingerprint studies of gold using LA-ICP-MS. J Anal Atom Spectrom 24:476–483

Kovacs R, Nishiguchi K, Utani K, Günther D (2010) Developement of direct atmospheric sampling for laser ablation-inductively coupled plasma-mass spectrometry. J Anal Atom Spectrom 25:142–147

Krauß R (2010) Zur Akkumulation von Prestigegütern im Westschwarzmeerraum während des 5. Jahrtausends v. Chr. In: Callmer J, Theune C, Biermann F, Struwe R, Jeute GH (eds) Zwischen Fjorden und Steppe; Festschrift für Johan Callmer zum 65. Geburtstag. Internationale Archäologie: Studia honoraria, Band 31. Verlag Marie Leidorf GmbH, Rahden/Westfahlen, p 289–525

Lehrberger G (1995) The gold deposits of Europe – an overview of the possible metal sources for prehistoric gold objects. In: Morteani G, Northover JP (eds) Prehistoric gold in Europe – mines, metallurgy and manufacture. Springer, Dordrecht, pp 115–144

Leroy S, Simon R, Bertrand L, Williams A, Foy E, Dillmann P (2011) First examination of slag inclusions in medieval armours by confocal SR-μXRF and LA-ICP-MS. J Anal Atom Spectrom 26 (5):1078–1087

Leusch V, Pernicka E, Armbruster B (2014) Chalcolithic gold from Varna – provenance, circulation, processing, and function. In: Meller H, Risch R, Pernicka E (eds) Metalle der Macht – Frühes Gold und Silber (Metals of Power – Early Gold and Silver) – 6th Archaeological conference of Central Germany 2013. Tagungen des Landesmuseums für Vorgeschichte Halle 11/I, p 165–182

Leusch V, Armbruster B, Pernicka E, Slavčev V (2015) On the invention of gold metallurgy: the gold objects from the Varna I Cemetery (Bulgaria)—technological consequence and inventive creativity. Cambridge Archaeol J 25:353–376

Lichardus J (ed) (1991) Die Kupferzeit als historische Epoche – Symposium Saarbrücken und Otzenhausen 1988. Saarbrücker Beiträge zur Altertumskunde 55, Rudolf Habelt GmbH, Bonn

Longerich HP, Jackson SE, Günther D (1996) Laser ablation inductively coupled plasma mass spectrometric transient signal data acquisition and analyte concentration calculation. J Anal Atom Spectrom 11:899–904

Mcdonald EH (2007) Handbook of gold exploration and evaluation. CRC Press, Cambridge

Morteani G (1995) Mineral economics, minerology, geochemistry and structure of gold deposits: an overview. In: Morteani G, Northover JP (eds) Prehistoric gold in Europe – mines, metallurgy and manufacture. Springer, Dordrecht, pp 97–113

Pernicka E (1986) Provenance determination of metal artifacts: methodological considerations. Nucl Instrum Methods 14:24–29

Pernicka E (1999) Trace element fingerprinting of ancient copper: a guide to technology or provenance? In: Young, S, Pollard A, Budd P, Ixer R (eds) Metals in antiquity. BAR International Series, 792:163–171

Pernicka E (2014a) Provenance determination of archaeological metal objects. In: Roberts BW, Thornton C (eds) Archaeometallurgy in global perspective, Methods and syntheses. Springer, New York, NY, pp 239–268

Pernicka E (2014b) On the authenticity of the gold finds from Bernstorf, community of Kranzberg, Freising district, Bavaria. Jahresschrift für mitteldeutsche Vorgeschichte 94:517–526

Pernicka E, Begemann F, Schmitt-Strecker S, Todorova H, Kuleff L (1997) Prehistoric copper in Bulgaria. Its composition and provenance. Eurasia Antiqua 3:41–180

Pickhardt C, Becker JS, Dietze H-J (2000) A new strategy of solution calibration in laser ablation inductively coupled plasma mass spectrometry for multielement trace analysis of geological samples. Fresenius J Anal Chem 368:173–181

Raub C (1995) The metallurgy of gold an silver in prehistoric times. In: Morteani G, Northover JP (eds) Prehistoric gold in Europe-mines, metallurgy and manufacture. Springer, Dordrecht, pp 243–529

Russo RE, Mao X, Liu H, Gonzalez J, Mao SS (2002) Laser Ablation in analytical chemistry—a review. Talante 57:425–451

Schlosser S, Kovacs R, Pernicka E, Günther D, Tellenbach M (2009) Fingerprints in Gold. In: Reindel M, Wagner GA (eds) New technologies for archaeology – natural science in archaeology. Springer, Berlin, pp 409–436

Schlosser S, Reinecke A, Schwab R, Pernicka E, Sonetra S, Laychour V (2012) Early Cambodian gold and silver from Prohear: composition, trace elements and technology. J Archaeol Sci 39:2877–2887

Schmiderer A (2008) Geochemische Charakterisierung von Goldvorkommen in Europa. Ph.D Dissertation, University of Halle

Todorova H (1991) Die Kupferzeit Bulgariens. In: Lichardus J (ed) Die Kupferzeit als historische Epoche – Symposium Saarbrücken und Otzenhausen 1988. Saarbrücker Beiträge zur Altertumskunde 55, Rudolf Habelt GmbH, Bonn, p 89–93

Wilson L, Pollard A (2001) The provenance hypothesis. In: Brothwell D, Pollard A (eds) Handbook of archaeological sciences. Wiley, Chichester, pp 507–517

Yovchev D (2014) Native gold and platinum in stream sediments from Dvoynitsa River and right tributaries of Kamchia River, Bulgaria. Geoscience 2014. Proceedings of the National Conference of the Bulgarien Geological Society, p 33–34

LA-ICP-MS Analysis of Prehistoric Copper and Bronze Metalwork from Armenia

8

David L. Peterson, John V. Dudgeon, Monica Tromp, and Arsen Bobokhyan

Abstract

Analysis of prehistoric copper and bronze in the Caucasus was performed previously on thousands of objects with arc optical emission spectroscopy (OES). While arc OES is no longer widely used in archaeometry, LA-ICP-MS has shown great promise for isotopic and chemical analysis of ancient copper and bronze artifacts. In order to explore the effectiveness of LA-ICP-MS for the characterization of materials in a large group of ancient copper-based metalwork from the South Caucasus, we analyzed 48 metal artifacts from the Horom necropolis and 16 from the Karashamb necropolis, at Idaho State University's Center for Archaeology, Materials and Applied Spectroscopy (CAMAS). These artifacts had been recovered from burials dating to the late second–early first millennium BC, a period noted for the use of a variety of copper alloy mixtures, including antimony bronze (which is very unusual at this early period in Europe and Asia). The metal artifacts from Horom had been previously analyzed by arc OES at the Institute of Archaeology and Ethnography in Yerevan, Armenia. This provided the opportunity to compare the performance of arc OES with LA-ICP-MS for analysis of variations in the use of copper alloys in ancient metal artifacts. In addition to LA-ICP-MS, EDS was used to analyze major elements, especially the proportion of copper in relation minor and trace elements that were measured with LA-ICP-MS. Besides unalloyed copper, the alloys detected by EDS and arc OES included mixtures with arsenic, tin, lead and antimony. More alloys were detected

D.L. Peterson (✉) • J.V. Dudgeon
Department of Anthropology, Center for Archaeology, Materials and Applied Spectroscopy, Idaho State University, 921 S. 8th Avenue Stop 8005, Pocatello, ID 83209-8005, USA
e-mail: davepeterson26@hotmail.com; dudgeon@isu.edu

M. Tromp
Department of Anatomy, Otago University, P.O. Box 913, Dunedin 9054, New Zealand
e-mail: monica.tromp@anatomy.otago.ac.nz

A. Bobokhyan
Institute for Archaeology and Ethnography, Armenian National Academy of Sciences, 15 Charensti Street, 375025 Yerevan, Republic of Armenia
e-mail: arsbobok@yahoo.com

L. Dussubieux et al. (eds.), *Recent Advances in Laser Ablation ICP-MS for Archaeology*, Natural Science in Archaeology, DOI 10.1007/978-3-662-49894-1_8

in the assemblage by LA-ICP-MS and EDS than with arc OES. This may be because copper levels were measured by EDS allowing the results for all elements to be normalized in proportion to copper. Normalization of results was therefore not possible with arc OES, which is another advantage of using LA-ICP-MS together with EDS.

8.1 Introduction

Laser ablation inductively coupled plasma mass spectrometry (LA-ICP-MS) has become increasingly utilized as an efficient, precise and accurate method for chemical analysis of trace elements in the μg/g (parts per million) and ng/g (parts per billion) range when applied with appropriate standard reference materials. Only microscopic portions of artifacts need to be vaporized in order to determine object composition, and it is therefore considered a virtually non-destructive archaeometric technique. Since copper and bronze artifacts are frequently among the items that collections managers most wish to preserve for future exhibition and study, the low impact of LA-ICP-MS compositional analysis has distinct advantages for research with these objects. However, for the analysis to be effective the material that an artifact is composed of must be more or less homogeneous, unless the goal is to identify compositional variation within an object. Copper-based artifacts include alloys with complex microstructures made up of distinct phases, which may vary according to the chemistry of the materials as well as the techniques used to work them (Scott 1991). Therefore, the results of LA-ICP-MS analysis of portions of artifacts that are typically only micrometers in width must be approached with caution. Despite these caveats, Dussubieux and colleagues have demonstrated the effectiveness of LA-ICP-MS for discriminating between North American native copper and European smelted copper on the basis of concentrations of As, Ag, Ni and Sb, as well as the method's utility for distinguishing artifacts made of brass from those made of copper—all without destructive sample preparation (Dussubieux et al. 2008). Dussubieux (2007) has also demonstrated the promise of LA-ICP-MS for successful analysis of copper alloys.

What requires further exploration is the performance of LA-ICP-MS in the analysis of a larger assemblage of ancient copper-based artifacts, which may include a variety of mixtures with As, Sn, Sb, Ag, Au, Pb and other metals in varying proportions and made with a variety of manufacturing techniques. This is needed to advance LA-ICP-MS analysis of ancient copper-based metalwork beyond the characterization of individual artifacts, and towards the broader examination of the technologies people used in particular regions and periods, as well as the examination of associated practices, like recycling, that are important to the archaeological assessment of past sociotechnical systems (Peterson 2012).

In order to explore the effectiveness of LA-ICP-MS for the characterization of materials in a relatively large assemblage of ancient copper and bronze metalwork, we undertook the analysis of 48 metal artifacts from the Horom necropolis in northwestern Armenia (Badalyan and Agekyan 1991; Badaljan et al. 1992, 1993, 1994). These objects were recovered from burials dating to the late second–early first millennium BC, a period noted for the utilization of an array of copper alloy mixtures (Chernykh 1992: 264–295; Pike 2002; Meliksetian et al. 2003a; Gevorkyan 2009). This group of artifacts was therefore well suited for an experiment testing the performance characteristics of LA-ICP-MS for the analysis of variations in copper alloys.

In this chapter, we discuss the methods and results of the experiment, and the potential of LA-ICP-MS analysis of ancient copper and bronze based on our analysis of metalwork from Horom. The analytical problems addressed include the results achieved utilizing several copper and bronze standards, matrix effects in the analysis of trace elements in copper and copper alloys with LA-ICP-MS, and calibration of the analyses with the allied use of silicon drift detector energy dispersive spectrometry (SDD-EDS) for quantifying major elements. The objects from Horom were analyzed previously by arc optical emission spectroscopy (OES) by Aram Gevorkyan (Institute for Archaeology and Ethnography, Yerevan, Armenia) in the 1990s, but the results are published here for the first time. In addition to presenting the LA-ICP-MS results, we compare the results of the two methods. Arc OES was the principal method of archaeometallurgical analysis in the South Caucasus and throughout the former Soviet Union until the 1990s, and longer in some places.[1] Evaluating the performance of LA-ICP-MS through comparison with the results of arc OES analysis was undertaken to demonstrate the potential of LA-ICP-MS for analysis of ancient copper and bronze in the region, and to begin to assemble a baseline comparison between the old and new technologies.[2]

8.2 General Background

The numerous copper ore deposits and associated evidence for prehistoric mining and metallurgical activities in the Armenian plateau have made archaeometallurgy an important focus of archaeological research in the region (Chernykh 1992; Gevorkyan 1980). Copper metalwork appeared in the South Caucasus by the Neolithic period. It was initially produced by cold working native copper as demonstrated by a sheet-bead bracelet from Aratashen, which dates to the first quarter of the sixth millennium BC (Meliksetian et al. 2007). Although copper smelting began in the South Caucasus in the early fourth or perhaps the late fifth millennium BC, regularization of copper and bronze production is not evident before the late fourth millennium BC. From then onward, metal producers throughout the Caucasus participated in geographically extensive networks in which metal, artifacts and technical knowledge circulated (Peterson 2012). These early metal networks initially emerged in the context of small villages and groups of mobile pastoralists associated with what are commonly referred to as the Early Bronze Age Maikop and Kura-Araxes archaeological cultures (Table 8.1).

Table 8.1 Archaeological periodization and chronology of the Armenian plateau for the periods discussed here (after Smith et al. 2009: Fig. 2)

Period		Approximate dates BCE
Iron I	Iron Ib	1000–800
	Iron Ia	1150–1000
Late Bronze Age	LB III	1300–1150
	LB II	1400–1300
	LB I	1500–1400
Middle Bronze Age		2400–1500
Early Bronze Age		3500–2400

The majority of Early Bronze Age metalwork in the region has been recovered from burials (Chernykh 1992, 2009; Kohl 2007; Peterson 2012). The Middle Bronze Age saw an increase in the consumption of metalwork in lavish burial mounds such as the Karashamb and Trialeti kurgans (Kohl 2007: 115–116; Kushnareva 1997; Oganesian 1992). Although the availability of more easily smelted oxide ores in the Caucasus is limited, heavy exploitation of primary, sulphidic copper ores is not apparent before the Late Bronze Age of the mid-second millennium BC.[3] Prehistoric metal consumption in the region reached its greatest heights during

[1] The majority of these analyses are summarized by Chernykh (1992).

[2] In addition to the OES results, additional points of comparison are available from XRF and neutron activation analysis of samples from other copper and bronze artifacts from Armenia that were collected previously by Meliksetian and colleagues (Meliksetian et al. 2003a, b; Meliksetian et al. 2007; Meliksetian and Pernicka 2010). However, these are discussed only in passing here since those analyses were not conducted on the same objects as those described in this chapter.

[3] These ores are more difficult to smelt into metallic copper than the weathered oxide ores that would have been encountered earlier in the upper levels of deposits.

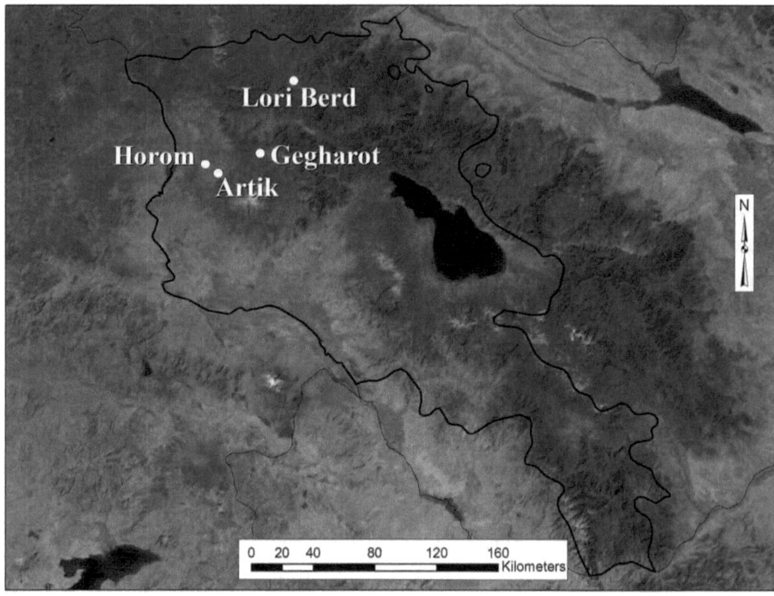

Fig. 8.1 Location of Horom and other major Late Bronze-Early Iron necropolises that are mentioned in the text, with the borders of present day Armenia outlined in *black* (prepared by Adam Clegg)

the Late Bronze II, Late Bronze III and Iron I periods, as evident from rich necropoli such as Artik, Lori Berd, and Horom (Fig. 8.1 and Table 8.1). The objects examined in this chapter originate from Horom, and date to these periods.

The Late Bronze and Early Iron Age sites in present day Armenia lie within an eastern focus of mining and metallurgical activity with ties to what is commonly referred to as the Kayakent-Korochoi culture in the area of present day Daghestan, in the northeastern Caucasus (Chernykh 1992: 278, 288; Pike 2002). Morphologically, Late Bronze Age metalwork associated with this eastern focus has been grouped into three principal diagnostic forms: shaft-hole pole axes with arched lunate blades (Chernykh 1992: Fig. 8.7, 9.9, 9.13, 9.15; Kohl 2007: Fig. 3.29); daggers and swords with a decorated and sharply tapering triangular blades, with hilts that were often cast separately and hafted onto the tang (Chernykh 1992: Figs. 100.1–100.3, 100.9–100.12; 101.1–101.8); and spearheads that frequently have a forged and split socket (Chernykh 1992: Fig. 101.9–101.12). Other ornaments and ceremonial items associated with this focus include elaborate fibulae, pendants, bracelets, belts, buttons, zoomorphic figurines, bronze vessels (Chernykh 1992: 264–295), and objects interpreted as representations of the solar system (Ouzounian 1984: Fig. 3). Points (dart or arrow heads) with a sharply tapering base account for a number of the implements recovered from the Horom necropolis (Fig. 8.2). By the second millennium BC, metal producers in the South Caucasus were able to smelt antimony, and utilized it in binary, ternary, and quaternary alloys with copper (Chernykh 1992: 290; Pike 2002; Meliksetian et al. 2003a; Gevorkyan 2009). Although people in the region utilized iron by 1200 BC, they continued to produce a variety of copper-based metalwork in the Late Bronze and Early Iron Age.

Knowledge of these developments in early mining and metallurgy on the Armenian plateau has relevance beyond the South Caucasus alone. It is also important to the understanding of the interregional relations that surrounded ancient metal making in neighboring areas. From the 1950s to early 1990s, Soviet researchers conducted over 35,000 analyses of ancient copper-based metalwork in investigations of early metallurgical developments across the Caucasus and Central Eurasia (Chernykh 1992: 16). These efforts unfolded within programs of 'spectral-chemical analysis' based on arc OES in

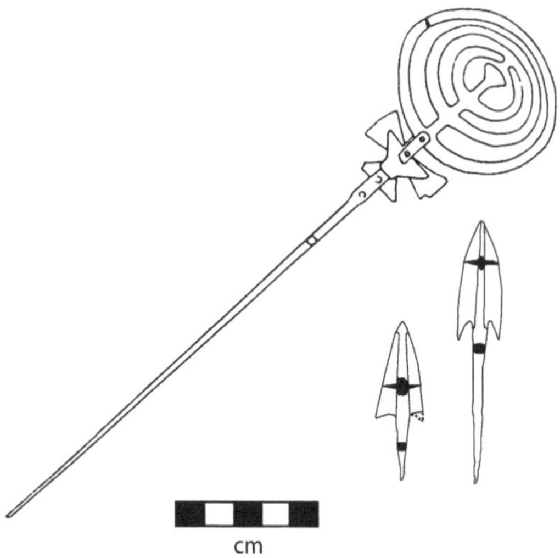

Fig. 8.2 Examples of metal objects from Horom included in this study. *Left*: 1021, a pin in a form interpreted to represent the solar system (see Ouzounian 1984). *Right*: 1043 and 1177, *arrow-* or *dart-heads* (prepared with the help of Bradley Paige)

branches of the Soviet Academy of Sciences, in Armenia and other former Soviet republics. Since then, the popularity of new techniques and the dwindling supplies for traditional arc OES have rendered it practically obsolete. The sharp contraction of institutional funding that followed the dissolution of the Soviet Union also resulted in a general decrease in archaeometallurgical research, which was once a hallmark of archaeology in the region. The South Caucasus Archaeometallurgy Project was initiated in 2009 to contribute to the revival of this important area of research.[4] Our research to date includes the present LA-ICP-MS analysis of Late Bronze to Early Iron Age copper-based artifacts from the Horom necropolis. Identification of the copper sources utilized in making ancient metalwork in the region requires further isotopic and chemical analysis of artifacts and ore deposits, using an appropriate technique such as multicollector-inductively coupled plasma mass spectrometry (MC-ICP-MS) or thermal ionization mass spectrometry (TIMS). This is planned for source analysis of copper-based artifacts from Armenia, utilizing ore samples we collected in 2009 and 2010. LA-ICP-MS analysis is an early step in the broader examination of the technologies and practices utilized in metal making during these periods.

8.3 The Horom Necropolis

Investigations were initiated at Horom with the salvage of 160 tombs for construction of the Nor Kyank reservoir in 1987–1989 (Badalyan and Agekyan 1991; Badaljan et al. 1993: 9). An Armenian-American team resumed excavations in the early 1990s (Badaljan et al. 1992, 1993, 1994). Horom covers an area of some 300 ha making it the largest prehistoric settlement in the Shirak Plain of northwestern Armenia. Stone architecture, including two citadels dating to the twelfth to ninth centuries BC, indicate careful planning under centralized leadership during the latest stages of prehistoric occupation of the region, which ended in its incorporation into the Urartian Empire (Badaljan et al. 1992: 31, 46). The necropolis is comprised of clusters of cromlechs scattered over several hundred hectares south and east of the settlement (Badaljan et al. 1992: 47; Badaljan et al. 1993: 8–12). The cromlechs typically contain an individual interment, and the most frequently encountered grave goods are ceramics and bronze metalwork. The large number of cromlechs that were investigated by salvage operations in the 1980s are mostly unpublished. Two that were opened by the Armenian-American team in 1992 to the southeast of southernmost citadel give a picture of the interior of the burials. Both had a central burial pit within a ring of stones that was sealed with a massive capstone (Badaljan et al. 1993: Figs. 7 and 8).

[4] In other projects using other techniques, copper, bronze, antimony and lead artifacts as well as Armenian copper ores have been analyzed at the Curt-Englehorn Zentrum, Mannheim (Meliksetian et al. 2003a, b; Meliksetian et al. 2007; Meliksetian and Pernicka 2010). Other ongoing archaeometallurgical research in the South Caucasus includes investigations of ancient gold mining in Georgia by Stöllner, Gambashidze, and Hauptmann (Hauptmann et al. 2010; Hauptmann and Klein 2009), and research recently initiated on the Sotk mine and neighboring sites in Armenia under the direction of Ernst Pernicka.

8.4 LA-ICP-MS Analysis: Goals and Methods

LA-ICP-MS analysis of copper and bronze artifacts from Horom was performed on samples collected earlier by one of us (Gevorkyan), which had been taken from objects recovered by the Nor Kyank salvage operations. The analysis was performed at the Center for Archaeology, Materials and Applied Spectroscopy (CAMAS) at Idaho State University.[5] These artifacts date to the thirteenth to ninth centuries BC, or the Late Bronze III-Iron I periods (Table 8.1).[6] This experiment had two general goals: (1) to examine the effectiveness of LA-ICP-MS for analyzing differences in the composition of copper and bronze artifacts both within and between assemblages, as an aid in characterizing ancient metal technologies and possibly sources of materials (e.g., the use of sulphide ores high in arsenic vs. oxide ores); and (2) comparison of the LA-ICP-MS results with those from arc OES performed by Gevorkyan on 48 of the 50 samples. Until the 1990s, arc OES was the chief instrumental method of copper-based artifact analysis in the South Caucasus. The comparison of LA-ICP-MS and arc OES results was proposed in order to determine the potential of relatively inexpensive LA-ICP-MS analysis for the ongoing investigation of ancient copper and bronze from Armenia. LA-ICP-MS analysis was performed using five standard reference materials (SRMs) with certified element concentrations (four bronze and one copper), in order to provide a broad range of elements and concentrations for calibrating and quantifying the results. For the comparison of techniques, it would have been useful to analyze these standards with arc OES as well. However, the only instrument available for this would have been the unit in the Institute of Archaeology and Ethnography in Yerevan, which Gevorkyan utilized for the analysis of the Horom assemblage. Unfortunately it no longer works, which prevented the analysis of certified standards as greater basis for comparing techniques. We are grateful to Laure Dussubieux for sharing the B10, B12, 51.13-4, 71.32-4 standards used in the experiment.[7] SRM 494 from the US National Institute for Standards and Technology was also used.

Laser ablation was performed on cut, ground, and polished surfaces of samples embedded in spectroscopic epoxy. The exposed surface of each sample ranged from 1.5 to 2.0 mm. While the analysis could have been performed on the surface of uncut objects, the samples had already been removed in the 1990s for arc OES and provided an opportunity to prepare them in this manner. Preparation of ground and polished samples with uniform surface geometry facilitated SDD-EDS determination of major element composition. It also aided in selecting the exact area of ablation and to visualize structural heterogeneity, which might affect the LA-ICP-MS analysis of material composition. Before ablation, element maps were collected for each sample utilizing the CAMAS laboratory's Bruker Quantax 200 SDD-EDS system in tandem with its FEI Quanta 200F ESEM, operated in backscattered electron imaging mode. Three non-overlapping 500 × 500 μm ablation sites were analyzed by SDD-EDS on each. Element mapping revealed that a 500 × 500 μm laser ablation raster pattern would be sufficient to subsume the heterogeneous domains in each sample and approximate a bulk sample analysis (Fig. 8.3). These heterogeneities consisted of different phases of the matrix materials, e.g., arsenic-rich phases in arsenic bronze and tin-rich phases in tin bronze. The size of these phases ranged from only a few micrometers to up to 300 μm across (Fig. 8.3). If LA-ICP-MS spot analysis of an area 100 μm in diameter had been performed within a single phase alone, it would

[5] By three of us: Dudgeon, Tromp, and Peterson.

[6] A closer examination of the dating and periodization of the Horom metalwork is beyond the scope of the present discussion.

[7] SRM B10 and B12 are from the Centre de Développement des Industries de Mise en Forme des Matériaux, France, and 51.13-4, 71.32-4 are from the Bureau of Analysed Samples Ltd, England.

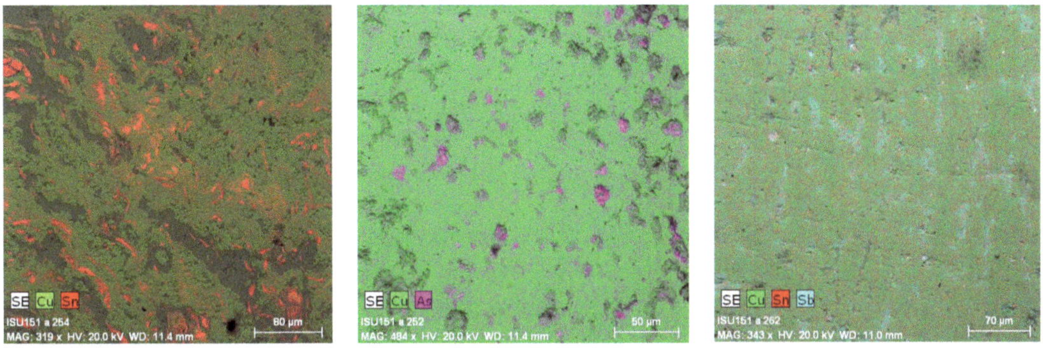

666 (Cu-Sn) 675 (Cu-As) 1040 (Cu-Sn-Sb)

Fig. 8.3 Examples of SEM-EDS element maps from tin bronze metalwork analyzed from Horom (for samples 666, 675, and 1040)

have diminished the accuracy of the determination of the alloy recipe, such as the tin to copper ratio in a tin bronze artifact. The 500 × 500 μm laser ablation raster pattern was large enough to analyze all the phases visible in the samples together, and to approximate a bulk analysis.[8]

At each of the three SDD-EDS analysis sites on each sample, three replicate or overlapping LA-ICP-MS analyses were performed in the raster patterns noted above. The laser device was a New Wave UP-213 laser ablation system operated in imaged aperture mode. Ultra high purity (UHP) helium was used as the laser sweep gas (0.343 l min^{-1}) in a standard volume laser cell. The sweep gas was blended with up to 0.70 l min^{-1} of UHP argon gas prior to entering the plasma torch, to minimize plasma flicker and increase the washout rate at the conclusion of each analysis. Sample raster patterns were pre-ablated in order to remove oxidation and residual surface contamination, using a 100 μm diameter spot size with a 100 μm s^{-1} laser speed at 60 % laser output power. Data acquisition was performed with an ablation pass within the pre-ablated area using an 80 μm diameter beam, moving at 50 μm s^{-1} operating at 70 % output power, at a pulse rate of 20 Hz. The ablated material was analyzed by a Thermo X-Series II inductively coupled plasma-mass spectrometer with a GCMS-based dual-inlet interface (typical operating parameters are found in Table 8.2). This interface permits simultaneous introduction of the laser ablation gas with a liquid internal standard (with 20 ppb each of Rh, Ru and Ir), to aid in correcting instrument drift, and to offset matrix suppression that can cause the signal for elements of interest to decrease. This may occur when more of a relatively soft, easily ablated material like metallic copper is ionized in the plasma formed by the laser, which if left uncorrected may lead to misleading results. The GC dual-inlet interface allows the mapping of signal suppression during an experiment (Fig. 8.4), permitting precise estimation of effects on intensity counts per second (ICPS) for both standards and unknowns. Signal variations noted in this experiment include the aforementioned matrix suppression for materials with higher ablation efficiency, as well as cyclical drift and secular reduction in signal from ablated material accumulating on the orifice of the sample cone. We believe that constant introduction of 2 % HNO_3 has positive effects on the rate of deposition and resultant signal suppression (see Fig. 8.4), although we have not performed

[8] Another strategy for dealing with heterogeneity in a situation in which it is not possible to remove samples for SEM-EDS visualization, such as analysis through the corroded surface of a bronze, might be to sample numerous points and to average the results. However, it is not possible to say how many points would be sufficient in a given case.

Table 8.2 LA-ICP-MS operating parameters for the present experiment

Laser ablation device	New Wave UP213 Nd:YAG
Wavelength	213 nm
Energy	0.2 mJ
Spot size	80 μm
Preablation	100 μm
Fluence	7.35 J cm^{-2}
Rep rate	20 Hz
Travel rate	50 μm/s
ICP-MS	**Thermo X-II Series ICP-QMS**
Detector	Discrete dynode electron multiplier
RF power	1400 W
Plasma Ar gas flow rate	10.0 l min^{-1}
Auxiliary Ar gas flow rate	0.70 l min^{-1}
Sample He gas flow rate	0.343 min^{-1}
Data acquisition mode	Time resolved, 1 acquisition/s
Dwell Time per analyte	10.0 ms
Channels	1
Separation	0.02 AMU

specific experiments to determine the effective improvement on long analysis routines with high ablation matrix load.

The data was calibrated with a slightly modified version of the approach described previously by Neff and Dudgeon (2006) and in Speakman and Neff (2005). Elements commonly found as trace elements, impurities, and alloy constituents in copper and bronze may segregate as cast metal cools, as in the case of lead (Pb), or may form different phases with copper, as arsenic (As) and tin (Sn) will do. In order to increase the precision and accuracy of spatially discrete compositional analysis across samples with varying concentrations of copper, we utilized a procedure developed by one of the authors (Dudgeon, as explained in Chap. 18) to externally measure the actual copper concentration at each ablation site by SEM-SDD-EDS (reported as normalized mass percent). At each ablation site, a minimum of 10 replicate SDD-EDS measures covering a suite of 15 major, minor, and trace element components (Al, Cr, Mn, Fe, Co, Ni, Cu, Zn, As, Ag, Sn, Sb, Au, Pb, Bi) was averaged, and the mean value of copper was used in the internal standard correction during ICP-MS signal calibration for analyses at the respective site. The analytes measured with LA-ICP-MS and the detection limits for each are given in Table 8.3.

For this experiment, detection limits (LOD) are calculated as three times the standard deviation of all experimental blank concentrations (sample chamber online, stage moving, laser at 0 % power, ICP-MS collecting ICPS data) *plus* the mean of all calculated experimental blank concentrations. This results in a more conservative estimate of LOD than using three times the standard deviation alone as is often done, and reflects the fact that for many elements the experimental blank ICPS remain above calibration blanks throughout the analysis. Experimental blanks for this experiment are the equivalent of QA/QC blanks in liquid aspiration analysis, and are not used in standard curve calibrations. Experimental blanks were performed after every 15 analyses (standards and unknowns), permitting further monitoring of the laser and ICP-MS operational conditions/washout efficiency over the course of the experiment. Use of this optimized laser ablation—liquid aspiration internal standard approach, we achieve tenfold or lower detection limits than the 0.01 % (100 ppm) LOD calculated by Gevorkyan (1980) for previously unpublished OES analyses (see Table 8.3) for all elements except copper (LOD = 2878.61 ppm) and tin (LOD = 213.45 ppm). As discussed below, broad comparability was discovered between LA-ICP-MS and arc OES for several elements (Fe, Sb, Sn, Bi), but with significant improvement of LOD using LA-ICP-MS (i.e., Mn, Au) (compare Tables 8.4 and 8.5). However, OES appears to perform better with lead, probably due to the high propensity of lead to segregate in copper and bronze, which would yield better results with bulk sampling method (OES) as opposed to a micro-sampling

Fig. 8.4 Sensitivity drift over time as measured on our LA-ICP-MS instrument setup for internal standard element rhodium. Trendline represents a moving average of ten 3-min acquisitions

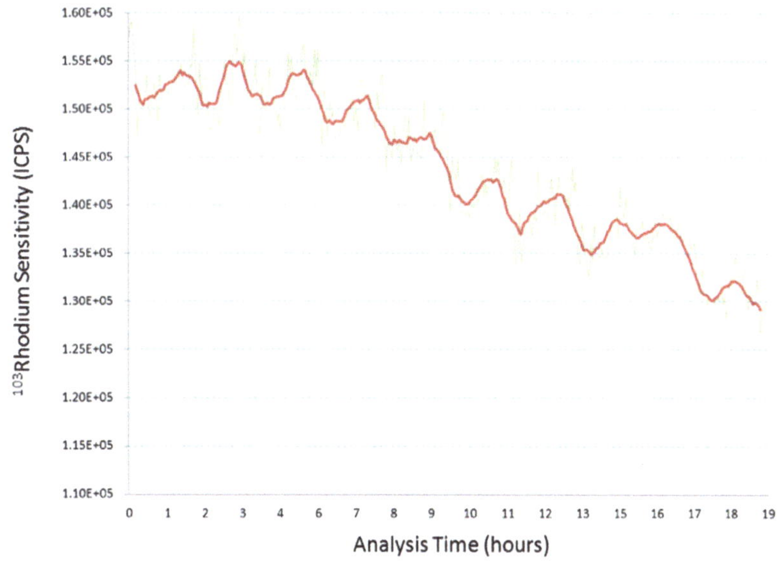

Table 8.3 Analytes measured by LA-ICP-MS in copper-based metalwork from Horom, and their detection limits

Analyte	Detection limit (ppm)
27Al	1.01
53Cr	3.11
55Mn	2.26
57Fe	12.89
59Co	0.18
62Ni	1.61
65Cu	2878.61
67Zn	0.77
75As	56.09
109Ag	1.35
117Sn	213.45
121Sb	2.88
197Au	0.02
204Pb	9.84
209Bi	0.19

technique (LA-ICP-MS). This is discussed further below.

8.5 Results of Arc OES Analysis of the Horom Metalwork

The 48 previously analyzed copper-based artifacts from Horom have been classified into two artifact categories: ornaments (n = 21) and implements (n = 27). The identified ornaments include flat copper sheets and bands, swastika-shaped plaques, rings, pins (including three associated with representations of the solar system—an example is shown in Fig. 8.2), sword pommels, and pendants. The implements are wire, points, and a dagger. Among the objects that were analyzed, the 23 points in the assemblage account for the majority of implements and the largest number of artifacts that can be grouped together by form (Table 8.4).

In every object, copper was identified as the base material (Table 8.4). No attempt was made to quantify copper. This was the norm for arc OES analysis of ancient copper-based materials. Therefore, it was not possible to normalize the arc OES results to the concentration of copper present in any sample. The results show the presence of 10 alloys among the objects, by the conventional association of the presence of more than 1 % of a major element in a copper-based material with "voluntary addition" as an alloy constituent (see Dussubieux et al. 2008: 650). Figure 8.5 summarizes the number of implements and ornaments identified for each alloy and unalloyed copper. Six of the objects register as unalloyed copper and three as leaded copper (Cu-Pb). Seven are arsenical copper or arsenic bronze (Cu-As), and another nine are leaded arsenical copper (Cu-As-Pb). Eleven are tin bronze

Table 8.4 Arc OES results for the Horom metalwork, in percent values (in all cases copper is the base material but was not measured)

No.	Object	Sn	Pb	Zn	Bi	Ag	Sb	As	Fe	Ni	Co	Mn	Au
899	Band	>12	1.1	0.035	0.018	0.005	0.12	0.4	0.3	0.25	0.14	>0.001	<0.003
904	Wire	0.2	0.14	0.06	0.008	0.025	0.15	0.4	0.09	0.08	0.013	>0.001	<0.003
905	Wire	>12	0.14	0.009	0.007	0.025	0.3	1.1	0.005	0.09	0.005	>0.001	<0.003
928	Band	6.1	0.14	0.12	0.003	0.0012	0.15	0.12	0.09	0.01	0.005	>0.001	<0.003
929	Plaque	10.5	2.3	0.045	0.007	0.025	0.025	0.052	0.11	0.023	0.02	>0.001	<0.003
930	Plaque	>12	1.6	0.12	0.022	0.025	0.063	0.062	0.07	0.012	0.027	>0.001	<0.003
934	Ring	>12	1.4	0.008	0.007	0.025	0.065	0.4	0.03	0.08	0.013	>0.001	<0.003
936	Pin	0.009	2.3	0.008	0.007	0.007	0.064	0.16	0.0045	0.0016	–	>0.001	–
1019	Pin	>12	0.15	0.02	0.009	0.12	0.12	2.75	0.12	0.035	0.004	0.031	0.0065
1020	Pin	>12	0.9	0.9	0.05	0.04	0.33	0.4	0.35	0.028	0.04	0.022	0.0065
1021	Pin	>12	1.1	0.02	0.026	0.25	0.15	0.9	0.25	0.028	0.0063	0.032	0.0065
1026	Point	0.02	2.5	0.25	0.0025	0.04	0.33	1.1	0.018	0.015	–	0.008	0.002
1027	Point	0.25	0.12	0.15	0.003	0.04	0.82	0.4	0.013	0.08	0.0012	0.005	0.002
1028	Point	0.9	2.5	0.18	0.0012	0.025	0.7	2.3	0.02	0.027	0.0012	0.005	0.002
1029	Belt	0.9	2.5	0.18	0.0035	0.025	0.7	2.3	0.02	0.027	0.0012	0.005	0.002
1030	Point	0.08	0.9	0.009	0.007	0.04	0.3	2.3	0.004	0.034	–	0.007	0.002
1031	Point	0.7	0.2	0.035	0.003	0.025	0.7	1.2	0.012	0.034	<0.001	0.007	0.002
1032	Point	0.085	0.35	0.009	0.007	0.04	0.22	1.1	0.08	0.04	0.004	0.004	0.002
1037	Pin	>12	0.9	0.016	0.003	0.025	0.15	0.4	0.025	0.035	0.013	0.008	<0.0003
1038	Pin	0.07	1.8	0.016	0.003	0.04	0.33	1.1	0.09	0.08	0.0012	0.007	0.00065
1039	Pin	>12	0.18	0.016	0.008	0.04	0.027	0.33	0.11	0.25	0.05	0.004	0.00065
1040	Pommel	>12	0.45	0.016	0.003	0.012	7–8	0.14	0.035	0.0045	0.028	0.004	–
1041	Pommel	>12	0.8	0.016	0.015	0.018	7–8	2.3	0.1	0.08	0.006	0.004	0.00065
1043	Point	0.18	0.35	0.016	0.0025	0.018	0.22	0.85	0.07	0.014	0.004	0.004	–
1048	Point	0.08	0.45	0.016	0.0035	0.04	0.33	1.6	0.007	0.08	<0.001	0.004	0.00065
1049	Point	0.016	0.9	0.016	0.009	0.04	0.33	1.6	0.011	0.16	–	0.004	<0.0003
1050	Point	0.08	0.18	0.025	0.003	0.04	0.33	0.85	0.013	0.018	–	0.005	0.00065

1051	Point	0.75	2.2	0.025	0.003	0.01	0.3	2.3	0.011	0.035	0.0012	0.004	0.00065
1054	Dagger	0.052	0.6	0.025	0.003	0.04	0.7	2.3	0.011	0.035	0.0012	0.004	0.00065
1079	Ring	0.12	2.5	0.009	0.014	0.025	0.06	2.3	0.37	0.08	0.004	0.022	0.002
1176	Point	0.03	0.45	0.007	0.0054	0.18	0.3	9.2	0.016	0.08	0.0025	0.008	0.0065
1178	Point	0.55	2.5	0.004	0.008	0.18	0.5	2.5	0.03	0.095	0.0025	0.008	0.0065
1179	Point	0.042	1.1	0.004	0.003	0.04	0.3	2.5	0.009	0.034	0.001	–	0.002
1180	Point	0.65	1.4	0.005	0.004	0.025	0.5	2.5	0.12	0.034	0.0032	0.01	0.002
1184	Sheet	0.03	0.18	0.01	<0.0001	<0.0002	–	0.6	0.15	*Trace*	–	–	–
1188	Point	0.08	0.17	0.009	0.0025	0.008	0.063	0.18	0.013	<0.001	–	0.006	–
1190	Pommel	>12	1.1	0.006	0.01	0.04	0.12	1.2	0.08	0.012	0.005	0.012	–
1192	Sheet	>12	0.06	0.005	–	0.092	0.3	0.16	0.02	0	0.0028	0.012	–
1193	Point	0.7	2.7	0.007	0.004	0.02	0.063	0.23	0.03	0.0017	0.001	0.006	–
1194	Point	6.2	0.35	0.006	0.0095	0.08	0.7	2.2	0.011	0.035	0.001	0.007	0.002
1195	Point	5	2.7	0.006	0.008	0.18	0.35	0.9	0.12	0.034	0.001	0.011	0.002
1196	Point	6.2	2.2	0.007	0.004	0.25	0.7	2.1	0.016	0.034	0.0028	0.005	0.002
1198	Point	0.026	0.2	0.005	0.0025	0.005	0.025	0.4	0.011	0.004	–	–	–
1210	Pendant	6.2	0.06	0.005	–	0.0035	0.063	0.18	0.03	0.012	0.033	0.01	–
1211	Pendant	6.1	0.06	0.007	*Trace*	0.005	0.063	0.18	0.036	0.012	0.0035	0.008	–
1212	Pendant	>12	0.35	0.006	–	0.025	0.13	0.4	0.3	0.06	0.0035	0.024	–
1214	Sheet	>12	0.35	0.005	0.002	0.005	0.3	0.9	0.1	0.012	0.0036	0.022	0.002
1217	Band	>12	0.45	0.007	0.022	0.18	0.12	0.35	0.36	0.035	0.027	0.012	0.002

Provided by A. Gevorkyan

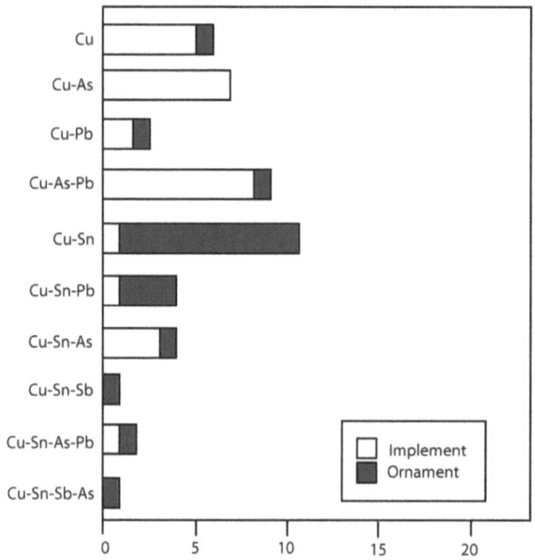

Fig. 8.5 Metal groups (unalloyed copper and various copper alloys) detected by arc OES in ornaments and implements from Horom

(Cu-Sn), and there are four each of tin-arsenic bronze (Cu-Sn-As) and leaded tin bronze (Cu-Sn-Pb). The most prevalent complex alloy is Cu-Sn-As-Pb, for which there are two examples.

One of the most remarkable aspects of the assemblage is the use of antimony within it. Arc OES detected the presence of over 7 % antimony in two sword pommels, in one as tin-antimony bronze (Cu-Sn-Sb, sample 1040) and in another as tin-antimony-arsenic bronze (Cu-Sn-Sb-As, sample 1041). The source of the antimony in these objects has not been identified. The best known prehistoric antimony mines near Horom are in the Zopkhito district of the Gornaya Racha in the Great Caucasus. There are at least four published calibrated radiocarbon dates for antimony mines in Zopkhito, which span a period from about 1800 to 1000 BC (Burchuladze and Togonidze 1987: 241) that overlap with the dating of the Horom objects. Zopkhito is therefore a strong candidate as the source of the antimony utilized in making these objects.

8.6 LA-ICP-MS Results

Small fragments of the 48 Horom artifacts left over from the arc OES analysis were analyzed by LA-ICP-MS. A total of 15 analytes and 3 liquid internal standard additions were collected during data acquisition (Table 8.2). Twelve of these analytes demonstrated adequate detection limits (^{55}Mn, ^{57}Fe, ^{59}Co, ^{62}Ni, ^{75}As, ^{109}Ag, ^{117}Sn, ^{121}Sb, ^{125}Te, ^{197}Au, ^{204}Pb, ^{209}Bi). The elements Mn, Fe, Co, Ni, As, Ag, Sn, Sb, Te, Au, Pb, Bi were previously analyzed by arc OES and provided the principle points of comparison between the techniques. Copper was also measured by SDD-EDS, and was used as an internal standard calibration for the LA-ICP-MS analysis. ICP-MS survey scans on the Horom artifacts determined that this suite of analytes comprised at least 99.73 % of the available masses measureable by ICP-MS. A combination of copper and bronze standards (SRM-494 and B10, B12, 51.13-4, 71.32-4) was analyzed to bracket the concentrations within the artifact samples for calibration. The SDD-EDS standardized LA-ICP-MS concentration results are given in Table 8.5. With the exception of four artifacts with high iron concentration (Horom 899, 928, 1029 and 1217; average 3150 ppm) and two with high nickel (Horom 899 and 1039; average 5780 ppm), all of the artifacts possessed high and variable concentrations of copper, arsenic, tin, antimony and lead. To visualize the major elements alloyed into the copper base, we compared the ratio of the four major non-copper elements and minor constituents (As, Sn, Sb, Pb) across artifact classes in the assemblage (Fig. 8.6).

The LA-ICP-MS data indicate that within the various artifact classes significantly different alloy recipes were used. In most cases alloying would have been achieved by the direct mixing of metals, but high levels of arsenic may have also been achieved through a co-smelting process (e.g., Lechtman and Klein 1999). The dominant relationship in this assemblage is the correlation between high arsenic (average 11,500 ppm) and point implements, the dagger and the wire artifacts. Conversely, the artifacts characterized as ornaments (i.e., pins, bands, rings and pommels) contained a higher proportion of tin (average 95,700 ppm) than the implement artifacts. The concentration of lead and antimony

Table 8.5 LA-ICP-MS results for metalwork from the Horom necropolis (all values ppm). Underlined values indicate the metal group

No.	Object	Metal group	^{55}Mn	^{57}Fe	^{59}Co	^{62}Ni	^{65}Cu	^{67}Zn	^{75}As	^{109}Ag	^{117}Sn	^{121}Sb	^{197}Au	^{204}Pb	^{209}Bi
899	Band	Cu + Sn	2.30	2804.00	1464.70	7327.09	929,561.00	149.68	2011.15	217.54	120,517.33	853.23	85.93	4440.38	25.52
904	Wire	Cu	0.71	1494.42	208.56	672.69	982,953.16	298.58	2699.35	237.22	1381.06	931.55	7.73	915.56	69.06
905	Wire	Cu + As + Sn	0.79	85.99	61.14	1482.43	937,733.71	61.07	11,113.39	320.69	41,717.09	1433.06	4.67	1403.25	49.48
928	Band	Cu + Sn	26.55	3646.23	147.53	297.76	881,921.57	1521.29	2312.47	207.24	169,956.85	1391.61	7.25	3101.80	34.21
929	Plaque	Cu + Sn + Pb	5.05	1129.74	376.16	245.83	931,770.30	800.64	363.18	342.21	87,213.34	236.65	1.47	10,736.96	185.48
930	Plaque	Cu + Sn + Pb	3.15	1637.58	406.27	393.51	940,124.72	794.65	388.19	335.81	97,697.07	372.27	1.89	12,452.42	200.32
934	Ring	Cu + Sn	0.19	627.56	341.63	1106.28	880,733.89	69.31	2531.16	156.32	95,391.02	351.08	7.37	7662.09	26.73
936	Pin	Cu + Sn	0.31	308.89	54.82	281.56	919,839.45	23.60	4243.51	330.85	75,027.11	689.64	15.98	1741.01	16.12
1019	Pin	Cu + Sn	2.33	1085.39	47.21	153.60	936,478.29	136.79	6187.65	285.01	45,684.99	345.97	1.69	535.17	25.94
1020	Pin	Cu + Sn + Pb	1.36	1539.95	95.40	242.90	890,432.18	97.16	4388.21	482.29	66,880.32	552.75	11.06	10,272.75	158.86
1021	Pin	Cu + Sn	2.51	468.55	32.28	277.27	879,069.15	83.58	7496.95	287.94	88,082.06	4407.69	3.12	2011.56	71.95
1026	Point	Cu + As(+Pb)	0.89	27.64	10.94	440.08	954,887.09	295.76	20,498.32	349.84	691.79	1442.89	18.14	8972.99	34.87
1027	Point	Cu	0.32	23.56	13.54	943.27	978,907.36	672.06	5757.61	282.13	3854.46	4327.42	10.52	3832.00	24.97
1028	Point	Cu + As + Pb	0.15	81.52	6.82	390.26	960,498.33	11.12	11,270.71	316.98	5855.17	2214.54	13.04	20,912.09	31.83
1029	Belt	Cu + Sn	6.75	3173.41	127.99	212.22	830,899.49	119.29	4916.21	134.02	219,979.10	6915.97	9.52	5263.90	23.34
1030	Point	Cu + As	0.22	15.36	3.54	508.03	962,540.36	1764.04	15,617.24	386.61	622.16	1269.85	21.15	7812.22	45.98
1031	Point	Cu + As	0.60	11.74	4.15	495.21	966,707.98	723.63	11,211.99	289.48	3410.59	2194.03	13.62	2262.96	24.34
1032	Point	Cu	0.41	692.11	60.67	726.32	994,670.20	76.82	4583.33	323.44	1141.28	774.81	4.83	1827.02	17.00
1037	Pin	Cu + Sn	1.09	222.46	193.01	578.47	899,943.32	20.32	4483.08	150.17	72,693.33	643.48	6.35	1243.56	24.88
1038	Pin	Cu + As(+Pb)	0.34	28.62	11.49	874.27	973,938.59	6.01	12,601.50	459.81	429.78	1777.16	9.53	9186.89	31.37
1039	Pin	Cu + Sn	0.62	833.46	495.27	4239.50	943,512.26	10.14	2900.07	48.10	41,957.37	116.29	11.10	1877.65	3.29
1040	Pommel	Cu + Sn + Sb + Pb	1.30	263.14	22.44	68.03	892,579.33	142.11	647.82	185.21	62,387.70	64,946.69	0.78	39,148.51	4.25
1041	Pommel	Cu + As + Sn + Sb + Pb	1.07	1040.48	92.61	791.86	793,705.55	63.10	10,165.22	290.84	116,862.72	29,174.59	15.57	10,297.41	28.05
1043	Point	Cu(+As)	2.46	942.94	6.77	227.27	997,669.00	21.10	9379.93	269.52	2515.09	1708.40	2.85	3913.46	30.57
1048	Point	Cu + As	0.76	29.28	3.81	516.57	973,519.72	8.83	11,286.42	334.24	1496.58	2076.66	20.27	2647.40	26.15
1049	Point	Cu + As	0.13	3.51	0.75	457.82	989,374.25	6.63	11,949.30	353.45	271.38	1308.32	7.32	3763.20	42.42
1050	Point	Cu(+As)	0.13	13.95	1.57	348.00	992,300.85	4.00	8916.22	269.74	204.10	927.72	11.43	635.31	16.89
1051	Point	Cu + As(+Pb)	0.11	6.56	19.98	673.42	971,892.37	11.79	16,234.80	267.65	432.27	1362.19	10.93	8595.12	33.42
1054	Dagger	Cu + As	0.49	43.28	17.83	463.67	958,265.79	37.48	17,519.21	224.39	735.37	2236.25	6.53	6943.61	33.67
1079	Ring	Cu + As + Sn	2.11	715.44	65.78	886.48	935,740.42	91.41	11,174.45	331.60	98,488.38	2538.08	17.35	3247.76	41.97
1176	Point	Cu + As	0.54	20.06	9.36	590.52	933,483.88	18.47	30,623.29	323.22	1589.57	886.25	13.30	1881.16	28.47
1178	Point	Cu + As + Pb	2.39	38.83	8.74	722.41	974,845.87	37.08	9768.41	400.57	3321.27	1687.25	9.78	10,306.02	26.62
1179	Point	Cu + As	0.45	14.16	1.70	385.74	967,193.74	14.73	11,860.15	451.09	422.82	1219.04	11.26	4053.80	23.45
1180	Point	Cu + As	0.46	136.79	32.78	279.28	937,453.31	18.36	19,525.51	279.72	3868.23	1777.74	5.39	2406.17	43.66
1184	Sheet	Cu	0.25	32.85	10.92	566.12	993,385.09	9.87	7482.79	354.85	6474.65	4819.62	31.26	7689.46	32.15
1188	Point	Cu	0.33	28.35	9.85	533.38	982,819.27	8.24	7804.63	373.63	6973.36	4812.61	30.01	8430.80	33.90

(continued)

Table 8.5 (continued)

No.	Object	Metal group	^{55}Mn	^{57}Fe	^{59}Co	^{62}Ni	^{65}Cu	^{67}Zn	^{75}As	^{109}Ag	^{117}Sn	^{121}Sb	^{197}Au	^{204}Pb	^{209}Bi
1190	Pommel	Cu + Sn	0.66	307.51	108.66	380.65	920,571.86	102.64	5513.10	226.56	74,844.25	525.77	1.57	6920.82	59.35
1192	Sheet	Cu + Sn	1.78	159.13	22.82	258.18	840,794.80	47.04	1337.30	168.21	234,783.99	1664.73	0.24	658.61	3.50
1193	Point	Cu + Sn	0.53	69.42	15.58	363.76	946,961.37	18.09	5747.98	240.76	22,837.89	1092.61	6.00	7755.90	13.68
1194	Point	Cu + Sn	3.04	168.25	22.17	308.66	873,406.64	60.50	1307.25	103.76	258,851.72	1777.80	0.21	425.86	1.20
1195	Point	Cu + Sn + Pb	0.44	51.25	15.84	461.72	969,661.22	34.35	7961.84	324.03	28,221.71	1720.79	10.21	12,422.85	21.26
1196	Point	Cu + Sn(+Pb)	0.47	60.71	21.73	397.41	945,914.18	18.64	9993.77	327.86	35,445.12	1835.37	9.45	9096.29	21.96
1198	Point	Cu + As + Sn	0.26	35.90	5.54	489.34	954,194.94	25.17	11,677.54	289.03	27,503.44	2159.48	9.05	3218.90	19.22
1210	Pendant	Cu + Sn	2.63	396.75	44.03	211.57	957,722.71	72.50	1313.26	146.78	76,088.45	546.69	0.06	666.54	0.67
1211	Pendant	Cu + Sn	1.09	425.83	32.58	133.40	937,046.71	77.21	1560.51	184.14	91,880.34	643.63	0.22	661.40	0.89
1212	Pendant	Cu + Sn	0.64	421.29	19.95	95.03	955,065.61	25.80	2111.73	125.24	81,812.32	626.70	0.03	574.85	0.48
1214	Sheet	Cu + Sn	4.34	1409.33	63.31	500.75	902,780.25	171.12	7162.61	233.46	113,137.67	2009.59	7.06	3118.51	28.90
1217	Band	Cu + Sn	2.48	2990.28	156.37	380.64	973,124.86	114.83	3169.75	79.83	159,877.26	630.57	13.41	2041.28	85.47

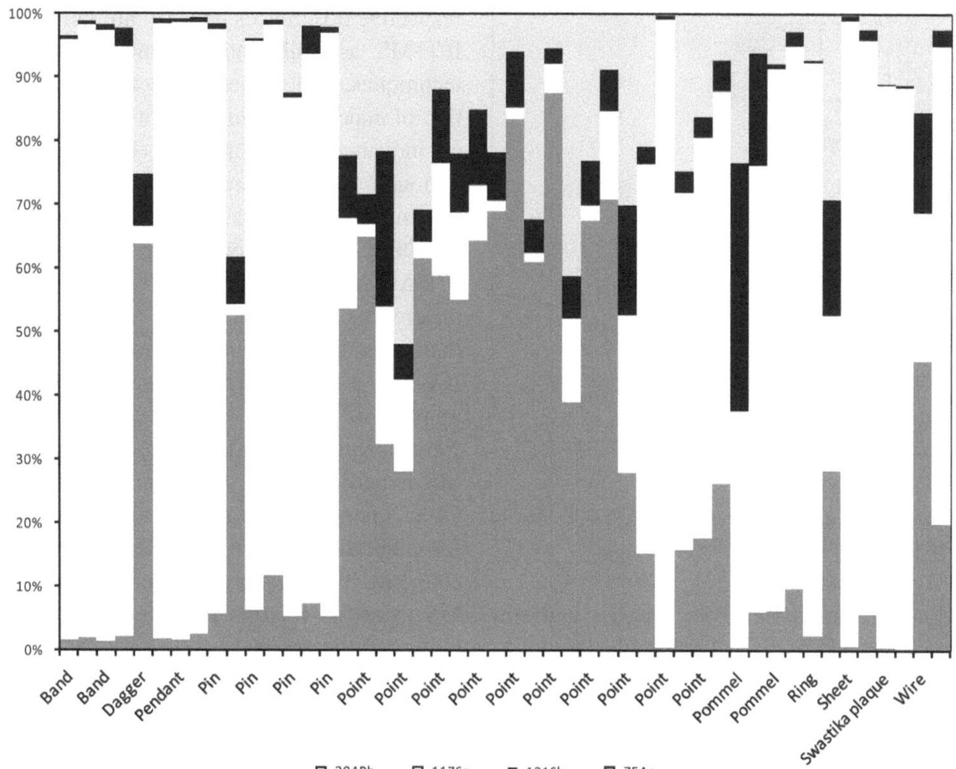

Fig. 8.6 Frequency of primary non-copper elements (e.g., As, Sn, Sb, Pb), by artifact form according to LA-ICP-MS results (Table 8.5)

in the Horom artifacts is not indicative of any pattern correlated with artifact type; rather, they appear to covary as lesser constituents within the copper—arsenic—tin modal recipe, suggesting a great deal of mixing of these materials through recycling.

8.7 Discussion

An important part of assessing the performance of LA-ICP-MS on the Horom metalwork is comparing the LA-ICP-MS results for the relative frequency of different materials in the tested samples with the arc OES results for the same samples. A comparison of the arc OES and LA-ICP-MS results shows a general pattern of agreement in the element constituents of the Horom artifacts when the analytes occur at relatively high concentrations, especially Fe, Sb, Sn, and Bi. At lower levels, the OES data break down into more categorical estimates of concentration, due to the higher detection limits (generally about 0.01 %) and the lower precision of arc OES spectroscopy (copper was not measured, and quantities of other elements were not normalized in relation to copper). Elements that are especially adversely affected by lower precision of estimation include manganese, gold, silver and zinc, which register at low to several hundred ppm using LA-ICP-MS.

Altogether, the results for the LA-ICP-MS analysis generally agreed with those that were gathered previously by arc OES on the composition of only 25 of the 48 objects analyzed with both techniques. We detected copper, arsenical copper, leaded arsenical copper, tin bronze, leaded tin bronze and tin arsenic bronze by LA-ICP-MS. We did not identify leaded copper by LA-ICP-MS as was found by arc OES. In

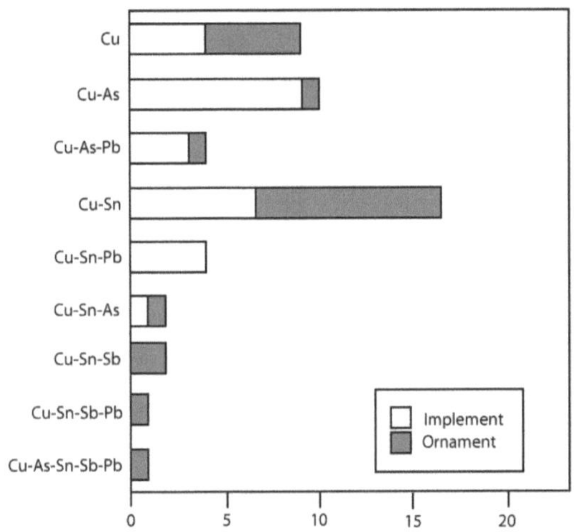

Fig. 8.7 Metal groups detected by LA-ICP-MS in ornaments and implements from Horom

addition, in the tin-antimony bronze and tin-antimony-arsenic bronze identified by arc OES, LA-ICP-MS detected the presence of alloy-level concentrations of lead; the composition of these two objects was therefore recategorized as leaded tin-antimony bronze and leaded tin-antimony-arsenic bronze. With the exception of these two objects, LA-ICP-MS tended to find lower levels of lead than arc OES in the same samples. This may indicate a weakness in the LA-ICP-MS method as compared to arc OES as a bulk sampling technique. Lead segregates in copper and bronze, and when analyzing a larger volume bulk sample, the arc OES may simply have encountered more of it. Alloy-level concentrations of antimony were identified with equal frequency by both techniques. Comparing Figs. 8.5 and 8.7, indication of a tendency to select tin alloys for ornaments is much greater in the LA-ICP-MS results than those for arc OES (LA-ICP-MS shows that 18 out of the 20 ornaments analyzed are made of alloys with tin), while the LA-ICP-MS results indicate a similar tendency to reserve unalloyed copper and arsenical copper for implements as did arc OES (four out of five unalloyed copper objects are implements, and implements account for all arsenical copper objects).

Thus, while there is notable disagreement on concentrations of lead, which we attributed to the respective differences between arc OES and LA-ICP-MS as bulk analysis and micro-analysis techniques, some general tendencies in the selection of materials for implements and adornments are indicated by both methods (Figs. 8.5 and 8.7), although the preference of tin alloys for ornaments, which was revealed originally with arc OES, is more strongly expressed in the LA-ICP-MS results (Fig. 8.7). In addressing the differences in performance, we would argue that other than the issue with lead noted, LA-ICP-MS is more reliable than arc OES for a number of reasons. Concentrations of copper were quantified by SEM-EDS for the LA-ICP-MS experiment, but never quantified with arc OES. Since copper was quantified by SDD-EDS for calibrating our LA-ICP-MS results for other elements, it was possible to examine the LA-ICP-MS values in relation to a measurement of the bulk material (copper). Measurement of copper was not part of arc OES analyses of ancient copper-based artifacts (Chernykh 1992; Gevorkyan 1980). As noted earlier, the LA-ICP-MS analysis was performed with five certified copper and bronze standards, which allow for the calibration of results for the unknown quantity of elements present in a test sample according to the known values for those elements in the certified standards when analyzed together in the same experiment, as we did in this case. No analogous form of standardization was used in the arc OES analysis. It would have been desirable analyze the standards used with LA-ICP-MS with the arc OES used to analyze the Horom metalwork, but this was not possible because it no longer works.

One of the most interesting aspects of the assemblage is the use of antimony as an alloying agent in two sword pommels. Pike (2002) has argued that antimony may have been used as a substitute for tin, which is not known to have been mined in the Caucasus before the late first millennium BC. However, antimony appears to have been more than just a tin substitute in the metalwork from Horom. In both items in which antimony was identified at alloying levels by both arc OES and LA-ICP-MS (samples 1040 and 1041), it is in mixtures that also contain high alloy levels of tin. Both objects are ornamental sword pommels in which tin and antimony may have been used for the combined aesthetic and functional effects of a light colored and hard material. Whether production was done locally or finished objects were imported cannot be determined through this analysis. However, as noted earlier, there is a strong likelihood that either antimony or the objects themselves are from present-day Georgia. The sources of tin used in prehistoric Eurasian metalwork are a topic of debate beyond the scope of the present discussion.

Of all the materials, the identification of arsenic with voluntary alloying is perhaps the most problematic, since arsenical copper with arsenic concentrations under 3 % can be created intentionally or unintentionally through co-smelting (Lechtman and Klein 1999). In all but one sample, the levels of arsenic detected by arc OES that may be associated with alloying range from 1.1 % to 2.75 %, and can be reasonably attributed to accidental production by co-smelting. The exception is sample 1176, a point that the OES results show as containing 9.2 % arsenic without signs of heavy corrosion that might lead to depletion of Cu and enrichment of As (Table 8.4). However, the OES results appear to be incorrect; the LA-ICP-MS results found that it contains 30,623 ppm arsenic, which is closer to the values for arsenic in the rest of the samples with value commensurate with alloying.

Another significant aspect of both the arc OES and LA-ICP-MS results is the potential evidence for recycling represented by the presence of Sn, As, Pb, and Sb in several objects at levels not far below the conventional threshold for intentional alloying. One explanation for this is that the materials were prepared by mixing alloys containing these metals with unalloyed copper through recycling.

8.8 Conclusions

Through analysis of late prehistoric metalwork from Horom, comparison of the results from arc OES (the traditional but now defunct method of compositional analysis in Eurasian archaeometallurgy) and LA-ICP-MS has shown that LA-ICP-MS is as effective or superior to arc OES in the detection of alloys and alloy recipes. We therefore conclude that LA-ICP-MS it is a highly viable technique for continuing analyses of the composition of ancient copper-based metalwork. Even so, we see two problems that should be addressed in future experiments. First, if possible, it would be important to measure the relative precision of arc OES and LA-ICP-MS through analysis of the same certified standards by both techniques. That could not be done in this case, since the arc OES used to analyze the metalwork from Horom is no longer operational and cannot be repaired. The second problem is a hindrance of LA-ICP-MS that is paradoxically also one of its strengths: it is a microanalysis technique. While this means microscopically small samples can be analyzed with no visible damage to artifacts, there are potential problems identifying the proportion of lead present throughout a sample, since lead tends to segregate in copper and bronze. We sought to overcome this by sampling in a relatively large raster pattern (500 × 500 μm), but it still appears that the results for lead are questionable.

For the purpose of this analysis, the objects were classified into two artifact types: ornaments and implements. While heuristically valuable for evaluating preliminary hypotheses, this scheme should be treated with care, since this type of classificatory arrangement is predicated on functional differences between the two artifact types. Compositional analysis alone is not enough to examine differences in function; this would be better served by combining compositional analysis with analyses of usewear and microhardness. This classification scheme may also be misleading for grave goods, which may have been made for a ceremonial function. However, there is utility in these categories from the standpoint of compositional analysis in relation to the selection of particular materials for specific object forms regardless of how the objects were used before burial. While all of the objects analyzed by arc OES and LA-ICP-MS were recovered as grave goods by the people who utilized the Horom necropolis, according to results with both arc-OES and LA-ICP-MS, the producers of these objects tended to systematically utilize alloys with tin to make ornaments, while they prepared the majority of implements (which are mostly points) with unalloyed copper and arsenic alloys. An important goal of compositional analysis of ancient copper-based metalwork is to determine the patterns in the selection of copper and alloys for particular kinds of artifacts. Our results are further indication of the benefits of LA-ICP-MS for analysis of variance in alloy composition in a large assemblage of copper-based artifacts.

Acknowledgments We are grateful to Aram Gevorkyan (Institute for Archaeology and Ethnography, Yerevan, Armenia) for providing the results of Arc OES analysis of metalwork from Horom presented in Table 8.4. Some of the results discussed here were presented at the Archaeometallurgy in Europe III conference (June 28–July 1, 2011) at the Deutsches-Bergbau Museum, Bochum, Germany (Peterson et al. 2011), and were the subject of a poster displayed at the 2011 Annual Meeting of the Society for American Archaeology in Sacramento, California (Meredith et al. 2011). This research was sponsored by the American Councils for International Education and the National Councils for Eurasian and East European Research, through a National Endowment for the Humanities International Collaborative Research Fellowship in 2009–2010 for *Collaborative Investigations of Early Mining and Metal Production on the Armenian Plateau, ca. 7000–800 BCE* (David Peterson, PI). Additional support was provided by Idaho State University's Faculty Research Committee (2009–2010) and Humanities and Social Sciences Research Committee (2010). Laure Dussubieux (Field Museum of Natural History, Chicago) graciously provided copper and bronze standards utilized in the LA-ICP-MS analysis. The objects analyzed were recovered in excavations at Horom directed by Ruben Badalyan and O. K. Agekyan in 1987–1989 (Badalyan and Agekyan 1991). Many thanks to Idaho State University Anthropology students Adam Clegg for preparation of Fig. 8.2, and Bradley Paige for Fig. 8.3.

References

Badaljan R, Edens C, Kohl PL, Tonikjan AV (1992) Archaeological investigations at Horom in the Shirak Plain of North-Western Armenia. Iran 30:31–48

Badaljan R, Edens C, Gorny R, Kohl PL, Stronach D, Tonikjan AV, Hamayakjan S, Mandrikjan S, Zardarjan M (1993) Preliminary report on the 1992 excavations at Horom, Armenia. Iran 31:1–24

Badaljan R, Kohl PL, Stronach D, Tonikjan AV (1994) Preliminary report on the 1993 excavations at Horom, Armenia. Iran 32:1–29

Badalyan RS, Agekyan OK (1991) Excavations of early Iron Age burials in the Horom cemetery situated on the territory of the Artik Reservoir. Abstracts of papers given at the scientific session devoted to field archaeological investigations in the Republic of Armenia, Yerevan (in Armenian)

Burchuladze AA, Togonidze GI (1987) Tblisi radiocarbon dates IV. Radiocarbon 29:239–262

Chernykh EN (1992) Ancient metallurgy in the USSR. Cambridge University Press, Cambridge

Chernykh EN (2009) Formation of the Eurasian steppe belt cultures: viewed through the lens of archaeometallurgy and radiocarbon dating. In: Hanks B, Linduff K (eds) Social complexity in prehistoric Eurasia: monuments, metals and mobility. Cambridge University Press, Cambridge, pp 115–145

Dussubieux L (2007) Laser ablation-inductively coupled plasma-mass spectrometry of ancient copper alloy artifacts. In: Glascock MD, Speakman RJ, Popelka-Filcoff, RS (eds) Archaeological chemistry: analytical methods and archaeological interpretation. ACS publication series 968. American Chemical Society, Washington, DC, p 336–348

Dussubieux L, Deraisme A, Frot G, Stevenson C, Creech A, Bienvenu Y (2008) LA-ICP-MS, SEM-EDS and EPMA analysis of Eastern North American copper-based artifacts: impact of corrosion and heterogeneity on the reliability of the LA-ICP-MS compositional results. Archaeometry 50:643–657

Gevorkyan A (1980) Iz istorii drevneyschey metallurgii Armyanskogo nagor'ya. Yerevan

Gevorkyan A (2009) Reflections on metal production in Armenia during the Late Bronze and Early Iron Age. Armazd 4:36–54

Hauptmann A, Klein S (2009) Bronze Age gold in Southern Georgia. Revue d'archéométrie 33:75–82

Hauptmann A, Bendall CHR, Brey GP, Japarize I, Gambasidze I, Klein S, Prange M, Stöllner TH (2010) Gold in Georgien. Analytische Untersuchungen an Goldartefakten und an Naturgold aus dem Kaukasus und dem Transkaukasus. In: Hansen S, Hauptmann A, Motzenbäcker I, Pernicka E (eds) Von Majkop bis Trialeti. Gewinnung und Verbreitung von Metallen und Obsidian in Kaukasien im 4. - 2. Jt. v. Chr. Habelt, Bonn, pp 139–160

Kohl PL (2007) The making of Bronze Age Eurasia. Cambridge University Press, New York, NY

Kushnareva KKh (1997) The Southern Caucasus in prehistory: stages of cultural and socioeconomic development from the eighth to the second millennium

B.C. University Museum Monograph 99. The University of Pennsylvania Museum, Philadelphia, PA

Lechtman H, Klein S (1999) The production of copper–arsenic alloys (arsenic bronze) by cosmelting: modern experiment, ancient practice. J Archaeol Sci 26:497–526

Meliksetian K, Pernicka E (2010) Geochemical characterisation of Armenian Early Bronze Age metal artefacts and their relation to copper ores. In: Hansen S, Hauptmann A, Motzenbäcker I, Pernicka E (eds) Von Majkop bis Trialeti. Gewinnung und Verbreitung von Metallen und Obsidian in Kaukasien im 4. - 2. Jt. v. Chr. Habelt, Bonn, pp 41–58

Meliksetian Kh, Pernicka E, Avetissyan P, Simonyan H (2003a) Chemical and lead isotope characterisation of Middle Bronze Age bronzes and Some Iron Age antimony objects (Armenia). In: Proceedings, second international conference archaeometallurgy in Europe, vol 2. Associazione Italiana di Metallurgia, Milan

Meliksetian Kh, Pernicka E, Badalyan R, Avetissyan P (2003b) Geochemical characterization of Armenian Early Bronze Age copper based artefacts and their relation to copper ores. Proceedings, second international conference archaeometallurgy in Europe, vol 1. Associazione Italiana di Metallurgia, Milan

Meliksetian Kh, Pernicka E, Badalyan R (2007) Compositions and some considerations on the provenance of Armenian Early Bronze Age copper artefacts. In: Proceedings, second international conference archaeometallurgy in Europe, Aquiliea, Italy, 17–21 June 2007. Associazione Italiana di Metallurgia, Milan

Meredith C, Peterson D, Tromp M, Dudgeon J, Gevorkyan A, Meliksetian Kh (2011) New standards in the analysis of archaeological metalwork using LA-ICP-MS: a case study from the South Caucasus archaeometallurgy project. Poster displayed at the 76th annual meeting of the society for American archaeology, Sacramento, 30 March–3 April 2011

Neff H, Dudgeon JV (2006) LA-ICP-MS analysis of ceramics and ceramic raw materials from the Gila River Indian Community, Arizona. Research report: Institute for Integrated Research in Materials, Environments, and Society, California State University, Long Beach, CA

Oganesian VE (1992) A silver goblet from Karashamb. Sov Anthropol Arch 30:84–102

Ouzounian JG (1984) Armenian astronomy in the Bronze Age. Archaeoastronomy 7:105–109

Peterson D (2012) Forging social networks: metallurgy and the politics of value in Bronze Age Eurasia. In: Hartley CW, Yazıcıoğlu GB, Smith AT (eds) The archaeology of power and politics in Eurasia: regimes and revolutions. Cambridge University Press, New York, NY, pp 283–301

Peterson D, Gevorkyan A, Meliksetian K, Bobokhyan A, Dudgeon J, Tromp M, Hovakimyan S, Vardanyan A, Meredith C, Schneyder T (2011) The South Caucasus Archaeometallurgy project: investigation of early mining and metal production on the Armenian Plateau. An interim report. In: Hauptmann A, Modaressi-Tehrani D, Prange M (eds) International conference, Archaeometallurgy in Europe III, Abstracts, Metalla 4. Duetsches Bergbau-Museum, Bochum, p 213

Pike AWG (2002) Appendix: Analysis of Caucasian metalwork – the use of antimonal, arsenical and tin bronze in the Late Bronze Age. In: Curtis J, Kruszynski M, Pike AWG, Searight A (eds) Ancient Caucasian and related materials in the British Museum. British Museum Occasional Paper 121. The British Museum, London, p 87–92

Scott D (1991) Metallography and microstructure of ancient and historic metals. Getty Museum, Malibu

Smith AT, Badalyan R, Avetisyan P (2009) The archaeology and geography of ancient Transcaucasian Societies I: the foundations of research and regional survey in the Tsaghkahovit Plain, Armenia. Oriental Institute Press, Chicago, IL

Speakman RJ, Neff H (eds) (2005) Laser ablation ICP-MS in archaeological research. University of New Mexico Press, Albuquerque, NM

Part III

Application to Vitreous Materials

Analysis of Vitreous Archaeological Materials by LA-ICP-MS

9

Bernard Gratuze

Abstract

This chapter introduces some of advantages of using LA-ICP-MS to analyze vitreous archaeological materials, highlighting the approaches utilized in the studies included in Part III of this volume.

9.1 Analysis of Vitreous Materials

As they are relatively homogeneous, vitreous matrices were among the first archaeological materials to be tested by Laser Ablation-Inductively Coupled Plasma-Mass Spectrometry, and the first applications of this technique to the characterization of archaeological artificial and natural glasses were published during the early 1990s (Gratuze et al. 1993; Telouk 1993). The application of LA-ICP-MS to archaeometric study of vitreous materials slowly increases in frequency up to the end of last millennium (Gratuze 1999; Tykot and Young 1996).

Since the beginning of the twenty-first century, studies using LA-ICP-MS for characterization of vitreous materials are increasingly numerous, and this technique tends nowadays to be one of the principal methods employed for analysis of these types of material (see Chaps. 4, 10–14 and 19). In its most broad meaning, the category of vitreous materials includes, as demonstrated in this volume, a large variety of matrices which comprise not only artificial silica glasses, but also lead glazes, glassy slags, and natural glasses. For example, in Chap. 10, Reepmeyer and colleagues refers to natural glasses such as obsidian, while in Chap. 14, L'Héritier et al. deal with metallurgical glassy iron slags, and Iñañez et al. in Chap. 19 study ceramic lead glazes. If in many cases, glass matrices are quite homogeneous, they can also include numerous small crystalline mineral phases and can therefore also be fairly heterogeneous.

At the beginning 1990s, quadrupole ICP-MS instruments were the only ones coupled with laser ablation, allowing only bulk analysis of minor and trace elements. At that time, LA-ICP-MS was often used as a complementary method for trace element analysis, major elements being separately determined by EPMA, SEM-EDX, PIXE or XRF. Nowadays, as described by Cagno et al. in Chap. 11, LA-ICP-MS can perform major to trace element analysis of almost all elements within a material during a single

B. Gratuze (✉)
Institut de Recherche sur les Archéomatériaux, Centre Ernest Babelon, CNRS/Université d'Orléans, 3D rue de la Férollerie, 45071 Orléans Cedex 2, France
e-mail: gratuze@cnrs-orleans.fr

L. Dussubieux et al. (eds.), *Recent Advances in Laser Ablation ICP-MS for Archaeology*, Natural Science in Archaeology, DOI 10.1007/978-3-662-49894-1_9

instrumental run. Furthermore, the coupling of laser ablation with all available plasma mass spectrometry techniques (multi- or single-collector sector field and time of flight instruments), allows a wide field of different characterization techniques. The different chapters in this volume demonstrate even that if LA-ICP-MS is primarily used for the full characterization of glass composition from major to trace elements, it can also produce valuable data for isotopic analysis, element mapping and profile analysis.

For example, in Chap. 4, van Elteren et al. demonstrate that elemental mapping—which was in the past mainly performed using EPMA—is now possible using LA-ICP-MS. This allows the characterization of chemical association within colorants and opacifiers and their mapping within an object. Depth profiling using single LA-ICP-MS spot analysis is highlighted in Chap. 12 by Gratuze. While more conventional methods require cutting a profile, DP-LA-ICP-MS allows generation of similar results with no visual damage to the object. In the near future, the rapid mass scanning allowed by TOF-ICP-MS will probably make it possible to fully characterize sub-surface mineral inclusions present in glass by combining elemental mapping and depth profile analysis. Lead isotopic analysis of ancient glazes using LA-MC-ICP-MS is described in Chap. 19 by Iñañez et al., while the potential of boron isotopic composition using this type of instrument was recently tested by Devulder on Roman glass (Devulder et al. 2015).

Due to the fact that LA-ICP-MS allows nearly non-destructive analysis, this technique is now widely accepted by curators and archaeologists for investigation of glass objects. LA-ICP-MS has thus made it possible to undertake studies on large population of objects rather than a few selected less valuable pieces. As shown by recent publications and theses, the amount of data generated using this technique is increasing exponentially. Hence the routine application of trace element determination using LA-ICP-MS for glass characterization coupled with isotopic analysis is opening new fields of investigation for glass provenance studies.

However, due to the rapidly increasing amount of available data produced by LA-ICP-MS, analysts now face two main challenges. The first one is evaluating and ensuring the compatibility of data produced by different laboratories. Different approaches are used to ensure data compatibility, one of which is the systematic publication of values obtained on glass standard reference materials such as NIST612 or the various Corning glasses (Purowski et al. 2014; Rehren et al. 2015; Varberg et al. 2015; Walton et al. 2012), as has often been done for EPMA analysis (Smirniou and Rehren 2011). A second approach, which has been tested here by Dussubieux (Chap. 13), is to conduct inter-laboratory comparisons on a set of selected glasses, as was done by Glascock (1999) for obsidian. The results obtained in Dussubieux's comparison study show that, despite some dispersion of values, the interpretation of the data published by the nine participating laboratories results in the same grouping of samples. In his paper on Pergamon glasses, when comparing data obtained at the Field Museum in Chicago and at the IRAMAT-CEB in Orleans, Rehren similarly concludes that "While no direct comparison of the performance of the two LAICPMS laboratories was done for these samples, there is no systematic difference visible for data from similar glasses analysed in the different labs; the data is assumed to be fully compatible." (Rehren et al. 2015: 272).

As shown by several recent papers (Rehren and Freestone 2015; Varberg et al. 2015), in the majority of cases, it is necessary to compare analytical data obtained directly on objects known to have different origins in order to be able to draw general conclusions about long distance glass trade. However, for diverse geopolitical and legislative reasons, it is often difficult to move archaeological objects between different countries. Maintaining high compatibility between data produced all over the world by different archaeometric laboratories will probably be one of the most important challenges in the future. To succeed, the multiplication of inter-comparison tests, the adoption of similar analytical protocols, and the development of new sets of reference materials will probably be the best solution.

The second challenge linked to the exponential increase of data volume which we will have to face is the development of common databases and statistical tools for the treatment of experimental data. While comparison of data points using binary plots or extended rare earth element spidergrams has been shown to be very efficient for relating glass objects of common origin (Jackson and Nicholson 2010; Shortland and Schroeder 2009; Wedepohl et al. 2011), this approach is not efficient for handling thousands of data points.

References

Devulder V, Gerdesc A, Vanhaeckea F, Degryse P (2015) Validation of the determination of the B isotopic composition in Roman glasses with laser ablation multi-collector inductively coupled plasma-mass spectrometry. Spectrochim Acta B 105(1):116–120

Glascock MD (1999) An inter-laboratory comparison of element compositions for two obsidian sources. IAOS Bull 23:13–25

Gratuze B (1999) Obsidian characterization by laser ablation ICP-MS and its application to prehistoric trade in the Mediterranean and the Near East: sources and distribution of Obsidian within the Aegean and Anatolia. J Archaeol Sci 26:869–881

Gratuze B, Giovagnoli A, Barrandon JN, Telouk P, Imbert JL (1993) Apport de la méthode ICP-MS couplée à l'ablation laser pour la caractérisation des archéomatériaux. Revue d'Archéométrie 17:89–104

Jackson CM, Nicholson PT (2010) The provenance of some glass ingots from the Uluburun shipwreck. J Archaeol Sci 37:295–301

Purowski T, Wagner B, Bulska E, Syta O, Dzierzanowski P (2014) Glassy faience from the Hallstatt C period in Poland: a chemicophysical study. J Archaeol Sci 50:288–304

Rehren T, Freestone I (2015) Ancient glass: from kaleidoscope to crystal ball. J Archaeol Sci 56:233–241

Rehren T, Connolly P, Schibille N, Schwarzer H (2015) Changes in glass consumption in Pergamon (Turkey) from Hellenistic to late Byzantine and Islamic times. J Archaeol Sci 55:266–279

Shortland AJ, Schroeder H (2009) Analysis of first millennium BC glass vessels and beads from the Pichvnari Necropolis, Georgia. Archaeometry 51:947–965

Smirniou M, Rehren T (2011) Direct evidence of primary glass production inlate Bronze Age Amarna, Egypt. Archaeometry 53(1):58–80

Telouk Ph (1993) Développements de la spectrométrie ICP-MS dans les domaines de l'archéologie et de la géologie, Utilisation de l'ablation laser. Ph.D Dissertation, University of Lyon

Tykot RH, Young SM (1996) Archaeological applications of ICP-mass spectrometry. In: Orna MV (ed) Archaeological chemistry: organic, inorganic, and biochemical analysis. American Chemical Society, Washington, DC, pp 116–130

Varberg J, Gratuze B, Kaul F (2015) Between Egypt, Mesopotamia and Scandinavia: Late Bronze Age glass beads found in Denmark. J Archaeol Sci 54:168–181

Walton M, Eremin K, Shortland A, Degryse P (2012) Analysis of Late Bronze Age glass axes from Nippur-a new cobalt colourant. Archaeometry 54(5):835–852

Wedepohl K-H, Simon K, Kronz A (2011) Data on 61 chemical elements for the characterization of three major glass compositions in late antiquity and the Middle Ages. Archaeometry 53(1):81–102

Contributions of LA-ICP-MS to Obsidian Sourcing in the Pacific 10

Christian Reepmeyer, Wallace Ambrose, and Geoffrey Clark

Abstract

This chapter examines results from the application of LA-ICP-MS to the identification of sources of obsidian artifacts from the Western Pacific. More than 700 analyses of obsidian samples collected at the Australian National University over the past years provide an accurate geochemical dataset for major obsidian source regions in the Western Pacific unambiguously discriminating sources and sub-sources. This dataset is employed to analyze social interaction in the 3000-year timeframe of human occupation of the western Pacific as reflected by variations in lithic raw material sources. Albeit not non-destructive, the minimal sample size necessary, and the precision and accuracy of analysis for a wide range of major, minor, trace and rare earth elements enables LA-ICP-MS to not only geochemically fingerprint archaeological artifacts to known source locations, but also to provide information about general geological contexts from which these objects derived. These additional data have been successfully employed in the past to identify locations of high probability for undetected obsidian outcrops.

10.1 Introduction

The combination of geochemical and technological analyses of lithic assemblages provides unique insight into cultural activities on the landscape, more specifically the nature of interaction between communities and their local environments. Obsidian is a prime material for understanding patterns of social interaction in the archaeological record because of the possibility to track the movement of this raw material between distant areas. This has been particularly

C. Reepmeyer (✉)
College of Society, Culture and Education, James Cook University, PO Box 6811Cairns, QLD 4870, Australia
e-mail: Christian.Reepmeyer@jcu.edu.au

W. Ambrose • G. Clark
Archaeology and Natural History, School of Culture, History and Language, Australian National University, H.C. Coombs Building (No. 9), Canberra 0200, ACT, Australia
e-mail: Wal.Ambrose@anu.edu.au; Geoffrey.Clark@anu.edu.au

L. Dussubieux et al. (eds.), *Recent Advances in Laser Ablation ICP-MS for Archaeology*, Natural Science in Archaeology, DOI 10.1007/978-3-662-49894-1_10

productive for the analysis of exchange systems and social interaction in various parts of the world, including the Pacific.

Obsidian is a natural volcanic glass that commonly develops when high silicate (rhyolitic) low temperature lava cools quickly during a volcanic eruption (Rothe 2005). High viscosity of alumino-silicates and the relative cooling rate limits the materialization of crystals in the melt and results in a homogeneous amorphous rock (Vinx 2005), which is very brittle and can easily be flaked, producing very sharp edges. In general, the overall chemical composition of the initial magma (high silicate, low Al, Ti, Mn) determines the quality of the resulting glass—the lower the silicate content, the more the glass tends to form crystals, which in turn reduces the quality of the resulting material. The rather complex nature of its genesis is the reason that obsidian deposits have a geographically discrete distribution: not every volcanic eruption, and indeed not all volcanic islands develop obsidian deposits. Furthermore, even when obsidian deposits did develop on islands, not all outcrops were exploited by humans in the past (Ambrose et al. 1981b; Ambrose and Johnson 1986).

As a consequence of its unique geochemical attributes and perceived importance as an item of cultural significance, obsidian has been a focus of archaeological research in the Pacific. Kononenko (2012) for example, has recently suggested that some small sharp flakes and, on occasion, specialized obsidian tools had been selected to perform specific tasks such as tattooing by piercing on the human body and/or for medical purposes. However, while obsidian produces a very sharp cutting edge, it is particularly fragile and this results in a less useful stone tool for domestic purposes than other, more durable raw materials (Torrence 2011). Consequentially, more mundane utilization as detected on a small number of artefacts (Kononenko et al. 2010) might only occur after artefacts lost their initial sharpness and were of no further use for these highly specialized tasks (see also Specht and Koettig 1981; Fullagar 1992, 1993). Ceremonial applications for certain mid-Holocene obsidian tools from New Guinea are also proposed by Araho et al. (2002). In the context of the Kula exchange system of south-eastern New Guinea, MacIntyre (1983:212) describes obsidian as "pasa"—a subsidiary decorative item. In contrast, other researchers have proposed a number of models where the value of obsidian in new colonies is explained by its role as a symbolic link to ancestral places or societies (Kirch 1988). Sheppard (1993) suggested, based on the usage and distribution patterns of obsidian artefacts at the Reefs/Santa Cruz sites in the Solomon Islands, that obsidian might have served as a concrete symbol of exchange at times when social relationships were being established or maintained within the expanding networks of Lapita colonies. It was transported primarily for this function and later usage patterns were not connected to its exchange value. Obsidian might have been an ideal symbolic material for creating social ties through gift-giving because of its physically attractive appearance and its relative rarity due to the limited number of sources and their discrete distribution (Torrence 2005, 2011). It has also been proposed that the social value of obsidian could have been exploited in profligate displays of conspicuous consumption to advertise wealth, status or prestige (Specht 2002). This scenario suggests that the distribution of obsidian may have been linked to its role in sustaining ancestral social practices.

This chapter focuses on the methods, calibration protocols, and data treatment employed at the Australian National University (ANU) to successfully fingerprint obsidian sources and artifacts in the Pacific using a combination of SEM-EDX (major elements) and LA-ICP-MS (minor and trace elements). Albeit not non-destructive, the minimal sample size necessary for analysis and the precision and accuracy possible in analyzing a wide range of major, minor, trace and rare earth elements enables LA-ICP-MS not only to geochemically fingerprint archaeological artifacts to known source locations, but also provides information about general geological contexts from which these artifacts derived. Sufficiently detailed trace element data, particularly for the rare earth

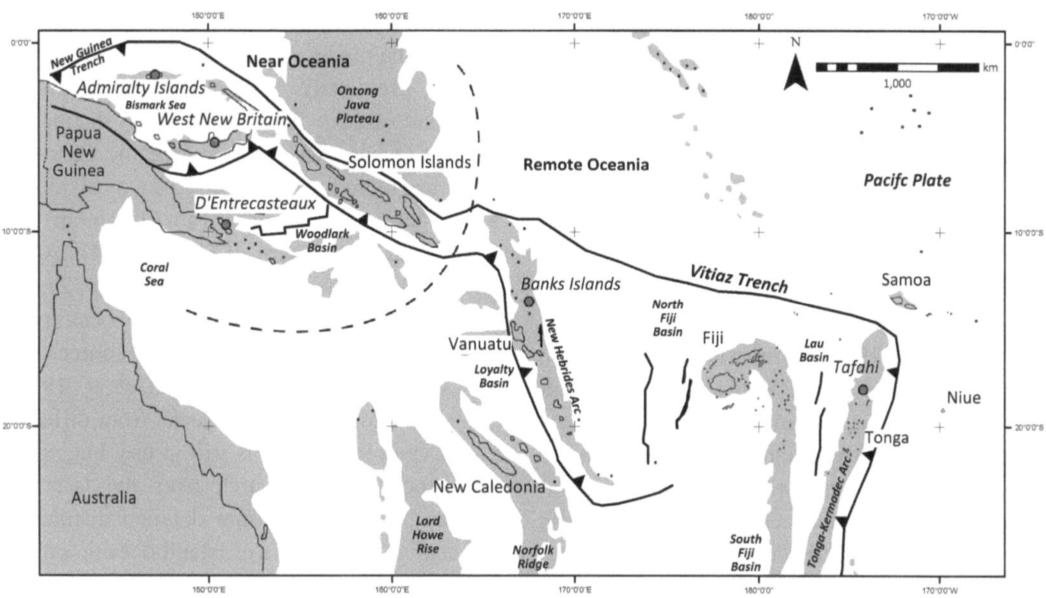

Fig. 10.1 Map of the Western Pacific with obsidian source areas and major geological boundaries indicated

elements, which are not available with bulk analyses such as PIXE-PIGME, give information about fractionation processes in magmatic melts, enabling researchers to identify tectonic origins of surface rocks on Pacific islands, track plate tectonic movements, and distinguish different geological formations of submarine and surface rocks. These additional data have been successfully employed in the past to identify locations of high probability for occurrence of undetected obsidian outcrops (Reepmeyer and Clark 2010; Reepmeyer et al. 2012).

10.1.1 Obsidian Sources in the Southwestern Pacific

Currently, five major centers of early prehistoric obsidian exploitation are known in the southwestern Pacific[1]: West New Britain (WNB) and the Admiralty Islands in the Bismarck Archipelago of Papua New Guinea (PNG), the D'Entrecasteaux Islands of southeastern PNG, the Banks Islands of Northern Vanuatu, and northern Tonga (Fig. 10.1). Most remarkable in the spatial distribution of this raw material in the Pacific is indeed its long-distance transportation, associated with the colonization of the Pacific beyond the northern Solomon Islands during the Lapita period in the second millennium BC. During colonization, obsidian from the Bismarck Archipelago was moved over distances of more than 3500 km into the western and central Pacific and also back into Southeast Asia (Bellwood and Koon 1989).

The present discussion concentrates on a sub-set of 87 samples collected from nine obsidian sources (Lou and Manus in the Admiralty Islands, Papua New Guinea (PNG); Kutau/Bao and Mopir in West New Britain, PNG; West and East Fergusson in the D'Entrecasteaux Islands, PNG; Gaua and Vanua Lava in the Banks Islands, Northern Vanuatu; and Tafahi, Northern Tonga) and the discrimination of regional compositional patterns based on the large array of major and trace elements analyzed by LA-ICP-MS.

[1] The numerous obsidian sources in New Zealand are excluded from this analysis, because of the late colonisation of the island and the limited transportation of New Zealand obsidian into the central and western Pacific.

10.1.2 Prior Chemical Characterization of Pacific Volcanic Glass

Since the beginning of archaeological geochemical research in the 1960s, the analysis of obsidian sources for provenance studies in the Pacific has seen the application of a wide range of different analytical methods (Ambrose 1976; Ambrose and Duerden 1982; Ambrose et al. 1981a, b; Bird et al. 1981a, b; Duerden et al. 1987; Sheppard et al. 1989; Summerhayes et al. 1998; Wall 1976; Ward 1979). With the improvement of instrumentation, starting with X-ray Fluorescence analysis (XRF) and Neutron Activation Analysis (NAA), Particle Induced X-ray Emission—Proton Induced Gamma-ray Emission (PIXE-PIGME, further description see below) quickly became the most widely used method in sourcing obsidian artifacts in the Pacific (Ambrose and Duerden 1982; Duerden et al. 1987; Fullagar et al. 1989; Green 1987; Leach and Davidson 1981). The improvement of analytical techniques (particular PIXE-PIGME) in the early 1980s, securing long term precision and accuracy of elemental measurements, absolute concentrations of F, Na, Al, Si, K, Ca, Ti, Mn, Fe, Rb, Sr, Y, Zr and Nb were expanded by nine ratios (Al/Na, Zr/Fe, Y/Fe, Sr/Fe, Rb/Fe, Mn/Fe, Ca/Fe, K/Fe, F/Na). Fluorine proved to be particularly helpful in distinguishing western Pacific obsidian sources from each other (Bird 1996; Bird et al. 1981a, b, 1997; Summerhayes et al. 1998).

Unfortunately, analyzing Fluorine (F) and Zinc (Zn) with LA-ICP-MS is problematic as the ionization potential of F exceeds that of the carrier gas argon, which means that F cannot effectively be ionized (Falkner et al. 1995:412), while the atomic weight of $^{64}Zn/^{70}Zn$ overlaps with those of isotopes of Nickel (^{64}Ni) and Germanium (^{70}Ge) (Goffer 2007). Nonetheless, the high precision and accuracy of LA-ICP-MS now gives the opportunity to not only distinguish obsidian sources and examine in detail the intra-source variation in elemental composition, but also to reveal the genesis of these sources. Based on a wide suite of analyzed isotopes, comparisons between different volcanic rock types can be made, which narrows the identification of previously unidentifiable artifacts to particular areas. The geological distinction between Oceanic and circum-Oceanic Cenozoic volcanics, including obsidian, based on alkalinity (K and Na content compared to Aluminum) is a well-researched pattern (Chayes 1964). Additionally, the higher content of Fe, Na, F, Zr, Nb and often Rb and Y, and lower concentrations of Al, Ca, Sr and Ba in Oceanic sources has been observed previously (Smith et al. 1977). This separation of volcanic rocks from different provinces in the Pacific is still in use, but extensive geochemical research over the last few decades has led to a more detailed distinction between volcanic rocks of Island oceanic provenance and volcanic rocks from Island arcs.

10.2 Methods

10.2.1 Sampling Method

Sampling methods used for the geochemical analysis reported here included a mixture of probabilistic and non-probabilistic sampling. All source samples available from western Pacific sources were included in the analysis. Flakes were taken from larger boulders and then cut with a wire saw, while smaller pebbles were cut directly with a wire saw. For archaeological assemblages, all artifacts were assessed macroscopically, and pieces with distinctive color, texture, luster, structure or matrix were selected, as well as a random sample of 1–5 % of all other pieces from analyzed assemblages.

10.2.2 Sample Preparation

The application of laser ablation radically reduces sample preparation compared to other techniques. However, a few requirements for analysis by LA-ICP-MS and EDXA-SEM must still be met:

1) A flat ablation surface is preferable;
2) For efficiency reasons, it is preferential to not analyze whole samples because limitations in sample chamber dimensions allow only a small number of artifacts to be put in the chamber at any time. So as to minimize changing sample-mounts during analysis (re-venting samples chambers in SEM vacuum can take up to several minutes), samples were sectioned from each artifact or source sample;
3) The same is imperative for the ablation chamber of the LA-ICP-MS. The chamber has light excess pressure of a helium/argon gas mix. After opening of the chamber a stable atmosphere lacking oxygen must be re-established.

Therefore, a 1 mm^3 piece of each sample was sectioned for analysis. Each of the samples were washed in an ultra-sonic bath for 10 min and then cut with a 200 μm diamond wire saw to minimize material loss. The wire was moisturized during sectioning with a mixture of tap water and detergent. The 1 mm^3 sectioned pieces were then embedded in epoxy resin. The constructed mount, 50 × 25 × 4 mm in size, can carry up to 40 samples, set in a 10 × 4 array. Excess resin was removed from the sample mount using wet and dry silicon carbide sandpaper and the surface was flat mirror polished. For the EDX analysis the samples were coated with a 30 nm thick carbon film to avoid charging.

10.2.3 Scanning Electron Microscopy with Energy Dispersive X-Ray Analysis (SEM-EDX)

The method of Scanning Electron Microscopy with Energy Dispersive X-ray analysis (SEM-EDX) combines two different analytical methods in one; an X-ray detector and an electron microscope. The source for both methods is a concentrated electron-beam, emitted from a tungsten filament cathode, which hits the sample. This beam reacts with the sample and emits electrons which are detected by a scintillator-photomultiplier producing so-called secondary electron images (Reed 2005). The same concentrated electron-beam reacts with the sample causing it to emit electromagnetic radiation in the form of X-rays. Different elements emit X-rays with distinctive energies which can be measured with a semiconductor detector. This X-ray spectrum allows quantitative analysis to a detection limit of about 50 ppm and is therefore useful for the analysis of major element composition (Goldstein et al. 1992; Tykot 2004).

Obsidian source samples and artifacts were examined with a JEOL JSM6400 Scanning Electron Microscope equipped with an Oxford ISIS Energy-Dispersive X-ray Analyzer at the Electron Microscopy Unit, Research School of Biological Sciences, ANU. The set-up employed an atmospheric-thin window (ATW), which is particularly sensitive for light elements. For processing the collected data, Oxford instruments Link ISIS 3.3 software was used. The size of the area analyzed is crucial for the correct analysis of the elemental composition of obsidian samples by SEM-EDX. Exciting an area of more than 200 μm with EDX can result in a loss of precision (Reed 2005). Due to detected inhomogeneity in low-silicate sources from several locations in the Western Pacific and particularly the northern Banks Islands, the occurrence of micro-phenocrysts has to be taken into consideration (Reepmeyer 2008). Analysis of relatively large areas using EDX can result in different compositional data compared to smaller areas analyzed by LA-ICP-MS. Therefore the decision was made to analyze the same area with EDX as was analyzed by LA-ICP-MS (86 μm in diameter).

Eight major elements (Na, Mg, Al, Si, Ca, K, Ti, Mn and Fe) were analyzed with SEM-EDX (Pollard et al. 2007). Additional analyses of P, S and Cl produced inconsistent results and these elements were excluded from later runs. The SEM–EDX was calibrated with 11 mineral standards (Albite 15 kv, MgO 15 kv, Sanidine [$KAlSi_3O_8$], CeP_5O_{14}, FeS_2, NaCl, Diopside, TiO_2 15 kv, Cr_2O_3 15 kv, pure Mn 15 kv, Fe_2O_3), against the NIST612 Standard Reference

Material and two high-silicate obsidian standards from Wekwok, Admiralty Islands and Kutau/Bao, West New Britain (Ambrose et al. 1981b; Duerden et al. 1987; Reepmeyer 2008; Summerhayes 2009). The calculated SiO_2 content was employed to calibrate results of ICP-MS analysis (Ambrose et al. 2009).

10.2.4 Laser Ablation-Inductively Coupled Plasma-Mass Spectrometry (LA-ICP-MS)

The system used in this study is an AGILENT 7500S Inductively Coupled Plasma Mass Spectrometer combined first with an EXCIMER and later a LAMBDA PHYSIK ArF laser ablation system (Longerich et al. 1996). The ArF laser, operating at a wavelength of 193 nm, is capable of ablating silicate, oxide and sulphide phases using an aperture to define pit diameters from about 20 to 200 μm. The pit diameter is controlled by the beam size, and only minimal residual melting occurs (Eggins et al. 1998). Beam diameters producing best ablation results depend on chemical and structural features of the sample (e.g., for clay >100 μm). Laser diameters of 81 μm (EXCIMER laser), and 86 μm (LAMBDA PHYSIK laser) were chosen because they produced count rates of 10^3–10^6 per second for most trace elements, allowing use of the same low count rate part of the detector system.

The utilized ablation chamber was specifically designed at the ANU to use a minimum amount of carrier gas for transport of the ablated material to the ICP-MS (Eggins et al. 1998). Ablation position was identified by digital imaging. Each sample was ablated in a He/Ar atmosphere, and power delivered to the sample surface was calibrated to approximately 20 mJ. The frequency of the laser pulse used for obsidian is 5 Hz, which results in a steady stream of volatized sample material (approximately 0.1 μm of material is ablated with each pulse). Each sample was ablated for at least 40 s with a drilling depth of about 30 μm (Ambrose et al. 2009).

10.2.5 Calibration Protocol

Because all materials ablate somewhat differently, giving different counts per second per ppm for each material, calibration was performed against the NIST612 Standard Reference Glass, measured in a round-robin fashion for a maximum of 10 analyses of unknowns (~15 min) (Ambrose et al. 2009). To control for time variability in signal strength, the count rates for all elements were ratioed to silica and the average of each measured element/Si ratio was multiplied by the SiO_2 content of the sample as determined by EDXA. Counts for 31 isotopes (^{31}P, ^{45}Sc, ^{49}Ti, ^{51}V, ^{55}Mn, ^{63}Cu, ^{85}Rb, ^{88}Sr, ^{89}Y, ^{90}Zr, ^{93}Nb, ^{95}Mo, ^{118}Sn, ^{133}Cs, ^{138}Ba, ^{139}La, ^{140}Ce, ^{144}Nd, ^{147}Sm, ^{153}Eu, ^{158}Gd, ^{162}Dy, ^{166}Er, ^{174}Yb, ^{175}Lu, ^{181}Ta, ^{186}W, ^{206}Pb, ^{207}Pb, ^{208}Pb, ^{232}Th and ^{238}U) were determined by calculating the mean concentration for each element from three analysis runs per sample (Lee and Sneddon 1994).

10.2.6 Data Treatment, Inter-method Comparability and Statistics

Provenance studies in the Pacific are largely based on the use of multivariate statistical analyses to identify similarities and dissimilarities between the chemical compositions of different sources. Several different methods have been employed: Principal components analysis (PCA), correspondence analysis (CA), various forms of cluster analyses (K-means, hierarchical, etc...), Principal component regression (PCR), discriminant analysis, Popper's Razor, and others (Baxter et al. 2006; Leach and Manly 1982). In this study, for a first assessment of the data structure, unsupervised multivariate statistical analyses (PCA and PCR) were performed (Juggins 2005; SPSS 2006) on log (base 10) transformed absolute concentrations (ppm) for all measured elements.

Multivariate statistical analyses are very helpful in re-arranging complex datasets to display underlying similarities and dissimilarities. They can, however, occasionally obscure information as their main aim is to reduce the complexity of multi-dimensional data. A number of statistical analyses employed in the past assume that the chance of correctly identifying a source rises with the amount of conformity between the chemical composition of the analyzed artifact and the source (Goffer 2007; Pollard et al. 2007; Wilson and Pollard 2001). While this assumption is correct, it is restricted by the amount of intra-source variation. If dissimilarity between intra-source variation and artifact is too high, this approach leads to possibly erroneous interpretations resulting in flawed source associations (e.g. Reepmeyer and Clark 2010; Weisler 1998). Multivariate statistical analyses are therefore always supported by the examination of several diagnostic elements and element ratios.

10.2.7 Mantle Normalized Trace Element Patterns

Supplementary tools applied in this work are mantle normalized trace element diagrams (spider-diagrams, Fig. 10.2a–e). These are tools commonly used by geochemists to display relative elemental abundances in specific rocks for the identification of fractionation processes in the melt (Sun and McDonough 1989). Mantle normalized trace element diagrams are typically arranged such that incompatible elements (elements which tend to not fractionate easily in the melt) are plotted to the left of the diagram and in decreasing order of their incompatibility to the right. Trace element distributions can be used to compare igneous rocks from different volcanic regions with each other. For example, it is well accepted now that island arc magmas are systematically depleted in Ta, Nb, Zr, and Ti (HFSE, high-field strength elements)[2] (Albarède 2003). This produces a trace element distribution which is easy to distinguish from Oceanic Island Basalt (OIB), whose genesis is not based on subduction of two or more tectonic plates, but rather on weak zones (hot spots) in Oceanic plates producing intra-plate volcanism. Igneous rocks formed by this process are on average less depleted in their trace elements (Albarède 2003).

10.3 Results

10.3.1 Admiralty Islands

Two major source locations in the Admiralty Islands (Manus Province) are known: southwest Manus and the Lou/Pam Islands (Ambrose et al. 1981a, b; Ambrose and Duerden 1982). The dataset available for this study includes five samples from Manus and 13 samples from Lou and Pam sources. The Lou sources contain 72–74 wt% Si-oxides, which is lower than the neighboring West New Britain obsidian sources. They are relatively high in Fe-oxides at around 2.0 wt%, and have a similar alkalinity as the West Fergusson sources in the D'Entrecasteaux Islands. In contrast, the Manus source contains a very high SiO_2 content of >77 wt% and an exceptionally low amount of Fe, with only 0.5 wt%. The trace element pattern shows a clear island arc geochemical composition (Fig. 10.2a) with a strong enrichment in incompatible elements, particularly Ba–Ta. A clear distinction between the Lou sources and the Manus source is observed in the concentrations of HFSEs, particular Ti and Zr, and selected lithophile elements (LILEs) such as Sr and Eu.

PCR on the source samples defined P (PCA Factor loading: Score one −0.294/Score two −2.228), Ti (−0.707/−0.292), V (−1.52/0.477), Co (−1.693/−0.171), Th (−0.538/0.799) and U (−0.027/0.235) as elements with a high ranking in discriminating these source locations from each other. PCA (with the first component explaining 97.7 % of the variance and the second component explaining 1 % of the variance) on the reduced elemental data shows an

[2] Large-ion lithophile elements (LILEs) and high-field-strength elements (HFSEs) are two categories of elements in geochemistry. Both groups have the tendency to stay in liquid phase during fractionation process in the melt. They are classified by the ratio of their electrostatic potential to their ionic radius, which is informative about the origin of rocks, e.g. abundance of LILEs would indicate continental crusts (Albarède 2003:18).

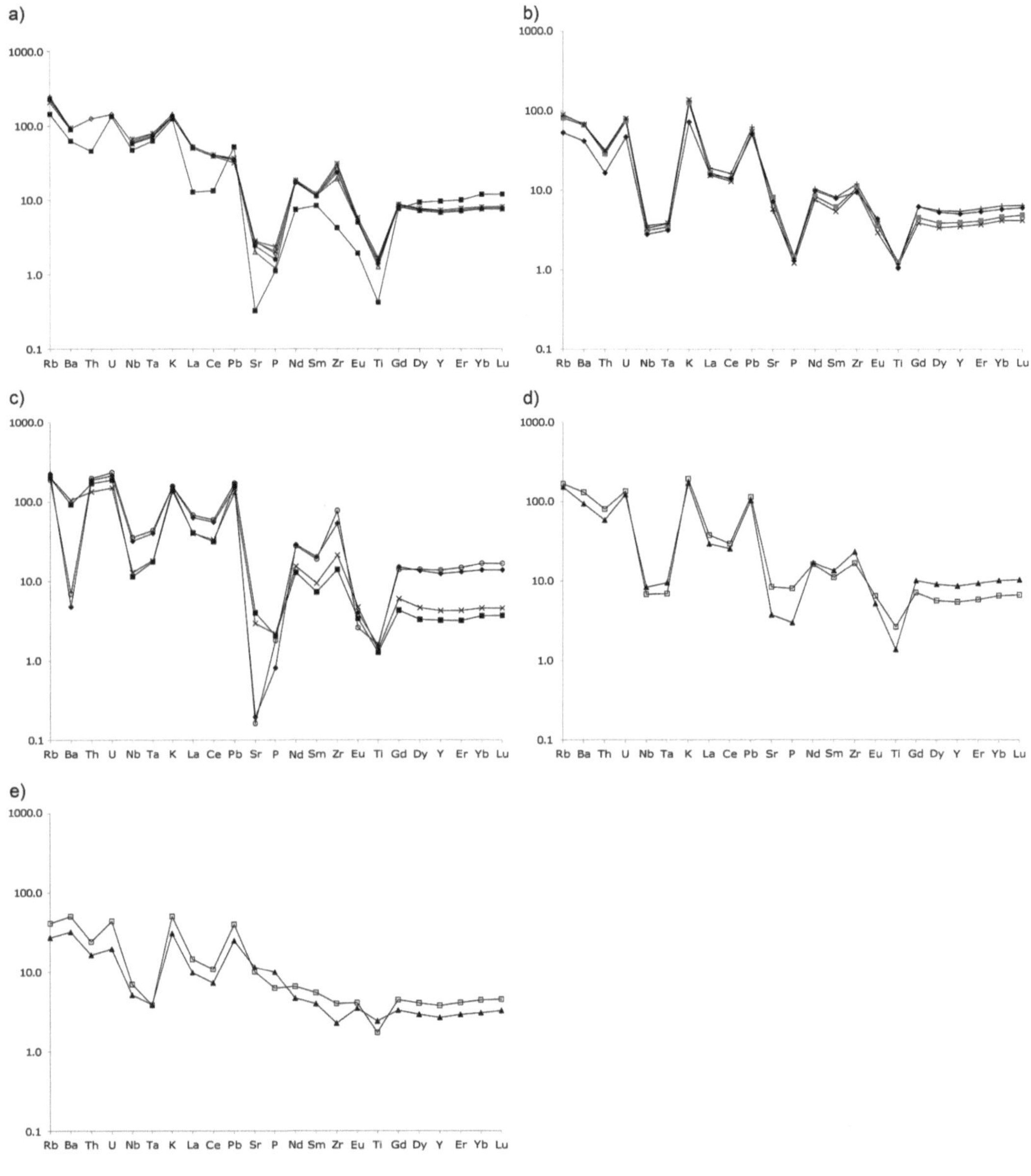

Fig. 10.2 Incompatible element diagrams, primitive mantle normalized (Sun and McDonough 1989), of Western Pacific obsidian sources (a) Admiralty Islands sources, b) West New Britain sources, c) D'Entrecasteaux Islands sources, d) Northern Vanuatu sources, e) Tonga sources)

unambiguous distinction between the Lou and Manus sources (Fig. 10.3). Unfortunately, the sub-sources of Lou and Pam are difficult to distinguish as there seems to be considerable variability in their geochemistry, as previously reported in the literature (Ambrose 1976; Bird 1996; Duerden et al. 1987). Focusing on selected trace elements such as Ga and Zr, but also on major elements like the silicate or K_2O content (Tables 10.1, 10.2, and 10.3), the sources Pam, Wekwok, Umrei/Umleang and Baun can tentatively be separated from each other.

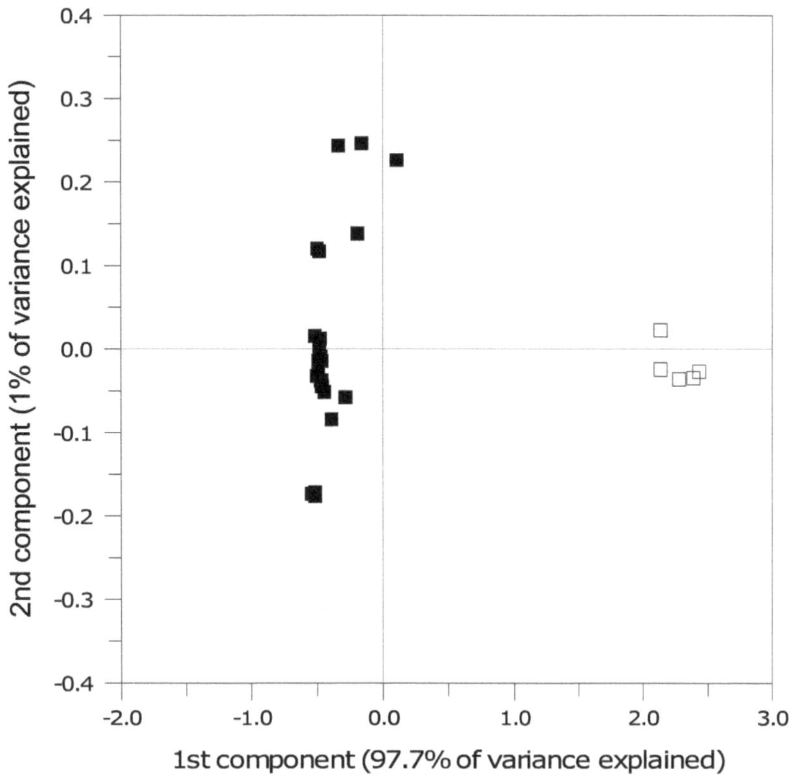

Fig. 10.3 Principal component analysis of source samples from the Lou sources (*filled square*) and Manus source (*open square*), Admiralty Islands

10.3.2 West New Britain

In West New Britain (WNB), five well-researched obsidian sub-sources in two distinct locations are known. Four sub-sources are located on the Willaumez Peninsula and adjacent Garua Island, previously grouped under the term "Talasea." These sources can be separated into four distinct chemical subgroups: Kutau/Bao, Gulu, Baki and Hamilton (Torrence 2004; Torrence et al. 1992, 1996). The second location is the Mopir obsidian source, situated approximately 50 km further east (Fullagar et al. 1991; Summerhayes and Allen 1993). The present dataset includes 20 samples from the Kutau/Bao sub-source, 1 sample from Gulu, 3 samples from the Baki sub-source, and 3 samples from the Mopir source. WNB sources are characterized by their high content of Si-oxides (>75.5 wt%). With the exception of Mopir (2.2 wt%), those from New Britain have, like most PNG sources, high alkalinity (~4 wt% K_2O), as well as low Fe content.

A PCR of 27 samples from four of the five West New Britain sub-sources indicates that K (−1.062/0.816), P (−0.174/−0.159), V (−1.938/−1.468), Mn (0.57/−0.282), Rb (−0.811/0.737), Sr (−0.22/−1.793), Th (−1.106/0.859) and U (−0.911/0.761) rank highest in distinguishing these sub-sources from one other. A PCA of the reduced dataset shows a discrete distribution of the four sub-sources—the first component represents 79.7 % of the data variance, the second 14.6 % (Fig. 10.4). Separating the Kutau/Bao and Gulu sub-sources from each other is difficult. At this stage it is tentatively suggested that the Gulu sources show a slight enrichment in Ga and a stronger depletion in Ti, Mn and Sr than the Kutau/Bao source (Fig. 10.2b and Table 10.2), but more source samples are needed to verify this trend.

Table 10.1 Concentrations measured in Western Pacific source samples by SEMEDXA

Source area	Source	n		Na_2O	MgO	Al_2O_3	SiO_2	K_2O	CaO	TiO_2	MnO	FeO	Totals
Admiralty Islands	Wekwok	5	%	4.84	0.50	13.77	73.18	4.03	1.12	0.32	0.08	1.97	99.38
			SD	0.00	0.37	0.42	0.53	0.02	0.07	0.03	0.00	0.02	
	Umrei	3	%	5.09	0.55	14.65	71.87	3.95	1.32	0.32	0.08	2.36	100.18
			SD	0.09	0.06	0.00	0.18	0.00	0.03	0.01	0.06	0.07	
	Baun	3	%	4.99	0.19	14.68	72.60	4.07	1.13	0.30	0.03	1.99	100.02
			SD	0.11	0.15	0.41	0.90	0.04	0.10	0.04	0.01	0.45	
	Pam	2	%	4.65	0.26	13.85	73.33	4.31	0.93	0.24	0.11	1.85	99.64
			SD	0.07	0.09	0.04	0.01	0.08	0.04	0.01	0.04	0.06	
	Manus	5	%	4.83	0.06	13.73	77.05	3.76	0.29	0.08	0.21	0.48	100.54
			SD	0.05	0.11	0.10	0.30	0.14	0.03	0.01	0.06	0.04	
West New Britain	Kutau/Bao	14	%	3.90	0.27	12.84	76.27	3.88	1.17	0.24	0.07	1.08	99.59
			SD	0.14	0.13	0.35	0.75	0.12	0.06	0.04	0.07	0.15	
	Gulu	1	%	3.60	0.01	12.35	76.10	4.14	1.12	0.23	0.11	1.09	98.56
	Baki	3	%	4.05	0.24	12.95	75.48	4.04	1.05	0.26	0.06	1.26	99.40
			SD	0.07	0.11	0.42	0.27	0.07	0.07	0.05	0.03	0.06	
	Mopir	3	%	4.39	0.22	12.05	77.46	2.19	1.25	0.24	0.07	1.14	99.21
			SD	0.05	0.05	0.10	0.18	0.04	0.04	0.02	0.06	0.10	
D'Entrecasteaux	Fagalulu	6	%	5.00	0.11	14.79	74.03	4.12	0.81	0.25	0.07	1.04	100.14
			SD	0.01	0.02	0.10	0.04	0.00	0.00	0.02	0.07	0.01	
	Kukuila	2	%	5.32	0.21	15.32	72.78	4.42	0.70	0.26	0.03	1.24	100.32
			SD	0.08	0.17	0.12	0.27	0.03	0.05	0.01	0.09	0.24	
	Sanoro	2	%	5.89		13.34	71.96	4.71	0.38	0.29	0.14	3.93	100.61
			SD	0.00		0.06	0.04	0.01	0.01	0.01	0.12	0.01	
	Dobu	2	%	6.31	0.08	14.70	71.05	4.77	0.36	0.31	0.11	2.95	100.63
			SD	0.03	0.08	0.47	0.45	0.06	0.01	0.01	0.04	0.19	
	Lamonai	3	%	6.27	0.18	14.53	70.90	4.87	0.34	0.38	0.05	3.18	100.86
			SD	0.11	0.08	0.47	0.60	0.09	0.03	0.08	0.07	0.12	

Vanuatu	Gaua	14	%	5.66	0.62	16.71	64.60	5.81	1.59	0.56	0.15	3.51	99.31
			SD	0.07	0.04	0.05	0.17	0.07	0.05	0.02	0.04	0.05	
	Vanua Lava	14	%	4.48	0.23	13.92	70.83	5.15	1.13	0.31	0.12	2.96	99.21
			SD	0.10	0.06	0.10	0.33	0.23	0.16	0.04	0.05	0.20	
Tonga	Tafahi A	4	%	3.43	0.54	11.71	74.38	1.52	3.10	0.36	0.14	4.47	99.88
			SD	0.07	0.09	0.09	0.31	0.02	0.09	0.06	0.07	0.12	
	Tafahi B	1	%	2.54	2.61	13.73	63.03	0.93	6.96	0.55	0.28	9.22	100.15
Standards	ANU2000	18	%	4.86	0.25	13.51	73.23	4.00	1.09	0.33	0.05	1.98	99.38
			SD	0.07	0.03	0.12	0.26	0.09	0.04	0.05	0.05	0.10	
	ANU9000	20	%	3.94	0.23	12.59	76.30	3.80	1.14	0.25	0.05	1.19	99.59
			SD	0.06	0.03	0.12	0.30	0.06	0.03	0.05	0.04	0.09	

10.3.3 D'Entrecasteaux Islands

Very little research has been done on the obsidian sources of the D'Entrecasteaux Islands. Previous research detected six sub-sources (Ambrose 1976:369; Bird et al. 1981a; Duerden et al. 1987; Smith 1974) separating West and East Fergusson Islands (Green and Bird 1989; White et al. 2006). West Fergusson sources were associated with obsidian outcrops on the Kukuia Peninsula and two sub-sources were detected there, Fagalulu and Kukuia (samples were taken from secondary deposits close to the village Ioupolo and beach deposits at the Igwageta village). Ambrose achieved a good discrimination of these sources based on geochemical analysis of U and additional density measurements (Ambrose 1976:371, Fig. 3; Summerhayes 2004).

The two source locations of West and East Fergusson have a distinctive element patterns, with the eastern sources containing generally lower SiO_2 concentrations, and showing a strong alkalinity with much higher Fe content. Additionally, the East Fergusson sources are very depleted in LILEs (Sr, Ba) and enriched in HFSEs (Ti, Zr). Both source locations are enriched in selected incompatible elements (Th, U) and Pb, suggesting these elements might be good indicators for finger-printing artifacts from the D'Entrecasteaux Islands (Fig. 10.2c).

Statistical analysis revealed a good possibility for discrimination of the two West Fergusson sources using V (−1.599/−1.238), Sr (−2.186/0.158), Y (0.926/−0.174), Mo (0.398/−1.372), Cs (0.176/−1.158), Gd (0.785/0.093), Dy (0.928/−0.022), Pb (0.116/−1.478), Th (0.232/−1.191) and U (0.261/−1.255). A PCA displays separation of these two source locations, with the first component explaining 95.3 % of the data variance and the second component 3.6 % (Fig. 10.5). Two outliers from both outcrops could indicate a second sub-source in this area. Both these samples showed significantly lower P, Sc, Ti, V, Mn, Ga, Rb, Zr, Nb, Ba, Pb and Th concentrations and higher Sn content.

Table 10.2 Concentrations measured in Western Pacific source samples by LA-ICP-MS

Source area	Source	n		P	Sc	Ti	V	Mn	Co	Cu	Ga	Rb	Sr	Y	Zr	Nb	Mo	Sn	Cs
Admiralty Islands	Wekwok	5	ppm	192.41	6.12	2072.97	5.86	472.16	1.44	3.57	63.06	152.37	60.75	33.04	294.10	43.73	3.65	3.31	2.04
			SD	7.74	0.35	96.56	0.24	21.99	0.09	0.36	6.45	9.59	5.13	2.41	22.83	1.70	0.15	0.26	0.11
	Umrei	3	ppm	193.88	5.42	1929.23	3.98	446.73	1.64	4.22	46.06	132.13	57.92	30.99	321.06	41.11	3.44	4.67	1.75
			SD	24.66	0.75	210.85	0.34	58.62	0.26	0.39	18.50	13.24	2.33	1.90	20.22	5.16	0.37	1.14	0.18
	Baun	3	ppm	152.16	5.51	1822.10	3.44	457.00	1.32	3.83	54.92	143.38	51.23	30.81	262.85	41.54	3.54	4.03	1.93
			SD	63.14	0.62	354.73	0.95	58.39	0.44	0.64	17.32	8.99	5.53	1.99	69.10	6.08	0.39	1.32	0.12
	Pam	2	ppm	115.50	6.12	1655.00	4.45	451.50	1.18	3.53	69.35	157.50	42.80	32.45	218.50	44.75	3.34	3.36	2.04
			SD	4.95	0.11	63.64	1.05	30.41	0.07	0.23	1.48	3.54	7.21	0.78	2.12	0.07	0.01	0.03	0.04
	Manus	5	ppm	106.60	7.08	543.80	0.33	916.80	0.06	0.34	50.16	91.70	6.84	43.74	47.46	33.82	2.63	2.14	1.41
			SD	1.14	0.14	21.58	0.05	33.24	0.01	0.03	2.08	3.24	2.56	1.43	1.40	0.52	0.06	0.07	0.04
West New Britain	Kutau/Bao	14	ppm	138.50	5.93	1581.00	6.05	444.00	0.66	3.63	10.73	52.90	163.70	17.40	117.40	2.23	3.23	1.84	1.55
			SD	10.10	0.26	41.00	0.74	48.00	0.09	1.11	0.10	2.70	20.80	0.86	5.10	0.06	0.12	1.89	0.10
	Gulu	1	ppm	116.00	5.66	1490.00	7.47	342.00	0.82	5.26	46.80	56.90	122.00	16.00	112.00	2.36	3.40	0.77	1.73
	Baki	3	ppm	132.46	7.77	1552.84	3.31	460.47	0.56	2.70	34.95	56.04	115.89	24.68	133.64	2.55	3.40	1.15	1.67
			SD	0.50	0.24	6.25	0.03	7.90	0.01	0.14	20.45	0.62	1.65	0.18	0.56	0.01	0.05	0.02	0.02
	Mopir	3	ppm	123.65	1.23	1370.27	2.26	589.20	0.47	2.68	11.27	33.88	151.58	22.88	105.13	2.00	3.76	0.88	1.31
			SD	0.24	0.02	5.60	0.04	5.39	0.02	0.02	0.09	0.30	0.91	0.20	0.73	0.02	0.05	0.02	0.02
D'Entrecasteaux	Fagalulu	6	ppm	197.00	3.87	1655.00	7.10	368.50	0.47	1.03	65.85	128.50	84.35	14.65	156.00	8.14	3.73	1.86	5.14
			SD	1.41	0.13	21.21	0.04	0.71	0.00	0.03	0.92	0.71	0.07	0.07	0.00	0.00	0.05	0.01	0.04
	Kukuila	2	ppm	206.40	4.72	1782.00	4.25	351.00	0.44	0.94	73.88	129.00	61.98	19.34	239.40	9.18	3.10	2.27	4.14
			SD	1.52	0.10	16.43	0.07	13.58	0.03	0.24	1.09	0.71	0.97	0.25	1.67	0.08	0.07	0.08	0.05
	Sanoro	2	ppm	77.35	6.20	1985.00	0.50	1015.00	0.19	1.85	25.95	143.50	4.09	56.25	601.50	23.00	5.05	5.70	5.69
			SD	0.78	0.05	7.07	0.01	7.07	0.01	0.08	0.07	0.71	0.02	0.49	3.54	0.14	0.06	0.09	0.01
	Dobu	2	ppm	157.10	7.11	2054.00	0.76	753.00	0.47	3.59	28.00	127.30	3.80	64.63	867.70	24.85	5.76	8.60	4.95
			SD	19.70	0.00	5.00	0.13	68.00	0.09	0.26		7.40	0.60	3.15	1.90	1.06	0.22	1.84	0.29
	Lamonai	3	ppm	170.70	7.35	2066.00	0.42	733.00	0.47	3.77		122.10	2.60	67.04	929.60	25.46	5.96	8.23	4.71
			SD	1.00	0.36	41.00	0.04	7.00	0.01	0.29		3.80	0.20	4.15	55.10	0.06	0.12	1.52	0.09
Vanuatu	Gaua	14	ppm	765.82	6.13	3452.07	21.42	1078.58	3.69	21.46		106.56	177.04	24.73	189.88	4.86	4.27	1.38	1.57
			SD	19.55	0.06	15.18	0.30	5.70	0.04	3.23		1.15	3.35	0.11	0.72	0.03	0.03	0.02	0.01
	Vanua Lava	14	ppm	284.48	9.88	1796.38	1.42	920.49	1.34	15.30		96.94	79.14	39.27	261.79	5.94	5.15	1.96	2.24
			SD	12.49	0.27	38.96	0.34	35.75	0.14	2.83		2.43	6.76	0.47	1.89	0.05	0.06	0.02	0.02
Tonga	Tafahi A	4	ppm	594.97	17.41	2252.12	19.17	907.15	6.03	51.82	11.22	26.13	212.16	17.17	44.77	4.98	2.51	0.57	0.67
			SD	14.19	0.13	47.09	0.61	8.52	0.04	1.71	0.74	0.60	22.93	0.43	1.31	0.12	0.09	0.01	0.02
	Tafahi B	1	ppm	946.88	34.56	3134.53	239.94	1426.08	21.79	90.48	13.08	17.27	241.79	12.15	25.43	3.67	1.53	0.37	0.49
Standards	ANU2000	18	ppm	183.59	6.53	2013.49	5.65	452.52	1.39	3.53	17.50	148.25	58.91	32.18	284.43	42.47	3.52	3.20	1.99
			SD	12.78	0.76	55.49	0.08	11.37	0.04	0.49	0.33	5.30	2.65	1.32	11.59	0.87	0.06	0.16	0.07
	ANU9000	20	ppm	138.76	6.47	1624.84	5.65	453.99	0.61	3.07	10.73	51.68	170.31	17.81	119.21	2.22	3.07	0.87	1.50
			SD	7.41	0.77	43.52	0.08	13.15	0.04	0.20	0.10	1.76	7.48	0.68	4.43	0.04	0.06	0.15	0.06

Table 10.3 Concentrations measured in Western Pacific source samples by LA-ICP-MS

Source area	Source	n		Ba	La	Ce	Nd	Sm	Eu	Gd	Dy	Er	Yb	Lu	Ta	W	206Pb	207Pb	208Pb	Th	U
Admiralty Islands	Wekwok	5	ppm	666.55	37.10	74.66	25.01	5.30	0.90	5.23	5.61	3.64	3.93	0.60	3.17	1.32	7.03	6.43	6.73	10.76	3.02
			SD	52.67	3.14	6.54	2.23	0.48	0.08	0.38	0.43	0.27	0.27	0.04	0.18	0.08	0.39	0.37	0.36	1.04	0.25
	Umrei	3	ppm	622.29	34.53	69.24	23.86	5.05	0.91	4.91	5.30	3.43	3.72	0.56	2.91	1.27	6.14	5.62	5.88	9.58	2.67
			SD	30.72	1.73	3.53	1.28	0.27	0.05	0.32	0.26	0.22	0.23	0.04	0.28	0.11	0.61	0.58	0.61	0.56	0.17
	Baun	3	ppm	624.35	34.66	69.40	23.30	5.00	0.85	4.83	5.27	3.40	3.73	0.56	2.98	1.29	6.77	6.21	6.45	10.16	2.82
			SD	20.99	1.10	2.69	1.06	0.28	0.06	0.35	0.28	0.21	0.28	0.04	0.30	0.11	0.56	0.55	0.51	0.26	0.09
	Pam	2	ppm	643.00	35.90	72.95	24.45	5.15	0.86	5.07	5.55	3.54	3.88	0.57	3.25	1.26	6.63	6.19	6.37	10.85	3.03
			SD	9.90	0.71	1.77	0.78	0.25	0.02	0.11	0.14	0.10	0.01	0.01	0.00	0.02	0.01	0.00	0.08	0.07	0.00
	Manus	5	ppm	439.00	8.80	23.66	10.07	3.75	0.32	4.57	6.86	4.78	5.86	0.88	2.58	1.09	10.20	9.25	9.67	3.89	2.81
			SD	23.65	0.82	1.51	0.59	0.14	0.05	0.17	0.22	0.16	0.21	0.03	0.06	0.02	0.19	0.20	0.18	0.21	0.08
West New Britain	Kutau/Bao	14	ppm	469.00	11.15	23.94	11.14	2.65	0.56	2.60	2.79	1.94	2.24	0.35	0.14	0.38	10.48	9.59	9.89	2.49	1.59
			SD	30.00	0.73	1.32	0.61	0.16	0.07	0.15	0.15	0.09	0.10	0.02	0.01	0.03	0.22	0.25	0.19	0.19	0.11
	Gulu	1	ppm	471.00	10.60	23.00	10.30	2.41	0.50	2.31	2.50	1.78	2.05	0.31	0.16	0.42	10.80	10.10	10.20	2.71	1.68
	Baki	3	ppm	477.62	13.00	28.70	14.03	3.62	0.68	3.69	4.07	2.81	3.14	0.48	0.16	0.40	12.19	11.01	11.56	2.60	1.66
			SD	6.04	0.27	0.46	0.41	0.06	0.00	0.07	0.08	0.05	0.04	0.00	0.00	0.00	0.12	0.09	0.13	0.02	0.02
	Mopir	3	ppm	292.62	10.78	25.17	13.33	3.52	0.73	3.69	3.88	2.58	2.85	0.45	0.13	0.44	9.86	9.05	9.38	1.40	0.99
			SD	2.12	0.03	0.26	0.12	0.04	0.01	0.04	0.05	0.01	0.05	0.01	0.01	0.01	0.04	0.10	0.15	0.01	0.01
D'Entrecasteaux	Fagalulu	6	ppm	649.00	28.05	56.50	17.50	3.25	0.57	2.57	2.44	1.54	1.81	0.27	0.72	1.30	32.30	29.80	30.80	14.55	3.98
			SD	2.83	0.07	0.00	0.00	0.02	0.00	0.01	0.01	0.02	0.01	0.00	0.01	0.01	0.71	0.28	0.42	0.07	0.01
	Kukuila	2	ppm	735.00	27.74	59.36	20.84	4.19	0.79	3.59	3.41	2.05	2.25	0.34	0.75	1.06	25.58	23.58	24.44	11.42	3.16
			SD	3.94	0.38	0.76	0.25	0.10	0.05	0.03	0.04	0.02	0.05	0.01	0.01	0.01	0.42	0.23	0.33	0.13	0.03
	Sanoro	2	ppm	33.30	43.60	99.30	39.10	9.09	0.68	8.98	9.95	6.27	6.80	1.02	1.66	1.59	29.10	26.75	27.80	16.00	4.50
			SD	0.00	0.28	0.57	0.42	0.01	0.00	0.10	0.06	0.01	0.04	0.01	0.01	0.00	0.00	0.21	0.14	0.14	0.01
	Dobu	2	ppm	49.00	49.92	111.94	41.10	9.16	0.52	8.86	10.75	7.31	8.35	1.25	1.77	1.43	32.24	29.93	30.81	17.47	5.04
			SD	1.00	4.26	8.40	4.38	0.91	0.11	0.68	0.63	0.29	0.06	0.01	0.05	0.08	1.36	1.52	1.41	0.94	0.16
	Lamonai	3	ppm	79.00	51.34	115.13	41.61	9.28	0.53	8.95	11.07	7.60	1.31	8.72	1.83	1.36	32.45	29.86	30.94	17.54	5.16
			SD	5.00	3.95	8.24	3.13	0.54	0.02	0.44	0.61	0.36	0.06	0.36	0.04	0.05	0.36	0.10	0.21	0.98	0.23
Vanuatu	Gaua	14	ppm	916.19	26.05	52.70	22.17	4.93	1.09	4.26	4.18	2.80	3.22	0.49	0.28	0.64	21.79	20.49	21.22	6.85	2.83
			SD	12.58	0.09	0.25	0.10	0.04	0.01	0.03	0.04	0.02	0.03	0.01	0.00	0.01	0.25	0.25	0.22	0.03	0.02
	Vanua Lava	14	ppm	657.09	20.29	45.49	23.10	6.00	0.87	6.01	6.67	4.48	4.98	0.77	0.39	0.73	19.97	18.63	19.17	4.99	2.57
			SD	7.61	0.15	0.39	0.26	0.06	0.03	0.08	0.08	0.06	0.07	0.01	0.01	0.01	0.20	0.17	0.19	0.04	0.02
Tonga	Tafahi A	4	ppm	350.46	10.00	19.27	8.90	2.44	0.69	2.66	2.98	1.97	2.18	0.33	0.16	0.87	7.69	6.86	7.37	2.05	0.92
			SD	8.70	0.26	0.48	0.20	0.06	0.03	0.10	0.06	0.08	0.06	0.01	0.00	0.01	0.17	0.23	0.20	0.08	0.03
	Tafahi B	1	ppm	223.60	6.75	12.97	6.34	1.77	0.59	1.96	2.15	1.40	1.51	0.24	0.16	0.32	4.90	4.32	4.62	1.40	0.41
Standards	ANU2000	18	ppm	644.43	36.12	72.35	24.43	5.15	0.88	5.18	5.42	3.54	3.83	0.58	3.08	1.28	6.80	6.20	6.47	10.65	3.01
			SD	37.99	1.74	3.56	1.32	0.24	0.04	0.20	0.26	0.16	0.18	0.03	0.12	0.04	0.30	0.28	0.28	0.82	0.21
	ANU9000	20	ppm	466.70	11.34	24.36	11.39	2.72	0.60	2.69	2.85	1.98	2.27	0.36	0.14	0.37	10.05	9.19	9.49	2.47	1.58
			SD	24.54	0.49	1.03	0.55	0.13	0.03	0.11	0.12	0.09	0.10	0.02	0.01	0.02	0.39	0.39	0.42	0.16	0.08

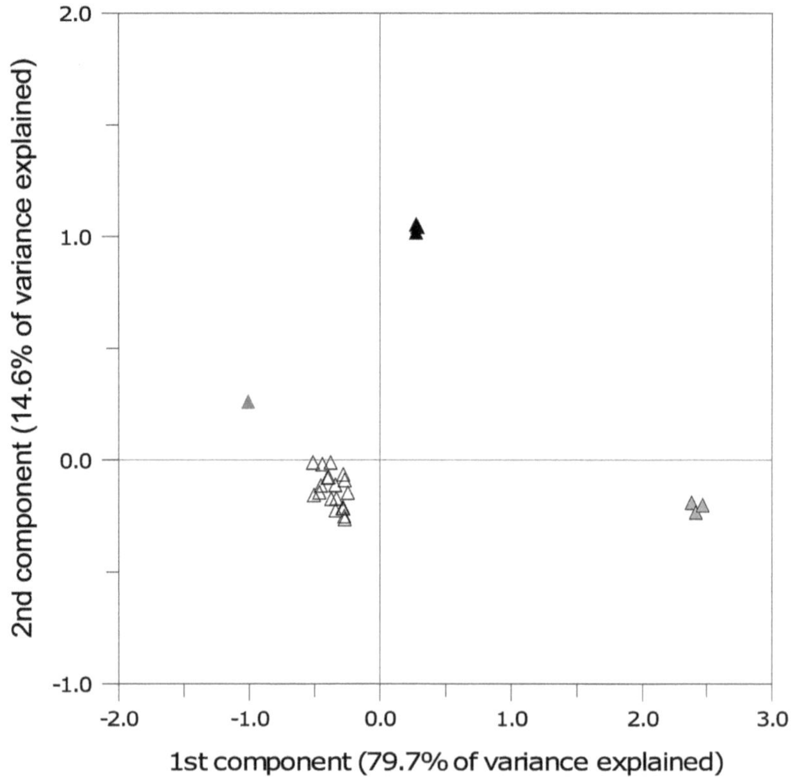

Fig. 10.4 Principal components analysis of source samples from the Kutau/Bao (*open triangle*), Baki (*filled triangle*), Gulu (*grey shaded triangle*) and Mopir (*grey shaded triangle* with *black border*) obsidian sources, West New Britain

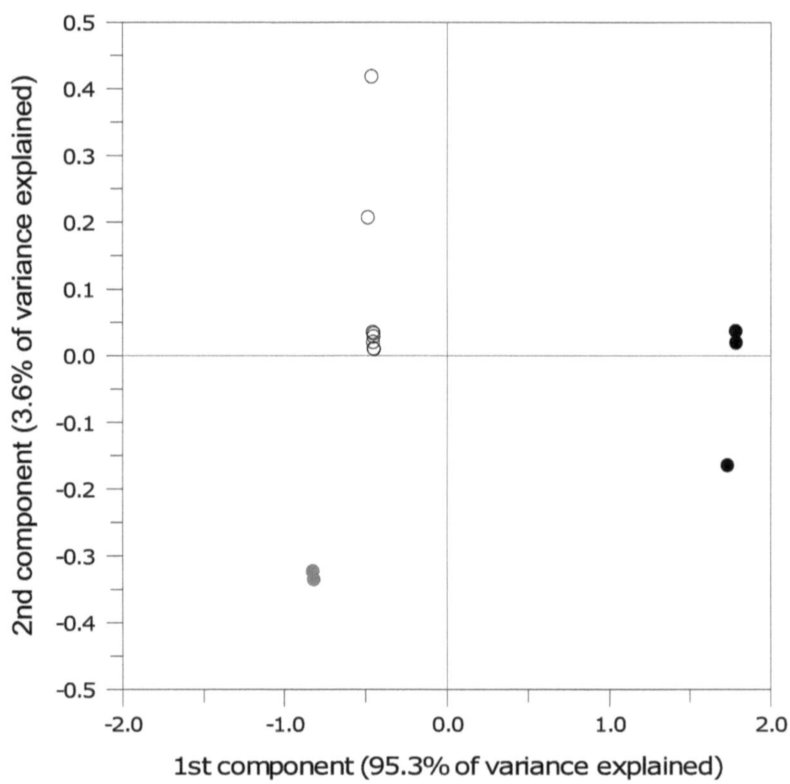

Fig. 10.5 Principal components analysis on West Fergusson (Igwageta/Iaupolu = *open circle*; Fagalulu = *grey shaded circle*) and East Fergusson sources (*filled circle*), D'Entrecasteaux Group

East Fergusson sources have been identified on Sanaroa Island, Dobu Island and at Lamonai. Whereas these sources are easy to separate from the West Fergusson sources using P, V, Ga, Sr and Ba, and indeed from all other Western Pacific obsidian sources, internal discrimination is ambiguous. All localities show a rather high variation in their chemical composition (SD of more than 5 %, which is higher than the expected machine variability), indicating difficulties in identifying specific outcrops.

10.3.4 Northern Vanuatu Obsidian Sources

In total, 28 samples collected from the two Vanuatu obsidian sources were analyzed to identify possible intra-source variation and to provide the basis for later comparison with artifacts from archaeological sites throughout southwest Remote Oceania. The two sources are readily distinguished from each other by different Si, Al and Na content. Vanua Lava material has, on average, a consistent SiO_2 content of more than 70.8 wt%, with 13.9 wt% Al_2O_3, 5.2 wt% K_2O and 4.5 wt% Na_2O; Gaua material ranges around only 64.6 wt% SiO_2, 16.7 wt% Al_2O_3, 5.8 wt% K_2O and 5.7 wt% Na_2O content. Both sources show high FeO concentrations of between 2.96 and 3.5 wt%. Consistent with the basic difference between these two sub-sources, a more silicate-rich (Vanua Lava) sub-source and silicate-poor (Gaua) sub-source, the titanium contents differ by a factor of two, the silicate-rich source being Ti poor (Fig. 10.2d). In addition, the more compositionally evolved silicate-rich Vanua Lava source has lower Rb and Sr content, but a distinctively higher Rb/Sr ratio. Especially distinctive is the high K_2O content of the northern Vanuatu sources, which matches the generally high alkaline geochemical composition of the Northern New Hebrides Arc (Barsdell et al. 1982; Raos and Crawford 2004).

PCR of the Vanuatu dataset identified K (−0.365/2.38), Rb (−0.283/−0.486), Zr (0.966/0.034), Sn (1.057/0.17), Er (1.415/0.220) and Yb (1.317.0.157) as particularly well suited to distinguish these samples from each other. PCA on the selected elements supports the detected differences in major element composition. The first two components explain 99.6 % of the variance, with the first component explaining 99 % of the total variance and the second component 0.6 % (Fig. 10.6). However, both sources show relatively high variability for Sr (expressed on PC2) which can be ascribed to fractionation processes in the melt, as plagioclase crystallization influences Sr distribution.

10.3.5 Tonga

Two chemically distinct sources on Tonga have been previously identified, both of which are located on the island of Tafahi in the northernmost part of the Tongan Island chain (Bird 1996; Duerden et al. 1987). In the present study, five source samples were analyzed from the island group, and two sub-sources were detected. Both are distinguishable from more Western Pacific obsidian sources by their low alkalinity (Fig. 10.2), especially low K_2O values (<1.5 wt%) (Smith and Price 2006:2325, Fig. 3) and high Fe, Ca and Mg content (Fig. 10.2e and Table 10.1). In comparison with the Oceanic obsidian sources of Samoa further north and the eastern Polynesian islands further east we can observe a higher depletion of all trace elements characteristic of island arc obsidian sources, compared to OIB trace element distribution (Albarède 2003) which is characteristic of Oceanic obsidian sources (Reepmeyer and Clark 2010; see Smith et al. (2003) for obsidian in the southern Tonga-Kermadec arc).

For the separation of these different sub-sources Zr (−1.22/0.725), Mo (−1.014/1.81), Ba (−1.068/−0.898, Nd (−0.811/−0.821), Sm (−0.724/−0.393) and Pb (−1.078/−0.751) were selected. A PCA (the first component accounts for 98.7 % of the variance and the second 1.1 %) shows an interesting distribution mainly on the first component, which is heavily influenced by W, Pb and U (Fig. 10.7). Additionally, strongly depleted values of Ba, Pb and Th

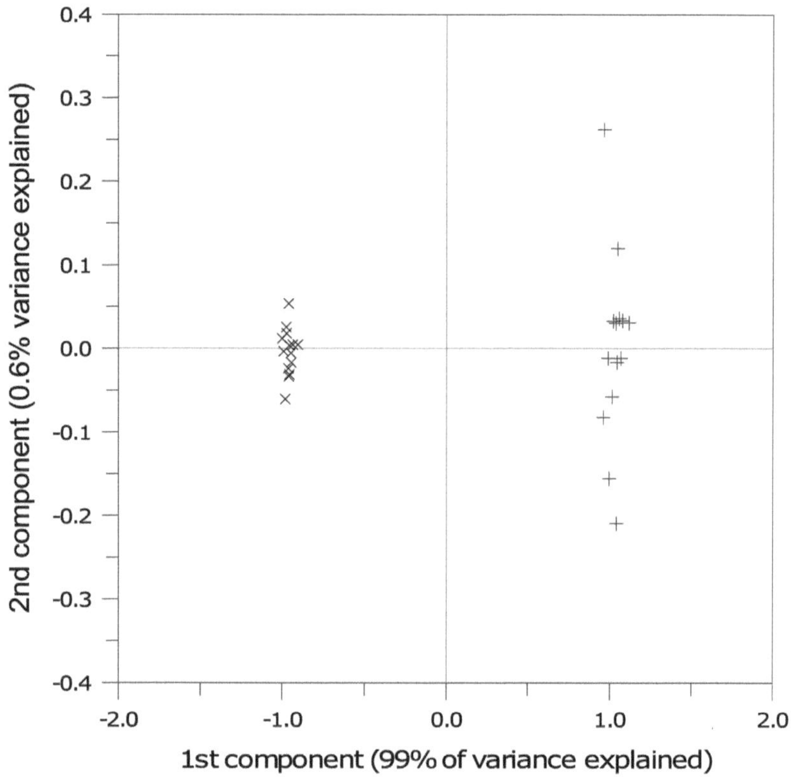

Fig. 10.6 Principal components analysis of selected elements for source samples from Vanua Lava (*plus symbol*) and Gaua (*multiplication symbol*), Banks Island, Vanuatux

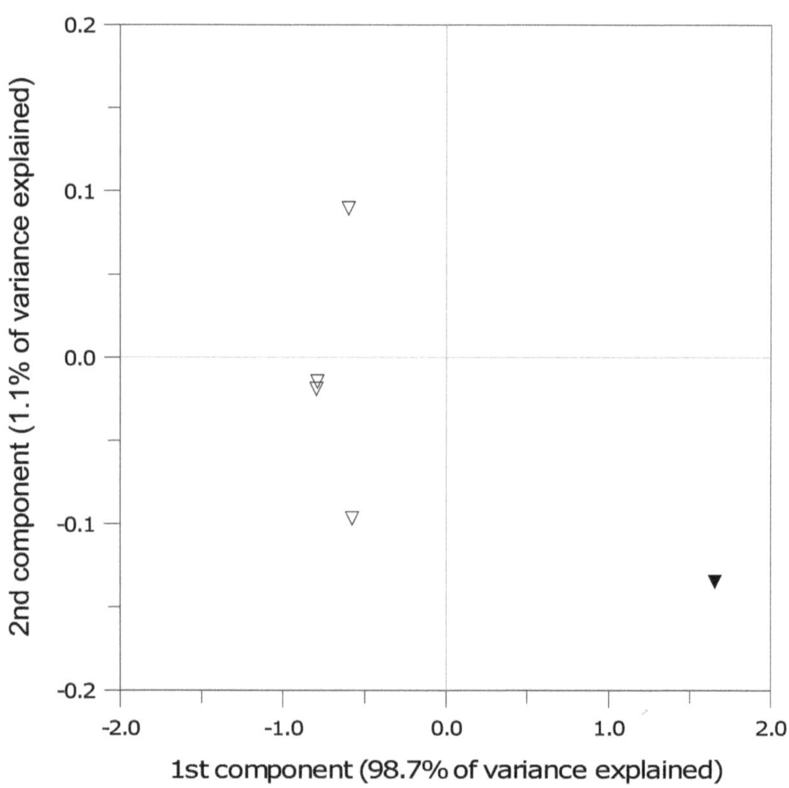

Fig. 10.7 Principal components analysis of selected elements for source samples from Tafahi, Tonga

support these samples as belonging to Tongan arc obsidian sources.

10.4 Discussion

Research on obsidian sources in the Western Pacific has shown that LA-ICP-MS is a useful tool for unambiguously finger-printing obsidian sources geochemically. Calibrating LA-ICP-MS with an external method has proven to be an efficient way to acquire high-resolution data and long-term stability in results (Longerich et al. 1996). Exemplary determination of inter-laboratory and inter-method comparability including analyses of Kutau/Bao obsidians from the Field Museum EAF (Golitko et al. 2010) based on a slightly different internal calibration protocol outlined in Gratuze et al. (2001) and NAA data from Missouri University Research Reactor (MURR) (Bird et al. 1997; Torrence personal communication) support the applicability of this calibration protocol (Fig. 10.8). Elements showing discrepancies in absolute concentration are limited to Nb, Sr, Sc and Th. Sc and Sr show additional high standard deviations which might suggest that there is an irregular distribution of certain trace elements in the matrix of Kutau/Bao obsidian. Nb and Th are both incompatible HFSEs with relatively low standard deviations. Inter-method comparability between NAA and ANU LA-ICP-MS is good for Th with only the Chicago ICP reporting higher values. Similar Nb values measured at both LA-ICP-MS laboratories show lower values than NAA results. At this stage it is unclear whether discrepancies in these two elements derive from machine induced errors or additional heterogeneity in the matrix of this obsidian sample.

While the ability to analyze small areas is a major advantage of micro-analytical techniques such as LA-ICP-MS in contrast to other methods such as NAA, XRF, PIXE-PIGME or solution ICP-MS, this brings with it new challenges to - inter-method comparability. One important observation stemming from analysis of the relatively low-silicate sources of northern Vanuatu is

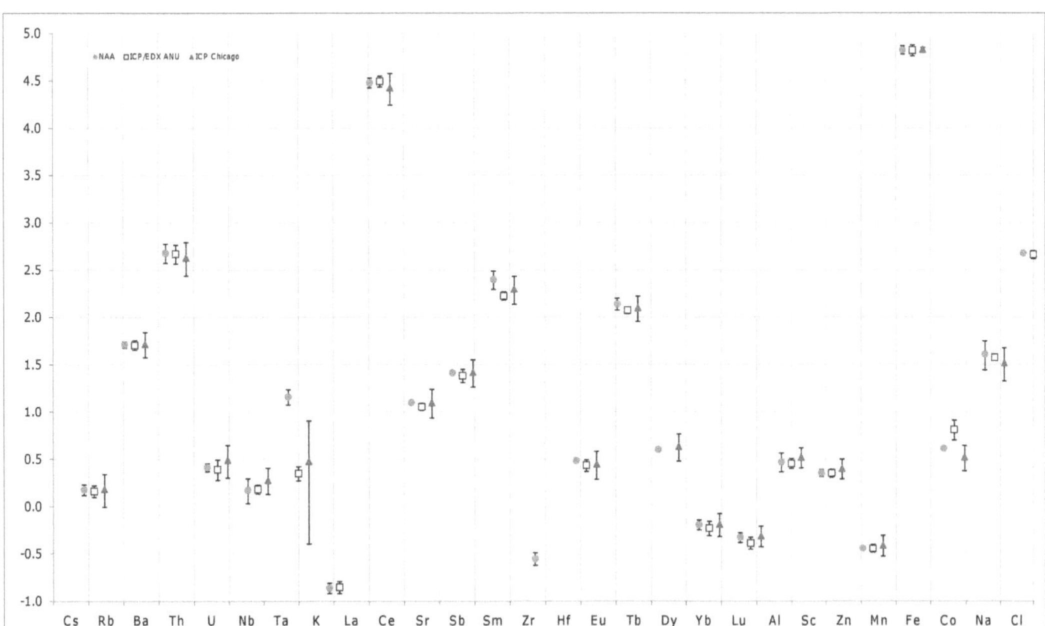

Fig. 10.8 Inter-laboratory comparison of the Kutau/Bao standard (ANU9000, Mt. Bao, W. New Britain) on selected elements (log10 transformed) with 2σ error bars

the occurrence of phenocrysts in the matrix of these raw materials. Some of these phenocrysts are macroscopically visible (mainly plagioclase, Ca-Na alumino-silicates), but several microphenocrysts also occur, mainly titano-magnite (Deer et al. 1992). Variable measured concentrations of Na, K, Ca, Ti and Fe result if differing volumes of glass, plagioclase and titano-magnetite are encountered in an ablation pit, especially if focused on large microphenocrysts (~100 μm). This indicates a problem with micro-analytical techniques in comparing data from true bulk analysis of rock specimens containing microphenocrysts where a detailed choice of the excited area is limited and therefore the average of a much larger sample volume is analyzed (Reepmeyer 2008).

High counts of crystallizations in obsidian sources, although problematic for inter-method comparability, provide further potential possibilities for fingerprinting Pacific obsidian sources. The analysis of microphenocrysts has been successfully applied in provenance studies in the Mediterranean (Acquafredda and Paglionico 2004; Kayani and McDonnell 1996), but is not well developed in the Pacific. For instance, the two different sources of Vanuatu volcanic glass show distinctive frequencies of microphenocrysts, with the Gaua source displaying microphenocrysts in only approximately 20 % of the sample, while microphenocrysts are present in 100 % of the material analyzed from the Vanua Lava source. No microphenocrysts can be found in samples from the Kutau/Bao source of West New Britain. However, other Bismarck Archipelago sources, for instance the Hamilton source on West New Britain, contain microphenocrysts (Torrence personal communication). Additionally, one piece from Wekwok, Admiralty Islands, displays small numbers of titano-magnetite microphenocrysts. At this stage the suggestion can be made that there is potential for fingerprinting obsidian sources in the Pacific by analyzing crystallizations in the material. Further research on a wider spectrum of obsidian sources is needed to identify the potential of microphenocrysts in sourcing artifacts.

Good precision and wide array of major, minor and trace elements analyzed by the combination of SEM-EDXA and LA-ICP-MS gives information not only for discriminating obsidian sources, but also their general geological setting. Primitive mantle normalized trace element diagrams show that all obsidian sources in the Western Pacific, besides the Admiralty Islands sources, have a geochemical composition typical of island arcs, confirming the origin of these sources in multiple subduction zones of the western Pacific Rim. Major-element distributions characterizing different island arcs, for example the low-K patterning of the Tongan-Kermadec arc, are mirrored in the composition of their obsidian sources. Selected by PCA factor loadings, four LREEs (La[−0.633/−0.956], Ce [−0.614/−0.872], Nd[−0.419/−0.880], Sm [−0.346/−0.585]) and four HFSEs (Zr[−0.57/ −1.569], Nb[−1.633/1.039], Ta[−1.777/0.946], Th[−0.883/−0.871]) prove to have the highest potential for discriminating the five major western Pacific source regions from each other, (Fig. 10.9). These elements are also useful for determining the origin of artifacts where a source cannot yet unambiguously be identified.

The potential of LA-ICP-MS to provide new data and understanding of social interaction can be seen in the study of artifacts from archaeological sites on Lakeba, in the Lau group of Eastern Fiji, which previously have been associated with obsidian outcrops on Vanuatu and Tonga (Reepmeyer and Clark 2010). New LA-ICP-MS results suggest that past provenancing of these artifacts to source locations in Vanuatu are incorrect, as these artifacts do not show trace element compositions typical of the New Hebrides arc. Additional radiogenic isotope data further suggests an unusual mixture of arc and OIB patterns excluding classical arc settings such as the Tongan-Kermadec arc (Reepmeyer and Clark 2010:12). These unusual geochemical patterns, however, are found in local formations on the main island of Fiji, Viti Levu, and a tentative source location, most likely on the southern side of the island, was proposed consequently.

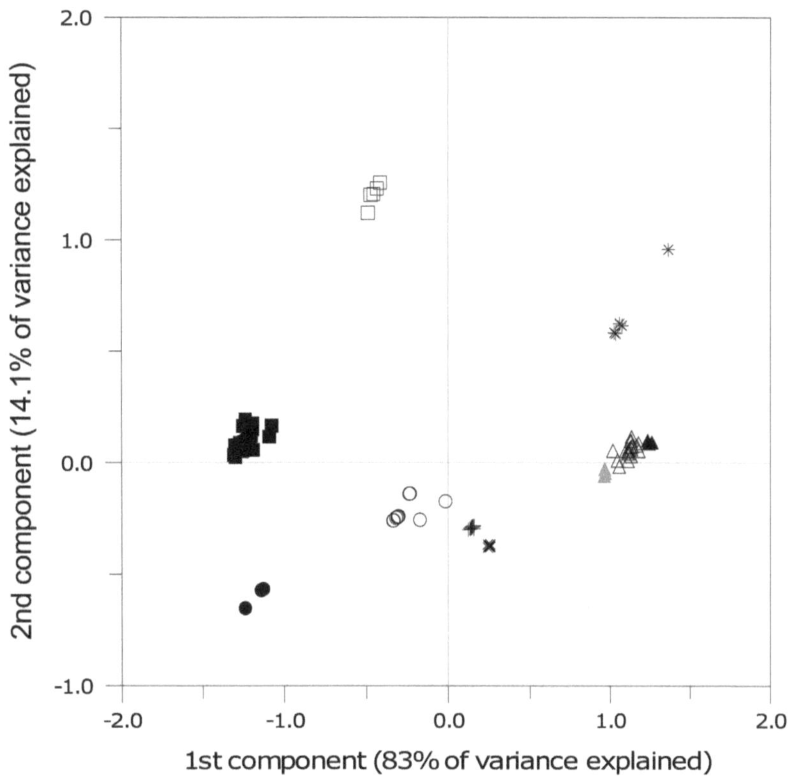

Fig. 10.9 Principal components analysis of selected elements for all source samples; Lou (*filled square*); Manus (*open square*); West New Britain (*open triangle*); Mopir (*grey shaded triangle*); West Fergusson (*open circle*); East Fergusson (*filled circle*); Vanua Lava (*plus symbol*); Gaua (*multiplication symbol*); and Tonga-Kermadec (*asterisk*) obsidian sources

10.5 Conclusion

LA-ICP-MS, particularly considering its low-cost, limited need for sample preparation, minimal size-limitations for analyzed artifacts, and rapidity of analysis, is an ideal geochemical technique for the analysis of small and heterogeneous archaeological assemblages. Benefiting from high precision and accuracy for the analysis of a wide range of major, minor, trace and rare-earth elements, and the resulting possibilities to compare geochemical data from archaeological contexts with the general geological background of the Pacific, enhances the potential for more targeted surveys aiming at the detection of utilized but previously unknown obsidian outcrops.

As shown, there is now a good understanding of the geochemistry of obsidian sources in the Western Pacific. Although not discussed in detail here, 510 obsidian artifacts from 21 sites in Vanuatu and neighboring islands (Reepmeyer 2009) were included in geochemical studies conducted at the ANU during the last decade, of which only 9 artifacts (1.8 %) could not be unambiguously provenanced. However, new research in the Pacific, advancing from the currently well-researched earliest colonization of the Pacific by Lapita people, increasingly focuses on small scale inter-island interactions between and within island groups. This shift of themes, from a narrative about the selection of obsidian sources utilized in the long-distance transport of raw material into Remote Oceania (which is generally agreed on, although in its details persistently contentious) to under-researched questions of development of social interaction in the post-colonization era, opens a new field for geochemical analysis of obsidian sources.

More comprehensive databases of hard evidence on a wider range of obsidian sources with a more detailed assessment of chronological changes in source use will further our understanding of patterns of obsidian transport over long distances and the perceived value of this raw material. In the introduction different propositions for the value of obsidian in prehistory have been described. However, there is currently no overwhelming evidence for any of the proposed options. Future research on extraction processes at sources and more detailed studies of use wear and site context of artifacts accompanying geochemical research on a wide range of archaeological materials will provide a deeper understanding of why past peoples transported and exchanged obsidian in the past.

References

Acquafredda P, Paglionico A (2004) SEM-EDS microanalysis of microphenocrysts of Mediterranean obsidians: a preliminary approach to source discrimination. Eur J Miner 16:419–429

Albarède F (2003) Geochemistry: an introduction. Cambridge University Press, Cambridge

Ambrose WR (1976) Obsidian and its prehistoric distribution in Melanesia. In: Barnard N (ed) Ancient Chinese bronzes and South East Asian metal and other archaeological artifacts. National Gallery of Victoria, Melbourne, pp 351–378

Ambrose WR, Duerden P (1982) PIXE analysis in the distribution and chronology of obsidian use in the Admiralty Islands. In: Ambrose WR, Duerden P (eds) Archaeometry: an Australasian perspective. Occasional papers in prehistory. RSPAS and ANU, Canberra, p 83–89

Ambrose WR, Johnson RW (1986) Unea: an Obsidian non-source in Papua New Guinea. J Polynesian Soc 95:491–497

Ambrose WR, Bird JR, Duerden P (1981a) The impermanence of obsidian sources in Melanesia. In: Leach BF, Davidson J (eds) Archaeological studies of Pacific stone resources. British Archaeological Reports: International Series, Oxford, pp 1–19

Ambrose WR, Duerden P, Bird JR (1981b) An archaeological Application of PIXE-PIGME analysis to Admiralty Islands obsidians. Nucl Instrum Method 191:397–402

Ambrose WR, Allen C, O'Connor S, Spriggs M, Oliveira NV, Reepmeyer C (2009) Possible obsidian sources for artefacts from Timor: narrowing the options using chemical data. J Archaeol Sci 36:607–615

Araho N, Torrence R, White JP (2002) Valuable and useful: mid-Holocene stemmed obsidian artefacts from West New Britain, Papua New Guinea. Proc Prehist Soc 68:61–81

Barsdell M, Smith IEM, Spoerli KB (1982) The origin of reversed geochemical zoning in the Northern New Hebrides volcanic arc. Contrib Mineral Petr 81:148–155

Baxter MJ, Cool HEM, Jackson CM (2006) Comparing glass compositional analyses. Archaeometry 48:399–414

Bellwood P, Koon P (1989) Lapita colonists leave boats unburned! The questions of Lapita links with Island South East Asia. Antiquity 63:613–622

Bird JR (1996) Pacific Obsidian Studies, Unpublished Report. ANSTO, Sydney

Bird JR, Ambrose WR, Russell LH, Scott MD (1981a) The characterisation of Melanesian obsidian sources and artefacts using the proton induced gamma ray emission (PIGME) technique, AAEC/E510. Australian Atomic Energy Commission, Lucas Heights (NSW)

Bird JR, Duerden P, Ambrose WR, Leach BF (1981b) Pacific obsidian catalogue. In: Leach BF (ed) Lithic resources of the Pacific Region, vol 104. British Archaeological Reports International Series (BAR), Oxford, pp 31–43

Bird JR, Torrence R, Summerhayes GR, Bailey G (1997) New Britain obsidian sources. Archaeol Ocean 32:61–67

Chayes F (1964) A petrographic distinction between Cenozoic volcanics in and around the open oceans. J Geophys Res 69:1573–1588

Deer WA, Howie RA, Zussman J (1992) An introduction to the rock-forming minerals, 2nd edn. Longman Scientific & Technical/Wiley, New York, NY

Duerden P, Clayton E, Bird JR, Ambrose WR, Leach BF (1987) Obsidian composition catalogue. In: Ambrose WR, Mummery JMJ (eds) Archaeometry: further Australasian studies. Occasional papers in prehistory. RSPAS and ANU, Canberra, pp 232–238

Eggins SM, Kinsley LPJ, Shelley JMG (1998) Deposition and element fractionation processes during atmospheric pressure laser sampling for analysis by ICP-MS. Appl Surf Sci 127:278–286

Falkner KK, Klinkhammer GP, Ungerer CA, Christie DM (1995) Inductively coupled plasma mass spectrometry in geochemistry. Annu Rev Earth Pl Sci 23:409–449

Fullagar R (1992) Lithically Lapita. Functional analysis of flaked stone assemblages from West New Britain Province, Papua New Guinea. In: Galipaud J-C (ed) Poterie Lapita et Peuplement. ORSTOM, Noumea, pp 135–143

Fullagar R (1993) Flaked stone tools and plant food production: a preliminary report on obsidian tools from Talasea, West New Britain, PNG. In: Anderson PC, Beyries S, Otte M, Plisson H (eds) Traces et Fonction: Les gestes retrouvés. ERAUL, Liège, pp 331–337

Fullagar R, Ambrose WR, Bird JR, Specht J, Torrence R (1989) Stocktaking the rocks: obsidian sources in West New Britain, Papua New Guinea. Proceedings of the AINSE conference on nuclear techniques of analysis, p 187–189

Fullagar R, Ivuyo B, Specht J, Summerhayes GR (1991) Obsidian sources at Mopir, West New Britain Province, Papua New Guinea. Archaeol Ocean 26:110–114

Goffer Z (2007) Archaeological chemistry. Wiley, Hoboken, NJ
Goldstein JI, Newbury DE, Echlin P, Joy DC, Fiori C, Lifshin E (1992) Scanning electron microscopy and X-ray microanalysis: a text for biologists, materials scientists, and geologists, 2nd edn. Plenum Press, New York, NY
Golitko M, Meierhoff J, Terrell JE (2010) Chemical characterization of sources of obsidian from the Sepik coast (PNG). Archaeol Ocean 45:120–129
Gratuze B, Blet-Lemarquand M, Barrandon J-N (2001) Mass spectrometry with laser sampling: a new tool to characterize archaeological materials. J Radioanal Nucl Chem 247:645–656
Green RC (1987) Obsidian results from the Lapita sites of the Reef/Santa Cruz Islands. In: Ambrose WR, Mummery JMJ (eds) Archaeometry: further Australasian studies. Occasional papers in prehistory. RSPAS and ANU, Canberra, pp 239–249
Green RC, Bird JR (1989) Fergusson Island obsidian from the D'Entrecasteaux group in a Lapita site of the Reef Santa Cruz group. NZ J Archaeol 11:87–99
Juggins S (2005) C2 data analysis, 1.4.2 edn. University of Newcastle, Newcastle
Kayani PI, McDonnell G (1996) An assessment of back-scattered electron petrography as a method for distinguishing Mediterranean obsidians. Archaeometry 38:43–58
Kirch PV (1988) Long-distance exchange and Island colonization: the Lapita case. Nor Archaeol Rev 21:103–117
Kononenko NA (2012) Middle and late Holocene skin-working tools in Melanesia: tattooing and scarification? Archaeol Ocean 47:14–28
Kononenko NA, Bedford S, Reepmeyer C (2010) Functional analysis of late Holocene flaked and pebble stone artefacts from Vanuatu, Southwest Pacific. Archaeol Ocean 45:13–20
Leach BF, Davidson J (1981) Archaeological studies of Pacific stone resources. British Archaeological Reports: International Series, Oxford
Leach BF, Manly B (1982) Minimum Mahalanobis distance functions and lithic source characterisation by multi-element analysis. NZ J Archaeol 4:77–109
Lee YI, Sneddon J (1994) Direct and rapid determination of potassium in standard solid glasses by excimer laser ablation plasma atomic emission spectrometry. Analyst 119:1441–1443
Longerich HP, Jackson SE, Günther D (1996) Laser ablation inductively coupled plasma mass spectrometric transient signal data acquisition and analyte concentration calculation. J Anal Atom Spectrom 11:899–904
MacIntyre M (1983) Changing paths: a historical ethnology of the traders of Tubetube. Research School of Pacific Studies, Australian National University, Canberra
Pollard M, Batt C, Stern B, Young SMM (2007) Analytical chemistry in archaeology. Manuals in Archaeology. Cambridge University Press, Cambridge
Raos AM, Crawford AJ (2004) Basalts from the Efate Island Group, central section of the Vanuatu arc, SW Pacific: geochemistry and petrogenesis. J Volcanol Geotherm Res 134:35–56
Reed SJB (2005) Electron microprobe analysis and scanning electron microscopy in geology, 2nd edn. Cambridge University Press, Cambridge
Reepmeyer C (2008) Characterising volcanic glass sources in the Banks Islands, Vanuatu. Archaeol Ocean 43:120–127
Reepmeyer C (2009) Obsidian sources and distribution systems emanating from Gaua and Vanua Lava on the Banks Islands of Vanuatu. Ph.D Dissertation, Australian National University
Reepmeyer C, Clark GR (2010) Post-colonization interaction between Vanuatu and Fiji reconsidered: the reanalysis of obsidian from Lakeba Island, Fiji. Archaeometry 52:1–18
Reepmeyer C, Clark GR, Sheppard PJ (2012) Obsidian source use in Tongan prehistory: new results and implications. J Island Coastal Archaeol 7:255–271
Rothe P (2005) Gesteine: Entstehung, Zerstörung, Umbildung, 2nd edn. Primus Verlag, Darmstadt
Sheppard PJ (1993) Lapita lithics: trade/exchange and technology. A view from the Reefs/Santa Cruz. Archaeol Ocean 28:121–137
Sheppard PJ, Hancock RGV, Pavlish LA, Parker R (1989) Samoan volcanic glass. Archaeol Ocean 24:70–74
Smith IEM (1974) Obsidian sources in Papua-New Guinea. Archaeol Phys Anthropol Ocean 9:18–25
Smith IEM, Price RC (2006) The Tonga-Kermadec arc and Havre-Lau back-arc system: their role in the development of tectonisc and magmatic models for the western Pacific. J Volcanol Geotherm Res 156:315–331
Smith IEM, Ward GK, Ambrose WR (1977) Geographic distribution and the characterization of volcanic glasses in Oceania. Archaeol Ocean 12:173–201
Smith IEM, Steward RB, Price RC (2003) The petrology of a large intra-oceanic silicic eruption: the Sandy Bay Tephra, Kermadec Arc, Southwest Pacific. J Volcanol Geotherm Res 124:173–194
Specht J (2002) Obsidian, colonising and exchange. In: Bedford S, Sand C, Burley D (eds) Fifty years in the field: essays in honour and celebration of Richard Shutler Jr's Archaeological career. New Zealand Archaeological Association Monograph, Auckland, pp 37–49
Specht J, Koettig M (1981) An obsidian flaking area near Talasea, West New Britain, Papua New Guinea. Archaeol Ocean 16:168–172
SPSS (2006) SPSS for Windows, 16.0 edition. SPSS, Chicago, IL

Summerhayes GR (2004) The nature of prehistoric obsidian importation to Anir and the development of a 3,000 year old regional picture of obsidian exchange within the Bismarck Archipelago, Papua New Guinea. Records of the Australian Museum 29 (Suppl):145–156

Summerhayes GR (2009) Obsidian network patterns in Melanesia – sources, characterisation and distribution. Bull Indo Pac Pre Hist Assoc 29:109–124

Summerhayes GR, Allen J (1993) The transport of Mopir obsidian to late Pleistocene New Ireland. Archaeol Ocean 28:144–148

Summerhayes GR, Bird JR, Fullagar R, Gosden C, Specht J, Torrence R (1998) Application of PIXE-PIGME to archaeological analysis of changing patterns of obsidian use in West New Britain, Papua New Guinea. In: Shackley MS (ed) Archaeological obsidian studies: method and theory. Plenum Press, New York, NY, pp 129–158

Sun S-S, McDonough WF (1989) Chemical and isotopic systematics of oceanic basalts: implications for mantle composition and processes. In: Saunders AD, Norry MJ (eds) Magmatism in the ocean basins, geological society special publications. Blackwell, Oxford, pp 313–345

Torrence R (2004) Pre-Lapita valuables in Island Melanesia. Records of the Australian Museum 29 (Suppl):163–172

Torrence R (2005) Valued stone – how so? In: Macfarlane I, Mountain M-J, Paton R (eds) Many exchanges: archaeology, history, community and the work of Isabel McBryde. Aboriginal History Monograph, Canberra, pp 357–372

Torrence R (2011) Finding the right question: learning from stone tools on the Willaumez Penisula, Papua New Guinea. Archaeol Ocean 46:29–41

Torrence R, Specht J, Fullagar R, Bird JR (1992) From pleistocene to present: obsidian sources in West New Britain, Papua New Guinea. Records of the Australian Museum 15(Suppl):83–98

Torrence R, Specht J, Fullagar R, Summerhayes GR (1996) Which obsidian is worth it? In: Davidson J, Irwin G, Leach BF, Pawley A, Brown D (eds) Oceanic culture history: essays in honour of Roger Green. New Zealand Journal of Archaeology Special Publication, Dunedin, pp 211–224

Tykot RH (2004) Scientific methods and applications to archaeological provenance studies. In: Martini M, Milazzo M, Piacentini M (eds) Proceedings of the International School of Physics "Enrico Fermi" course CLIV. IOS Press, Amsterdam, pp 407–432

Vinx R (2005) Gesteinsbestimmung im Gelände. Spektrum Akademischer Verlag, Munich

Wall T (1976) Use of the research reactor Moata and associated facilities for the source identification of obsidian artefacts. In: Barnard N (ed) Ancient Chinese bronzes and South East Asian metal and other archaeological artifacts. National Gallery of Victoria, Melbourne, pp 337–350

Ward G (1979) Prehistoric settlement and economy in a tropical small island environment: the Banks Islands, Insular Melanesia. Australian National University, Canberra

Weisler MI (1998) Hard evidence for prehistoric interaction in Polynesia. Curr Anthropol 39:521–532

White JP, Jacobsen H, Kewibu V, Doelman T (2006) Obsidian traffic in the Southeast Papuan Islands. J Island Coastal Archaeol 1:101–108

Wilson L, Pollard A (2001) The provenance hypothesis. In: Brothwell D, Pollard A (eds) Handbook of archaeological sciences. Wiley, Chichester, pp 507–517

The Role of LA-ICP-MS in the Investigation of Archaeological Glass

11

Simone Cagno, Kevin Hellemans, and Koen Janssens

Abstract

A number of analytical techniques (e.g. SEM-EDX, XRF) are able to provide glass scientists with quantitative information on major and minor components of historical glass. However, at present, only LA-ICP-MS can deliver precise and accurate data on major, minor and trace (down to ppm or sub-ppm levels) elements with minimal or no sample preparation. The μm-sized sampling area of LA-ICP-MS makes it compatible, conservation-wise, with most cases in which analysis of archaeological glass is required. The capabilities of this technique are highlighted in this chapter by means of a summary of recent research on black-appearing Roman glass, for which a large set of LA-ICP-MS data have been assembled during the last few years.

11.1 LA-ICP-MS as a Tool for Chemical Analysis of Historical Glass

In recent years, glass scientists, conservators and archaeologists have become increasingly interested in the use of Laser Ablation-Inductively Coupled Plasma-Mass Spectrometry (LA-ICP-MS) for the analysis of historical glass samples (Barca et al. 2013; Bertini et al. 2011; Cagno et al. 2010; Dussubieux et al. 2009; Gratuze 2013; Mirti et al. 2009; Panighello et al. 2012). In contrast to other techniques such as Scanning Electron Microscopy—Energy Dispersive X-ray Spectroscopy (SEM-EDX) (Janssens 2013a) or X-Ray Fluorescence (XRF) (Janssens 2013b), LA-ICP-MS is capable of providing quantitative information on both major components and trace elements, often in a single run. Although the technique is destructive by nature, most samples can readily be analyzed without any kind of sample preparation. In this regard, LA-ICP-MS can be considered to be less destructive in comparison to other (micro-) analytical techniques that often require embedding, polishing and coating of the sample material in a preparatory step. The term *micro-destructive* is often used in this context, since in most cases, ablation at single

S. Cagno (✉) • K. Hellemans • K. Janssens
Department of Chemistry, University of Antwerp, Groenenborgerlaan 171, 2020 Antwerpen, Belgium
e-mail: simone.cagno@uantwerpen.be; kevin.hellemans@uantwerpen.be; koen.janssens@uantwerpen.be

L. Dussubieux et al. (eds.), *Recent Advances in Laser Ablation ICP-MS for Archaeology*, Natural Science in Archaeology, DOI 10.1007/978-3-662-49894-1_11

points or in rasters of 10–100 μm wide lines is required, thus significantly limiting resulting damage to the object. Obviously, the effect of glass alteration (i.e. the presence of a Si-gel superficial layer, depleted in cations) has to be considered in virtually every ancient glass find. To address this issue, pre-ablation, or manual polishing of the first few μm of the area of interest is employed. An additional advantage of LA-ICP-MS is the possibility of ablating lateral or vertical profiles in the sample, thus allowing the study of heterogeneity of the bulk glass and/or phenomena such as glass weathering (van Elteren et al. 2013).

To be suitable for LA-ICP-MS, glass artifacts essentially need at least one (macroscopically) flat surface and they have to fit in the ablation cell. This used to be a problem in the past, due to the limited size of most ablation cells, but recently the introduction of two-volume cells has allowed increase of cell size to about 20 by 20 cm, which is large enough for most objects. That said, practically, it is still very common to analyze glass samples embedded in resin disks for a number of reasons:

1) comparison of results with SEM-EDX is possible (the same samples can be analyzed with both techniques);
2) a flat and polished surface is already available, and possible surficial alteration is removed;
3) a much larger number of samples can be introduced into the ablation chamber at once and thus can be analyzed in the same run, increasing efficiency.

LA-ICP-MS has always had difficulties to overcome in order to produce quantitative results. Due to the strong matrix dependency of ablation efficiency, matrix-matched calibration is required to accurately determine composition. Fairly good results have been reported with the use of liquid standards (Aeschliman et al. 2003; Boue-Bigne et al. 1999; Halicz and Günther 2004), unfortunately this requires a highly modified set-up and is as such less desirable for general use. Other strategies include the fabrication of home-made matrix-matched standards (Becker et al. 2007; Fitzpatrick et al. 2008; Odegard et al. 2005). In more recent attempts to provide accurate quantification, Claverie et al. (2013) used a spinning plate to simultaneously ablate standard and sample and thus create the ability to perform standard addition in a laser ablation ICP-MS experiment; similarly to liquid standard calibration, the highly modified setups required for these approaches exclude these techniques from broad use. Luckily, for silicate glasses there are a large number of secondary reference materials available, such as those provided by the Corning Museum of Glass, designed to match the composition of the different types of historical glass, like soda-lime silica glass (CMG A-B), lead-silica glass (CMG C) and potash lime glass (CMG D) (Brill 1999). Thus, a close match can be found for nearly all glass samples, although these secondary reference materials have been reported to show low accuracy and precision for some elements (Vicenzi et al. 2002; Wagner et al. 2012). Other frequently used standards are the ones provided by the National Institute of Standards and Technology (NIST 61X–trace elements in glass series).

Over the years, many quantification protocols have been proposed and investigated on different samples and on different setups. Dussubieux et al. (2009) and Bertini et al. (2013) presented interesting inter-laboratory comparisons for the analysis of glass beads, showing good agreement in composition. Bertini et al. (2013) discussed in detail the effect of heterogeneity in standards and of fractionation, as well as the difficulties associated with quantification of light, volatile elements. In both articles, the quantification routine was based on the work done by Gratuze et al. (2001) and Longerich et al. (1996), commonly referred to as the Internal Standard Independent (ISI) method. This routine requires the availability of a known internal standard element and as a direct result, additional analyses with another technique to determine the concentration of this element.

Preferably, one would like to acquire all the necessary information from a single LA-ICP-MS experiment, as additional techniques are not always available. Sum normalization has been proposed to address this problem (Gagnon et al. 2008; Halicz and Günther 2004; Van Elteren and Tennent 2009), and has proven to yield reliable results given that the laser wavelength used is in the UV range, as shorter wavelengths tend to yield better reproducibility and ablation efficiency (Bertini et al. 2013; Guillong et al. 2003; Günther and Heinrich 1999, Horn et al. 2001). In contrast, for lasers with longer wavelengths, where the laser energy is sometimes dissipated through the sample, fractionation effects cannot be disregarded (Gaboardi and Humayun 2009).

In the last few years we have shown the benefit of using LA-ICP-MS for glass analysis in several cases; in particular for the discrimination between Façon-de-Venise glass and genuine Venetian production (Cagno et al. 2012a; Janssens et al. 2013; De Raedt et al. 1999), or the identification of different traditions in early medieval Tuscan and Altarist glass making (Cagno et al. 2008, 2010, 2012b, c) and the investigation of eighteenth century non-durable glass (Hellemans et al. 2014). In this chapter, after a discussion of our analytical approach and quantification method, we will highlight the capabilities of LA-ICP-MS as a tool for analysis of major, minor and trace elements by means of a case study on black-appearing Roman glass.

11.2 Analytical Approach

In the last few years we have performed LA-ICP-MS during several analytical sessions performed by means of different instruments, comprising both solid state and excimer lasers. Given that reliable analytical and quantification procedures were used, the results can be considered as intercomparable. Precision and accuracy can be determined by repeatedly measuring glass standards. A brief overview of the instrumental setups employed is presented below; for a more detailed description, the reader should refer to the publications cited in this section. For the reasons cited in the introduction, acrylic resin blocks, containing about 10 samples each, were inserted one by one in a laser ablation sample cell on the XYZ-translation stage. The exact position of the sample was observed with a CCD camera under PC control, which allows a selective ablation of the chosen areas of the investigated samples in the preferred pattern (line, points, raster). Helium was preferred as a carrier gas, while Argon was used as plasma and auxiliary gas.

Line ablation was chosen in order to maximize transient signal stability and also identify possible micro-inhomogeneities in the glass across the profile. Each measurement cycle consisted of pre-cleaning of the sample surface with a laser beam (no signal was registered during this step) followed by the laser ablation accompanied by registering of signals. Before or after ablation, a blank was always registered for each sample with the helium and argon flow only. Before quantification the blank signal was subtracted for each element individually. Pre-cleaning of the surface of the samples was generally performed by ablating the selected line with a laser beam diameter of 100 μm before the actual laser sampling of the selected line with a laser beam diameter of 80 μm. The most important ICP-MS instrument settings (such as carrier gas flow rate and RF power) were optimized, aiming at maximum signal intensities ($^{24}Mg^+$, $^{115}In^+$, $^{232}U^+$) and minimal background intensity (m/z = 8.5 and 220) and low oxide ion ($^{232}Th^{16}O^+/^{232}Th^+$) and doubly charged ($^{232}Th^{2+}/^{232}Th^+$) ion formation, based upon the signal intensities obtained when ablating NIST SRM 610 with the same pattern chosen for the samples. Different isotopes of particular elements were measured where possible. In this way, if interferences occur, adequate correction or selection of an interference-free isotope is possible.

At the Faculty of Chemistry of Warsaw University, a quadrupole ICP-MS ELAN 9000 (Perkin Elmer, SCIEX, Canada) equipped with a LSX-200+ (CETAC, USA) laser ablation system was used. The LSX-200+ combines a stable, environmentally sealed 266 nm UV laser (Nd: YAG, solid state, Q-switched) with a high sampling efficiency, variable 1–20 Hz pulse repetition rate and maximum energy up to 6 mJ/pulse (Van der Linden et al. 2009).

At the Ghent University Department of Analytical Chemistry, two different lasers coupled to two ICP-MS instruments were used. First, a GeoLas 193 nm ArF excimer-laser ablation system (MicroLas, Gottingen, Germany) coupled to a PerkinElmer SCIEX DRCplus quadrupole-based ICP-MS instrument (Concord, Ontario, Canada) in vented mode was used. An output energy of 0.1 mJ, yielding an energy density of ~13 J/cm^2 on the sample surface was used throughout all of the measurements. The ablation cell was coupled to the ICP torch via a 3 mm internal diameter Tygon tubing and argon was used as carrier gas (Cagno et al. 2012a).

A New Wave Research UP-193HE ArF excimer laser was also employed in the same laboratory, coupled with a quadrupole-based Thermo Corporation X-Series 2 ICP-MS. This second mass spectrometer was also used in combination with the GeoLas ablation system described previously (Cagno et al. 2012b).

At the Natural History Museum in London, a VG Elemental PQ3 ICP-MS with 'S-option' together with a VG Elemental Microprobe II 266 nm laser ablation system was used. Data were collected using the peak-jumping time resolved method (Cagno et al. 2012a).

In the next section we describe the two most commonly used procedures for the quantification of LA-ICP-MS data (ISI and SN method). The evaluation of these methods was performed at the LA-ICP-MS laboratory at the University of Antwerp, by means of a New Wave NWR193 ArF laser and an Agilent 7700 ICPMS (Table 11.1).

Table 11.1 Instrumental parameters for New Wave NWR193 ArF laser and Agilent 7700 setup

Laser	Type	NWR 193 ArF Excimer Laser
	Line length	1 mm
	Beam diameter	100 μm
	Repetition rate	10 Hz
	Scan speed	5 μm/s
	Laser warm-up	10 s
	Washout delay	60 s
	Carrier gas flow rate	0.3 L/min (He)
	Irradiance	2.30 GW/cm^2
	Fluence	11.45 J/cm^2
ICP-MS	Type	Agilent 7700X
	Forward power	1350 W
	Plasma gas flow rate	15 L/min (Ar)
	Auxiliary gas flow rate	0.9 L/min (Ar)
	Integration time	50 ms/mass
	Detector	Pulse Detector + Faraday Cup

11.2.1 Quantification Protocols

Internal Standard Independent (ISI) Method In its simplest form, quantification for LA-ICP-MS would require a standard STD to determine the concentration of the elements in an unknown material X. From this standard, sensitivity factors S_i for every element i can be calculated from the background corrected average signal of element i for the standard STD ($I_{i,STD}$) and the concentration $c_{i,STD}$ of element i in this standard STD:

$$S_i = \frac{I_{i,STD}}{c_{i,STD}} \tag{11.1}$$

Hence, the concentration $c_{i,X}$ of an element i in an unknown X could then be calculated as

$$c_{i,X} = \frac{I_{i,X}}{S_i} \tag{11.2}$$

in which $I_{i,X}$ represents the background corrected average signal of element i obtained from unknown X. Reference material concentrations are taken from the literature (Gao et al. 2002; Vicenzi et al. 2002).

In reality however, it became evident that this equation did not yield accurate results, as drift and variable ablation efficiency are not corrected for by this method. Thus, the use of an internal standard was introduced, commonly determined by another technique such as SEM-EDX. By assuming that every sample has at least one element that is homogeneously distributed, the signal intensity for the internal standard is a good indicator for both drift and variations in ablation efficiency. As such, the intensities used in the Eqs. (11.1) and (11.2) can be replaced by normalized intensities, by dividing I_i by the background corrected signal of the internal standard element I_{IS} (^{43}Ca in this study) and multiplying by its concentration c_{IS} (Heinrich et al. 2003; Longerich et al. 1996):

$$S_i = \left\langle \frac{I_{i,STD}\, c_{IS,STD}}{I_{IS,STD}\, c_{i,STD}} \right\rangle : \tag{11.3}$$

$$c_{i,X} = \left\langle \frac{I_{i,X}\, c_{IS,X}}{I_{IS,X}\, S_i} \right\rangle \tag{11.4}$$

The method described here is commonly referred to as the ISI method and provides satisfactory results.

Sum Normalization (SN) Method In the case of glass analysis however, sum normalization can also be employed. This technique, described in many different forms (Gagnon et al. 2008; Gratuze et al. 2001; Halicz and Günther 2004; Van Elteren and Tennent 2009), assumes that (1) as glass is almost exclusively comprised of oxides in known oxidation states and (2) all oxides can be detected by the ICP-MS, the sum of the concentration of all oxides should equal 100 %.

If c_t represents the cumulative oxide concentration (in wt%) of n oxides:

$$c_t = \sum_i^n c_i \tag{11.5}$$

The sum normalized concentration $c_{i,X}^{norm}$ can then be calculated as:

$$c_{i,X}^{norm} = 100 \times \frac{c_i}{c_t} \tag{11.6}$$

With this adaptation in the quantification protocol, there is no more need for the determination of the concentration of the internal standard by means of another technique. Described in more detail in Van Elteren and Tennent (2009), it suffices to use an initial guess of the IS concentration ($c_{IS,X}$) in the calculation; after the calculation has been done, $c_{IS,X}$ is replaced with $c_{IS,X}^{norm}$. We used an adapted version of this technique by employing a new normalized intensity:

$$c_{i,X} = \left\langle \frac{I_{i,X}}{I_{IS,X}} \frac{1}{S_{i,X}} \right\rangle \tag{11.7}$$

As the concentration of the internal standard is a constant, it is cancelled out in the sum normalization. By using this new normalized intensity, the replacement of $c_{IS,X}$ is no longer required.

In most cases, due to matrix-dependent ablation efficiency, the response factors from a single standard are often insufficient to accurately quantify a single sample set, especially when that sample set contains a variety of glass types. Therefore, multiple standards are usually analyzed together with the samples. Here, two different protocols are reported. The first option is to select the standard that most closely compositionally matches the sample, as done by Wagner et al. (2008). The second option is to use a linear fit between normalized intensity and concentration (Gratuze et al. 2001; Van Elteren and Tennent 2009). The former relies on the strongly matrix-dependent ablation rate and fractionation, but is prone to error as heterogeneity in standards would create a bias on the results. The latter is a more robust method, especially when outliers are removed, but relies on linearity over a large concentration range, which may not always be present, so that accuracy is sacrificed for robustness. In our quantification protocol, we used the latter, as the loss of accuracy is negligible, and also keeping in mind that the primary goal of the analysis of historical glass is classification.

To further the accuracy of the results, a combination of different standards based on the Euclidean distance between the sample x and all available standards (index j) can be employed to obtain an average sensitivity factor (Hellemans et al. 2015, unpublished results):

$$S_{i,x} = \sum_j^k r_{x,j} S_{i,j} \tag{11.8}$$

Where:

$$r_{x,j} = \frac{1}{\sqrt{\sum_i (c_{i,j} - c_{i,x})^2}} \times \left(\sum_{j=1}^{k} \frac{1}{\sqrt{\sum_i (c_{i,j} - c_{i,x})^2}} \right)^{-1} \quad (11.9)$$

Where:

- $S_{i,j}$: the sensitivity factors calculated from Eq. (11.3)
- $c_{i,x}$: initial guess of the concentration of oxide i, obtained from a quantification with all standards
- $c_{i,j}$: concentration of oxide i in standard j

11.2.2 Comparison SN and ISI Methods

In Fig. 11.1 and Table 11.2, the quantitative analysis results are summarized for CMG reference glasses B-D using the ISI method (a) and the SN method (b), while NIST SRM 610 was used as a single calibration standard. As can be seen from the results, even though the matrix of the "unknowns" differs from that of the standard (potash glass vs. soda-lime glass), the calculated concentrations match the reported values to an acceptable extent, with two notable outliers (MnO_2, Sb_2O_5) for the CMG C standard. The concentration of these two oxides has already been reported to vary greatly in the CMG C standard (Wagner et al. 2012).

The results of both quantification methods are comparable, and the same outliers are visible. The results clearly show the similarity in accuracy between ISI and sum normalization, so both techniques may be considered equally reliable, although it must be kept in mind that any error in the standard composition may create a significant bias. Quantification accuracy can be further improved by adapting the calibration standard set to include additional secondary glass standards. This is especially necessary in the analysis of potash glass. Due to the very low content of potassium in the NIST SRM610 and the isobaric interference of ArH^+, the determination of potassium should be done with a standard containing potassium in the same range of concentration as samples.

Indeed, from Table 11.2 it becomes apparent that using only NIST SRM610 as a calibration standard underestimates the potassium oxide concentration by a factor of two, whereas a combined calibration set yields more accurate results. Furthermore, the outliers MnO_2 and Sb_2O_5 are underestimated in every quantification method, suggesting that these are true outliers and not quantification artifacts. When a heterogeneous standard is used, the sensitivity factors calculated from this standard might induce a positive or negative bias in the final results. The plots of Fig. 11.1 may be used to detect outliers. By using these plots we may decide to exclude certain standards from the calibration, or simply exclude problematic isotopes from quantification.

11.3 Black-Appearing Roman Glass

11.3.1 Background

During the Roman era, a specific type of intensely colored, black-appearing glass was used for different purposes. While it represented only a small fraction of the entirety of Roman glass production, the use of black-appearing glass enjoyed fluctuating popularity throughout the Roman Imperial era and consisted of two main types of artifacts: glass vessels and jewelry.

The production of vessels in black-appearing glass first became fashionable across the entire Empire from about AD 30 to AD 70–80 (Cosyns and Fontaine 2009). The reappearance of black glass vessels from AD 170–180 to AD 230–250 did not result in Empire-wide distribution, but only regional distribution in an area comprising the north-western provinces *Gallia Belgica*, *Germania Inferior* and *Germania Superior* (Cosyns and Hanut 2005; Cosyns et al. 2006). A third re-emergence of black glass vessel production took place in the course of the fourth and fifth centuries AD, and was limited to the south-eastern Mediterranean (Cosyns 2011).

Black glass jewelry was commonly manufactured through the entire duration of the Imperial era, especially from the second half of

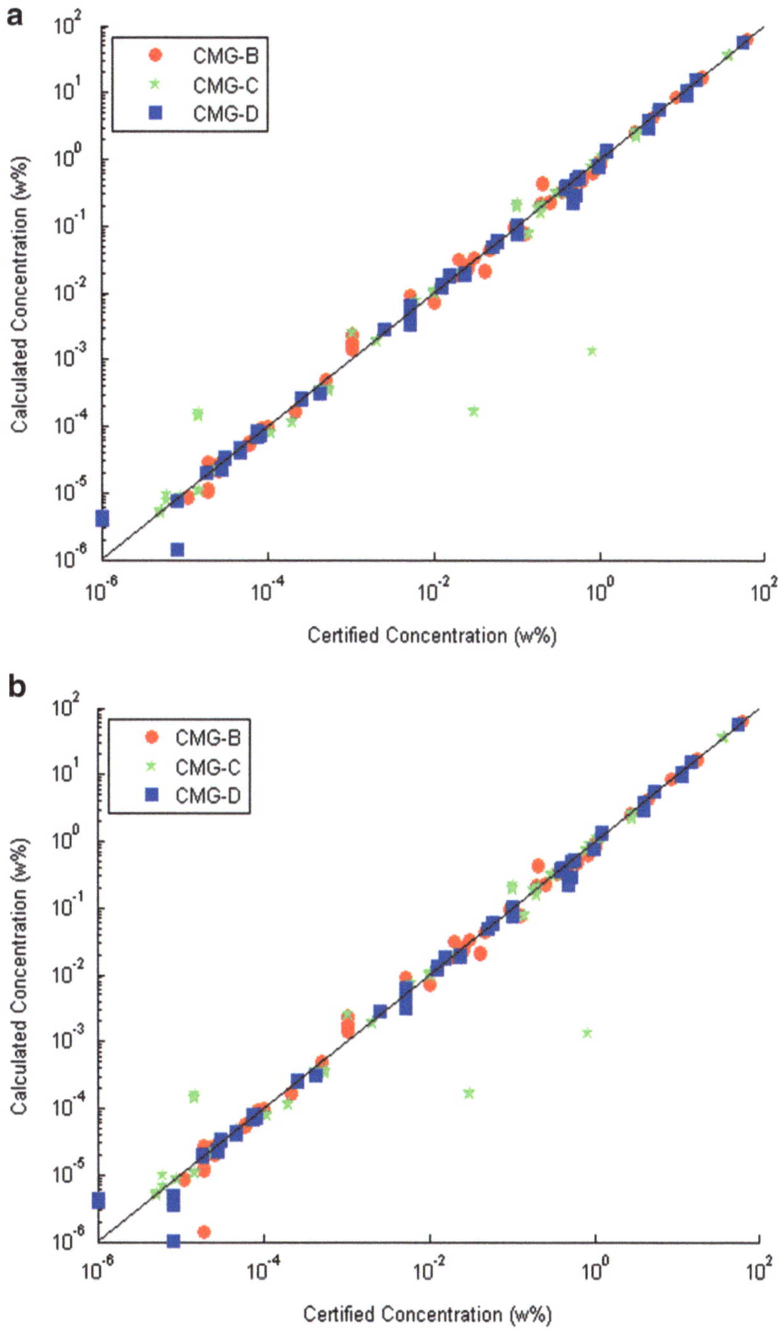

Fig. 11.1 ISI quantification of reference materials using NIST SRM610 (*a*) and SN quantification of reference materials using NIST SRM610 (*b*)

the second century AD onwards (Cosyns 2011). Empire-wide production of glass bangles, finger-rings, beads, pendants, hairpins, and gems gradually disappeared during the fifth century AD in the western part of the Empire, except for black glass beads decorated with an applied trail in vividly colored opaque glass, which remained in use until the mid-sixth century AD (Cosyns 2011). Within the eastern part of the Empire, continuity of a wide range of jewelry production in black glass can be observed into the Byzantine era (Cosyns 2011). This changeable age- and location-specific distribution makes black glass a valuable type of archaeological find that can be used to explore in more

Table 11.2 Calculated composition of Corning Museum of Glass Standards B, C and D compared with published values (Wagner et al. 2012) by means of (a) Sum Normalization Method using NIST SRM610 as a calibration standard, (b) Internal Standard Independent Method using NIST SRM610 as a calibration standard, (c) Internal Standard Independent Method using NIST SRM610 and two CMG glasses as calibration standards with the determined CMG standard excluded

	Analyte											
	CMG B				CMG C				CMG D			
Source	Literature	(a)	(b)	(c)	Literature	(a)	(b)	(c)	Literature	(a)	(b)	(c)
Na_2O	17.0	17 ± 1	16.4 ± 0.9	18 ± 1	1.07	1.07 ± 0.01	1.02 ± 0.01	1.09 ± 0.06	1.20	1.37 ± 0.05	1.26 ± 0.04	1.35 ± 0.06
MgO	1.03	0.93 ± 0.02	0.92 ± 0.04	0.32 ± 0.03	2.76	2.52 ± 0.03	2.39 ± 0.01	0.60 ± 0.08	3.94	4.01 ± 0.06	3.61 ± 0.09	1.23 ± 0.06
Al_2O_3	4.36	4.3 ± 0.1	4.2 ± 0.9	4.0 ± 0.3	0.87	0.88 ± 0.02	0.82 ± 0.01	0.78 ± 0.02	5.30	5.8 ± 0.1	5.23 ± 0.03	5.06 ± 0.07
SiO_2	61.6	63 ± 1	61.4 ± 0.6	65 ± 3	34.9	35 ± 1	33.7 ± 0.4	35 ± 2	55.2	58.8 ± 0.2	53.9 ± 0.9	55.9 ± 0.7
P_2O_5	0.82	0.64 ± 0.02	0.62 ± 0.01	0.3 ± 0.2	0.14	(7.5 ± 0.2) × 10^{-2}	(7.03 ± 0.01) × 10^{-2}	(2.3 ± 0.3) × 10^{-2}	3.93	3.2 ± 0.1	2.88 ± 0.07	1.24 ± 0.04
Cl	0.20	0.48 ± 0.04	0.44 ± 0.02	0.28 ± 0.02	0.10	0.21 ± 0.01	0.20 ± 0.01	0.12 ± 0.01	0.40	0.41 ± 0.02	0.37 ± 0.01	0.23 ± 0.01
K_2O	1.00	0.4 ± 0.1	0.4 ± 0.1	1.0 ± 0.1	2.84	1.24 ± 0.05	1.07 ± 0.05	2.53 ± 0.03	11.3	6.1 ± 0.3	5.05 ± 0.06	11.3 ± 0.5
CaO	8.56	8.9 ± 0.2	8.6	8.6	5.07	5.5 ± 0.2	5.1	5.1	14.8	16.7 ± 0.2	14.8	14.8
Sc_2O_3	8.24 × 10^{-5}	(8.8 ± 0.9) × 10^{-5}	(8.8 ± 0.06 × 10^{-5}	(1.0 ± 0.1) × 10^{-4}	4.67 × 10^{-5}	(4.3 ± 0.8) × 10^{-5}	(4.2 ± 0.4) × 10^{-5}	(4.2 ± 0.8) × 10^{-5}	7.44 × 10^{-5}	(7.5 ± 0.8) × 10^{-5}	(7 ± 1) × 10^{-5}	(7.7 ± 0.8) × 10^{-5}
TiO_2	8.90 × 10^{-2}	(9.5 ± 0.5) × 10^{-2}	(9.3 ± 0.1) × 10^{-2}	(8 ± 1) × 10^{-2}	0.79	0.79 ± 0.04	0.70 ± 0.01	0.60 ± 0.03	0.38	0.40 ± 0.01	0.35 ± 0.01	0.32 ± 0.01
V_2O_5	3.00 × 10^{-2}	(3.5 ± 0.2) × 10^{-2}	(3.32 ± 0.06) × 10^{-2}	(3.2 ± 0.3) × 10^{-2}	6.00 × 10^{-3}	(7.8 ± 0.3) × 10^{-3}	(6.97 ± 0.07) × 10^{-3}	(6.8 ± 0.2) × 10^{-3}	1.50 × 10^{-2}	(1.97 ± 0.05) × 10^{-2}	(1.71 ± 0.01) × 10^{-2}	(1.64 ± 0.05) × 10^{-2}
Cr_2O_3	5.00 × 10^{-3}	(8.9 ± 0.2) × 10^{-3}	(8.9 ± 0.1) × 10^{-3}	(5.9 ± 0.2) × 10^{-3}	1.00 × 10^{-3}	(2.50 ± 0.03) × 10^{-3}	(2.40 ± 0.02) × 10^{-3}	(1.60 ± 0.05) × 10^{-3}	2.50 × 10^{-3}	(2.9 ± 0.1) × 10^{-3}	(2.68 ± 0.03) × 10^{-3}	(1.80 ± 0.09) × 10^{-3}
MnO	0.25	0.23 ± 0.03	0.23 ± 0.05	0.19 ± 0.07	0.82	(1.34 ± 0.04) × 10^{-3}	(1.24 ± 0.02) × 10^{-3}	(1.01 ± 0.07) × 10^{-3}	0.55	0.56 ± 0.01	0.51 ± 0.01	0.48 ± 0.01
Fe_2O_3	0.34	0.32 ± 0.02	0.32 ± 0.08	0.24 ± 0.07	0.34	0.32 ± 0.01	0.29 ± 0.01	0.22 ± 0.01	0.52	0.52 ± 0.01	0.47 ± 0.01	0.40 ± 0.01
CoO	4.60 × 10^{-2}	(4.5 ± 0.7) × 10^{-2}	(4.34 ± 0.04) × 10^{-2}	(4.9 ± 0.2) × 10^{-2}	0.18	0.18 ± 0.01	0.17 ± 0.01	0.20 ± 0.01	2.30 × 10^{-2}	(1.98 ± 0.05) × 10^{-2}	(1.81 ± 0.02) × 10^{-2}	(2.06 ± 0.07) × 10^{-2}
NiO	0.10	(9.2 ± 0.2) × 10^{-2}	(9.2 ± 0.2) × 10^{-2}	(9.7 ± 0.2) × 10^{-2}	2.00 × 10^{-2}	(2.11 ± 0.08) × 10^{-2}	(1.94 ± 0.02) × 10^{-2}	(2.1 ± 0.1) × 10^{-2}	5.00 × 10^{-2}	(5.16 ± 0.09) × 10^{-2}	(4.62 ± 0.04) × 10^{-2}	(5.0 ± 0.2) × 10^{-2}
CuO	2.66	2.55 ± 0.09	2.53 ± 0.08	1.9 ± 0.5	1.13	1.16 ± 0.04	1.06 ± 0.02	0.82 ± 0.04	0.38	0.38 ± 0.01	0.34 ± 0.01	0.29 ± 0.01
ZnO	0.19	0.21 ± 0.01	0.21 ± 0.04	0.20 ± 0.01	5.20 × 10^{-2}	(5.3 ± 0.2) × 10^{-2}	(4.94 ± 0.08) × 10^{-2}	(4.7 ± 0.1) × 10^{-2}	0.10	0.10 ± 0.01	(9.72 ± 0.01) × 10^{-2}	(9.5 ± 0.5) × 10^{-2}
Rb_2O	1.00 × 10^{-3}	(1.5 ± 0.2) × 10^{-3}	(1.28 ± 0.07) × 10^{-3}	(1.5 ± 0.3) × 10^{-3}	1.00 × 10^{-2}	(1.01 ± 0.05) × 10^{-2}	(9.5 ± 0.2) × 10^{-3}	(8.8 ± 0.4) × 10^{-3}	5.00 × 10^{-3}	(5.3 ± 0.3) × 10^{-3}	(4.80 ± 0.01) × 10^{-3}	(4.7 ± 0.3) × 10^{-3}
SrO	1.90 × 10^{-2}	(1.83 ± 0.05) × 10^{-2}	(1.79 ± 0.04) × 10^{-2}	(1.79 ± 0.09) × 10^{-2}	0.29	0.32 ± 0.02	0.29 ± 0.01	0.30 ± 0.01	5.70 × 10^{-2}	(6.6 ± 0.1) × 10^{-2}	(5.62 ± 0.07) × 10^{-2}	(6.0 ± 0.2) × 10^{-2}

Y_2O_3	6.02×10^{-5}	$(5.3 \pm 0.2) \times 10^{-5}$	$(5.4 \pm 0.3) \times 10^{-5}$	$(6.9 \pm 0.8) \times 10^{-5}$	5.44×10^{-4}	$(3.6 \pm 0.2) \times 10^{-4}$	$(3.3 \pm 0.1) \times 10^{-4}$	$(4.4 \pm 0.2) \times 10^{-4}$	4.70×10^{-5}	$(4.5 \pm 0.5) \times 10^{-5}$	$(4.1 \pm 0.2) \times 10^{-5}$	$(5.2 \pm 0.6) \times 10^{-5}$
ZrO_2	2.50×10^{-2}	$(2.5 \pm 0.2) \times 10^{-2}$	$(2.4 \pm 0.1) \times 10^{-2}$	$(2.6 \pm 0.3) \times 10^{-2}$	5.00×10^{-3}	$(5.5 \pm 0.4) \times 10^{-3}$	$(5.12 \pm 0.07) \times 10^{-3}$	$(5.5 \pm 0.4) \times 10^{-3}$	1.25×10^{-2}	$(1.37 \pm 0.06) \times 10^{-2}$	$(1.21 \pm 0.05) \times 10^{-2}$	$(1.32 \pm 0.06) \times 10^{-2}$
Nb_2O_5	2.56×10^{-5}	$(2.4 \pm 0.5) \times 10^{-5}$	$(2.09 \pm 0.01) \times 10^{-5}$	$(2.7 \pm 0.7) \times 10^{-5}$	1.06×10^{-4}	$(8.5 \pm 0.5) \times 10^{-5}$	$(7.3 \pm 0.2) \times 10^{-5}$	$(1.00 \pm 0.04) \times 10^{-4}$	8.00×10^{-5}	$(7.9 \pm 0.3) \times 10^{-5}$	$(6.84 \pm 0.05) \times 10^{-5}$	$(8.98 \pm 0.4) \times 10^{-5}$
SnO	4.00×10^{-2}	$(2.25 \pm 0.05) \times 10^{-2}$	$(2.12 \pm 0.03) \times 10^{-2}$	$(2.7 \pm 0.1) \times 10^{-2}$	0.19	0.16 ± 0.01	0.15 ± 0.03	0.18 ± 0.01	0.10	$(8.16 \pm 0.26) \times 10^{-2}$	$(7.09 \pm 0.06) \times 10^{-2}$	$(9.43 \pm 0.32) \times 10^{-2}$
Sb_2O_5	0.46	0.38 ± 0.01	0.36 ± 0.01	0.6 ± 0.2	3.00×10^{-2}	$(1.3 \pm 0.4) \times 10^{-4}$	$(1.51 \pm 0.03) \times 10^{-4}$	$(8 \pm 3) \times 10^{-5}$	0.97	0.8 ± 0.3	0.74 ± 0.01	0.58 ± 0.01
BaO	0.12	$(7.8 \pm 0.4) \times 10^{-2}$	$(7.6 \pm 0.3) \times 10^{-2}$	$(7 \pm 2) \times 10^{-2}$	11.4	11.7 ± 0.5	10.6 ± 0.2	9.7 ± 0.7	0.51	0.31 ± 0.01	0.27 ± 0.01	0.28 ± 0.01
Ce_2O_3	1.92×10^{-5}	$(1.6 \pm 0.6) \times 10^{-5}$	$2.0 \pm 0.1 \times 10^{-5}$	$(2.0 \pm 0.7) \times 10^{-5}$	5.39×10^{-6}	$(4.2 \pm 0.8) \times 10^{-6}$	$(4.8 \pm 0.3) \times 10^{-6}$	$(4.6 \pm 0.9) \times 10^{-6}$	3.00×10^{-5}	$(3.5 \pm 0.1) \times 10^{-5}$	$(3.21 \pm 0.06) \times 10^{-5}$	$(3.6 \pm 0.1) \times 10^{-5}$
PbO	0.61	0.46 ± 0.03	0.46 ± 0.01	0.5 ± 0.1	36.7	38.5 ± 0.9	35.0 ± 0.4	35 ± 3	0.48	0.24 ± 0.01	0.21 ± 0.01	0.23 ± 0.01
ThO_2	9.72×10^{-5}	$(9.7 \pm 0.4) \times 10^{-5}$	$(9.5 \pm 0.3) \times 10^{-5}$	$(1.1 \pm 0.1) \times 10^{-4}$	2.46×10^{-5}	$(2.4 \pm 0.4) \times 10^{-5}$	$(2.14 \pm 0.06) \times 10^{-5}$	$(2.5 \pm 0.3) \times 10^{-5}$	7.82×10^{-5}	$(8.5 \pm 0.6) \times 10^{-5}$	$(7.4 \pm 0.3) \times 10^{-5}$	$(8.4 \pm 0.5) \times 10^{-5}$
UO_2	2.56×10^{-5}	$(10 \pm 10) \times 10^{-6}$	$(2.65 \pm 0.03) \times 10^{-5}$	$0 \pm 4 \times 10^{-5}$	8.92×10^{-6}	$(1.0 \pm 0.2) \times 10^{-5}$	$(8.2 \pm 0.2) \times 10^{-6}$	$(10 \pm 1) \times 10^{-6}$	1.82×10^{-5}	$(2.2 \pm 0.1) \times 10^{-5}$	$(1.84 \pm 0.02) \times 10^{-5}$	$(2.0 \pm 0.1) \times 10^{-5}$

The reported error corresponds to 1 standard deviation

Table 11.3 Summary statistics for compositional groups of black-appearing Roman glass; mean values and 1 standard deviation for major, minor and trace element contents in % w/w and μg/g (Cagno et al. 2014)

Group	Na_2O	MgO	Al_2O_3	SiO_2	P_2O_5	SO_3	Cl	K_2O	CaO	MnO	Fe_2O_3
BG1.1	19 ± 1	2.6 ± 0.5	1.9 ± 0.5	63 ± 1	0.6 ± 0.2	0.4 ± 0.2	1.0 ± 0.2	1.8 ± 0.4	8 ± 1	0.4 ± 0.1	1.6 ± 0.2
BG1.2	18 ± 2	2.8 ± 0.5	1.9 ± 0.5	63.6 ± 0.9	0.2 ± 0.1	0.4 ± 0.1	0.9 ± 0.1	1.9 ± 0.4	9 ± 2	0.5 ± 0.2	1.8 ± 0.3
BG1.3	20.7± 0.5	5.1± 0.6	2.0± 1.0	63± 2	0.3± 0.1	0.4± 0.1	0.8± 0.2	2.4± 0.5	3.8± 0.3	0.1± 0.1	1.4± 0.4
BG2	18 ± 3	0.7 ± 0.2	1.9 ± 0.2	68 ± 2	0.2 ± 0.1	0.2 ± 0.1	0.8 ± 0.1	0.9 ± 0.1	7 ± 1	2.6 ± 0.4	0.5 ± 0.3
BG3.1	18 ± 2	0.6 ± 0.2	1.8 ± 0.2	64 ± 2	0.1 ± 0.1	0.3 ± 0.1	0.7 ± 0.1	0.7 ± 0.2	6.0 ± 0.7	0.3 ± 0.1	8 ± 2
BG3.2	16.8 ± 0.7	0.6 ± 0.1	1.9 ± 0.2	64 ± 2	0.1 ± 0.1	0.2 ± 0.1	0.7 ± 0.1	0.7 ± 0.1	6.3 ± 0.4	0.3 ± 0.1	9 ± 3
BG4.1	17 ± 1	1.2 ± 0.2	2.1 ± 0.3	61.0 ± 1.5	0.1 ±0.1	0.3 ± 0.1	0.7 ± 0.1	0.8 ± 0.3	6.2 ± 0.4	1.5 ± 0.2	8 ± 2
BG4.2	17 ± 4	0.9 ± 0.3	1.9 ± 0.5	61 ± 2	0.1 ± 0.1	0.2 ± 0.1	0.7 ± 0.2	0.7 ± 0.2	6 ± 1	1.2 ± 0.4	9 ± 3
BG4.3	18 ± 2	0.9 ± 0.2	1.6 ± 0.2	62 ± 1	0.1 ± 0.1	0.3 ± 0.1	0.9 ± 0.2	0.7 ± 0.3	7 ± 1	0.8 ±0.2	8 ± 2

Group	B	Ti	V	Cr	Co	Ni	Cu	Zn	As	Rb	Sr	Y	Zr	Nb	Ag	Cd	Sn	Sb	Ba	La	Ce	Hf	Pb	Th	U
BG1.1	250 ± 70	800 ± 200	19 ± 5	18 ± 7	5 ± 2	8 ± 2	50 ± 20	30 ± 10	2 ± 2	7 ± 4	450 ± 90	3 ± 1	60 ± 20	2 ± 1	nd	2 ± 1	11 ± 9	300 ± 300	250 ± 60	5 ± 2	12 ± 2	1 ± 0	100 ± 100	1 ± 1	1 ± 1
BG1.2	240 ± 40	1300 ± 300	22 ± 7	10 ± 10	10 ± 10	10 ± 10	200 ± 200	40 ± 10	5 ± 5	7 ± 4	800± 200	10 ± 2	140 ± 30	4 ± 1	nd	nd	20 ± 20	500 ± 300	370 ± 70	11 ± 2	16 ± 4	2 ± 2	300 ± 300	2 ± 2	1 ± 1
BG1.3	190 ± 20	1500 ± 200	23 ± 7	27 ± 3	10 ± 10	30 ± 20	60 ± 40	40 ± 20	1 ± 1	15 ± 5	280 ± 80	10 ± 10	100 ± 20	6 ± 1	nd	2 ± 1	7 ± 8	nd	70 ± 30	9 ± 2	22 ± 9	3 ± 1	20 ± 10	2 ± 1	nd
BG2	200 ± 100	400 ± 200	31 ± 9	8 ± 2	40 ± 40	19 ± 2	200 ± 200	30 ± 10	5 ± 2	11 ± 2	700 ± 300	8 ± 3	40 ± 10	1 ± 0	5 ± 5	6 ± 6	40 ± 40	40 ± 40	400 ± 100	8 ± 2	13 ± 2	2 ±2	100 ± 100	1 ± 1	1 ± 1
BG3.1	160 ± 20	470 ± 80	12 ± 3	11 ± 1	40 ± 20	50 ± 10	200 ± 200	50 ± 20	40 ± 10	12 ± 3	390 ± 50	6 ± 1	60 ± 20	1 ± 1	6 ± 6	8 ± 8	60 ± 40	3000 ± 2000	250 ± 50	7 ± 1	11 ± 1	1 ± 1	600 ± 600	1 ± 1	1 ± 1
BG3.2	170 ± 20	800 ± 100	13 ± 2	nd	40 ± 20	nd	300 ± 300	40 ± 20	30 ± 10	11 ± 2	630 ± 60	16 ± 1	140 ± 10	3 ± 0	nd	nd	nd	4000 ± 1000	290 ± 30	14 ± 1	16 ± 2	nd	500 ± 500	nd	nd
BG4.1	171 ± 30	2200 ± 500	50± 20	50 ± 10	30 ± 10	36 ± 5	1000 ± 1000	110 ± 60	50 ± 10	9 ± 5	600± 100	14 ± 6	200± 100	5 ± 1	60 ± 40	80 ± 70	300 ± 300	200± 200	700± 400	13 ± 4	17 ± 3	4 ± 1	4000 ± 4000	2 ± 2	1 ± 1
BG4.2	130 ± 20	900 ± 100	22 ± 3	nd	15 ± 8	nd	120 ± 80	30 ±20	31 ±8	8 ±2	660 ± 70	16 ± 1	130 ± 20	3 ± 0	nd	nd	nd	400 ± 300	320 ± 60	13 ± 1	15 ± 1	nd	200 ± 200	nd	nd
BG4.3	110 ± 20	600 ± 200	20 ± 6	18 ± 5	40 ± 40	30 ± 30	300 ± 300	200 ± 100	7 ± 1	7 ± 3	450 ± 90	6 ± 1	60 ± 20	2 ± 1	13 ± 4	20 ± 10	60 ± 50	70 ± 70	270 ± 90	6 ± 1	10 ± 1	2 ± 1	800 ± 500	1 ± 1	1 ± 1

detail the relation between primary and secondary glass workshops in the Roman Empire.

11.3.2 Analyses of Black Roman Glass

In the framework of black Roman glass investigation (Cosyns 2011), over 400 archaeological glass samples were analyzed with several analytical techniques, including non-destructive analyses on museum artefacts (Cagno et al. 2015), but mostly by LA-ICP-MS in combination with SEM-EDX, as in the early stages of investigation only information about major components was sought (Van der Linden et al. 2009; Cagno et al. 2013, 2014). The quantification of these samples was performed with the ISI method, using ^{43}Ca as an internal standard, the Ca-concentration being pre-determined via SEM-EDX. NIST SRM610 and NIST SRM612 were used as external standards.

This approach was chosen because quantitative SEM-EDX results were already available, and mainly trace element information was needed in the latest investigation stage. For this reason, the use of NIST SRM610 and 612 was preferred, having a similar matrix as the historical glass (soda-lime-silica), and allowing us to overcome the known issues with some trace components in the CMG series (Vicenzi et al. 2002; Wagner et al. 2012). Additionally, because of the previously determined major elements, we were also able to keep the same quantification strategy for all samples, even though some major elements (e.g., Fe) differed by several orders of magnitude with respect to concentrations in NIST SRM 610. For those elements, concentrations were taken from SEM-EDX quantification.

All of the glass fragments analyzed have a silica-soda-lime composition (Table 11.3); four main groups were identified that approximate the

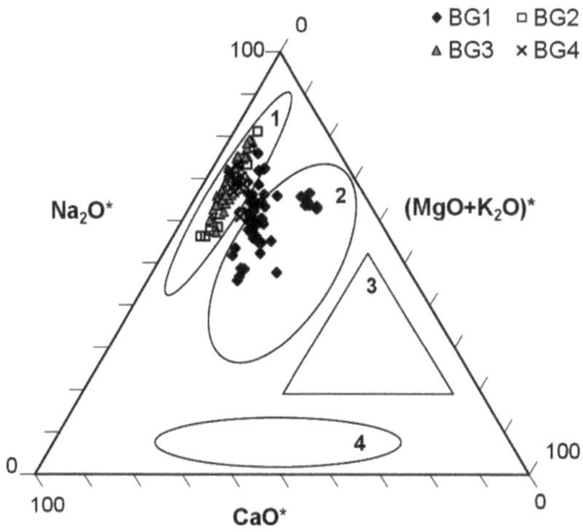

Fig. 11.2 Ternary plot of normalized (*) Na_2O, MgO + K_2O and CaO contents in the analyzed black glass finds; the normalization was done against the sum of the Na_2O, MgO + K_2O and CaO contents. The four main groups of historical glass are represented by the *ellipses* and *triangle*: (1) mineral-soda glass typical of the first millennium BC–first millennium AD; (2) vegetable-ash glass of the medieval period; (3) mixed alkali glass of late Bronze Age and post-medieval period; (4) potash glass of the medieval period (Cagno et al. 2014)

main ancient glass groups found in the classification initially proposed by Sayre and Smith (1961). These were named BG1 (soda ash glass with natural Fe), BG2 (natron glass with additional Mn), BG3 (natron glass with addition of Fe), and BG4 (High Iron Manganese Titanium (HIMT) glass with elevated levels of Fe) (Freestone 1994; Mirti et al. 1993). These groups are shown in Fig. 11.2 in the ternary plot introduced by Gratuze (2004) in order to classify historical glass. Roman mainstream natron glass (BG2 and 3) represents the majority of samples in the analyzed set. Such glass was manufactured by means of lime-rich sand (for instance from the Belus river, as cited by Pliny) and mineral soda (natron) obtained from inland lakes (for instance in Wadi al Natrun, Egypt) (Picon and Vichy 2003). To this base glass, coloring agents such as Mn or Fe (see below) were added in order to obtain the desired color.

The HIMT glass group (BG4) is comprised of glass from the fourth century onwards and its base composition is similar to analyses of HIMT glass reported in the literature (Foster and Jackson 2009; Freestone 1994; Freestone et al. 2005). Group BG4 was identified when the sample set was expanded with late Roman glass (Cagno et al. 2013, 2014). HIMT glass is a late Roman composition originating from the eastern Mediterranean (but not from Syro-Palestine) that is richer in Na, Mn, Ti, and Fe than Syro-Levantine glass due to the different geological origin of the sand used to produce it, and is of overall lower quality than Syro-Levantine glass.

The BG1 glass, instead, is produced by means of ash of halophytic plants (e.g., *Salsola Soda*) as a fluxing agent. This composition is not typical of Roman glass, however, it is reported that during the Roman period in areas not under Roman domination, glass was produced in this manner (Mirti et al. 2009). A few samples are also found with intermediate composition between soda ash and natron glass (Van der Linden et al. 2009). Thus, it is possible that a mixing process occurred, e.g. that recycled glass with a Low Magnesium Glass (LMG, Sayre and Smith 1961) composition was added to a batch containing vegetable ash and sand.

Concerning the coloring agents used to obtain a black appearing glass, two main groups are apparent when observing the glass in transmission with the help of a light source:

1) glass with a deep green hue, colored with iron (BG1, BG3, BG4)
2) glass with an intense purple hue, colored with manganese (BG2)

The dark green group is by far the largest and covers all the periods and areas studied, while the black/purple glass samples represent a characteristic production mainly related to pre-/early-Roman bracelets dated to between the first century BC and first century AD, and early Roman counters of the first to second centuries AD.

The use of Mn in moderate to high amounts (2–4 % MnO) is observed as a coloring agent for obtaining dark purple glass (group BG2). Such MnO contents are distinctly higher than those considered as natural in Roman (non-purple)

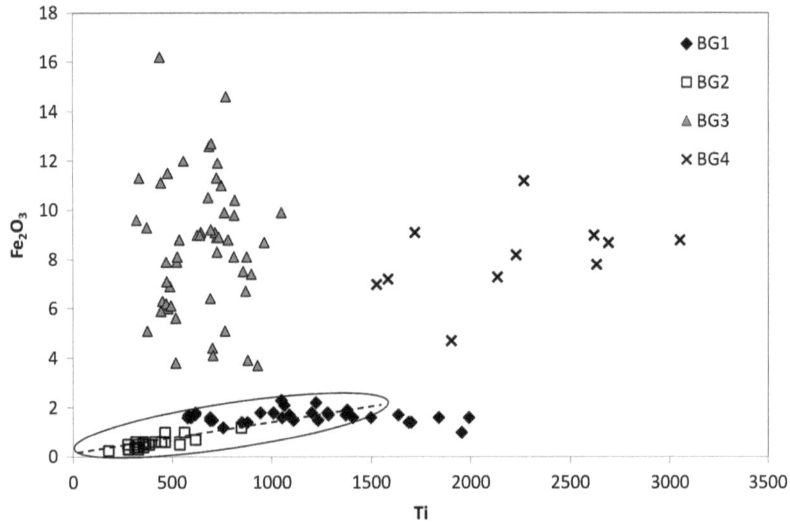

Fig. 11.3 Fe_2O_3 (% w/w) vs. Ti (μg/g) content in the analyzed samples (Cagno et al. 2014)

glass (Aerts et al. 2003; Freestone et al. 2005; Silvestri et al. 2008) and can be considered as deliberately produced (e.g. by adding Mn-minerals or Mn-rich glass to the batch). The iron content of this last group of glass (below 1 %) corresponds to that of "naturally colored" glass (Foster and Jackson 2009). Iron-colored glass represents about 80 % of the analyzed samples. The BG1 group contains glass with Fe_2O_3 concentrations in the range 1.5–2.5 %, while the BG3 and BG4 glass groups comprise black appearing glass with elevated iron contents (5–10 % Fe_2O_3). For a few samples, and especially the BG1 group, the quantity of coloring agent that is detected is by itself not enough to give the glass its dark hue; thus other factors must be considered in order to explain the black appearance of these objects. Typically this involves the overall thickness and shape of the glass artifact in question.

Our hypothesis is that for BG3 glass, the introduction of iron in the glass batch is unlikely to have taken place in the primary Levantine workshops, devoted to the production and trade of widely marketable raw glass (and thus not to specialties with changing popularity such as black glass), but was probably done in a second step, coincident with the reworking and shaping of the raw glass in Western workshops (Van der Linden et al. 2009, Cagno et al. 2014) in areas where a large amount of dark appearing glass is found (e.g. the Northwestern provinces).

In Fig. 11.3, the lack of correlation between Fe and Ti concentrations in BG3 and BG4 glass, in contrast to BG1 group and mainstream Roman glass, confirms that the two elements originate from different sources—Ti is contained in the raw glass, while a pure source of Fe, such as magnetite or hammer scale (Rehren et al. 2012) is added as a colorant (Cagno et al. 2014). The appearance of a change in the procedure used to obtain deeply colored artifacts from the second half of the second century AD (i.e. from glassmaking and coloring in a single step to two separate steps), and continuing into the late Roman period, is confirmed. Moreover, inside the four macro-groups of black-appearing Roman glass, a number of subgroups can be identified on the basis of trace element content (see Table 11.3): the evaluation of the provenance and age of the glass finds included in these smaller sub-groups allows interesting conclusions to be drawn about the history of black Roman glass. We will discuss the subgroups in the following paragraphs.

When observing Ti and Zr concentrations plotted in Fig. 11.4, it is clear that samples with Ti concentrations higher than 1000 μg/g either belong to the BG1 or BG4 groups. Within the BG1 group, two subgroups can be identified. The difference between them has to be sought in the

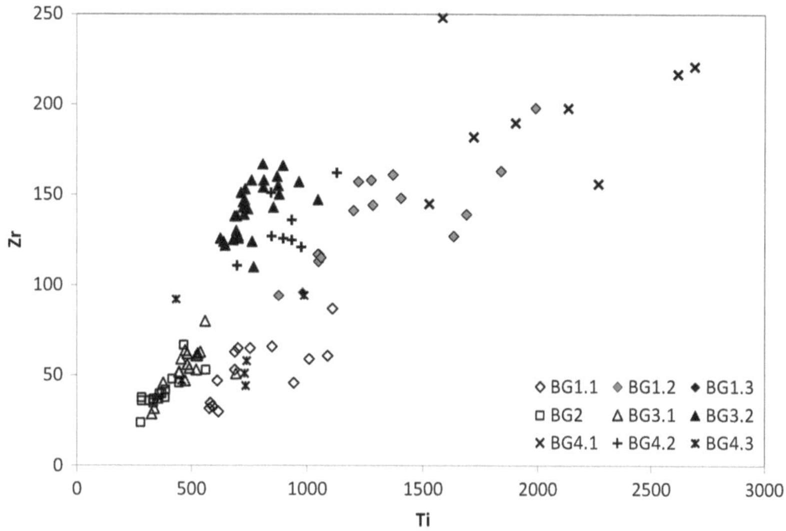

Fig. 11.4 Ti and Zr (μg/g) content in the analyzed glass samples (Cagno et al. 2014)

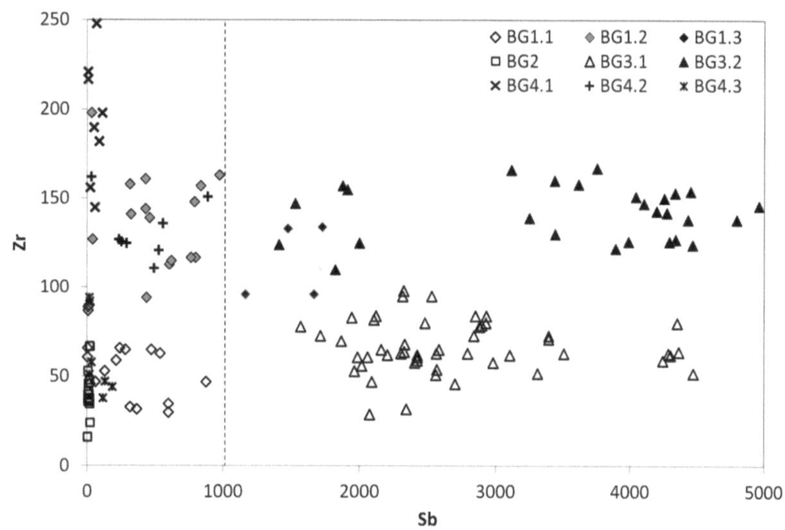

Fig. 11.5 Zr and Sb (μg/g) content in the analyzed samples (Cagno et al. 2014)

different silica sources used for the production of glass, purer in the first case, and with a higher percentage of mineral impurities in the second one. As can be seen in Fig. 11.4, group BG1.1 has Zr concentrations between 30 and 80 μg/g, and Ti concentrations between 600 and 1100 μg/g. Group BG1.2, in contrast, features much higher Zr-values (100–170 μg/g) comparable with the BG4 (HIMT) group, and higher Ti contents (1000–2000 μg/g). Besides Zr and Ti, group BG1.2 shows higher contents of a number of tri- and tetravalent silica impurities (see Table 11.3). For the BG3 group two subgroups can be observed: one (BG3.1) with Ti (400–700 μg/g) and Zr (40–80 μg/g) values similar to dark purple glass (BG2), and another one (BG3.2) with higher contents of both elements (Ti 700–1000 μg/g, Zr 100–170 μg/g), closer to BG4 (HIMT) glass. The difference between these two groups appears to be connected with the diverse geological origin of the sands used to manufacture the glass.

The distinction between HIMT and Syro-Levantine LMG glass is also evident in the trace element patterns: the BG4 glass samples have different Zr/Ti ratios than the BG3 finds, and are more similar to the earlier BG1.2 glass. In particular, BG4.2 glass features extremely high concentrations of Ti (up to 4000 μg/g) and Zr (up to 500 μg/g) that are compatible with "strong HIMT" glass (Silvestri et al. 2005). In Fig. 11.5 it can be seen that values of Sb from 1400 up to 6000 μg/g are found in the BG3 group. Such values are not found in any of the other groups, where Sb concentrations are in most cases 0 or in the 0–1000 μg/g range. The high amount of Sb in the BG3 group appears to be due to the presence of Sb in the naturally colored glass that was used as starting material to obtain black glass, consistent with the values reported by Foster and Jackson (2009).

11.4 Conclusion

The major advantage of LA-ICP-MS for glass analysis is its ability to perform major to trace element analysis of almost all elements in the periodic table within the same run; this requires micro-destructive sampling giving rise to craters of dimensions 10–100 μm, i.e. invisible to the naked eye. Moreover, the method allows one to perform lateral scans as well as depth scans, which can be extremely useful in the study of glass alteration. The case study presented here demonstrates how historical glass analysis can benefit from application of LA-ICP-MS. Here, trace analysis has shown a finer structure in our dataset that would not have been visible by measuring only major and minor components (e.g. by SEM-EDX or XRF). From this we can deduce that the role of LA-ICP-MS in glass investigation can be that of an important complement to routine glass analysis. Moreover, the discussion on quantification methods and the literature cited demonstrate that LA-ICP-MS can also be utilized as a standalone technique for glass analysis, when the sum-normalization method can be used.

Acknowledgements This research was supported by the Interuniversity Attraction Poles Programme-Belgian Science Policy (IUAP VI/16). The text also presents results of GOA "XANES meets ELNES" (Research Fund University of Antwerp, Belgium), and FWO (Brussels, Belgium) projects no. G.0704.08 and G.01769.09. We are indebted to the HERCULES program (Brussels) under grant AUHA1103 that made the purchase of the Laser Ablation sampling unit possible.

References

Aerts A, Velde B, Janssens K, Dijkman W (2003) Change in silica sources in Roman and post-Roman glass. Spectrochim Acta B 58:659–667

Aeschliman DB, Bajic SJ, Baldwin DP, Houk RS (2003) Spatially-resolved analysis of solids by laser ablation-inductively coupled plasma-mass spectrometry: trace elemental quantification without matrix-matched solid standards. J Anal Atom Spectrom 18:872–877

Barca D, Miriello D, Pecci A, Barba L, Ortiz A, Manzanilla LR, Blancas J, Crisci GM (2013) Provenance of glass shards in archaeological lime plasters by LA-ICP-MS: implications for the ancient routes from the Gulf of Mexico to Teotihuacan in Central Mexico. J Archaeol Sci 40(11):3999–4008

Becker JS, Matusch A, Depboylu C, Dobrowolska J, Zoriy MV (2007) Quantitative imaging of selenium, copper, and zinc in thin sections of biological tissues (Slugs-Genus arion) measured by laser ablation inductively coupled of plasma mass spectrometry. Anal Chem 79:6074–6080

Bertini M, Shortland A, Milek K, Krupp EM (2011) Investigation of Iron Age north-eastern Scottish glass beads using element analysis with LA-ICP-MS. J Archaeol Sci 38:2750–2766

Bertini M, Izmer A, Vanhaecke F, Krupp EM (2013) Critical evaluation of quantitative methods for the multi-elemental analysis of ancient glasses using laser ablation inductively coupled plasma mass spectrometry. J Anal Atom Spectrom 28:77–91

Boue-Bigne F, Masters BJ, Crighton JS, Sharp BL (1999) A calibration strategy for LA-ICP-MS analysis employing aqueous standards having modified absorption coefficients. J Anal Atom Spectrom 14:1665–1672

Brill RH (1999) Chemical analyses of early glasses, vol 2, Tables of analyses. Corning Museum of Glass, New York, NY

Cagno S, Janssens K, Mendera M (2008) Compositional analysis of Tuscan glass samples: in search of raw material fingerprints. Anal Bioanal Chem 391:1389–1395

Cagno S, Mendera M, Jeffries T, Janssens K (2010) Raw materials for medieval to post-medieval Tuscan glassmaking: new insight from LA-ICP-MS analyses. J Archaeol Sci 37:3030–3036

Cagno S, Favaretto L, Mendera M, Izmer A, Vanhaecke F, Janssens K (2012a) Evidence of early medieval soda ash glass in the archaeological site of San Genesio (Tuscany). J Archaeol Sci 39:1540–1552

Cagno S, Badano Brondi M, Mathis F, Strivay D, Janssens K (2012b) Study of medieval glass fragments from Savona (Italy) and their relation with the glass produced in Altare. J Archaeol Sci 39:2191–2197

Cagno S, De Raedt I, Jeffries T, Janssens K (2012c) Façon de Venise glass from early 17th century London: its composition and a comparison with known compositions of the same age. Proceedings of the SPIE8422, Integrated approaches to the study of historical glass, 842205, 7 November 2012, doi:10.1117/12.975212

Cagno S, Cosyns P, Van der Linden V, Schalm O, Izmer A, Deconinck I, Vanhaecke F, Nowak A, Wagner B, Bulska E, Nys K, Janssens K (2013) Composition data of a large collection of black-appearing Roman glass. Open J Archaeom 1. doi:10.4081/4964

Cagno S, Cosyns P, Izmer A, Vanhaecke F, Nys K, Janssens K (2014) Deeply colored and black-appearing Roman glass: a continued research. J Archaeol Sci 42:128–139

Cagno S, Cosyns P, Ceglia A, Nys K, Janssens K (2015) The use of vitrum obsianum in the Roman Empire: some new insights and future prospects. Periodico di Mineralogia 84:465–482

Claverie F, Malherbe J, Bier N, Molloy JL, Long SE (2013) Standard addition method for laser ablation ICPMS using a spinning platform. Anal Chem 85:3584–3591

Cosyns P (2011) The typological, chronological, contextual, technological and archaeometric investigation of the production and use of black glass in the Roman Empire during the 1st to 5th centuries AD. Ph.D Dissertation, Vrije Universiteit Brussel

Cosyns P, Fontaine S (2009) Le verre noir romain du 1er siècle: un aperçu. In: Janssens D, Degryse P, Cosyns P, Caen J, Van't dack L (eds) Annales du 17e Congrès de l'Association Internationale pour l'Histoire du Verre, Antwerp, 2006. Asp/Vubpress/Upa, Antwerp

Cosyns P, Hanut F (2005) Black glass of second to third-century date in northern Gaul: a preliminary survey. In: Annales of the 16th AIHV Congress, London 2003. AIHV, Nottingham, pp 113–118

Cosyns P, Janssens K, Van der Linden V, Schalm O (2006) Black glass in the Roman Empire: a work in progress. In: Creemers G, Demarsin B, Cosyns P (eds) Roman Glass from Germania Inferior. Interregional comparisons and recent finds, international colloquium, 13th May 2005 Atuatuca 1, Tongeren, Hasselt, p 30–41

De Raedt I, Janssens K, Veeckman J (1999) Compositional distinctions between 16th century 'Façon-de-Venise' and Venetian Glass Vessels, excavated in Antwerp, Belgium. J Anal Atom Spectrom 14:493–498

Dussubieux L, Robertshaw P, Glascock MD (2009) LA-ICP-MS analysis of African glass beads: laboratory inter-comparison with an emphasis on the impact of corrosion on data interpretation. Int J Mass Spectrom 284:152–161

Fitzpatrick AJ, Kurtis Kyser T, Chipley D, Beauchemin D (2008) Fabrication of solid calibration standards by a sol-gel process and use in laser ablation ICPMS. J Anal Atom Spectrom 23:244–248

Foster E, Jackson CM (2009) The composition of naturally coloured late Roman vessel glass from Britain and the implications for models of glass production and supply. J Archaeol Sci 36:189–204

Freestone IC (1994) Chemical analysis of 'raw' glass fragments. In: Hurst HR (ed) Excavations at Carthage, vol II, 1: The Circular Harbour, North Side. British Academy, Oxford

Freestone IC, Wolf S, Thirlwall M (2005) The production of HIMT glass: elemental and isotopic evidence. In: Annales of the 16th AIHV Congress, London 2003. AIHV, Nottingham, pp 153–157

Gaboardi M, Humayun M (2009) Elemental fractionation during LA-ICP-MS analysis of silicate glasses: implications for matrix-independent standardization. J Anal Atom Spectrom 24:1188–1197

Gagnon JE, Fryer BJ, Samson IM, Williams-Jones AE (2008) Quantitative analysis of silicate certified reference materials by LA-ICPMS with and without an internal standard. J Anal Atom Spectrom 23:1529–1537

Gao S, Liu X, Yuan H, Hattendorf B, Günther D, Chen L, Hu S (2002) Determination of forty two major and trace elements in USGS and NIST SRM glasses by laser ablation-inductively coupled plasma-mass spectrometry. Geostand Newslett 26: 181–196

Gratuze B (2004) Etude des perles protohistoriques en verre du tumulus de Courtesoult (Haute-Saône). IRAMAT, CNRS, Orléans

Gratuze B (2013) Glass characterisation using laser ablation inductively coupled plasma mass spectrometry methods. In: Janssens K (ed) Modern methods for analysing archaeological and historical glass, vol 1. Wiley, Chichester, pp 199–232

Gratuze B, Blet-Lemarquand M, Barrandon JN (2001) Mass spectrometry with laser sampling: a new tool to characterize archaeological materials. J Radioanal Nucl Chem 247:645–656

Guillong M, Horn I, Günther D (2003) A comparison of 266 nm, 213 nm and 193 nm produced from a single solid state Nd:YAG laser for laser ablation ICP-MS. J Anal Atom Spectrom 18:1224–1230

Günther D, Heinrich CA (1999) Comparison of the ablation behaviour of 266 nm Nd:YAG and 193 nm ArF excimer lasers for LA-ICP-MS analysis. J Anal Atom Spectrom 14:1369–1374

Halicz L, Günther D (2004) Quantitative analysis of silicates using LA-ICP-MS with liquid calibration. J Anal Atom Spectrom 19:1539–1545

Heinrich CA, Pettke T, Halter WE, Aigner-Torres M, Audetat A, Günther D, Hattendorf B, Bleiner D, Guillong M, Horn I (2003) Quantitative multi-element analysis of minerals, fluid and melt inclusions by laser-ablation inductively-coupled-plasma mass-spectrometry. Geochim Cosmochim Acta 67:3473–3497

Hellemans K, Vincke A, Cagno S, Herremans D, De Clercq W, Janssens K (2014) Composition and state of alteration of 18th-century glass finds found at the Cistercian nunnery of Clairefontaine, Belgium. J Archaeol Sci 47:121–133

Horn I, Guillong M, Günther D (2001) Wavelength dependant ablation rates for metals and silicate glasses using homogenized laser beam profiles—implications for LA-ICP-MS. Appl Surf Sci 182:91–102

Janssens K (2013a) Electron microscopy. In: Janssens K (ed) Modern methods for analysing archaeological and historical glass, vol 1. Wiley, Chichester, pp 129–154

Janssens K (2013b) X-ray based methods of analysis. In: Janssens K (ed) Modern methods for analysing archaeological and historical glass, vol 1. Wiley, Chichester, pp 79–128

Janssens K, Cagno S, De Raedt I, Degryse P (2013) Transfer of glass manufacturing technology in the sixteenth and seventeenth centuries from Southern to Northern Europe. In: Janssens K (ed) Modern methods for analysing archaeological and historical glass, vol 1. Wiley, Chichester, pp 537–562

Longerich HP, Jackson SE, Günther D (1996) Laser ablation inductively coupled plasma mass spectrometric transient signal data acquisition and analyte concentration calculation. J Anal Atom Spectrom 11:899–904

Mirti P, Casoli A, Appolonia L (1993) Scientific analysis of Roman glass from Augusta-Praetoria. Archaeometry 35:225–240

Mirti P, Pace M, Malandrino M, Ponzi MN (2009) Sasanian glass from Veh Ardašīr: new evidences by ICP-MS analysis. J Archaeol Sci 36:1061–1069

Odegard M, Skar O, Schiellerup H, Pearson NJ (2005) Preparation of a synthetic titanite glass calibration material for in situ microanalysis by direct fusion in graphite electrodes: a preliminary characterisation by EPMA and LA-ICP-MS. Geostand Geoanal Res 29:197–209

Panighello S, Orsega EF, van Elteren JT, Šelih VS (2012) Analysis of polychrome Iron Age glass vessels from Mediterranean I, II and III groups by LA-ICP-MS. J Archaeol Sci 39:2945–2955

Picon M, Vichy M (2003) D'Orient en occident: l'origine du verre à l'époque romaine et durant le haut moyenâge. In: Foy D, Nenna MD (eds) Échanges et commerce du verre dans le monde antique. Actes du colloque international de l'AFAV (Aix-en-Provence/ Marseille 2001). Éditions Monique Mergoil, Montignac, pp 17–31

Rehren T, Cholakova A, Zivanovic M (2012) The making of black glass in Late Roman Doclea, Montenegro. New Antique Doclea 3:75e90

Sayre EV, Smith RW (1961) Compositional categories of ancient glass. Science 133:1824–1826

Silvestri A, Molin G, Salviulo G (2005) Roman and Medieval glass from the Italian Area: bulk characterization and relationships with production technologies. Archaeometry 47:797–816

Silvestri A, Molin G, Salviulo G (2008) The colourless glass of Iulia Felix. J Archaeol Sci 35:331–341

Van der Linden V, Cosyns P, Schalm O, Cagno S, Nys K, Janssens K (2009) Deeply coloured and black glass in the northern provinces of the Roman Empire: differences and similarities in chemical composition before and after 150 AD. Archaeometry 51:822–844

Van Elteren JT, Tennent NH (2009) Multi-element quantification of ancient/historic glasses by laser ablation inductively coupled plasma mass spectrometry using sum normalization calibration. Anal Chim Acta 644:1–9

Van Elteren JT, Izmer A, Sala A, Orsega EF, Selih VS, Panighello S, Vanhaecke F (2013) 3D laser ablation-ICP-mass spectrometry mapping for the study of surface layer phenomena–a case study for weathered glass. J Anal Atom Spectrom 28:994–1004

Vicenzi EP, Eggins S, Logan A, Wysoczanski R (2002) Microbeam characterization of corning archeological reference glasses: new additions to the Smithsonian microbeam standard collection. J Res Natl Inst Stand Technol 107:719–727

Wagner B, Nowak A, Bulska E, Kunicki-Goldfinger J, Schalm O, Janssens K (2008) Complementary analysis of historical glass by scanning electron microscopy with energy dispersive X-ray spectroscopy and laser ablation inductively coupled plasma mass spectrometry. Microchim Acta 162:415–424

Wagner B, Nowak A, Bulska E, Hametner K, Günther D (2012) Critical assessment of the elemental composition of Corning archeological reference glasses by LA-ICP-MS. Anal Bioanal Chem 402:1667–1677

Glass Characterization Using Laser Ablation-Inductively Coupled Plasma-Mass Spectrometry Methods

12

Bernard Gratuze

Abstract

The accurate determination of archaeological glass sample composition is challenging due to possible heterogeneity and post-production chemical alteration. Here, depth profiling using the time resolved signal generated by LA-ICP-MS (DP-LA-ICP-MS) is applied to two case studies. The first study concerns corroded Late Bronze Age glass beads. While many varieties of ancient glass beads are corrosion resistant, some recipes are particularly impacted by devitrification, and depth profiling allows for analysis of un-corroded material reflecting original glass composition. In the second case study on a seventeenth century composite glass vessel, depth profiling allowed for measurement of composite layers including a thin ruby gold glass layer and white opaque glass filigree embedded in colorless lead glass. The minimal destructiveness of this technique combined with other advantages of LA-ICP-MS—the large number of elements that can be measured and low detection limits—make this technique ideal for analysis of many kinds of compositionally complex glass objects.

12.1 Introduction

LA-ICP-MS is now a well established analytical method for the characterization of ancient glass (e.g. Robertshaw et al. 2014; Varberg et al. 2015; Walder 2013), however its use is generally limited to the measurement of the bulk composition of objects. The focus of this paper is the application of LA-ICP-MS to the quasi-non-destructive determination of concentration profiles in ancient glass using time resolved signals (Heinrich et al. 2003; Mason and Manks 2001; Woodhead et al. 2008). This approach will be called here depth profile or DP-LA-ICP-MS. LA-ICP-MS will refer to standard single spots analyses. We will illustrate DP-LA-ICP-MS with two case studies: measurement of the composition of heavily corroded Late Bronze Age glass beads, and the characterization of seventeenth century composite glass vessels.

B. Gratuze (✉)
Institut de Recherche sur les Archéomatériaux, Centre Ernest Babelon, CNRS/Université d'Orléans, 3D rue de la Férollerie, 45071 Orléans Cedex 2, France
e-mail: gratuze@cnrs-orleans.fr

L. Dussubieux et al. (eds.), *Recent Advances in Laser Ablation ICP-MS for Archaeology*, Natural Science in Archaeology, DOI 10.1007/978-3-662-49894-1_12

During the Late Bronze Age, a new type of glass referred as LMHK (Low Magnesium High Potassium) or mixed potash-soda glass (Henderson 1988, 1989) appeared in northern Italy. Recently, Angelini et al. (2004) identified a new type of LMHK glass containing more potash and less soda than those previously studied. While typical LMHK glass appears to be quite resistant to corrosion, this new glass type is significantly impacted by devitrification. As the thickness of the resulting corroded layer is unknown, analyses of this glass using LA-ICP-MS will require uptake times of variable duration that is difficult to estimate. The realization of concentration profiles appears to be the easiest solution for the characterization of these devitrified glasses.

In 2010, 14 whole glass pieces attributed to Bernard Perrot, a famous Italian glass worker that moved to Orléans (France) in 1662, were quasi non-destructively analyzed with either Particle-induced X-ray emission (PIXE), or for very small samples, LA-ICP-MS (Biron et al. 2010). Among them, some objects contained gold ruby glass. The results obtained for gold concentrations on gold ruby glass show a large dispersion: in some glass samples, no gold was detected (mostly with PIXE) while in glass analysed using LA-ICP-MS, gold concentrations were far below the usual gold concentrations found in this type of glass (Biron et al. 2011). The recent discovery in Orléans of a broken piece of glass produced by Bernard Perrot shows that the red glass which seems to occur at the surface of the object is actually embedded inside a colourless lead glass layer. The presence of this lead glass explains the difficulties in obtaining a representative analysis of the glass encountered during PIXE and LA-ICP-MS measurements. Our results show that in such cases DP-LA-ICP-MS can be very helpful for characterizing the different types of glass making up such complex glass objects.

12.2 Background

Only a few analytical methods are commonly used to determine the composition of ancient glass. A review of recent publications dealing with that subject (e.g. Barca et al. 2009; Carmona et al. 2010; Giussani et al. 2009; van der Werf et al. 2009; Walton et al. 2009; Wedepohl et al. 2011) shows that the most common methods remain Electron Probe Microanalysis (EPMA) and Scanning Electron Microscopy with Energy or Wavelength Dispersive X-ray Spectroscopy (SEM-EDX or WDX). Next most common are other X-ray methods: X-ray fluorescence, particle induced X-ray and gamma ray emission (PIXE and PIGE) and liquid mode or laser ablation ICP-MS.

LA-ICP-MS presents many advantages over the other analytical techniques mentioned above. Ancient glass artifacts are often weathered, thus the corroded surface of these objects has a different composition when compared to the intact glass underneath. Depending on the type of glass and on the intensity of the weathering, this perturbed layer may be several hundreds of μm thick. Most X-rays methods only measure material to a depth of a few microns, and thus typically do not reach intact glass. Moreover, these methods are highly sensitive to surface effects and need to be applied on very flat surfaces. Embedding of the glass sample in resin and polishing is often necessary, which is destructive, time consuming and expensive. LA-ICP-MS analysis can be done on unprepared objects. Removal of the corrosion layer is easily carried out with a pre-ablation of the area to be analyzed, regardless of the thickness of the corrosion layer. It is possible to get reliable results down to 1 mm below the surface. As the laser sampling is invisible to the naked eye, it is thus possible to analyze a greater number of objects and to get a better representativeness of compositional variation within a population of such glass objects.

Most of the energy dispersive X-rays methods suffer matrix effects and interferences. For example, the detection limits of bismuth or gold in a glass are impacted when lead is also present in high concentrations. LA-ICP-MS allows the measurement of a large number of chemical elements (present either as major, minor or trace components) with only a few chemical or spectrometric interferences. It is therefore possible to obtain reliable compositions for all kinds of glass matrices. LA-ICP-MS also offers lower detection limits than any energy or wavelength dispersive X-rays method.

Although the spatial resolution of LA-ICP-MS analysis is not as good as that of EPMA, this method is able to characterize all the different parts of composite objects such as millefiori and mosaic glasses (Nenna and Gratuze 2009). Lastly, LA-ICP-MS offers the possibility to measure concentration profiles for stratified glass samples. This is particularly important for glass objects made with different layers (coloration by surface treatment, inserted metal foil, etc.). Full characterization of these types of objects can be undertaken without any sample preparation and with no resulting visible damage. One of the disadvantages of LA-ICP-MS is related to the maximum size of the object that can be analyzed; however, this problem can be overcome either by using large or very small ablation cells (sealed to the object) or by carrying out direct atmospheric sampling (Fricker et al. 2011; Glaus et al. 2012; Golitko et al. Chap. 3; Kovacs et al. 2010).

12.3 Analytical Protocol

Analyses of glass objects conducted at the Centre Ernest-Babelon of the IRAMAT (Orléans) are carried out with an Element XR mass spectrometer from Thermofisher Instrument and a VG UV laser ablation device. The mass spectrometer offers the advantage of being equipped with a three stage detector system: a conventional high performance discrete dynode, a dual mode (counting and analog modes) secondary electron multiplier (SEM) with a linear dynamic range of over nine orders of magnitude, and a single Faraday collector (combined with the SEM). This last collector allows an increase of the linear dynamic range by an additional three orders of magnitude. A wide cross-over range (>2 orders of magnitude) between the different detector modes allows an accurate automated cross-calibration. This feature is particularly important for laser ablation analysis as dilution of sample is impossible compared to ICP-MS with liquid sample introduction. For glass, it is possible to analyze major, minor, and trace elements in a single run regardless of their concentrations and their isotopic abundance. For example, it is possible to analyze high lead glass (Pb > 20 %) and glass in which this element is present at the ppm (parts per million) level with the same parameters.

Routinely, the concentrations of 52 elements are determined in glass objects:

- The main major and minor ancient glass constituents which are often present at the percent level or above: silicon, sodium, calcium, potassium, aluminum, magnesium, lead, chlorine, phosphorus and iron,
- most of the coloring and opacifying agents and their associated impurities: cobalt, copper, antimony, tin and manganese but also zinc, nickel, arsenic, barium, chromium, vanadium, gold and silver,
- 29 other trace elements ranging from lithium to uranium. Among them the most important for glass classification studies are rubidium, strontium, cesium, zirconium, uranium, thorium and the rare earth elements.

Glass object are placed in the ablation cell together with standard reference materials (SRM). At the Centre Ernest-Babelon, three ablation cells of 5, 8 and 12 cm diameter and 4 cm height have been designed specifically for archaeological objects, ranging from small beads or glass fragments to entire glass objects such as bracelets, earrings or small flasks. The objects and SRMs are sampled by a laser beam generated by a Nd:YAG pulsed laser with a wavelength of 266 nm. The laser operates at a maximum energy of 3–4 mJ and at a maximum pulse frequency of 15 Hz. The diameter of the ablation crater ranges from 60 to 100 μm, and its depth is around 250 μm, depending of the ablation duration and the laser pulse frequency. Standard parameters are 70 s of ablation (20 s for pre-ablation and 50 s for analysis) and 6–8 Hz laser pulse frequency. A pre-ablation time of 20 s is set in order to eliminate the transient part of the signal and ensure that possible surface contamination or corrosion does not affect the results of the analysis.

Pre-ablation parameters (duration and laser pulse frequency) can be adjusted according to the thickness of the corrosion layer at the surface of the glass in order to reach uncorroded glass.

An argon gas flow carries the ablated aerosol to the injector inlet of the plasma torch, where the matter is dissociated, atomized and ionized. Typical flow rate values range from 1.15 to 1.35 L/min depending on the laser cell size. The ions are then injected into the vacuum chamber of a high resolution mass spectrometer, which filters the ions depending upon their mass-to-charge ratio. The ions are then collected by the channel electron multiplier or the Faraday cup.

Measurements are carried out in peak jump acquisition mode, with four points per peak for counting and analogue detection modes and ten points per peak for Faraday detection. Automatic detection mode is used for most of elements; only sodium, aluminum and potassium are systematically detected with the Faraday detector. Silicon, used as the internal standard, is measured using the ^{29}Si isotope. The mass scanning velocity of double-focusing high resolution mass spectrometers is slower than for quadrupole systems. With our analytical parameters the scanning time necessary to measure 52 selected isotopes is about 2.6 s, thus only 20 scans of the total mass range are done for each ablation. For glass analysis, most isobaric interferences can be resolved by selecting an isotope that is not impacted by interference, even if its isotopic abundance is low. It is thus possible to work in low resolution mode for all the measured isotopes. This allows for a reduction in the duration of each scan while still working with the highest sensitivity.

SRMs are used to calculate the response coefficient factor K_y as defined by Gratuze (1999) and thus convert signal intensities generated by the mass spectrometer for each element into fully quantitative data. To prevent matrix effects, the composition of these SRMs has to be as close as possible to that of the samples. Three different types of standards are used to measure major, minor and trace elements (Table 12.1):

- SRM610, manufactured by the National Institute for Standards and Technology.
- Corning glasses B, C and D.
- An in-house standard glass with composition determined by Fast Neutron Activation Analysis.

SRM610 is a soda-lime-silica glass doped with trace elements with concentrations in the range of 500 ppm. Certified values have been published by several authors for this glass. A recent compilation was proposed by Jochum et al. (2011), however, we use the values reported by Hollocher and Ruiz (1995), Norman et al. (1996) and Pearce et al. (1997). SRM610 is used to calculate the K_y response coefficient factors for all elements except for magnesium, chlorine, phosphorus, potassium and iron, which are present at low levels in this glass, as well as aluminum, whose concentration in SRM610 is not reliable enough to generate acceptable results.

The Corning glass series was designed to match compositions of ancient glasses (Brill 1999: 544; Bronk and Freestone 2001; Dussubieux et al. 2009; Verità et al. 1994; Vicenzi et al. 2002; Wagner et al. 2012). Corning B is a mineral soda-lime glass, C a lead-barium-potash glass, and D is a lime-potash glass. These glasses also contain coloring agents such as cobalt, copper, antimony, and other elements. This set of reference materials is mainly used to calculate the response coefficient factors for sodium, magnesium, aluminum, phosphorus, potassium, calcium, manganese, iron, cobalt, copper, antimony and lead concentrations. The in-house standard is used to calculate K_y for chlorine but also for some major elements such as sodium, magnesium, aluminum, phosphorus, potassium, calcium, manganese and iron.

Each analysis consists of a blank measurement followed by two ablations located at different places on the object. At the beginning of the analytical sequence, all the SRMs are analyzed once or more (two or three times for SRM 610) before starting archaeological glass analysis. To improve reproducibility and to correct eventual

Table 12.1 Measured isotopes, detection mode and reference material used for standardization

Used isotope	Detection mode	Reference materials used for K calculations	Main interferences
Na23	Faraday	SRM610, Corning B and D and laboratory glass standard	
Al27, K39	Faraday	Corning B and D and laboratory glass standard	
Cl35	Pulse counting or analogue	laboratory glass standard	
Mg24, P31, Fe57	Pulse counting or analogue	Corning B and D and laboratory glass standard	
Ni60	Pulse counting or analogue	SRM610, Corning B	
Si29, Ca44, Ti47, Mn55, Co59, Cu63, Sr88, Sn118, Sb121	Pulse counting or analogue	SRM610, Corning B and D and laboratory glass standard	Negligible Si28O16 on Ca44
Ba137, Pb204	Pulse counting or analogue	SRM610, Corning B, C and D	For profile measurement Pb206 or 208 are used instead of Pb204
Li7, B11, V51, Cr52, Zn66, As75, Rb85, Y89, Zr90, Nb93, Ag107, In115, Cs133, La139, Ce140, Pr141, Nd146, Sm147, Eu153, Gd157, Tb159, Dy163, Ho165, Er166, Tm169, Yb172, Lu175, Hf178, Ta181, Au197, Bi209, Th232, U238	Pulse counting or analogue	SRM610	Sn115 on In115, needs correction
Other occasionally analysed elements: Be9, Se78, Mo95, Cd111, W182	Pulse counting or analogue	SRM610	Kr78 on Se78 correction is made with background measurement

instrumental drift or fluctuation in ablation efficiency, the SRM 610 and, at random, one other reference material from the above list are systematically analyzed every six or eight samples. At the end of the analytical sequence all the SRMs are analyzed again. Corning A glass (a soda plant ash glass type) is also analyzed as an unknown sample from time to time to check the accuracy of the results (Table 12.2). At the end of the analytical session, an average response coefficient factor K_y is calculated from all the analyzed SRMs. If instrumental drift is observed, the session is divided in different segments in order that the standard deviation on the average coefficients remains below 10 % relative standard deviation.

Concentrations are calculated using net average intensity count rates measured for each isotope. The following formula is used to calculate concentrations for all elements assuming that the sum of their concentrations in weight percent in glass is equal to 100 %:

$$\%\mathrm{Y_mO_n} = \frac{\mathrm{I_Y} \times \alpha_\mathrm{Y}}{\mathrm{I_{Si}} \times \mathrm{K_Y}} \Big/ \sum \frac{\mathrm{I_X} \times \alpha_\mathrm{X}}{\mathrm{I_{Si}} \times \mathrm{K_X}} \tag{12.1}$$

I_y, I_x and I_{Si} are the net intensity counts rates, corrected for isotopic abundance, measured for elements Y, X and silicon (the internal standard). α_y and α_x are the conversion factors from element to oxide for elements Y and X. K_y and K_x are the response coefficient factors for elements Y and X, calculated as follows:

$$\mathrm{K_Y} = \frac{\mathrm{I_{Y_{std}}} \times [\mathrm{Conc}]_{\mathrm{Si_{std}}}}{\mathrm{I_{Si_{std}}} \times [\mathrm{Conc}]_{\mathrm{Y_{std}}}} \tag{12.2}$$

I_{ystd} and I_{Sistd} are the net intensity counts rates, corrected for isotopic abundance, measured for elements Y and silicon in the standard material and $[Conc]_{Ystd}$ and $[Conc]_{Sistd}$ are the concentrations of element Y and Si in the standard material.

For a soda or a potash lime glass, with an average composition of 70 % silica, 20 % soda

Table 12.2 Typical average values measured in Corning A glass

Oxide	Average	Standard deviation	Oxide	Average	Standard deviation
Values in %			Values in ppm		
B_2O_3	0.20	0.01	SrO	964	18
Na_2O	15.0	0.2	Y_2O_3	0.45	0.05
MgO	2.43	0.01	ZrO_2	49	3
Al_2O_3	0.74	0.03	Nb_2O_3	0.67	0.05
SiO_2	66.8	0.3	Ag	17	1
P_2O_5	0.13	0.01	Cs_2O	0.21	0.03
Cl	0.13	0.01	La_2O_3	0.4	0.1
K_2O	3.01	0.05	CeO_2	0.6	0.5
CaO	5.08	0.14	PrO_2	0.1	0.0
TiO_2	0.68	0.04	Nd_2O_3	0.2	0.1
MnO	1.00	0.01	Sm_2O_3	0.04	0.02
Fe_2O_3	1.10	0.03	Eu_2O_3	0.02	0.02
CoO	0.17	0.01	Gd_2O_3	nd	nd
CuO	1.26	0.04	Tb_2O_3	0.01	0.00
SnO_2	0.17	0.003	Dy_2O_3	0.05	0.00
Sb_2O_3	1.54	0.04	Ho_2O_3	0.01	0.00
BaO	0.42	0.02	Er_2O_3	0.02	0.00
PbO	0.067	0.005	Tm_2O_3	0.01	0.00
Values in ppm			Yb_2O_3	0.03	0.01
Li_2O	103	4	Lu_2O_3	0.02	0.03
V_2O_5	61	3	HfO_2	1.1	0.1
Cr_2O_3	30	2	Ta_2O_3	0.12	0.01
NiO	231	6	Au	0.2	0.1
ZnO	496	22	Bi	8.1	0.4
As_2O_3	30	2	ThO_2	0.34	0.03
Rb_2O	98	4	UO_2	0.27	0.04

or potash and 10 % lime, the experimental detection limits calculated on the basis of a peak intensity equal to three times the standard deviation of the average value of the background intensity ranges from 0.07 % to 0.002 % for minor elements and from a few ppb to 10 ppm for the other elements (Table 12.3). Accuracy and reproducibility are difficult to estimate when dealing with archaeological materials. These factors can only be assessed by using the SRMs and by measuring the difference between certified values and calculated values for the accuracy and the deviation measured on the average response coefficient factor K for reproducibility. A good estimate of these two analytical parameters is given by the average content values and their associated standard deviations for Corning A glass (Table 12.2).

Table 12.3 Range of detection limits achieved for glass characterization with the analytical protocol developed with the Element XR

Average range of ldd values	Elements
Below 10 ppb	Pr, Tb, Ho, Tm, Lu, Ta, Th, U
Between 10 and 25 ppb	Y, Nb, In, La, Ce, Nd, Sm, Dy, Er, Yb, Hf, Bi
Between 50 and 100 ppb	V, Au
Between 100 and 500 ppb	Co, As, Rb, Sr, Zr, Ag, Sb, Cs, Eu
Between 500 and 1000 ppb	Ni, Zn, Gd
Between 1 and 10 ppm	Li, B, P, Ti, Cr, Mn, Cu, Sn Ba
Between 25 and 50 ppm	Mg, Fe, Pb
Between 200 and 700 ppm	Na, Al, K, Ca, Cl

The protocols for LA-ICP-MS and DP-LA-ICP-MS are not fundamentally different: however, it is important to note that no standard analytical protocol is used for DP-LA-ICP-MS analysis. The laser repetition rate and the ablation duration depend of the thickness of the analyzed glass structure. A blank measurement is not carried out independently, but is done during the first 15 s of acquisition. DP-LA-ICP-MS is usually carried out on a shorter list of isotopes even if it can be also performed on the full list of isotopes. The list of analyzed isotopes depends on the questions to be addressed (characterization of metal foil between two glass layers or of surface layers, analysis of highly corroded glasses, etc.).

Profile measurements are conducted with the same parameters on samples and on SRMs. The above calculation method is then applied to obtain oxide concentrations for each time slice. A time slice is an individual scan of the whole mass range. It lasts from 0.2 to 0.3 s depending on the analytical parameters. If the value of each K_y factor does not vary in time for the different reference materials, an average K_y factor is used to calculate the concentrations for the whole profile. Depending on the duration of the profile, chemical fractionation may occur for some elements such as zinc or lead causing a time dependence of K_y. For these elements, the K_y factors are calculated for each time slice. In both cases, we assume that the sum of the concentrations of all the elements, in weight percent, is equal to 100 % for each time slice. From a general point of view, when using profile analysis, the glass compositions are calculated when the signals of the elements of interest reach a plateau.

12.4 DP-LA-ICP-MS Applied to Ancient Glass

12.4.1 Estimation of the Analyzed Depth for Profile Analysis

In order to estimate the depth reached during analysis, ablation was performed on glasses with different colors and compositions and for different durations (1, 5 and 10 min) with laser frequencies of 5 and 10 Hz. Two different diaphragm apertures (0.08 and 0.12 mm) were used. The results show that the analyzed depth varies according to the glass type (color and composition) and the diaphragm apertures. Depths of 0.50 and 0.74 mm were reached, respectively, for apertures of 0.08 and 0.12 mm, with a laser frequency of 5 Hz and a duration of 5 min. Slight variation of the crater diameter was observed for the same diaphragm aperture, depending on glass type. During the first seconds the relationship between the depth and the number of laser shots is not linear as at first the laser is focused on the surface and then defocuses progressively. After one minute, a linear relationship between the ablation duration and the ablation depth is observed. For ablation durations longer than 5 min, an average ablation rate varying between 0.3 and 0.5 μm/shot according to the glass type and the diaphragm aperture is obtained. Working at 5 Hz with an aperture of 0.12 mm, an average ablation rate varying between 0.41 and 0.49 μm/shot is reached for five minutes of ablation. It corresponds to an analyzed depth of 0.65 to 0.74 mm. The measured average ablation rate after one minute of ablation appears to be more variable: from 0.64 to 1.0 μm/shot, i.e. a depth which varies from 0.2 to 0.3 mm after one minute of ablation. With the same parameters the analyzed depth reaches from 1.2 to 1.5 mm after 10 min. Our measurements show that the ablation rate is not affected by the laser shot rate: for the same type of glass, the depths reached while working at 5 Hz for 10 min or 10 Hz for 5 min are equal. In corroded glass the ablation rate is much more important and depending on its structure, its thickness and its cohesion, the corrosion layer will be crossed over more or less rapidly.

12.4.2 High Potassium (LMHK) Glass Beads of the Late Bronze Age in Western Europe

During the Late Bronze Age, glass beads with a particular mixed alkali glass composition, referred as LMHK glass by Henderson (1988, 1989), are found through all Western Europe from the British Isles (Raftery and Henderson 1987) down to the north of Greece (Nikita

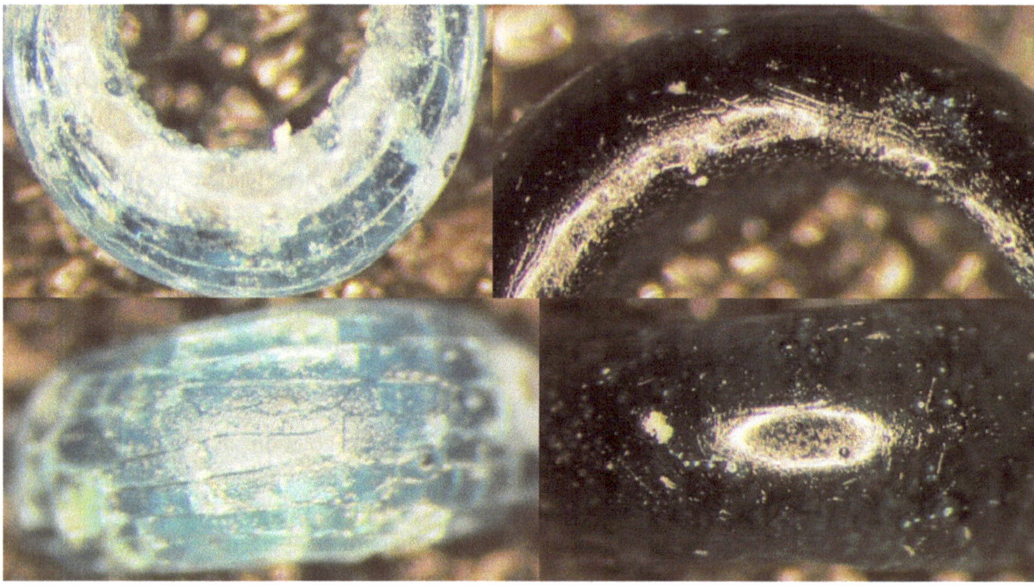

Fig. 12.1 Mathay glass beads. The cracked surface of the highly corroded potash glass bead on the *left hand side* consists mainly of hydrated amorphous silica. The classical soda-potash glass bead on the *right hand side* shows no sign of weathering

et al. 2009), including Bohemia (Venclová et al. 2011), France (Azemar et al. 2000; Gratuze et al. 1998; Gratuze and Billaud 2003; Guilaine et al. 1991), Belgium (Cosyns et al. 2005), Germany (Hartmann et al. 1997), Switzerland (Henderson 1993) and the north of Italy (Angelini et al. 2004; Biavati and Verità 1989; Bellintani et al. 1998; Brill 1992). Among the different typologies encountered for this type of material, three of them are largely predominant outside Italy: small annular copper blue beads, barrel shaped beads decorated with a continuous spiral of white opaque glass, and globular glass beads decorated with three or four prominent horns formed by stratified blue and white glass layers (Bellintani and Stefan 2009). As shown by numerous studies, this particular glass probably originates from the region of Frattesina di Fratta Polesine, Rovigo province, northern Italy (Biavati and Verità 1989).

During the same period, another glass composition, represented by only a few specimens, has been identified. In this glass, soda is practically entirely replaced by potash. This particular composition, which represents a high potash type of LMHK glass, has been found in northern Italy (Angelini et al. 2004) and elsewhere (Ploin et al. 2012; Venclová et al. 2011). While the classic LMHK glass appears to be quite resistant to corrosion, its potash rich counterpart seems to suffer strongly from weathering effects as shown in Fig. 12.1. Their corroded aspect makes the high potash LMHK glass beads easily recognizable from other beads by simple visual examination (Fig. 12.1). The fragility of these beads, which are often nearly entirely corroded, may explain why they are less represented in most studied Bronze Age sites.

The recent discovery of the Mathay deposit (Doubs, France) gives us the opportunity to study a large representative population of the glass beads exported by northern Italian glass workshops. At this site, more than one thousand glass beads, together with copper and gold ornaments, were discovered in a ceramic jar. Among these glass beads, 127 show evidence of strong corrosion. Twenty-two of these beads were selected for DP-LA-ICP-MS analysis after careful examination of their surface with a binocular microscope.

The results show that for a large part of these beads, the average signal measured using the

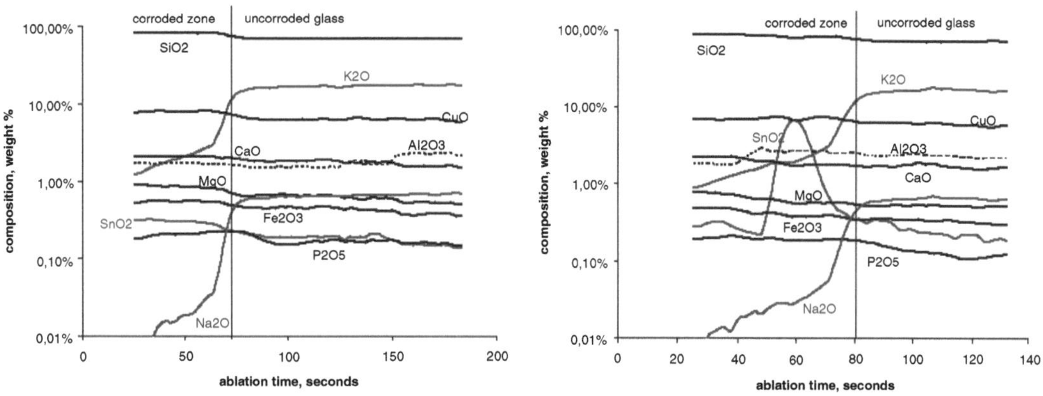

Fig. 12.2 (**a** and **b**) Concentration profiles measured on two high potash glass beads from Mathay. A strong depletion in potash and soda is observed in (**a**) at the beginning of the ablation, although a 20 s ablation uptake time was employed. In (**b**), enrichment in tin oxide is observed at the interface between the corroded zone and the uncorroded glass

standard laser ablation time, which is about 60 s after 20 s of uptake time, is not representative of the whole glass composition but reflects instead the composition of the corroded glass at the surface of the beads (Fig. 12.2a). In some cases, a chemical enrichment of elements such as tin or phosphorus occurs at the interface between the weathering layer and the pristine glass (Fig. 12.2b). Thus, due to the presence of this weathered layer, which has a highly variable thickness, standard LA-ICP-MS analysis will first underestimate potash and soda content, and may secondly overestimate the content of other elements such as tin. However, a good estimation of the glass composition can be derived from the concentration profiles measured for most of these beads. Table 12.4 gives the major constituent compositions of these beads measured first by standard LA-ICP-MS and then by DP-LA-ICP-MS.

12.4.3 Seventeenth Century Lead Gold Ruby Glass

In 2010, PIXE and LA-ICP-MS analyses were carried out on 14 whole glass pieces attributed to Bernard Perrot (Biron et al. 2010). This Italian glass worker, who settled in Orléans (France) in 1662, is known for the use or invention of different glass recipes to produce lead glass, gold ruby glass and white porcelain glass. The present analysis focuses on a particular type of gold ruby glass attributed to Bernard Perrot. This red glass was manufactured using gold and arsenic rather than gold and tin chloride, as described for the Purple of Cassius made by Kunckel (von Kerssenbrock-Kosigk 2008). All measurements with PIXE and LA-ICP-MS were conducted in a quasi-non-destructive manner, directly on the surface of the objects without any sample preparation. The detection limits for gold with both methods are below the ppm level.

Gold concentrations show wide variability: they range from no gold (gold concentrations below the detection limits) to 287 ppm for PIXE measurements and from 60 to 284 ppm for LA-ICP-MS measurements. This variability in gold concentration was attributed to the possible presence of a thin colorless glass layer over the red glass (Biron et al. 2011). The recent discovery in Orleans of a broken piece of glass with ruby red and porcelain white decorations, which probably comes from a glass cup produced by Perrot (Geyssant 2010; Geyssant et al. 2010: 54, 104, 148–149) confirms this hypothesis. Although it appears that the red glass occurs at the surface, observation of the broken edge of the object shows that this glass is embedded in a colorless glass layer. The thickness of this colorless layer, which varies from 100 to 300 μm (Fig. 12.3c) prevents the measurement of gold

Table 12.4 Composition of the surface (corroded zone) and the interior (uncorroded glass) of the glass beads derived from profiles shown in Fig. 12.2

	Left profile		Right profile		Composition of potash glass beads	
	Surface	Interior	Surface	Interior	Average	Standard deviation
Na_2O	0.02	0.66	0.03	0.66	1.1	*0.6*
MgO	0.86	0.62	0.62	0.52	0.70	*0.17*
Al_2O_3	1.69	1.70	2.59	2.26	1.72	*0.47*
SiO_2	84.1	70.8	82.7	71.4	71.3	*1.5*
P_2O_5	0.21	0.17	0.19	0.12	0.24	*0.10*
Cl	0.052	0.039	0.002	not detected	0.04	*0.04*
K_2O	1.91	17.1	1.83	16.9	16.3	*1.2*
CaO	2.12	1.79	1.91	1.67	2.23	*0.74*
MnO	0.0092	0.0077	0.0081	0.0070	0.0095	*0.0028*
Fe_2O_3	0.56	0.44	0.41	0.32	0.45	*0.11*
CoO	0.0033	0.0027	0.0029	0.0024	0.0027	*0.0008*
CuO	8.10	6.45	7.13	5.91	5.53	*1.33*
ZnO	0.013	0.009	0.012	0.006	0.010	*0.004*
As_2O_3	0.0094	0.0063	0.0088	0.0046	0.052	*0.015*
Rb_2O	0.017	0.020	0.015	0.019	0.023	*0.006*
SrO	0.012	0.010	0.012	0.010	0.013	*0.006*
ZrO_2	0.0015	0.0013	0.0016	0.0014	0.0017	*0.0006*
Ag	0.0012	0.0018	0.0014	0.0016	0.0017	*0.0011*
SnO_2	0.31	0.19	2.54	0.21	0.30	*0.24*
Sb_2O_3	0.0033	0.0026	0.0034	0.0021	0.0023	*0.0014*
BaO	0.0068	0.0059	0.0068	0.0059	0.0071	*0.0019*
Au	0.0001	0.0001	0.0001	0.0001	0.0001	*0.0001*
PbO	0.0005	not detected	not detected	not detected	0.0058	*0.0075*

All values are expressed in mass percent

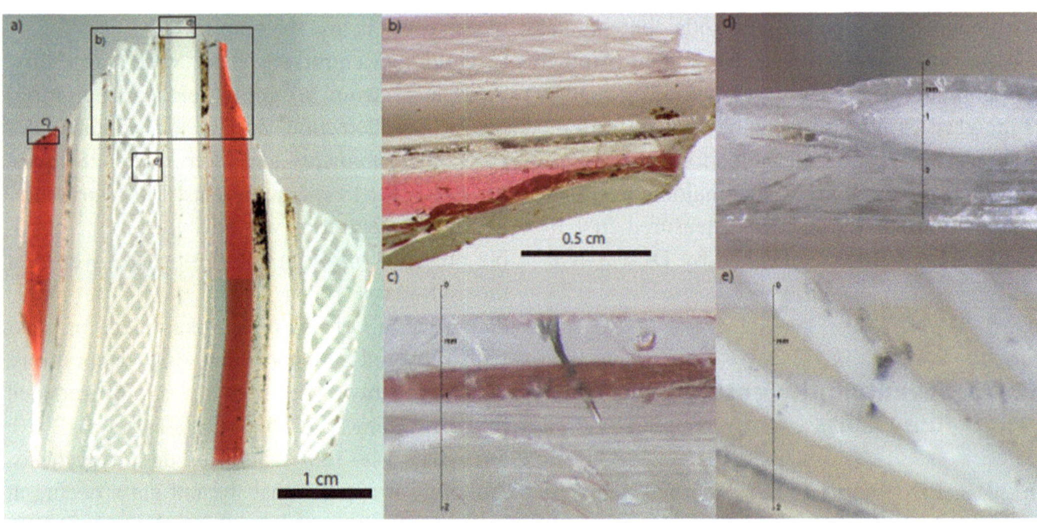

Fig. 12.3 (**a**) Perrot's glass fragment; (**b**) and (**d**) detail of the red and white glasses; (**c**) and (**e**) ablation craters on the colorless and red glasses and on the colorless and white glasses

when using PIXE and seriously alters the average gold signal measured with conventional LA-ICP-MS, as the resulting measured values are obtained from an unknown mixture of colorless and gold ruby glasses. By carrying out measurement of concentration profiles it is possible to determine the composition of both glasses: the colorless layer and the embedded red ruby (or in other cases, white porcelain) glass. As this object gave us the possibility to measure the composition of the different glasses (red, white and colourless) either directly or via the concentration profile, it was used to test the validity of compositions determined using depth-profiling.

In the present case, a much greater glass thickness must be characterized than for the late Bronze Age glasses analysed in the previous example (about 1 or 2 mm instead of a few tenth of mm). The use of a longer ablation time and a higher laser repetition rate (10 Hz instead of 5 Hz) to allow measurement at greater depths however leads to the occurrence of chemical fractionation relative to silicon for some elements. In Fig. 12.4a, it is apparent that the response coefficient factor for elements such as lead, zinc, gold, tin, antimony and arsenic tend to increase over the course of ablation, while coefficients for alkaline elements (Na, K) decrease. Those of aluminum, magnesium, and calcium tend to be fairly stable for the duration of ablation (Fig. 12.4b). In order to assess the impact of chemical fractionation, concentration profiles measured for red and white glasses (Fig. 12.3c, e) have been plotted in two different ways: with and without fractionation corrections (Figs. 12.5 and 12.6 for red glass and 12.7 to 12.9 for white glass). Concentrations calculated from corrected profiles were then compared to those determined directly on the different glasses by conventional LA-ICP-MS (Tables 12.5 and 12.6).

In Figs. 12.5, 12.6, 12.7, 12.8 and 12.9, which show the corrected and uncorrected profiles measured for red and white glasses, we observe that the uncorrected profiles (normal line) for oxides of elements such as sodium, phosphorus, potassium, zinc, arsenic, antimony, gold and lead do not show a concentration plateau but a continuous increase or decrease of their elemental concentrations. Concentration plateaus are however observed for these elements on the corrected profiles (bold line). However, for some other elements such as magnesium, calcium, manganese, cobalt, zirconium and barium, both profiles are identical and the differences between the two curves are mainly due to the fact that it is assumed that the sum of all oxides is equal to 100 %. It is also apparent in Fig. 12.7 that no real plateaus are observed for the white glass for antimony and arsenic oxides. This is probably due to the fact that white glass is not homogeneous but contains small opacifier crystals made of antimony, arsenic and probably lime. Some areas with higher concentrations of opacifier crystals (brighter white spots) can be observed on Fig. 12.3e.

Another interesting feature is evident when sodium oxide profiles are compared (Figs. 12.5

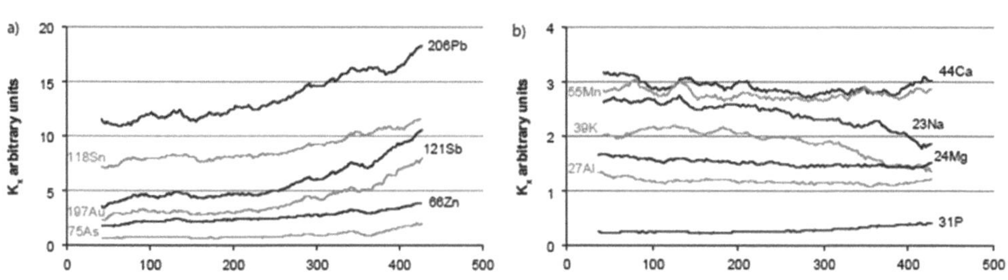

Fig. 12.4 Evolution of average response coefficient factors (calculated from NIST610 and Corning B, C and D) for different elements during ablation. Coefficients for lead, tin, antimony, gold, zinc, arsenic and phosphorus tend to increase while those of sodium and potassium decrease. Others coefficients such as those of calcium, manganese, magnesium and aluminium exhibit greater stability over the course of ablation

Table 12.5 Compositions of the colorless glass component of the Bernard Perrot cup calculated from depth-profiling (DP-LA-ICP-MS) and obtained directly by single spot analysis (LA-ICP-MS)

	Surface colorless glass layer				Colorless glass						
	1st red profile	2nd red profile	White filigree profile	White rod profile	Direct LA-ICP-MS analysis	1st red profile	2nd red profile	White filigree profile	White rod profile	After the red rod (1st red profile)	Between the white filigree
Na_2O	0.28	0.26	0.56	0.60	0.28	0.35	0.31	0.85	0.90	0.20	0.78
MgO	0.075	0.082	0.11	0.12	0.075	0.066	0.075	0.094	0.11	0.058	0.11
Al_2O_3	0.27	0.23	0.24	0.26	0.20	0.22	0.20	0.19	0.23	0.22	0.19
SiO_2	65.2	64.9	65.0	64.8	63.2	61.9	62.6	62.1	62.8	61.4	61.6
P_2O_5	0.065	0.074	0.066	0.071	0.075	0.058	0.073	0.061	0.075	0.085	0.094
Cl	0.20	0.19	0.32	0.28	0.20	0.17	0.17	0.29	0.21	0.28	0.39
K_2O	12.7	12.0	12.6	12.4	13.1	14.1	14.2	13.2	14.9	12.7	12.6
CaO	0.59	0.62	0.80	0.84	0.52	0.51	0.56	0.64	0.83	0.23	0.55
MnO	0.082	0.092	0.10	0.12	0.077	0.081	0.087	0.092	0.13	0.069	0.082
Fe_2O_3	0.099	0.10	0.13	0.14	0.086	0.098	0.099	0.13	0.14	0.079	0.12
CoO	0.0001	0.0001	0.0001	0.0001	0.0001	0.0001	0.0001	0.0001	0.0002	0.0001	0.0001
CuO	0.0036	0.0025	0.0039	0.0045	0.0037	0.0040	0.0030	0.0042	0.0050	0.0040	0.0045
ZnO	0.0020	0.0020	0.0024	0.0029	0.0019	0.0020	0.0019	0.0021	0.0032	0.0029	0.0033
As_2O_3	0.048	0.052	0.043	0.041	0.049	0.049	0.052	0.043	0.039	0.064	0.058
Rb_2O	0.0056	0.0053	0.0064	0.0069	0.0056	0.0059	0.0058	0.0068	0.0080	0.0047	0.0055
SrO	0.0023	0.0026	0.0032	0.0034	0.0021	0.0019	0.0022	0.0027	0.0030	0.0018	0.0029
ZrO_2	0.030	0.027	0.027	0.028	0.024	0.024	0.026	0.024	0.023	0.025	0.023
AgO	0.0024	0.0021	0.0023	0.0021	0.0025	0.0029	0.0026	0.0027	0.0022	0.0051	0.0046
SnO_2	0.0026	0.0028	0.0040	0.0040	0.0022	0.0020	0.0022	0.0038	0.0035	0.0029	0.0040
Sb_2O_3	0.055	0.047	0.041	0.046	0.044	0.061	0.048	0.046	0.047	0.076	0.12
BaO	0.0068	0.0054	0.0065	0.0068	0.0057	0.0056	0.0047	0.0055	0.0066	0.0050	0.0046
Au	0.0000	0.0000	0.0000	0.0000	0.0001	0.0000	0.0000	0.0000	0.0000	0.0005	0.0000
PbO	20.3	21.3	19.9	20.3	22.0	22.2	21.4	22.2	19.5	24.5	23.3

Composition of the thin colorless glass layer present on the filigree and red rods was measured only by DP-LA-ICP-MS. All values are expressed in mass percent

Table 12.6 Compositions of the red and white glasses of the Bernard Perrot cup obtained directly by single spot analysis (LA-ICP-MS) and calculated from two different depth profiles (DP-LA-ICP-MS)

	Red glass			White glass			
	Direct LA-ICP-MS analysis	1st profile	2nd profile	Direct LA-ICP-MS analysis	1st (top) white filigree	2nd (under) white filigree	Single white tube
Na_2O	0.72	0.57	0.52	3.06	3.15	1.24	3.19
MgO	0.14	0.11	0.13	1.09	1.12	0.90	0.99
Al_2O_3	0.34	0.31	0.29	0.86	0.69	0.56	0.82
SiO_2	59.7	59.8	59.9	60.5	59.2	57.8	63.3
P_2O_5	0.14	0.15	0.18	0.20	0.27	0.33	0.20
Cl	0.50	0.50	0.48	0.22	0.21	0.35	0.16
K_2O	13.5	13.1	13.7	8.90	9.3	10.9	11.0
CaO	1.12	0.92	1.08	5.55	5.83	5.10	5.29
MnO	0.059	0.053	0.064	0.069	0.056	0.056	0.069
Fe_2O_3	0.14	0.12	0.13	0.49	0.47	0.42	0.49
CoO	0.0004	0.0003	0.0003	0.0004	0.0003	0.0003	0.0003
CuO	0.0065	0.0057	0.0047	0.011	0.010	0.010	0.011
ZnO	0.0064	0.0035	0.0035	0.020	0.024	0.031	0.021
As_2O_3	1.09	1.28	1.04	0.27	0.20	0.19	0.17
Rb_2O	0.0089	0.0073	0.0069	0.0074	0.0052	0.0052	0.0062
SrO	0.0033	0.0030	0.0034	0.025	0.028	0.027	0.026
ZrO_2	0.017	0.019	0.017	0.018	0.020	0.022	0.024
AgO	0.0031	0.0039	0.0031	0.0006	0.0014	0.0022	0.0012
SnO_2	0.0018	0.0018	0.0019	0.0078	0.0065	0.0076	0.0075
Sb_2O_3	0.035	0.053	0.038	5.81	6.12	7.68	4.20
BaO	0.011	0.0094	0.0083	0.015	0.014	0.013	0.015
Au	0.021	0.023	0.023	0.0000	0.0000	0.0000	0.0000
PbO	22.4	23.0	22.4	12.8	13.3	14.4	10.0

All values are expressed in mass percent

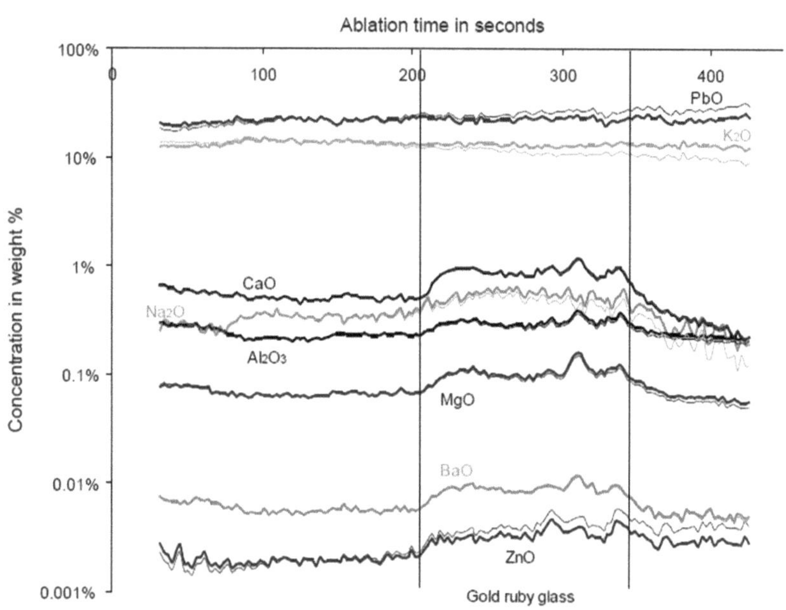

Fig. 12.5 Corrected (*bold line*) and uncorrected (*normal line*) profiles derived from the ablation data measured on the colorless and red glasses (crater on Fig. 12.3c) for some major and minor oxides

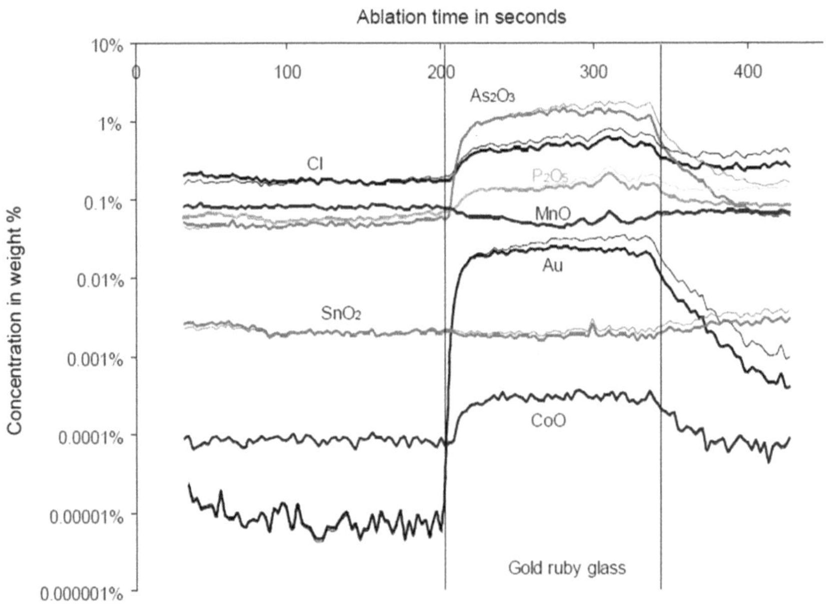

Fig. 12.6 Corrected (*bold line*) and uncorrected (*normal line*) profiles derived from the ablation data measured on the colorless and red glasses (crater on Fig. 12.3c) for some minor and trace oxides

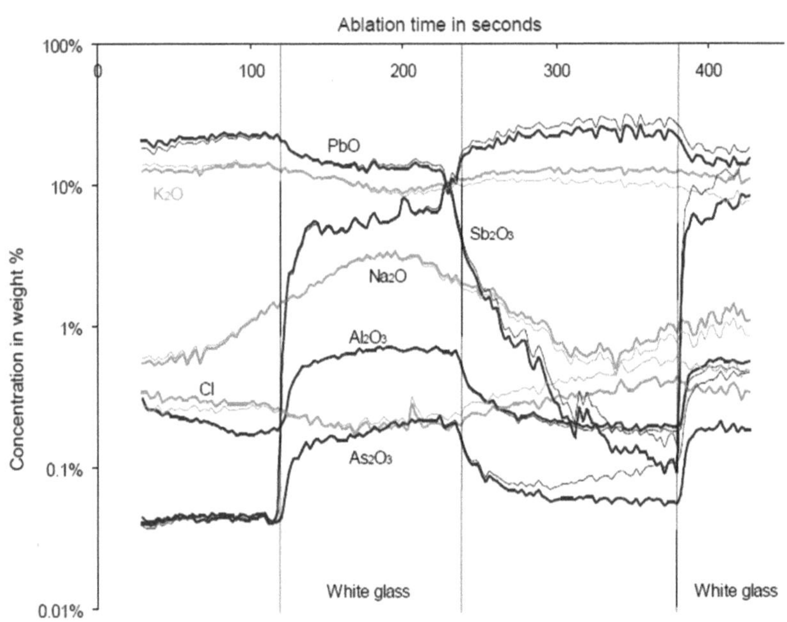

Fig. 12.7 Corrected (*bold line*) and uncorrected (*normal line*) profiles derived from the ablation data measured on the colorless and white glasses (crater on Fig. 12.3e) for some major and minor oxides

and 12.7). In Fig. 12.5, one can observe a small layer of glass with lower sodium content, then a first plateau which corresponds to the colourless glass and a small increase in concentration when the red glass is reached. Information measured after the red glass is altered by the memory effect of the cell. In Fig. 12.7, we observe again a small layer of glass with lower sodium content but after we have a continuous increase of sodium which reaches a maximum concentration in the middle

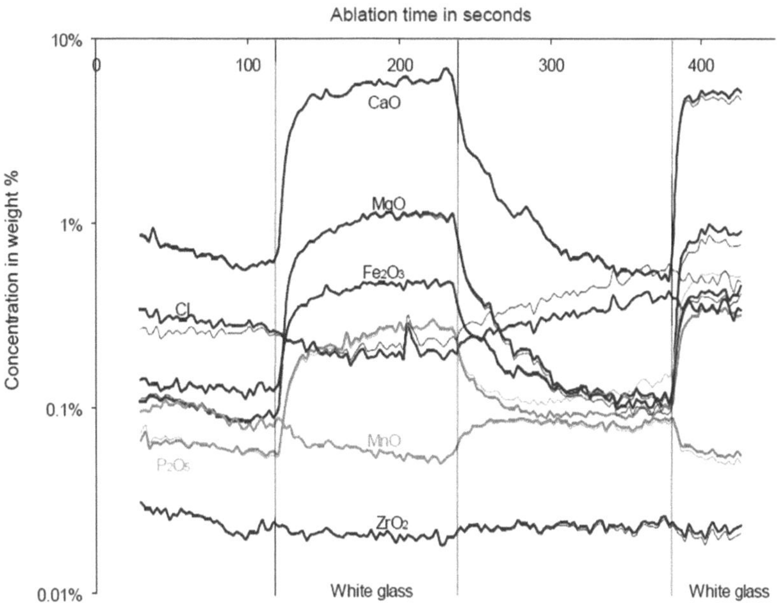

Fig. 12.8 Corrected (*bold line*) and uncorrected (*normal line*) profiles derived from the ablation data measured on the colorless and white glasses (crater on Fig. 12.3e) for others minor oxides

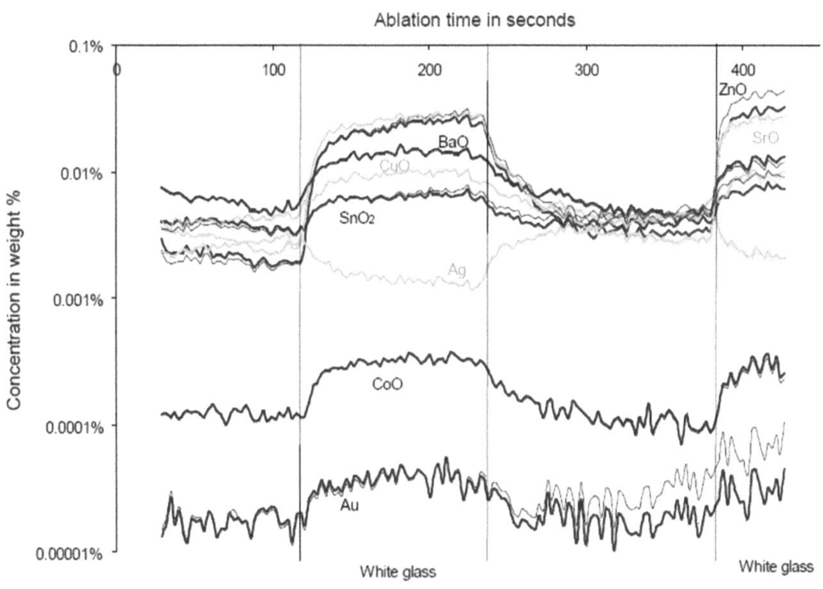

Fig. 12.9 Corrected (*bold line*) and uncorrected (*normal line*) profiles derived from the ablation data measured on the colorless and white glasses (craters on Fig. 12.3e) for trace oxides

of the white glasses and then slowly decreases. Sodium is the only element which exhibits this particular profile. All the others elements show a sharper change in concentration from one glass layer to the other, even when memory effect is present. The presence of a small glass layer of different composition is confirmed when we observe other oxides profiles such as those of aluminum, tin, and barium for example. These particularities provide some information on the making of the object. Firstly, Bernard Perrot probably used preformed glass rods made of red and white glass already embedded in colorless glass, which were then applied on a colorless glass core. Secondly, the colourless glass of the rods probably originates from a different batch than the colorless glass of the cups, as small concentration plateaus are observed for the colourless glass. Thirdly, the preparation of the white filigree probably took a longer time than the red rod, as sodium has significantly diffused from the white glass to the colorless glass.

These profiles were measured at a laser frequency of 10 Hz. A lower laser frequency may have generated a more detailed profile, but would not have permitted removal of material to as great a depth and sensitivity would have been negatively impacted, jeopardizing the reliable measurement of some elements. The use of a time of flight mass spectrometer might help in obtaining a more detailed profile. Using a double-focusing mass spectrometer with the operating conditions reported here, it takes 2.7 s to carry out mass scanning from sodium to lead. Thus not all elements are recorded at exactly the same depth, and it may thus mask the interface between different layers. Time of flight mass spectrometers have shorter mass scanning times (less than 1 ms), it is possible to record precise data for each laser shot. Such a system combined with a low memory cell would allow for a perfect definition of glass structure. The memory effect depends directly on the total volume of the cell. It is thus possible to reduce this effect to below two seconds instead of a few tens of seconds in the present case, by using smaller ablation cells (Fricker et al. 2011).

12.5 Conclusion

The accurate determination of archaeological glass sample composition is challenging due to possible heterogeneity. The surfaces of ancient artifacts are typically weathered and do not reflect internal composition. Consequently, most available non-destructive analytical methods, such as EPMA, SEM-EDS, PIXE or XRF, will produce non-representative results when carried out on an unprepared sample. If the corrosion layer is too thick, even LA-ICP-MS conducted in a conventional manner (line or spot ablations) may not be able to reach pristine glass and will produce data that will not reflect the original glass composition. In other cases, glass objects may be composed of different glass layers of varying composition. The determination of inner glass composition will then be possible only on cut slices of the object.

Compared to these methods, depth profiling by LA-ICP-MS offers the possibility to analyse sub-surface unaltered glass with minimum damage to the object, as shown by the concentration profiles generated here for late Bronze Age glass beads and red and white inlays glass from the seventeenth century. As laser sampling causes no visual damage to the object, DP-LA-ICP-MS analysis can be carried on a larger number of objects than would be possible with more conventional methods which require cutting a profile.

Compared to EPMA, XRF and PIXE, DP-LA-ICP-MS offers the same advantages as conventional LA-ICP-MS, for instance the ability to measure a greater number of trace elements with lower detection limits. It thus gives the possibility to compare chemical compositions on a large range of chemical elements and thus permits establishment with greater confidence the concordance or the difference between the compositions of two objects. As was shown for the profile generated on the seventeenth century glass made by Bernard Perrot, profile analysis can also provide information on glass making by identifying the different glass structures present in an object (rod and filigree in this case). One of the only limiting factors for LA-ICP-MS analysis is the case of heterogeneous opaque glass, where EPMA or SEM-EDX offer indispensable complementarity information.

References

Angelini I, Artioli G, Bellintani P, Diella V, Gemmi M, Polla A, Rossi A (2004) Chemical analyses of Bronze Age glasses from Frattesina di Rovigo, Northern Italy. J Archaeol Sci 31(8):1175–1184

Azemar R, Billaud Y, Costantini G, Gratuze B (2000) Les perles protohistoriques de l'Aveyron. Cahiers d'Archéologie Aveyronnaise 14:75–88

Barca D, Abate M, Mirocle Crisci G, De Presbiteris D (2009) Post-medieval glass from the Castle of

Cosenza, Italy: chemical characterization by LA-ICP-MS and SEM-EDS. Per Mineral 78(2):49–64

Bellintani P, Stefan L (2009) Nuovi dati sul primo vetro europeo: il caso di Frattesina. In: Atti del Primo Convegno Interdisciplinare sul Vetro nei Beni Culturali e nell'Arte di Ieri e di Oggi, Parma 27–28 novembre 2008. Aula Magna, Università degli Studi di Parma, Parma, pp 71–86

Bellintani P, Biavati A, Verità M (1998) Alcune considerazioni su materiali vetrosi da constesti dell'età del Bronzo media e recente dell'Italia settentrionale. In: La Guardia R (ed) Il vetro dall'antichità all'età contemporanea: aspetti tecnologici, funzionali e commerciali, Atti 2e Giornate Nazionali di Studio AIHV-Camitato Nazionale Italiano (Milano, 14–15 dicembre 1996). Edizioni Et, Milano, pp 15–24

Biavati A, Verità M (1989) The glass from Frattesina, a glassmaking center in the late Bronze Age. Rivista della Staz Sper Vetro 4:295–299

Biron I, Gratuze B, Pistre S (2010) Etude en laboratoire d'objets en verre attribués à Bernard Perrot. In: Gorget C, Klinka I (eds) Bernard Perrot, Secret et chefs-d'œuvre des verreries royales d'Orléans. Musée des Beaux-Arts d'Orléans and Somogy, Orléans and Paris, pp 87–93

Biron I, Gratuze B, Lehuede P, Pistre S (2011) Etude en laboratoire d'objets en verre attribués à Bernard Perrot. Bulletin de l'AFAV 2011:19–25

Brill RH (1992) Chemical analysis of some glasses from Frattesina. J Glass Stud 34:11–22

Brill RH (1999) Chemical analyses of early glasses. Corning Museum of Glass, New York

Bronk H, Freestone IC (2001) A quasi non-destructive microsampling technique for the analysis of intact glass objects by SEM-EDXA. Archaeometry 43 (4):517–527

Carmona N, Ortega-Feliu I, Gómez-Tubío B, Villegas MA (2010) Advantages and disadvantages of PIXE/ PIGE, XRF and EDX spectrometries applied to archaeometric characterisation of glasses. Mater Charact 61:257–267

Cosyns P, Warmenbol E, Bourgeois J, Degryse P (2005) Pre-Roman glass beads in Belgium. In: Nenna MD (ed) Annales du 16e Congrès de l'Association Internationale pour l'Histoire du Verre, London 2003. AIHV, Nottingham, pp 323–326

Dussubieux L, Robertshaw P, Glascock MD (2009) LA-ICP-MS analysis of African beads: laboratory inter-comparison with an emphasis on the impact of corrosion on data interpretation. Int J Mass Spectrom 284 (1-3):152–161

Fricker M-B, Kutscher D, Aeschlimann B, Frommer J, Dietiker R, Bettmer J, Günther D (2011) High spatial resolution trace element analysis by LA-ICP-MS using a novel ablation cell for multiple or large samples. Int J Mass Spectrom 307(1–3):39–45

Geyssant J (2010) Secret du verre rouge transparent de Bernard Perrot et comparaison avec celui de Jehann Kunckel. In: Gorget C, Klinka I (eds) Bernard Perrot, Secret et chefs-d'œuvre des verreries royales d'Orléans. Musée des Beaux-Arts d'Orléans and Somogy, Orléans and Paris, pp 51–54

Geyssant J, Gorget C, Tétart-Vittu F (2010) Présentation, particularités et catalogue des œuvres exposées. In: Gorget C, Klinka I (eds) Bernard Perrot, Secret et chefs-d'œuvre des verreries royales d'Orléans. Musée des Beaux-Arts d'Orléans and Somogy, Orléans and Paris, pp 95–184

Giussani B, Monticelli D, Rampazzi L (2009) Role of laser ablation–inductively coupled plasma–mass spectrometry in cultural heritage research: A review. Anal Chim Acta 635:6–21

Glaus R, Koch J, Günther D (2012) Portable laser ablation sampling device for elemental fingerprinting of objects outside the laboratory with laser ablation inductively coupled plasma mass spectrometry. Anal Chem 84(12):5358–5364

Gratuze B (1999) Obsidian characterisation by laser ablation ICP-MS and its application to prehistoric trade in the Mediterranean and the Near East: sources and distribution of obsidian within the Aegean and Anatolia. J Archaeol Sci 26(8):869–881

Gratuze B, Billaud Y (2003) La circulation des perles en verre dans le Bassin méditerranéen de l'Age du Bronze moyen jusqu'au Hallstatt. In: Foy D, Nenna MD (eds) Echanges et commerce du verre dans le monde antique, actes du colloque de l'AFAV, Aix-en-Provence Marseille 7–9 juin 2001. Monique Mergoil, Montagnac, pp 11–15

Gratuze B, Louboutin C, Billaud Y (1998) Les perles protohistoriques en verre du musée des Antiquités nationales. Antiquités nationales 30:13–24

Guilaine J, Gratuze B, Barrandon JN (1991) Les perles de verre du chalcolithique et de l'Age du Bronze. Analyses d'exemplaires trouvés en France. In: Chevillot C, Coffyn A (eds) L'Age du Bronze Atlantique, Actes du 1er colloque du parc archéologique de Beynac. Edition de l'association des musées du Sarladais, Beynac et Cazenac, pp 255–266

Hartmann G, Kappel I, Grote K, Arndt B (1997) Chemistry and technology of prehistoric glass from Lower Saxony and Hesse. J Archaeol Sci 24 (6):547–559

Heinrich CA, Pettke T, Halter WE, Aigner-Torres M, Audetat A, Gunther D, Hattendorf B, Bleiner D, Guillong M, Horn I (2003) Quantitative multi-element analysis of minerals, fluid and melt inclusions by laser-ablation inductively-coupled-plasma mass-spectrometry. Geochim Cosmochim Ac 67 (18):3473–3497

Henderson J (1988) Electron probe microanalysis of mixed-alkali glasses. Archaeometry 30(1):77–91

Henderson J (1989) The earliest glass in Britain and Ireland. In: Feugère M (ed) Le verre préromain en Europe occidentale. Monique Mergoil, Montagnac, pp 13–24

Henderson J (1993) Chemical analysis of the glass and faience from Hauterive-Champréveyres, Switzerland. In: Rychner-Faraggi AM (ed) Hauterive-Champréveyres 9, Métal et Parure au Bronze Final. Musée Cantonal d'Archéologie Archéologie neuchâteloise, Neuchâtel, pp 111–117

Hollocher K, Ruiz J (1995) Major and trace element determination on NIST glass standard reference material 611, 612, 614 and 1834 by inductively coupled plasma mass spectrometry. Geostandard Newslett 19(1):27–34

Jochum KP, Weis U, Stoll B, Kuzmin D, Yang Q, Raczek I, Jacob D-E, Stracke A, Birbaum K, Frick D-A, Günther D, Enzweiler J (2011) Determination of reference values for NIST SRM 610-617 glasses following ISO Guidelines. Geostandard Geoanal Res 34 (4):397–429

Kovacs R, Nishiguchi K, Utani K, Günther D (2010) Development of direct atmospheric sampling for laser ablation-inductively coupled plasma-mass spectrometry. J Anal Atom Spectrom 25:142–147

Mason PRD, Mank AJG (2001) Depth-resolved analysis in multi-layered glass and metal materials using laser ablation inductively coupled plasma mass spectrometry (LA-ICP-MS). J Anal Atom Spectrom 16:1381–1388

Nenna MD, Gratuze B (2009) Étude diachronique des compositions de verres employés dans les vases mosaïqués antiques: résultats préliminaires. In: Janssens K, Degryse P, Cosyns P, Caen J, Van't dack L (eds) Annales of the 17th Congress of the International Association for the History of Glass, 2006, Antwerp. VUP University Press, Brussels, pp 199–205

Nikita K, Henderson J, Nightingale G (2009) An archaeological and scientific study of Mycenaean glass from Elateia-Alonaki, Greece. In: Janssens K, Degryse P, Cosyns P, Caen J, Van't dack L (eds) Annales of the 17th Congress of the International Association for the History of Glass, 2006, Antwerp. VUP University Press, Brussels, pp 39-46

Norman MD, Pearson NJ, Sharma A, Griffin WL (1996) Quantitative analysis of trace elements in geological materials by laser ablation ICPMS: instrumental operating conditions and calibration values of NIST glasses. Geostandard Newslett 20(2):247–262

Pearce NJG, Perkins WT, Westgate JA, Gorton MP, Jackson SE, Neal CR, Chenery SP (1997) A compilation of new and published major and trace element data for NIST SRM 610 and NIST SRM 612 glass reference materials. Geostandard Newslett 21(1):115–144

Ploin S, Koenig M-P, Gratuze B (2012) Les perles en verre de l'âge du Bronze et du premier âge du Fer d'Alsace et de Lorraine. In: Arveiller V, Cabart H (eds) Le verre en Lorraine et dans les régions voisines, Actes des 26ème rencontres de l'AFAV, Metz 2011, Monographie Instrumentum 42. Monique Mergoil, Montagnac, pp 11–36

Raftery B, Henderson J (1987) Some glass beads of the Later Bronze Age in Ireland. In: Frey O-H, Haevernick TE, Matthäus H, Braun C (eds) Glasperlen der vorrömischen Eisenzeit I. von Zabern, Mainz, pp 39–53

Robertshaw P, Wood M, Haour A, Karklins K, Neff H (2014) Chemical analysis, chronology, and context of a European glass bead assemblage from Garumele, Niger. J Archaeol Sci 41:591–604

Van der Werf I, Mangone A, Giannossa L-C, Traini A, Laviano R, Coralini A, Sabbatini L (2009) Archaeometric investigation of Roman tesserae from Herculaneum (Italy) by the combined use of complementary micro-destructive analytical techniques. J Archaeol Sci 36(12):2625–2634

Varberg J, Gratuze B, Kaul F (2015) Between Egypt, Mesopotamia and Scandinavia: Late Bronze Age glass beads found in Denmark. J Archaeol Sci 54:168–181

Venclová N, Hulínský V, Henderson J, Chenery S, Šulová L, Hložek J (2011) Late Bronze Age mixed-alkali glasses from Bohemia, Skla typu mixed alkali mladší doby bronzové v Čechách. Archeologické rozhledy LXIII–2011:559–585

Verità M, Basso R, Wypyski MT, Koestler RJ (1994) X-ray microanalysis of ancient glassy materials: a comparative study of wavelength dispersive and energy dispersive techniques. Archaeometry (36) 2:241–251

Vicenzi EP, Eggins S, Logan A, Wysoczanski R (2002) Microbeam characterization of Corning archeological reference glasses: new additions to the Smithsonian microbeam standard collection. J Res Natl Inst Stand 107(6):719–727

von Kerssenbrock-Kosigk D (2008) Glass of the alchemists: lead crystal–gold ruby, 1650–1750. Corning Museum of Glass, New York

Wagner B, Nowak A, Bulska E, Hametner K, Günther D (2012) Critical assessment of the elemental composition of Corning archeological reference glasses by LA-ICP-MS. Anal Bioanal Chem 402:1667–1677

Walder H (2013) Stylistic and chemical investigation of turquoise blue glass artifacts from the Contact Era of Wisconsin. Midcontinent J Archaeol 38 (1):119–142

Walton MS, Shortland A, Kirk S, Degryse P (2009) Evidence for the trade of Mesopotamian and Egyptian glass to Mycenaean Greece. J Archaeol Sci 36 (7):1496–1503

Wedepohl K-H, Simon K, Kronz A (2011) Data on 61 chemical elements for the characterization of three major glass compositions in Late Antiquity and the Middle Ages. Archaeometry 53(1):81–102

Woodhead J, Hellstrom J, Paton C, Hergt J, Greig A, Maas R (2008) A guide to depth profiling and imaging applications of LA-ICP-MS. In: Sylvester P (ed) Laser ablation ICP–MS in the earth sciences: current practices and outstanding issues. Mineralogical Association of Canada, Vancouver, BC, pp 135–145

A Proficiency Test for LA-ICP-MS Laboratories Analyzing Ancient Glass

13

Laure Dussubieux

Abstract

LA-ICP-MS is now one of the most common methods used for glass analysis in archaeology, and has the potential to build large databases for comparison. However, the actual comparability of measurements generated in different laboratories has not yet been addressed systemically. Here, the results of a round-robin comparison test on two modern glass samples with compositions comparable to ancient glasses is presented. The test involved nine laboratories using a diversity of mass spectrometers and lasers, as well as different calibration protocols and standard reference materials. Measured values differ significantly between labs for some elements, reflecting differences in calibration and instrumentation. However, these differences do not cause overlap between the glass samples that might confuse provenance in archaeological studies, but are significant enough to make finer-grained assignments to individual glass batches difficult or impossible when comparing across different laboratories.

13.1 Introduction

In a 2007 paper, Shortland et al. reported analysis of glass samples from Egypt and Mesopotamia dated from the Late Bronze Age period with LA-ICP-MS. The glass samples from both regions were manufactured with similar recipes involving low alumina sand and soda from halophytic plant ashes; however distinct trace element patterns reflecting different geologic environments for the raw materials available in these two different regions appeared. Elements such as Ti, Zr, La and Cr were found particularly useful for separating the Mesopotamian and the Egyptian produced glasses. Subsequently, Walton et al. (2009) used the Shortland et al. (2007) dataset to provenance glass beads found in Mycenaean contexts and could establish by comparison that this Mycenaean material was in fact made from Egyptian glass for some beads and from Mesopotamian glass for others.

L. Dussubieux (✉)
Integrative Research Center, Elemental Analysis Facility, Field Museum of Natural History, 1400 S. Lakeshore Drive, Chicago, IL 60605, USA
e-mail: ldussubieux@fieldmuseum.org

L. Dussubieux et al. (eds.), *Recent Advances in Laser Ablation ICP-MS for Archaeology*, Natural Science in Archaeology, DOI 10.1007/978-3-662-49894-1_13

These two studies illustrate the capacity of LA-ICP-MS to address provenance issues for ancient glass, but also the usefulness of building datasets of glass compositions for comparative purposes; however, a careful examination of the data published in these two articles raises the issue of the comparability of different datasets obtained by different research teams. If the concentrations of the diagnostic elements for sourcing Egyptian and Mesopotamian Late Bronze Age glass vary in the same range in the Shortland et al. (2007) and the Walton et al. (2009) datasets, other elements, specifically Sc, with respective average concentrations of 1.4 ± 0.5 ppm and 12 ± 3 ppm, exhibit wide discrepancies between the two publications. Is this difference real? Is this due to a problem of transcription of the results, of calculation of the concentrations or of calibration of the instrument? In this case, such a discrepancy between results from two laboratories has no consequence on the data interpretation, levels for Sc being similar in the Egyptian and the Mesopotamian glass; however, each LA-ICP-MS laboratory analyzing ancient glass should be able to show that the results they produced are comparable with the results produced by other laboratories conducting similar types of measurements.

Inter-laboratory comparison performed on a regular basis is a way to assess the quality of the results generated by a given laboratory by comparing them with the results obtained by other laboratories when measuring identical samples. This paper reports on an inter-laboratory comparison involving nine LA-ICP-MS laboratories that occasionally or regularly analyze ancient materials, and specifically ancient glass. Two glass samples were tested. The results provided by the different laboratories are compared to each other through the calculation of similarity coefficients. The results will also be compared to the concentrations measured by the organizing laboratory, the Elemental Analysis Facility (EAF) at the Field Museum. The results of this inter-laboratory comparison test will be discussed in the context of archaeological research.

13.2 Inter-comparison Test

13.2.1 Factors Impacting Inter-laboratory Comparability

Different factors can explain discrepancies between results produced by different laboratories; some of them were highlighted in two recent articles (Bertini et al. 2013; Wagner et al. 2012). These factors fall into three different categories: laser type and settings, standard reference materials, and data acquisition and data reduction.

Ideally, all the isotopes to be measured should be ablated and transported to the plasma with the same efficiency so as the composition of the object and the composition of the ablated material in the plasma are the same. In reality, inter-element fractionation can modify the composition during the ablation and transportation processes. This phenomenon depends on a wide range of parameters including laser wavelength, pulse duration, laser frequency, laser energy and gas carrier. To summarize (and to simplify), shorter wavelength lasers produce smaller particles. In the plasma, they will be atomized and ionized more efficiently, limiting fractionation (Kuhn and Günther 2004). In addition, shorter pulse durations (e.g., in femtosecond lasers) create ablation without heat dissipation through the sample avoiding any preferential vaporization (Russo et al. 2002). Fractionation has also been linked to ablation depth (Eggins et al. 1998; Košler et al. 2005). Selecting a depth to diameter ratio <6 for the ablation crater reduces fractionation for elements that are most sensitive to this problem (Mason and Mank 2001). The choice of the gas carrier in the laser, generally He or Ar, will affect elemental fractionation as well (Eggins et al. 1998; Günther and Heinrich 1999; Wang et al. 2006).

To obtain quantitative results, standard reference materials (SRM) are measured between groups of samples with unknown compositions. Among the most commonly used SRMs are the 61X glass series from the National Institute for Standard and Technology (NIST) and Corning glasses A, B, C and D. Other SRMs more specifically developed for geological applications are also sometimes used. Most of these SRMs have compositions that are fairly different from the compositions of ancient glass (Corning glasses being an exception). As different glass matrices lead to different ablation efficiencies (Bertini et al. 2013), laboratories using different sets of SRMs will obtain slightly different results. Also, heterogeneous distribution of some elements has been observed in the NIST 61X SRM series (see Jochum et al. 2011) and in the Corning glasses (Vicenzi et al. 2002; Wagner et al. 2012) that are among the most widely used SRMs for ancient glass analysis Additionally, for those SRMs that have been in circulation for a number of years, several sets of concentrations are available. Generally speaking, most laboratories use the values that were most recently published at the time they began their analyses, and consequently, newer laboratories tend to use different sets of SRM concentrations compared to older laboratories. For an older laboratory, updating concentration values for one or more of its SRMs implies re-calculating all previously measured compositions. This is a daunting task for long-established laboratories that have accumulated thousands of measurements, and it is therefore rarely done.

Lastly, data acquisition and data reduction are important steps that can influence the final concentrations measured in the samples in different ways. Acquisition mode, scanning mode, dwell time and other acquisition parameters have an influence on measurement error. Most of the laboratories included in this study have adopted the approach described in Gratuze (1999) for data reduction. In this approach, all elements are measured with LA-ICP-MS and one of them is used as internal standard. The composition of the sample is then calculated, assuming that the sum of all the major and minor constituents, expressed as oxides, is 100 %; however, another approach exists. It involves the measurement of the internal standard (Si or Ca in the case of glass) using an independent analytical method. The concentration of the internal standard, obtained for each sample and standard, is used in a second step to calculate the concentrations of all the elements measured with LA-ICP-MS (e.g. Saitowitz 1996). In cases where some constituents of the samples are not measured with LA-ICP-MS (e.g. S, Cl, etc.) but are present in significant quantities, the two approaches will result in different concentrations for measured elements.

13.2.2 Samples

In the 1970s, Corning produced four different glasses matching the composition of four different types of ancient glass: glass A is similar to soda plant ash glass, glass B is similar to a mineral soda glass, glass C is a lead-barium glass, and glass D is a lime-potash glass. They have been widely used by LA-ICP-MS (and other) laboratories for standardization and quality control, but being manufactured from pure modern raw materials, their trace element content remains much lower than those generally encountered in ancient glass (Vicenzi et al. 2002; Wagner et al. 2012). As a consequence, they have been deemed unsuitable for this inter-laboratory comparison test.

The selection of glass samples for this inter-laboratory comparison test was made according to a certain set of criteria. First, the samples had to be free or very cheap. They had to be reasonably homogeneous, at least as homogeneous as ancient material can be. Trace element concentrations in the samples and in ancient glass had to be in the same range. The glass samples had to be available in sufficient quantities so as to be divided before being distributed to 10 different laboratories. After testing many potential samples, two with maximally dissimilar compositions were selected: a group of dark blue "snake" beads (called Blue through this paper) and a green wine bottle (Green).

Snake beads were manufactured in Czechoslovakia around the beginning of the twentieth century, and were molded using the Prosser technique (Francis 1999: 24). As the beads selected for this study, although all similar, might have been made from different glass batches, it was decided to crush them and to melt them in a crucible placed into a furnace heated at 900 °C. This temperature was the maximum the furnace could reach. The result was a block of glass with visible heterogeneities such as bubbles and inclusions. One of the participating teams (Dudgeon and Tromp) produced a map of the composition of their Blue glass sample with a scanning electron-microscope attached to an energy dispersive spectrometer. It revealed calcium-enriched areas measuring 10–-20 microns. Blue has a soda-alumina rich composition with significant quantities of potash (~4 %). This glass was colored with cobalt.

The second sample was a modern bottle of Italian red wine. Wine bottles are made of recycled glass and contain significant levels of trace elements. They also receive a surface treatment to make them more resistant to scratches. This treatment unfortunately results in tin enrichment of the outside surface of the container that one participant (Hector Neff) noticed when this part of the bottle was ablated. It was therefore recommended that participants sample the glass on the broken edge of their respective fragments. Green is a soda lime glass with magnesia and alumina concentration around 2 %. The color of Green is due to iron, which is quite likely accidently introduced with the sand used to produce the glass.

13.2.3 List of Participating Laboratories

The EAF organized this inter-laboratory comparison test and participated in it. The two fragments of glass were also shipped to 12 other participants. Nine laboratories (Table 13.1) returned results. One of the laboratories asked that its results be removed after a first draft of this paper was circulated among the participants. Therefore, this report will only discuss the results obtained by eight laboratories in addition of the results generated by the EAF. In Table 13.1, laboratories appear by alphabetical order of the

Table 13.1 List of participating laboratories with instrumentation and standard reference materials used

Participating individuals	Institution	Type of ICP-MS	Type of laser	Standard reference materials
Donatella Barca	Università della Calabria, Arcavacata di Rende, Italy	Quadrupole-ICP-MS	213 nm	NIST 612
John Dudgeon Monica Tromp	Idaho State University, Pocatello, ID, USA	Quadrupole-ICP-MS	213 nm	GSC-1G, GSD-1G, GSE-1G, BHVO-2G, BCR-2G, BIR-1G
Laure Dussubieux	Field Museum of Natural History, Chicago, IL, USA	Quadrupole-ICP-MS	213 nm	NIST 612 and NIST 610 Corning B and D
Bernard Gratuze	Institut de Recherche sur les ArchéoMATériaux Centre Ernest-Babelon, Orléans, France	HR-magnetic sector-ICP-MS	266 nm	NIST 610, Corning B, C, D and a in-house glass standard (mainly for Cl)
Hector Neff	University of California—Long Beach, CA, USA	Time of Flight-ICP-MS	213 nm	NIST 614, 612 and 610 Corning B and D
Klaus Simon	Georg-August-Universität Göttingen, Germany	Quadrupole-ICP-MS	193 nm	NIST 610
Johannes Van Elteren	National Institute of Chemistry, Ljubljana, Slovenia	Quadrupole-ICP-MS	213 nm	NIST 610 and 612 Corning B, C and D Society of Glass Technology 2, 3, 4 and 5 P&H Developments Ltd. DLH6, 7 and 8
Barbara Wagner	University of Warsaw, Warsaw, Poland	Quadrupole-ICP-MS	213 nm	NIST 610
Marc Walton	Formerly Getty, Los Angeles, CA, USA	Time of Flight -ICP-MS	213 nm	NIST 610, 612, 614
Timothy Ward	Millsaps College, Jackson, MS, USA	Quadrupole-ICP-MS	213 nm	NIST 612

individuals that carried out the measurements. To preserve confidentiality, a number ranging from one to nine was randomly attributed to all the participating laboratories with the exception of the EAF.

Most of the participating laboratories are equipped with quadrupole mass spectrometers. Time-of-Flight ICP-MS instruments were used by two participants, and a High Resolution Magnetic Sector ICP-MS by one laboratory. All included Laser ablation systems operate at 213 nm with two exceptions: one laboratory is equipped with a 266 nm laser, and another one with a 193 nm laser. Standard reference materials 610 and/or 612 manufactured by the NIST are used by almost all the laboratories for external standardization. Other standards used by several laboratories are NIST SRM 614, and Corning glasses A, B, C and D. One laboratory used Society of Glass Technology SRMs 2, 3, 4, 5 and P&H Developments Ltd. SRMs DLH6, 7, 8. Another participating laboratory uses USGS Microanalytical Reference materials GSC-1G, GSD-1G, GSE-1G BHVO-2G, BCR-2G, and BIR-1G for quantification.

The lists of elements provided by each laboratory also differ, ranging from 28 to 58 measured and/or quantified elements. One laboratory did not provide concentrations for major elements (lab 2) and another one measured a very limited number of trace elements (lab 1). Elements that were only measured by a single laboratory will not be taken into account, as no comparison is possible. The participants were asked to carry out three series of 10 measurements on each sample on different days to test repeatability and reproducibility. Not all laboratories performed the reproducibility test.

13.3 Results

13.3.1 Heterogeneity of Samples

Before assessing the results of inter-laboratory comparison, one major issue that needs to be dealt with is the potential heterogeneity of Blue and Green glasses. To assess the degree of heterogeneity in these samples, the repeatability of the results obtained at the Field Museum is reported. The repeatability (or precision) of a measurement depends on the stability of the signal (for a given element) provided by the instrument, but also of the homogeneity of the analyzed sample. Repeatability can be calculated as the relative standard deviation (RSD) on ten successive measurements carried out on a sample or SRM. For a SRM, assumed to be "ideally" homogeneous, the repeatability will only reflect the stability of the instrument. A sample can be considered homogeneous when the RSD calculated for a given element in the sample is close to the RSD for the same element in the SRM(s). However, it is important to keep in mind that heterogeneity is not the only factor affecting the repeatability of measurement. Low concentrations near the limits of detection can adversely affect repeatability as well.

For the SRM NIST 610, repeatability is generally 5 % or less. For Blue and Green, it varies in a wide range (Fig. 13.1). It is near 1 % for some major elements such as potassium or

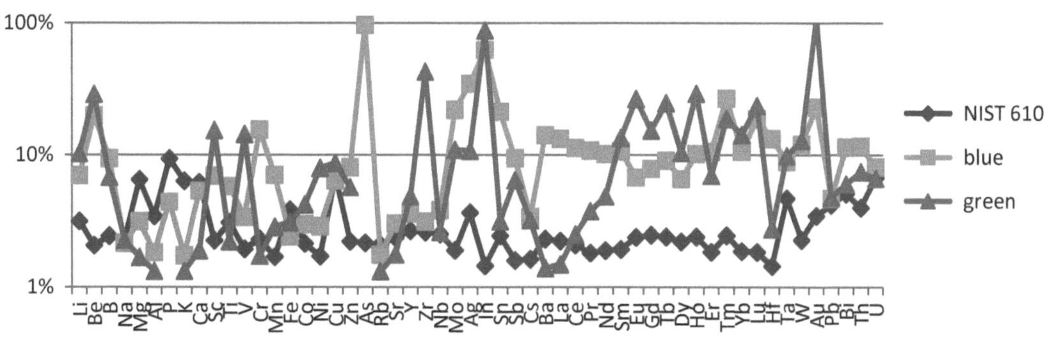

Fig. 13.1 RSD calculated from 10 measurements on SRM NIST 610, Blue, and Green

calcium (in Green), but can be as high as 96 % for arsenic (in Blue) or 109 % for gold (in Green). These two elements have concentrations very near their respective limits of detection. From a general point of view, the repeatability obtained for Blue and Green does not appear significantly worse than that of some archaeological glass samples tested in Dussubieux et al. (2009), which were generally less than 20 %, at least for elements not involved in coloring and opacifying the glass.

13.3.2 Comparison of Results from Different Laboratories

Direct comparison of the results (see Tables 13.2 and 13.3) is of course difficult. To assess the homogeneity of the results provided by the nine participating laboratories, similarity coefficients were calculated. To measure the similarity of the results produced by two given laboratories, the following equation was used (Hunt et al. 1998):

$$d_{(A,B)} = \frac{\sum_{i=1}^{n} R_i}{n} \qquad (13.1)$$

where:

$d_{(A,B)} = d_{(B,A)}$ = similarity coefficient for comparison between laboratory A and laboratory B.
i = element number
n = number of elements in calculation

$$R_i = \frac{X_{iA}}{X_{iB}},\ if\ X_{iA} > X_{iB} \qquad (13.2)$$

$$R_i = \frac{X_{iB}}{X_{iA}},\ if\ X_{iB} > X_{iA} \qquad (13.3)$$

X_{iA} = concentration of element i as measured by laboratory A
X_{iB} = concentration of element i as measured by laboratory B

A similarity coefficient is one or less. A value of one is obtained when two identical sets of data are compared and in Tables 13.4 and 13.5, when the composition for a given sample obtained by a laboratory is compared with itself. Values ranging downwards from one reflect increasing dissimilarity between two sets of measurements (Tables 13.4 and 13.5).

For Green, similarity coefficients range from 0.70 to 0.90, and for Blue, from 0.72 to 0.91. The highest similarity coefficients occur for both glass samples when laboratories 1 and 2 are compared. These two laboratories only have 15 elements in common and the comparison is based on far fewer elements than for the other laboratories.

For a more detailed analysis of the results produced by each laboratory, comparisons were carried out with results obtained at the EAF. As Blue and Green do not have known concentrations, and were not measured in advance by multiple independent laboratories employing well-established methods for the trace analysis of glass, it is impossible to assess the accuracy of the results provided by the different participants. Instead, all results were compared to the EAF measurements using the following equation:

$$X_i = \frac{X_i - X_i^{EAF}}{X_i^{EAF}} \qquad (13.4)$$

X_i = value provided by a given laboratory for element i
X_i^{EAF} = value measured at the EAF for element i

The magnitude of x_i is an indication of how well a given laboratory performed compared to the EAF. Values for x_i that are negative indicate an overestimation of the concentration by the EAF compared to the other laboratories. Ideally, all x_i values for an element i should be close to zero, which would mean that the EAF and all the participants provided very similar values for a given element, increasing the likelihood that

Table 13.2 Compositions provided by the participating laboratories for Green

wt%	1	2	3	4	5	7	8	9	EAF
Li	20.9		43.3	20.2	13.5		23.7	21.1	19.7
Be			1.07	1.48				1.28	2.26
B	106		97.0		112		112		92.0
S					215				
Cl					208				
Na_2O	11.6 %		13.3 %	12.2 %	14.3 %	11.7 %	12.2 %	12.6 %	13.6 %
MgO	2.12 %		2.10 %	2.02 %	1.80 %	2.63 %	2.14 %	1.66 %	2.18 %
Al_2O_3	1.74 %		1.52 %	1.51 %	2.35 %	2.02 %	2.02 %	1.41 %	2.19 %
SiO_2	73.2 %		72.9 %		69.2 %	72.3 %	72.1 %	74.6 %	70.7 %
P_2O_5	0.02 %		0.07 %		0.01 %	0.03 %	0.03 %		
K_2O	0.69 %		0.74 %	0.61 %	0.54 %	0.82 %	0.63 %	1.10 %	0.72 %
CaO	9.6 %		8.6 %	9.2 %	11.1 %	10.2 %	10.2 %	7.9 %	10.0 %
Sc		2.68	7.47	42.2	2.0	5.43		24.0	1.48
Ti	296	413	290	339	350	478	325	311	306
V	10.6	10.9	12.5	11.0	11.8	9.7	11.2	8.1	12.1
Cr	863	929	940	887	959	810	852	935	933
Mn	159	162	194	160	191	153	153	126	179
Fe	2474		2600	2487	2673	1630	2509	1995	2604
Co	7.86	8.28	10.40	7.79	8.29	7.09	6.95	8.02	8.34
Ni	<65	10.3	11.5	12.7	11.5	9.6	10.9	34.6	9.49
Cu	14.4	15.2	22.0	20.3	14.8	15.0	13.2	13.7	16.9
Zn	52.2	58.1	51.4	62.1	64.8	64.8	56.3	43.2	45.5
Ga			8.08	15.86	2.19	1.78		28.2	
Ge				0.87	0.76	0.64		7.15	
As	9.85	9.55	12.5	7.13	9.08	11.6	8.27		
Se			26.10						
Rb	25.6	27.3	30.7	26.9	31.1		26.8	33.0	30.5
Sr	102	118	113	103	130		121	84	122
Y		4.28	3.49	3.41	4.77	4.81	4.67	2.58	4.82
Zr	101	115	96	110	148		127	67	134
Nb		2.72	2.47	2.42	2.78	2.83	2.81	2.07	3.33
Mo			1.32	1.34	1.26	1.22	1.32	2.13	1.57
Ag			0.50	0.48	0.43	0.36	0.61	0.51	0.37
Cd	<12	3.25	0.32	0.15	0.26	1.14		1.12	
In			0.15	0.12	0.10	0.10			0.06
Sn	22.1	22.0	29.4	20.4	23.2	20.4	34.5	26.6	22.6
Sb	8.35	12.5	17.6	14.0	11.9		14.3	13.2	11.1
Cs				1.89	1.95	1.58	1.81	1.69	1.93
Ba	377	419	492	395	465	463	441	509	427
La		5.41	5.32	4.65	6.29	5.86	6.42	6.05	5.50
Ce		20.3	24.5	19.7	22.6	22.0	23.4	18.6	22.3
Pr		1.07	1.12	1.00	1.20		1.33	0.95	1.24
Nd		4.40	3.20	3.72	4.68	4.36	4.83	3.57	4.27
Sm		0.94	0.56	0.64	0.74	0.76	0.80	0.75	0.77
Eu		0.29	0.17	0.22	0.21	0.20	0.25	0.26	0.20
Gd		0.71	0.49	0.66	0.75	0.70	1.06	1.05	0.63
Tb		0.11	0.10	0.10	0.12	0.11	0.14	0.12	0.12
Dy		0.76	0.44	0.53	0.69	0.64	0.70	0.52	0.64
Ho		0.15	0.12	0.11	0.14	0.14	0.15	0.08	0.14

(continued)

Table 13.2 (continued)

wt%	1	2	3	4	5	7	8	9	EAF
Er		1.84	1.07	1.36	1.81	1.74	1.93	1.10	1.66
Tm		0.07	0.06	0.06	0.06	0.07	0.06	0.07	0.08
Yb		0.58	0.33	0.36	0.43	0.47	0.51	0.37	0.45
Lu		0.07	0.07	0.07	0.07	0.07	0.08	0.06	0.08
Hf		3.34		2.61	3.84	4.43	3.37	2.01	3.32
Ta		0.34		0.21	0.30		0.28	0.18	0.30
W				0.53	0.49		0.30	0.49	0.48
Au			0.04				0.04	0.02	0.06
Tl				0.07	0.09	0.04		0.22	
Pb	136	150	242	161	182	144	150	185	200
Bi	0.48		0.92	0.74	0.77	0.60	0.78	6.13	0.61
Th		1.40	1.10	1.20	1.69	1.84	1.88	0.78	1.31
U		0.84	1.06	0.94	1.01	0.88	1.06	0.90	0.84
Te								4.08	
Re								0.03	

those values are close to the "real" concentration in the sample. If all x_i values are different from zero but close to each other, this means that all the laboratories provided similar values, these values being different from the ones measured at the EAF. It is likely in this case that the value obtained by the EAF is further away from the real glass concentration compared to the values provided by the other participants. In reality, the x_i values are quite different from one laboratory to another for a given element (Fig. 13.2). More implications for the x_i values obtained for a given element are discussed below.

A few general trends can be inferred from the comparison of the results from the different participants and from the EAF. For most of the laboratories and most of the elements, the relative difference between the participants and the EAF results are within ±50 %. This is particularly true for trace elements such as the lanthanides (from La to Lu), with most of the values within 25 % of those measured at the EAF. However, for some elements, a more significant dispersion of the results is evident. Generally speaking, dispersion of the results can be attributed in some cases to low concentrations that are close to detection limits. Gold is present at very low concentrations in Blue and Green and concentrations and limits of detection for this element are in the same range, as shown in Table 13.6. Table 13.6 gives the range of the limits of detection reported by seven of the laboratories involved in this inter-laboratory comparison test. Limits of detection vary considerably, and the extreme values differ by two to four orders of magnitude.

While heterogeneity of the samples certainly accounts for some of the dispersion evident in the results obtained by the different laboratories, other factors may also contribute to variance in measured concentrations. For an element such as indium, an interference of its more abundant isotope (^{115}In) with one of the isotopes of tin, an element which is always far more abundant than indium in glass samples, creates a need for correction that may introduce a large error for the measurement of indium concentrations. Some elements such as scandium are systematically over- or underestimated by a given laboratory in the two samples. Other elements such as Li, Na, or Ba (Fig. 13.3a–d) display the same pattern, suggesting that for some elements calibration problems exist. A critical aspect of the calibration of the measurements, as mentioned earlier, is the selection of the appropriate values for the concentration of each element present in the utilized SRM(s). For NIST SRM 610, 612 or 614, certified values are provided by the

Table 13.3 Compositions provided by the participating laboratories for Blue

wt%	1	2	3	4	5	7	8	9	EAF
Li	12.5		42.1	15.4	18.7		14.1	12.1	13.6
Be			3.40	3.64				2.97	3.26
B	548		450	374	659		541		521
S					310				
Cl					988		1521		814
Na_2O	12.2 %		13.2 %	11.9 %	14.6 %	12.4 %	12.9 %	13.0 %	14.1 %
MgO	0.17 %		0.19 %	0.17 %	0.14 %	0.17 %	0.18 %	0.13 %	0.21 %
Al_2O_3	9.76 %		10.8 %	8.99 %	13.0 %	11.7 %	12.5 %	10.9 %	9.09 %
SiO_2	70.5 %		68.5 %	63.6 %	66.7 %	71.2 %	67.3 %	68.5 %	70.0 %
P_2O_5	0.12 %		0.25 %		0.14 %	0.19 %	0.16 %	0.17 %	0.17 %
K_2O	4.32 %		4.45 %	3.87 %	3.39 %	3.10 %	4.40 %	4.80 %	4.45 %
CaO	1.10 %		1.11 %	1.26 %	1.14 %	1.20 %	1.15 %	1.11 %	1.18 %
Sc		3.35	8.20	40.9	2.44	6.30		24.2	2.43
Ti	126	167	149	214	156	259	153	219	111
V	2.80	2.76	3.53	3.18	3.51	2.86	3.02	2.54	3.37
Cr	14.4	17.6	15.3	13.2	18.4	14.4	13.0		18.6
Mn	417	402	470	399	512	443	402	333	564
Fe	2067		2200	2144	2334	1519	2130	1904	1944
Co	5564	5699	5800	5801	6417	5613	5886	7363	4644
Ni	105	104	112	107	123	102	98.2	42.4	105
Cu	256	279	332	371	294	215	235	203	323
Zn	387	401	354	376	473	406	392	336	429
Ga			13.5	13.1	11.8	10.2		13.1	
Ge				1.41	1.68	1.35		3.74	
As	12.9	11.4	14.1	3.47	10.8	13.5	9.22	5.24	10.5
Se			33.0		1.52				
Rb	110	103	117	99	122		100	120	104
Sr	44.8	47.6	56.4	39.5	49.5		42.9	31.1	36.5
Y		3.42	3.98	2.94	3.59	4.17	3.19	2.47	2.48
Zr	21.5	21.3	21.1	22.6	24.5		23.2	14.4	15.3
Nb		2.65	2.83	2.50	2.89	3.38	2.81	3.06	2.52
Mo			0.26	0.44	0.19	0.24		1.66	0.28
Ag			0.21	0.71	0.17	0.33		0.19	0.13
Cd	31.5	30.1	35.1	18.2	34.3	31.9		34.2	
In			0.27	0.98	0.19	0.19			0.04
Sn	41.7	38.4	55.0	34.4	45.4	40.9	46.1	51.3	34.8
Sb	2.51	3.94	5.44	4.25	3.75		4.34	4.63	3.20
Cs				2.78	3.21	2.69	2.83	3.38	2.62
Ba	117	126	172	115	131	141	113	140	94.3
La		2.09	2.79	1.98	2.83	2.45	2.10	2.31	1.55
Ce		4.41	5.83	4.24	5.87	5.10	4.52	4.09	3.73
Pr		0.53	0.67	0.50	0.65		0.57	0.40	0.45
Nd		2.18	2.09	1.79	2.58	2.35	2.17	1.50	1.56
Sm		0.78	0.62	0.56	0.61	0.69	0.63	0.71	0.46
Eu		0.12	0.13	0.14	0.13	0.14	0.12	0.29	0.10
Gd		0.64	0.58	0.53	0.59	0.70	0.57	0.50	0.42
Tb		0.10	0.14	0.11	0.12	0.12	0.13	0.10	0.08
Dy		0.72	0.64	0.55	0.69	0.70	0.72	0.52	0.48
Ho		0.13	0.14	0.11	0.11	0.13	0.13	0.10	0.08

(continued)

Table 13.3 (continued)

wt%	1	2	3	4	5	7	8	9	EAF
Er		0.38	0.30	0.30	0.30	0.35	0.34	0.22	0.23
Tm		0.05	0.06	0.08	0.04	0.06	0.05	0.03	0.04
Yb		0.61	0.38	0.33	0.33	0.42	0.38	0.30	0.28
Lu		0.08	0.07	0.07	0.05	0.06	0.06	0.07	0.04
Hf		0.70		0.79	0.89	1.05	0.91	0.61	0.57
Ta		0.55		0.45	0.58		0.52	0.47	0.41
W				0.61	0.54		0.57	0.72	0.45
Au			0.09		0.05		0.09	0.07	0.08
Tl				0.13	0.15	0.12		0.19	
Pb	32.5	31.0	48.1	39.3	35.9	30.2	31.0	41.5	37.1
Bi	0.48		0.97	0.70	0.71	0.72	0.80	1.97	0.53
Th		0.94	0.97	0.79	1.48	1.05	0.97	0.68	0.65
U		0.64	0.84	0.85	0.82	0.72	0.78	0.77	0.63
Te								0.01	
Re									

Table 13.4 Similarity coefficients for Green

	1	2	3	4	5	EAF	7	8	9
1	1								
2	0.90	1							
3	0.79	0.76	1						
4	0.89	0.81	0.83	1					
5	0.83	0.87	0.78	0.81	1				
EAF	0.88	0.88	0.80	0.82	0.89	1			
7	0.86	0.87	0.75	0.78	0.84	0.87	1		
8	0.87	0.88	0.79	0.84	0.88	0.86	0.87	1	
9	0.79	0.74	0.74	0.76	0.70	0.76	0.70	0.75	1

Table 13.5 Similarity coefficients for Blue

	1	2	3	4	5	EAF	7	8	9
1	1								
2	0.91	1							
3	0.82	0.82	1						
4	0.83	0.81	0.78	1					
5	0.84	0.85	0.82	0.77	1				
EAF	0.86	0.78	0.74	0.77	0.77	1			
7	0.88	0.86	0.85	0.78	0.83	0.72	1		
8	0.89	0.90	0.86	0.87	0.86	0.80	0.88	1	
9	0.77	0.75	0.75	0.77	0.74	0.75	0.73	0.79	1

manufacturer for the concentrations of a small number of elements. For other values, laboratories rely on published data. Jochum et al. (2011) report new values for the NIST SRM 610-617 glasses that differ as much as 10 % from the Pearce et al. (1997) values. Two laboratories using two different sets of assumed concentrations for the same SRM will produce results that will differ at least in the same proportion as the published concentration values.

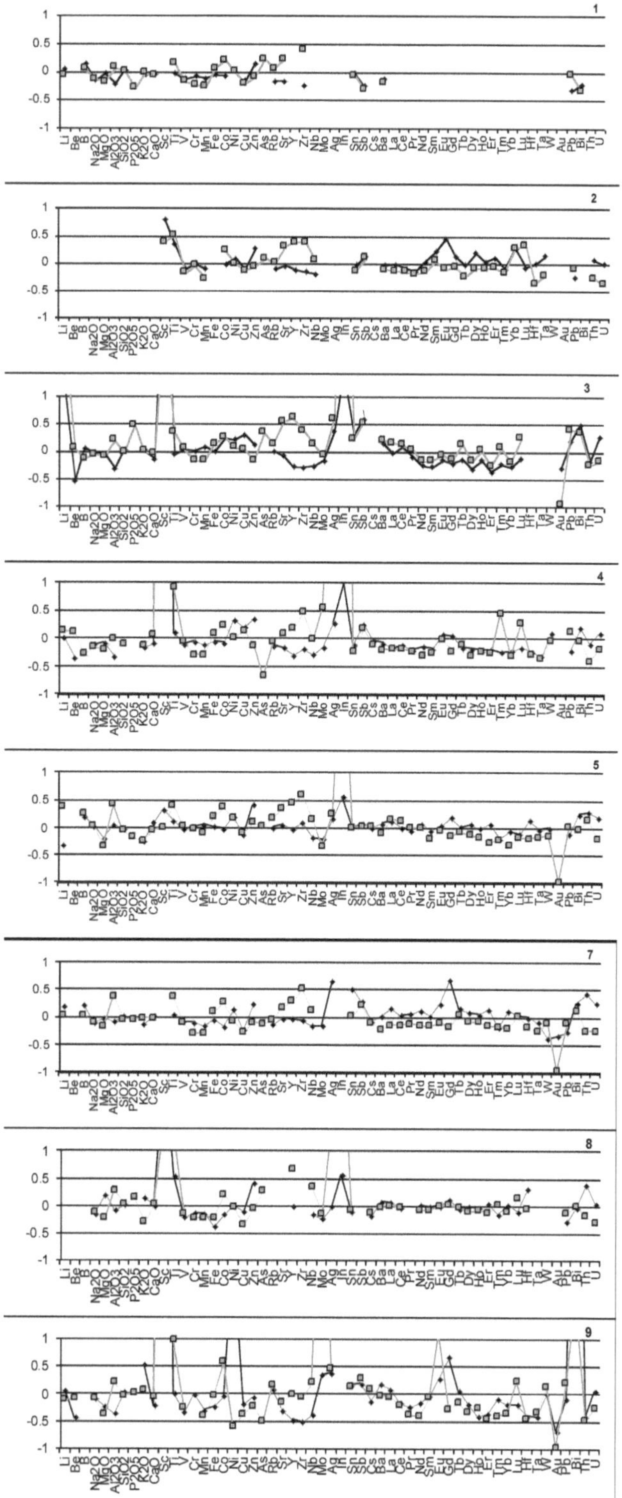

Fig. 13.2 Relative difference between results from each participating laboratory and those from the EAF. *Grey square* = Blue, *black diamond* = Green

Table 13.6 Comparison between the range of limits of detection from the different participating laboratories and the average measured concentrations for the same elements in Green and Blue

	LOD		Green	Blue
ppm	Min	Max	Average	
Li	0.002	0.1	23	18
Be	0.01	0.04	2	3
B	0.01	2	104	515
Sc	0.05	2	12	13
Ti	0.01	4	345	275
V	0.01	0.5	11	3
Cr	0.06	10	901	16
Mn	0.3	1	164	438
Fe	3	24	2371	2030
Co	0.001	6	8	5865
Ni	0.006	2	14	100
Cu	0.4	5	16	279
Zn	0.08	4	55	395
As	0.2	3	10	10
Rb	0.01	0.2	29	109
Sr	0.002	0.1	112	44
Y	0.002	0.1	4	3
Zr	0.0005	0.2	112	20
Nb	0.002	1	3	3
Mo	0.0007	0.2	1	1
Ag	0.003	0.3	0.5	0.3
In	0.0009	12	0.1	0.3
Sn	0.06	1	25	43
Sb	0.02	0.7	13	4
Cs	0.0006	0.3	2	3
Ba	0.001	0.6	443	128
La	0.0008	0.2	6	2
Ce	0.001	0.06	22	5
Pr	0.0006	0.7	1	0.5
Nd	0.002	0.4	4	2
Sm	0.003	0.5	0.7	0.6
Eu	0.001	0.9	0.2	0.1
Gd	0.002	0.1	0.8	0.6
Tb	0.0005	0.3	0.1	0.1
Dy	0.002	0.2	0.6	0.6
Ho	0.0005	0.5	0.1	0.1
Er	0.001	0.5	1.6	0.3
Tm	0.001	0.6	0.1	0.1
Yb	0.003	0.3	0.4	0.4
Lu	0.0005	0.6	0.07	0.1
Hf	0.004	0.2	3	0.8
Ta	0.0007	0.03	0.3	0.5
W	0.002	0.01	0.5	0.6
Au	0.03	0.03	0.04	0.1
PbO	0.03	0.3	172	36
Bi_2O_3	0.005	0.2	1	0.9
Th	0.002	0.4	1	0.9
U	0.002	0.2	0.9	0.8

13.4 Archaeological Implications

The following discussion will explore the impact of the dispersion between laboratories noted in this comparative study on archaeological interpretation. Statistical analysis was performed on rare earth element (REE) compositions of the two different glasses provided by seven of the participating laboratories and the EAF. In Blue and Green glasses, the concentrations for the REEs are sufficiently close to one another that they permit a good simulation of trying to separate two groups of ancient glass with different compositions. These elements are very helpful as they usually characterize the silica source used for glass production, and are generally unaffected by the colorants potentially added to a glass.

Statistical routines obtained from Missouri University Research Reactor implemented through Gauss Runtime 8.0 were used to perform a principal component analysis (PCA) and hierarchical cluster analysis on logged (base 10) REE concentrations. Projection of PCA1 and PCA2 on a bi-plot display two clusters, one for each glass sample (Fig. 13.4). For each glass, one laboratory provided results that fall outside the 95 % confidence ellipse for all measurements on each sample. However, a different laboratory falls outside of the 95 % confidence interval for Green and Blue.

Cluster analysis separates the two glasses (Fig. 13.5). If Blue and Green were two archaeological glasses, it is quite likely that all the laboratories involved in the inter-laboratory comparison test would have been able to place them accurately into their respective chemical group, each chemical group being related to a

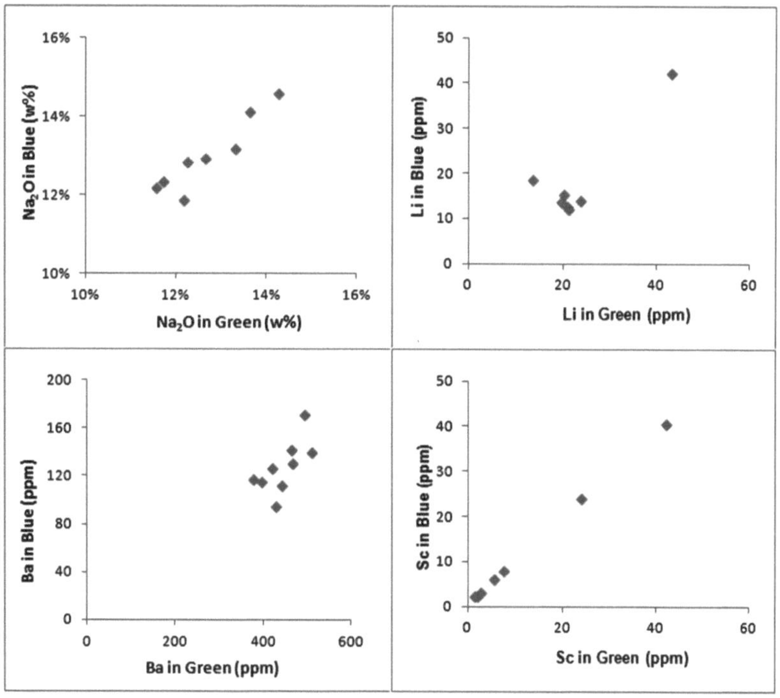

Fig. 13.3 (a–d) Concentrations of different constituent elements in Green and Blue

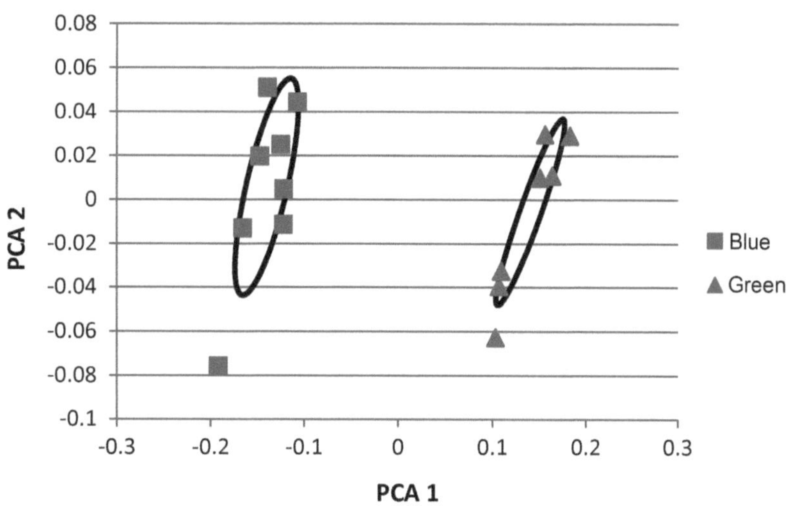

Fig. 13.4 Bi-plot for PCA1 and PCA2 with 95 % confidence ellipses for measurements on Green and Blue

glass producing workshop or region. However, it would be much more difficult for these laboratories to characterize different glass samples taken from a single batch. If a batch can be defined as the quantity of glass produced in one operation, this word designates a rather complex notion as far as ancient glass is concerned. Freestone et al. (2000) report on the chemical analysis of glass produced at a large-scale glass making center in Israel, where several furnaces able to melt nine tons of glass at once were discovered. Samples taken from a single furnace presented a large range of concentrations for oxides such as soda (~11 to 16 %) inversely correlated with silica, suggesting that either the batch was heterogeneous or that several batches with slightly different proportions of sand and soda-rich flux were used to charge the furnace

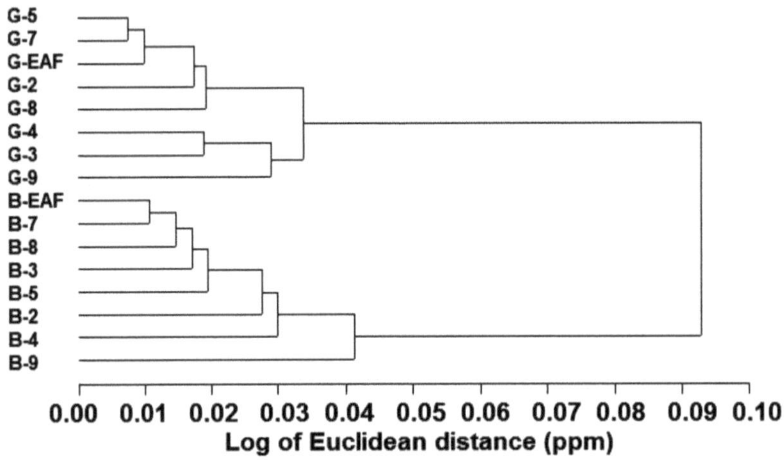

Fig. 13.5 Dendrogram for the compositions of Blue and Green as determined by the different participating laboratories

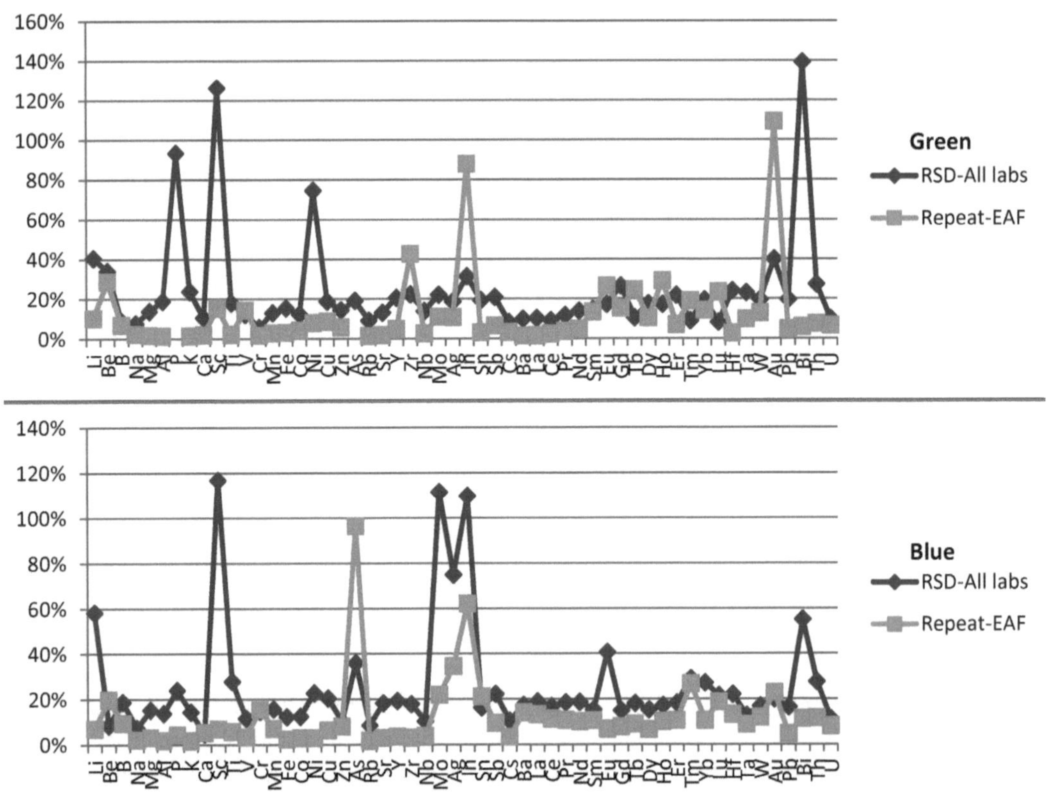

Fig. 13.6 (**a** and **b**) RSD calculated from the concentrations for each element as measured by each participating laboratory compared to the RSD calculated from the measurements carried out at the EAF on the samples sent to each participating laboratory

through the firing process. On the other hand, Freestone et al. (2009) identified a group of objects that had such similar compositions (the variation of the concentration for all measured elements was within experimental error) that not only did the author attribute them to a single workshop, but also it was assumed that they were manufactured from a single glass batch.

If the RSD calculated from the results from the different laboratories are compared to the RSD calculated for the same elements measured on the same pieces sent to each laboratory by the EAF, generally, a larger dispersion of the results is observed for the results obtained by the different laboratories (Fig. 13.6a, b). The variation in the results provided by the different participating laboratories would probably make it impossible to identify glass samples from a single batch with a very homogeneous composition across different studies.

13.5 Conclusion

This is the first inter-laboratory comparison test for laboratories involved in the measurement of the elemental composition of ancient glass using LA-ICP-MS with a focus on trace element determination. Firstly, it should be acknowledged that this test was not ideal, as the two samples tested presented heterogeneity that might impact the results reported by the different laboratories. Secondly, a list of elements should have been provided to the laboratories so as the all range of elements that can be analyzed using LA-ICP-MS could be compared.

This test indicates that results produced by nine different LA-ICP-MS laboratories concur quite well. From the results provided by these different laboratories, it should be possible to interpret the resulting data reliably in terms of group attributions. However, some information might be lost due to the observed dispersion between laboratories. It would for example likely be impossible to identify different objects manufactured from a perfectly homogeneous batch of glass.

This experiment shows that there is a need to develop a more uniform protocol between laboratories, and for those laboratories that use the same standards for quantification, consensus values for their composition should be used instead of different values published in different journals at different times. In the future, another improved inter-laboratory comparison test should be organized, to insure that the recommendations resulting from this study result in improved comparability between laboratories measuring ancient glass with LA-ICP-MS.

References

Bertini M, Izmer A, Vanhaecke F, Krupp EM (2013) Critical evaluation of quantitative methods for the multi-elemental analysis of ancient glasses using laser ablation inductively coupled plasma mass spectrometry. J Anal Atom Spectrom 28:77–91

Dussubieux L, Robertshaw P, Glascock MD (2009) LA-ICP-MS analysis of African glass beads: laboratory inter-comparison with an emphasis on the impact of corrosion on data interpretation. Int J Mass Spectrom 284:152–161

Eggins SM, Kinsley LPJ, Shelley JMG (1998) Deposition and element fractionation processes during atmospheric pressure laser sampling for analysis by ICP-MS. Appl Surf Sci 127–129:278–286

Francis P Jr (1999) Beads of the world. A Schiffer book for collectors, 2nd edn. Schiffer Publishing, Atglen, PA

Freestone IC, Gorin-Rosen Y, Hughes MJ (2000) Primary glass from Israel and the production of glass in Late Antiquity and the Early Islamic period. In: Nenna MD (ed) La route du verre: ateliers primaires et secondaires du second millénaire av. J.-C. au Moyen Âge, Travaux de la Maison de l'Orient Mediterranéen 33. Maison de l'Orient Mediterranée, Lyon, pp 65–83

Freestone IC, Price J, Cartwright CR (2009) The batch: its recognition and significance. In: Janssens K, Degryse P, Cosyns J, Caen J, Van't dack L (eds) The Annales of 17th Congress of the AIHV, Anvers, 2006. University Press of Antwerp, Brussels, pp 130–135

Gratuze B (1999) Obsidian characterization by laser ablation ICP-MS and its application to prehistoric trade in the Mediterranean and the Near East: sources and distribution of obsidian within the Aegean and Anatolia. J Archaeol Sci 26:869–881

Günther D, Heinrich CA (1999) Enhanced sensitivity in laser ablation-ICP mass spectrometry using helium-argon mixtures as aerosol carrier. J Anal Atom Spectrom 14:1363–1368

Hunt JB, Clift PD, Lacasse C, Vallier TL, Werner R (1998) Interlaboratory comparison of electron probe microanalysis of glass geochemistry. In Saunders AD, Larsen HC, Wise SW Jr (eds) Proceedings of the Ocean Drilling Program, Scientific Results, vol 152, pp 85–91

Jochum KP, Weis U, Stoll B, Kuzmin D, Yang Q, Raczek I, Jacob DE, Stracke A, Birbaum K, Frick

DA, Günther D, Enzweiler J (2011) Determination of reference values for NIST SRM 610-617 glasses following ISO guidelines. Geostandard Geoanal Res 35 (4):397–429

Košler J, Wiedenbeck M, Wirth R, Hovorka J, Sylvester P, Míková J (2005) Chemical and phase composition of particles produced by laser ablation of silicate glass and zircon—implications for elemental fractionation during ICP-MS analysis. J Anal Atom Spectrom 20(5):402–409

Kuhn H-R, Günther D (2004) Laser ablation-ICP-MS: particle size dependent elemental composition studies on filter-collected and online measured aerosols from glass. J Anal Atom Spectrom 19:1158–1164

Mason PRD, Mank AJG (2001) Depth-resolved analysis in multi-layered glass and metal materials using laser ablation inductively coupled plasma mass spectrometry (LA-ICP-MS). J Anal Atom Spectrom 16:1381–1388

Pearce NJG, Perkins WT, Westgate JA, Gorton MT, Jackson SE, Neal CR, Chenery SP (1997) A compilation of new and published major and trace element data for NIST SRM 610 and SRM 612 glass reference materials. Geostand Newslett XXI: 114–115

Russo RE, Mao X, Gonzalez JJ, Mao SS (2002) Femtosecond laser ablation ICP-MS. J Anal Atom Spectrom 17:1072–1075

Saitowitz SJ (1996) Glass beads as indicators of contact and trade in Southern Africa ca AD 900–AD 1250. Ph.D. Dissertation, University of Cape Town

Shortland A, Rogers N, Eremin K (2007) Trace element discriminants between Egyptian and Mesopotamian Late Bronze Age glasses. J Archaeol Sci 34 (5):781–789

Vicenzi EP, Eggins S, Logan A, Wysoczanski R (2002) Microbeam characterization of Corning archaeological reference glasses: new additions to the Smithsonian microbeam standard collection. J Res Natl Inst Stand 107(6):719–727

Wagner B, Nowak A, Bulska E, Hametner K, Günther D (2012) Critical assessment of the elemental composition of Corning archaeological reference glasses by LA-ICP-MS. Anal Bioanal Chem 402:1667–1677

Walton M, Shortland A, Kirk S, Degryse P (2009) Evidence for the trade of Mesopotamian and Egyptian glass to Mycenaean Greece. J Archaeol Sci 36 (7):496–1503

Wang Z, Hattendorf B, Günther D (2006) Vaporization and ionization of laser ablation generated aerosols in an inductively coupled plasma mass spectrometer—implications from ion distribution maps. J Anal Atom Spectrom 21:1143–1151

Characterization of Slag Inclusions in Iron Objects

14

Maxime L'Héritier, Stéphanie Leroy, Philippe Dillmann, and Bernard Gratuze

Abstract

Tracing the origins of iron artifacts generates great interest in iron metallurgy. Chemical analysis of non-metallic slag inclusion (SI) entrapped in iron produced in bloomery furnaces is an efficient mean for provenance studies as, within their composition, SI do not only carry the conditions of production but also the chemical signature of the iron ore initially used. The analytical protocol is a three stage process: (a) metallographic identification and selection of SI, (b) characterization of their main component using SEM-EDX, (c) traces element determination using LA-ICP-MS. LA-ICP-MS analysis allows to quantify up to 39 trace elements in the fayalitic-rich matrix of SI with detection limits at the ppm level. Compositional ratios are then compared with one another and to the chemical signature of previously analysed regions of production. The methodology was applied on iron armatures from the cathedral of Bourges to study their fabrication, their installation in the building and their provenance. Results obtained allow a better understanding of the role of these armatures and the supply of this construction site.

14.1 Introduction

Since the Iron Age, iron based materials have played an essential role in the technical and economic history of ancient societies, and tracing the origins of iron artifacts to understand their circulation and the distribution of ancient markets and commercial routes has always been among the major questions that generate great interest among historians and archaeologists in iron metallurgy. For example, recent archaeometallurgical studies suggest that in the late Iron

M. L'Héritier (✉)
Department of History, Université Paris 8, EA 1571 Histoire des Pouvoirs, Savoirs et Sociétés (HPSS), 2, rue de la liberté, 93 526 Saint Denis, France
e-mail: maxime.l_heritier@univ-paris8.fr

S. Leroy • P. Dillmann
Laboratoire Archéomatériaux et Prévision de l'Altération (LAPA), IRAMAT LMC, UMR 5060 CNRS and NIMBE UMR 3685 CEA/CNRS, CEA/Saclay, 91191 Gif-sur-Yvette Cedex, France

B. Gratuze
Institut de Recherche sur les Archéomatériaux, Centre Ernest Babelon, CNRS/Université d'Orléans, 3D rue de la Férollerie, 45071 Orléans Cedex 2, France
e-mail: gratuze@cnrs-orleans.fr

L. Dussubieux et al. (eds.), *Recent Advances in Laser Ablation ICP-MS for Archaeology*, Natural Science in Archaeology, DOI 10.1007/978-3-662-49894-1_14

Age, iron production was organised around major political and economic centres which concentrated post-reduction forging activities thanks to the provision of raw blooms (Berranger and Fluzin 2007). Metallographic analyses of iron bars coming from Roman shipwrecks off the coasts of Saintes-Maries-de-la-Mer in the Mediterranean Sea proved that these semi-products were classified and might have been selected according to their quality (Pagès et al. 2011). In the Middle Ages, historical sources also testify that iron products could be traded locally or internationally. As one example among many, Spanish iron is found in the construction records of several cathedrals of the French kingdom: in Rouen and Amiens, cities which are close to the coasts and even in Troyes, an inland city far from the usual maritime trading routes of Spanish iron, although iron was also produced locally (L'Héritier et al. 2010). Was it chosen for its better quality? Verna (2001) states that high quality *fer de Foix* ("iron from Foix") was traded over several hundreds of kilometres and for example entered into the construction of the now collapsed Lion tower in Arles.

To date, these written sources have been the only means to trace the provenance of iron products, but in the past 10 years, several studies have attempted to trace their origins using trace elements analysis of slag inclusions (SI) embedded in the metallic matrix, which carry the chemical signature of the iron ore initially used. After a brief review of these previous studies, this chapter will detail the analytical protocol used to perform provenance analysis on iron objects with an emphasis on the role of LA-ICP-MS. An example from medieval world will be used to illustrate the results that can be obtained using these new analytical techniques, by considering the iron supply of an emblematic monument of the thirteenth century, the cathedral of Bourges. Medieval buildings used tons of iron armatures, thus providing a great deal of well-dated iron artifacts for study (L'Héritier et al. 2010).

14.2 Background

A succession of chemical, physical and mechanical transformations are necessary to obtain a usable object from iron ore. We will focus here on the bloomery process, which is the main process used until the end of the Middle Ages. In the bloomery furnace, where iron ore is mixed with charcoal at high temperature (>1200 °C), most iron ore compounds are not removed, or not entirely, and become rejects known as slag, which have to be evacuated. Yet, as the entire reduction process takes place below the melting temperature of the metal (1535 °C), some non-metallic second phase particles, known as slag inclusions (SI), can remain entrapped in the metallic matrix of the object. These SI are composed of a partly fayalitic matrix sometimes containing wüstite dendrites. Within their composition, slag and SI carry information not only about the conditions of production (charcoal, lining, flux), but also the chemical signature of the iron ore initially used. Many studies of iron materials have therefore tried to obtain quantitative chemical information about the SI (Hedges and Salter 1979; Tylecote 1970).

The earliest analytical studies on SI focused on major (and minor) element composition with SEM-EDS (Scanning Electron Microscope-Energy Dispersive Spectroscopy) (Blakelock et al. 2009; Buchwald 2005; Buchwald and Wivel 1998; Dillmann and L'Héritier 2007; Gimeno Adelantado et al. 2003; Hedges and Salter 1979; Paynter 2006). In very favourable cases, i.e. very particular archaeological and geological contexts, it is possible to distinguish objects coming from very specific ores. Leroy (1997) used the very different compositions of two ores found in Lorrain—the siderolitic ore called "fer fort," and the oolithic ore better known as "minette," to trace their respective use in the region. A similar example is found in Scandinavia, where Buchwald and Wivel (1998) were able to differentiate Danish iron artifacts from those from other Scandinavian regions such as Norway and Sweden. However, in most cases, scattering due to contamination of the major elements by flux, charcoal or furnace linings is too strong, even between two smelts using the same ore (Dillmann and L'Héritier 2007; Kronz 2003; Lychatz and Janke 2000), to use these major elements as relevant geographic markers.

Research gradually turned to trace elements, which are more numerous than major elements and less likely to be polluted (Coustures et al. 2003, 2006; Desaulty et al 2009; Leroy 2010; Leroy et al. 2011a; Schwab et al. 2006). They can be classified in three main groups: lithophile elements which tend to be concentrated in the silicate phase during iron ore smelting in the bloomery furnace (the slag), siderophile elements which are preferentially retained in the iron phase, and volatile elements which will evaporate. Coustures et al. (2003) suggested that highly lithophile elements can be considered for provenance study, as their ratios are not likely to vary from the ore to the slag during the whole process of iron production, meaning that samples with a same origin should display a linear correlation. This suggestion of conservation of ratios from ore to slag and SI has been confirmed experimentally (Desaulty et al. 2009; Leroy 2010). A few lithophile trace elements (Ba, Zr, Sr, Rb) should however be discarded from provenance studies, as they are likely to be contaminated by furnace linings, flux, or charcoal when present in small quantities in the initial ore, as shown by these experiments.

A crucial point of any provenance study therefore resides in the choice of the best suited analytical method to quantify trace elements. INAA (Instrumental Neutron Activation Analysis) or solution ICP-MS are relevant methods currently used for macroscopic analyses of ore and slag, with detection limits below 1 ppm for all of these elements (Desaulty et al. 2008; Mariet et al. 2008). Their small size (<1 mm) and the fact that SI are embedded in the iron matrix prevent any easy adaptation of these techniques for their analysis, however. Several micro-analytical methods such as EPMA (Electron Probe Micro-Analysis), μ-XRF (X-ray fluorescence) and LA-ICP-MS, which permit targeted investigations at the scale of SI inclusions, were tested to quantify trace elements concentrations in SI.

Schwab et al. (2006) used EPMA to quantify trace elements in iron objects coming from the oppidum of Manching (Germany) and compare them with known sources of ores (iron crust, bog ores and bean ores). The analytical set-up for SI (WDS with an accelerating voltage of 20 kV, a beam current of 40 nA and counting times of 10 s) allowed a quick quantification of major and trace elements at the same time. Iron contents seem to be overestimated in small inclusions, although this problematic size limit is not mentioned by the author. Moreover, typical detection limits for this method are between 50 and 500 ppm depending on the element and matrix, which is clearly above the concentration of many trace elements. The author himself regards the sensitivity of the EPMA as a limiting factor. In this study, only Ce, Ba, Sr, and Zr could be accurately quantified along with minor elements such as V, Cr, and Ti, most of which are likely to be impacted by contamination, and are therefore not fully appropriate for provenance studies. Therefore, results of this EPMA study cannot be considered as totally relevant: even though most objects seem to have a SI composition compatible with bog ores, it is clearly not possible to reach a firm provenance assignment based on EPMA.

Other provenance studies have relied on LA-ICP-MS for trace element analysis in SI (Coustures et al. 2003, 2006; Desaulty et al. 2009; Leroy 2010). In each study, a Nd: YAG (neodymium-doped yttrium aluminium garnet) laser operating at 266 nm was used for SI ablation. The size of the resulting crater varied between 80 μm and 300 μm after a 50 to 160 s-long analysis. Therefore, only SI exceeding this size could be analysed. More than 30 trace elements including the rare earth group can be quantified by LA-ICP-MS, with detection limits below 1 ppm. However, due to interferences caused by the recombination of matrix major elements (notably O, Si, Fe...) or argon, some very light elements such as scandium (^{45}Sc versus $^{29}Si^{16}O^{+}$) are hardly detectable with this method and therefore cannot be accurately quantified (Desaulty et al. 2009).

These three studies provided relevant results in the field of iron provenance: highlighting differences between various origins was possible thanks to the large number of quantified elements. However, despite the measurement accuracy, the reliability of the results differs from one author to another, depending on the gathered analytical corpora but also on the choice of relevant elements to characterize each provenance. For example, whereas Desaulty and Leroy analysed on average ten SI per sample, Coustures et al. only analysed 13 SI coming from five different iron bars, which seems to be insufficient for estimating dispersion (Desaulty et al. 2009), and no metallographic analysis was carried out beforehand to understand whether the studied metallic matrices were homogeneous or not. They also used elemental pairs behaving like quasi-isotopes (e.g. Zr/Hf), which should be avoided as their elemental ratios usually do not vary from one region to another and are therefore not relevant for a particular region (Desaulty et al. 2009). Moreover, whereas pointing out dissimilarities between samples from different origins is sometimes easy using bivariate plot diagrams, validating a provenance hypothesis is much more difficult, as many ratios have to be considered. Other treatments using multivariate statistical analysis were therefore developed to check the similarities and incompatibilities between several ensembles of elements (Leroy 2010; Leroy et al. 2011a; Disser 2014).

Although LA-ICP-MS gives very good results for quantifying trace elements in SI, to have a sufficient signal with the analytical set-ups used so far, only SI larger than 80–100 μm could be analysed. Finding such SI is usually easy in iron by-products or poorly manufactured objects such as construction iron. On the other hand, it can remain a challenge in highly manufactured objects such as tools, weapons or armour. To attempt to overcome this limitation, Leroy tested confocal μ-XRF using Synchrotron Radiation on medieval Italian armour samples, which required a non-destructive analysis of much smaller SI (<30 μm) (Leroy et al. 2011b). The main fluorescence peaks are Rb, Sr, Y, and Zr for trace elements, whose detection limits could be estimated at around 10 ppm (for an energy of 22 KeV) by comparison with LA-ICP-MS and ICP-MS. No other trace elements are likely to be detected unless present in very high quantities (e.g. Ba ≈ 4000 ppm). Interferences may occur between peaks with very close energies (e.g. Kβ (Rb) and Kα(Y) or Kβ(Sr) and Kα(Zr)), thus preventing the detection of Y and Zr if Rb or Sr respectively are too concentrated. Despite the small numbers of trace elements which can be quantified with this method—among which most are likely to be contaminated—it might prove relevant in certain specific cases. For example, Leroy was able to reject the hypothetical Lombard provenance for some of the studied armours, as their SI compositions do not match the region's chemical signature.

Table 14.1 sums up the characteristics of the analytical methods tested so far for trace element analysis in SI. Among them, LA-ICP-MS is clearly the best suited to quantify lithophile trace elements at the ppm levels in SI from iron objects, although size is likely to be an issue for small inclusions (<80 μm). However, more sensitive—and more recent—devices should partly be able to overcome this last issue.

Table 14.1 Characteristics of analytical methods tested for trace element analysis in slag inclusions (advantages in bold, hindrances in italic)

Analytical method	Number of analysed trace elements	Detection limits	Size of analysis	Time of analysis
EPMA	*4*	*50–500 ppm*	Not mentioned (a few microns?)	**10 s**
μ-XRF	*4*	≈10 ppm	**5–30 μm depending on Z**	*10 min in average*
LA-ICP-MS	**>30**	**<1 ppm**	80–300 μm	50–160 s

14.3 Protocol

Prior to LA-ICP-MS trace element analysis, a rigorous protocol (Fig. 14.1) has to be followed to carry out a reliable provenance study (Desaulty et al. 2009; L'Héritier et al 2013; Leroy et al. 2011a). It consists in (a) performing a complete metallographic analysis on each artefact to consider the history of the artefact, in particular to evaluate if its forging required voluntary welding of iron from different origins for technical reasons or simply because of recycling and (b) selecting relevant SI for provenance studies, i.e. produced during the smelting stage. First, one (or several) cross-sections should be chosen on the artefact to evaluate its potential heterogeneity, especially when the object was likely to have been forged from different metals (e.g. tools with a steel point or edge). After mounting in epoxy resin and polishing using SiC abrasive paper (grade 80 to 1200) and diamond polishing (medium 9, 3 and 1 μm), a preliminary observation is performed on the sample using an optical microscope. This allows distinguishing the different kinds of inclusions and possible welding lines, suggesting different zones of interest on the artifact. Metallographic etching such as Nital or Oberhoffer etchings can also be used at this stage to better identify these potential features by revealing the metallic structure (carbon and phosphorus content).

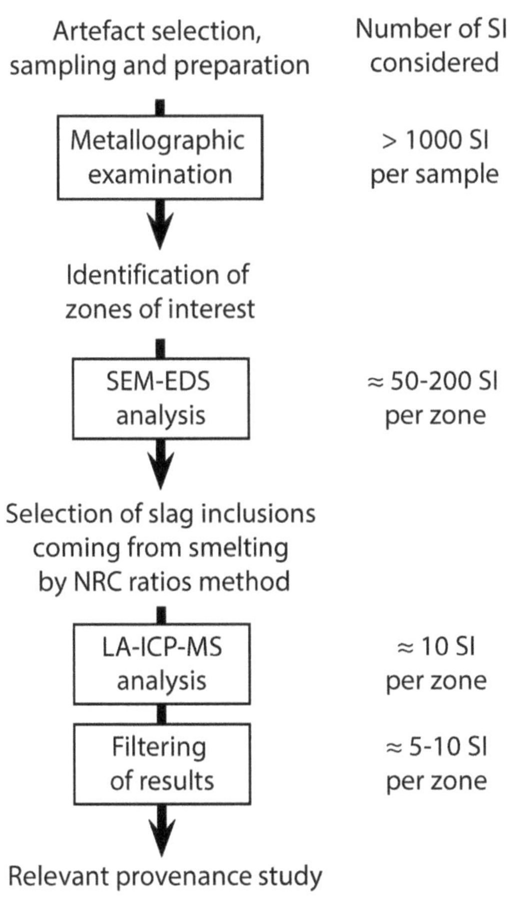

Fig. 14.1 Schematic diagram of the protocol used for provenance analysis

Next, major composition analyses of the SI are performed using Energy Dispersive Spectrometry (EDS) coupled to a Scanning Electron Microscope (SEM). As proposed by Dillmann and L'Héritier (2007), at least 50 SI per zone of interest on the sample should be analysed to select only those which are characteristic of the reduction system (ore, fuel, lining, flux), i.e. which come from the smelting stage. These SI can easily be identified as their Non-Reduced Compounds (NRC) ratios (i.e. Al_2O_3, SiO_2, MgO, CaO, K_2O) are constant throughout the sample. SI whose compositions vary from these ratios are likely to have undergone phase partitioning during the *chaine opératoire* or to come from the use of additives (antioxidants) during the forging stage. In both cases, these cannot be considered for provenance studies and should consequently be discarded from any trace element analysis. Moreover, if several constant NRC ratios are found within the same object, matching with previously identified zones of interest, it is likely that the object was forged out of iron pieces coming from different smelts and therefore potentially having different origins. Each zone of interest with a different signature should then be considered separately during trace element analysis. The combination of optical metallographic analysis with SI major element analysis by SEM-EDS allows (a) identification of iron objects likely to have been recycled (i.e. to have several origins), (b) selection within each homogeneous section of the artefact of up to a dozen SI with a sufficiently large size (>80 μm wide) whose compositions are related to the ore's chemical signature.

Analyses performed at the Centre Ernest-Babelon of the IRAMAT (Orléans) are carried out with an Element XR mass spectrometer from Thermofisher Instrument and a VG UV microprobe ablation device (quadrupled frequency Nd: YAG operated at 266 nm). The concentrations of 39 elements are routinely determined in SI, among them the main lithophile elements present in SI including most of the rare earth group and some siderophile elements mainly present in metallic iron. Among these elements, only 12 (mainly the rare earth and some related elements) are routinely used to determine the geochemical signature of the ore. Others elements are however useful either to obtain an estimation of the metallic iron present in the ablation (Mo, Sb, W, Cu, Ni...), or to distinguish objects coming from very specific ores or to check contamination during the reduction process (Ti, V, Ba, Zr, Sr, Rb...). Silicon is measured on the ^{28}Si isotope and is used as internal standard.

While new laser ablation systems have laser beam sizes as small as 4 μm, the 266 nm laser ablation device used in this study does not allow ablation below 20 to 30 μm. Moreover, for SI with average silica content (<40 %), an 80 μm ablation is usually required to generate sufficient signal. One of the first problems encountered with SI analysis using LA-ICP-MS is linked to the size (diameter and thickness) of the studied material. These objects are often too small to carry out line analysis and their thickness is unknown when using spot analysis. The ablation of pure SI is therefore nearly impossible, and the ablated material will generally be a combination of siliceous slag and metallic iron. As SI contain iron as well, distinguishing between iron oxides and silicates contained in the SI and metallic iron from the surrounding matrix is thus not possible and the classical internal standard procedure used to calculate the composition of glass and metal is consequently not applicable to SI.

Therefore, a compromise has to be made to obtain reliable results representing the geochemical signature of the ores. Assuming that the concentrations of lithophile elements of interest are negligible in the metallic phase, relative concentrations of these elements, rather than absolute values, can however be calculated using the silicon values determined by SEM-EDS and a semi-quantitative internal standard calculation procedure. Our aim is not to determine absolute rare earth concentration but to obtain the geochemical signature of the sample assuming that rare earth element ratios remain unchanged from the ore to slag and SI as shown by Desaulty and Leroy (Desaulty et al. 2009; Leroy 2010). To check the representativeness of the ablated material, the relative proportion of metallic iron can be estimated by measuring the signal of siderophile elements such as Mo and Sb and by comparing the intensity of the raw signal values obtained for silicon with the one obtained for glass SRM610 used as a reference material. Inclusions with high Mo and Sb signal and low Si signal should then be rejected.

LA-ICP-MS analysis of SI operates as follows. The objects are put in a large ablation cell (12 cm in diameter and 4 cm height) together with a glass disc of SRM610. The 266 nm Nd: YAG laser is operated at 5–7 Hz with a beam diameter of between 30 and 80 μm depending on the size of the SI and its silica content (usually 80 μm). Argon carrier gas at a flow rate of 1.35 L/min is used along with a cool gas at 16 L/min and auxillary gas flow at 1.07 L/min. There is 15 s of uptake time prior to SI analysis, and 3 min prior to each blank between two different samples. All isotopes except ^{28}Si are measured in dual mode with four points per peak; ^{28}Si is measured using a Faraday cup with ten points per peak (Table 14.2). Scan time for 39 isotopes is 1.96 s, with 25 scans for a total analysis time of 49 s.

For each sample, between 9 and 12 SI are selected per zone of interest. To improve reproducibility and to correct instrumental drift or changes in ablation efficiency, SRM610 standard is measured at least once per hour, approximately every four or five samples. SI analyses are carried out in a random order by groups of three. Each analysis consists of a blank measurement

Table 14.2 Measured isotopes, detection mode and main interferences

Used isotope	Detection mode	Main interferences
Si28 (used as internal standard)	Faraday, 10 points per peak	$Fe56^{++}$ (negligible)
Be9, Sc45, Ti47, V51, Cr52, Co59, Ni60, Cu63, Zn66, Ga69, Ge72, As75, Rb85, Sr88, Y89, Zr90, Nb93, Mo95, Pd108, Sn118, Sb121, Cs133, Ba137, La139, Ce140, Pr141, Nd146, Sm147, Eu153, Gd157, Tb159, Yb172, Hf178, Ta181, W182, Pb208, Th232, U238	Dual mode: pulse counting or analogue, 4 points per peak	

followed by three ablations of different SI from the same sample. Then, the next iron sample is analysed. When three SI have been measured on each sample, we start again on the first one and so on, until totalling 9 to 12 SI analyses per sample or zone of interest. This procedure allows us to check instrumental drifts, thus insuring that if two sets of SI from two different samples show different behaviours, they could really be interpreted in terms of different ores or provenance without being due to analytical artifacts. At the end of the analytical sequence, the reference materials are analysed again.

The glass standard reference material SRM610 manufactured by NIST (National Institute of Standards and Technology) is used to calculate the response coefficient factor K_Y as defined by Gratuze (1999) and thus convert signal data into quantitative values. Concentrations in SRM610 are taken from Pearce, Norman and Hollocher (Hollocher and Ruiz 1995; Norman et al. 1996; Pearce et al. 1997). Concentrations are then calculated using the silicon content determined by SEM-EDS. For measurement carried out directly on archaeological slag an additional reference material, Corning glass B (Brill 1999), is used.

The following formula is used to calculate the relative concentrations for all elements:

$$[Y] = [Si]_{SEM-EDS} \times \frac{I_Y}{I_{Si} \times K_Y} \quad (14.1)$$

where I_Y and I_{Si} are the net intensity counts rates, corrected for isotopic abundance, measured for elements Y and silicon.

K_Y is the response coefficient factor for element Y, calculated as follows:

$$K_Y = \frac{I_{Y_{N610}} \times [Si_{N610}]}{I_{Si_{N610}} \times [Y_{N610}]} \quad (14.2)$$

where I_{YN610} and I_{SiN610} are the net intensity counts rates, corrected for isotopic abundance, measured for elements element Y and silicon in the standard material SRM610 and $[Y_{N610}]$ and $[Si_{N610}]$ are their concentrations.

For slag analysis, the complete analytical protocol developed for glass is used (see Gratuze, Chap. 12). Iron concentrations are determined by using the ^{57}Fe isotope of iron, and Corning B glass is used to calculate K_{Fe}. As the analysed material (slag which can contain more than 50 % iron) differs strongly in composition from this reference material (Corning B iron concentration is only 0.23 %), the validity of our protocol and the absence of matrix effects for slag were checked by analysing a set of slag samples using both LA-ICP-MS and ICP-MS in liquid mode (liquid ICP-MS measurements were carried out by ALS Minerals). Data obtained on these slag samples using both methods shows a good agreement between laser ablation values and liquid mode values (Table 14.3). If we then compare the element concentrations determined with the slag inclusion protocol and the full analytical protocol, a good agreement is observed between rare earth element ratios.

The trace elements determined in slag inclusions are mainly concentrated in the silica phase, and for a determined laser spot size, their detection limits are strongly dependent on both the size of the analysed SI and its silica content. Silica content of SI can vary from 10 % up to 70 %, however it is more often in the range of 20 to 40 %. The net signal intensity measured for silica is between 2 and 30 times lower than the

Table 14.3 Comparison of LA-ICP-MS (Centre Ernest-Babelon) and ICP-MS (ALS Minerals) values for slag samples (all values in ppm)

	Y		Nb		La		Ce		Nd		Sm		Eu		Tb		Yb		Hf		Ta		Th		U	
Sample	Liq.	LA	Liq.	LA	Liq.	LA	Liq.	LA	Liq.	LA	Liq.	LA	Liq.	LA	Liq.	LA	Liq.	LA	Liq.	LA	Liq.	LA	Liq.	LA	Liq.	LA
PE 3701	130	143	7.4	7.3	83	98	73	68	63	72	13	14	3.1	3.2	2.0	2.3	5.0	6.5	2.6	2.8	0.4	0.4	8.1	11	5.3	5.5
PE 3702	77	82	13	13	58	65	99	89	53	58	10	11	2.4	2.5	1.5	1.7	4.0	5.3	5.3	6.9	0.8	1.0	10	15	5.1	5.2
PE 3703	124	141	12	12	80	96	125	115	75	86	14	16	3.2	3.6	2.1	2.6	6.3	8.9	4.4	5.8	0.8	0.9	11	16	6.9	6.5
MA 0003	47	46	8.4	11	55	54	161	133	73	70	17	17	3.6	3.8	2.2	2.2	6.0	7.2	5.7	6.7	0.5	0.8	32	38	8.7	7.6
MA 0014	13	14	2.5	4.5	18	25	51	64	19	25	4,3	5.1	0.8	1.1	0.5	0.6	1.5	1.8	2.5	3.5	0.1	0.3	8.0	11	3.0	3.3
BL 0012	9	10	2.0	2.2	11	11	26	26	11	11	2.3	2.5	0.5	0.5	0.3	0.4	0.9	1.3	2.2	1.6	0.1	0.2	4.2	5.3	1.2	1.3
US 1206	80	101	7.6	11	53	72	146	155	70	89	17	21	4.5	4.9	3.2	4.0	8.8	13	5.1	6.3	0.5	0.8	27	38	11	11
US 1215	84	100	8.6	9.3	61	68	167	133	79	83	19	21	4.7	4.3	3.3	3.8	9.6	13	5.1	5.9	0.5	0.7	34	41	13	11
BE 3501	131	105	20	15	87	75	185	139	85	79	20	19	4.9	4.6	3.5	2.9	11	12	9.1	8.5	1.6	1.3	29	22	17	11

Table 14.4 Range of detection limits achieved for a slag inclusion whose silicon net signal intensity is approximately 12 % of the one measured for SRM610 glass reference material

Average range of LOD values	Elements
Below 15 ppb	Th, U
Between 30 and 100 ppb	Y, Nb, La, Ce, Pr, Tb, Yb, Hf, Ta, W
Between 150 and 500 ppb	V, Zr, Mo, Pd, Sb, Nd, Sm, Eu
Between 500 and 1000 ppb	Ga, Rb, Sr, Cs
Between 1 and 5 ppm	Be, Co, Ni, Cu, Zn, Ge, As, Sn, Ba, Gd
Between 6 and 15 ppm	Sc, Cr, Pb
Between 100 and 150 ppm	Ti

one measured for the SRM610 standard glass. The detection limits given in Table 14.4 are calculated for SI containing an average value of 30 % silica and with a net Si signal intensity on the order of 12 % of the one measured for SRM610. In that case, detection limits range from a few ppb to 15 ppm for all elements except titanium, whose detection limit on ^{47}Ti is 130 ppm. This high value is due to the high background value measured at mass 47 as well as low isotopic abundance (7 % of ^{47}Ti), while the other elements are measured on positions with lower background and with more abundant isotopes. The achieved detection limits could be up to five times lower or higher, depending on the size of the SI and on its silica content.

Precision has been estimated by considering both the standard deviation of the sensitivity coefficient K_Y calculated for SRM610 (which is below 5 % relative standard deviation for most elements) and the measurement error on the peak intensity (which is equal to the ratio of the square root of the signal intensity divided by the signal intensity). Precision ranges from 5 to 10 % depending on the intensity of the measured signal. As the aim of this work is to obtain the reliable geochemical signature of the sample rather than the exact concentration values of elements, the uncertainty of silica measurement by SEM-EDS has not been taken into account. For a given SI, silica concentration can be considered as a constant. Variations in its content are not likely to affect element ratios, but only real element concentrations and thus the accuracy of the determined value.

14.4 Archaeological Application: The Tie-Rods of Bourges Cathedral

The cathedral of Bourges was rebuilt in its current gothic form after 1195, under the reign of Bishop Henri de Sully, who donated money for its reconstruction. According to art historical studies, the eastern half of the nave up to the 6th bay was erected before 1214 (Branner 1962). Then, after a break in construction which is clearly visible thanks to changes in the masonry, the western parts were probably begun under Simon de Sully. According to Ribault (1995), the nave was finished before 1232, yet more recently Epaud (2011) proved that it was only roofed in 1256, endorsing Branner's first assumptions about a later completion date. Among the ferrous armatures that were used inside the building, five meter long tie-rods were placed in the side-attic over the aisles at the triforium level. These tie-rods are used to attach the nave pillars to the corresponding abutments and are inserted above the transverse arches spanning the vaults (Fig. 14.2). They weigh more than 45 kg each and therefore needed the welding together of several blooms, which is partly visible from the existing forging marks on their surface. Despite their amazing dimensions, these tie-rods are not very deeply anchored inside the masonry: only 17 cm deep with a 5 cm long hook (Férauge and Mignerey 1996). Moreover, their distribution is not totally regular: on the northern side, the first tie-rod still in place corresponds to bays 2 and 3, whereas on the southern side, they start at bays 5 and 6 (Fig. 14.3). Considering the substantial restoration work that this part of the church underwent in the eighteenth and nineteenth centuries, Férauge and Mignerey (1996) question their installation date, not knowing whether to classify them as "iron armatures used *a priori* for a temporary consolidation" or "iron armatures used *a*

Fig. 14.2 View of the northern aisle tie-rods

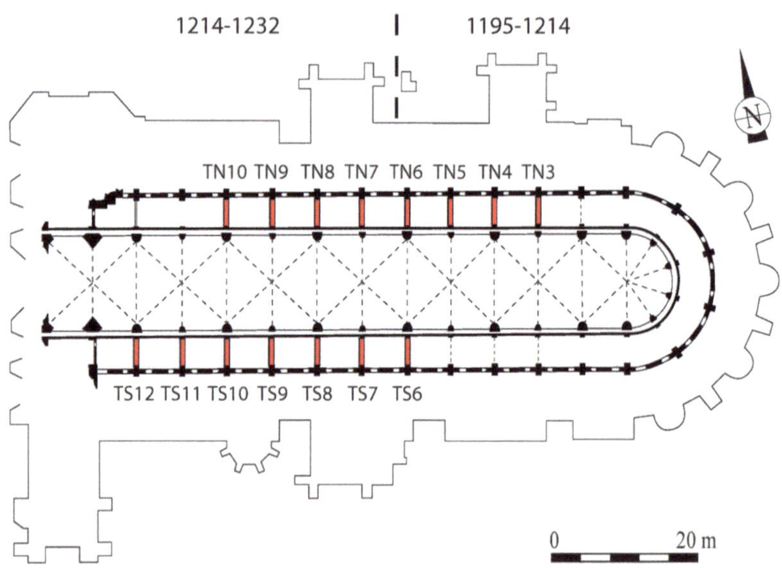

Fig. 14.3 Location of the tie-rods in Bourges cathedral (plan at the triforium level)

posteriori for a definitive consolidation," two totally different groups architecturally speaking.

Thanks to the collaboration of architect P. Ponsot and curator J.-P. Blin, some sampling was carried out on these tie-rods to understand (a) if these 45 kg bars were forged by welding blooms from the same or from different origins and (b) if their supply matches the known construction phases, particularly the so-called "Branner break" between the 5th and 6th bays.

Forty samples were taken from the 13 available bars (Table 14.5) totalling up to five samples per bar to check their homogeneity. Most analysed samples are too small (about 5 mm long) to give an idea of the structural homogeneity of the metallic matrix and of the forging techniques which were used by the smith, yet all of them but two contained sufficient SI to perform major and trace element analysis. Larger samples could only be studied on tie-rod TN7. One of them

(TN7H) displays a welding line, which proves that several pieces of metal were indeed welded together to forge this tie-rod (Fig. 14.4).

Major element analysis by SEM-EDS shows that most SI come from the smelting stage and that they are therefore relevant for provenance studies. This feature also proves that little or no flux was added by the smiths during forging. However, SI coming from different samples of the same tie-rod sometimes show two distinct constant NRC ratios (Fig. 14.5). It is in particular true for tie-rod TN7, where SI have a different composition on each side of the welding line, thus giving evidence of blooms from diverse workshops.

Trace elements analysis emphasizes this feature, when comparing the chemical signatures of several samples from the same tie-rod (12 trace elements were considered: Y, Nb, La, Ce, Nd, Sm, Eu, Tb, Yb, Hf, Th, U). Indeed, for TN3, TN4, TN6, TN7, TS8, TS9 and TS11, at least one bivariate plot diagram shows at least two distinct

Table 14.5 List of the samples coming from Bourges cathedral tie-rods

Tie-rod reference	Location (bay)	Number of samples (analysed by LA-ICP-MS)
TN3	North 2–3	4
TN4	North 3–4	4
TN5	North 4–5	3
TN6	North 5–6	4
TN7	North 6–7	3
TN8	North 7–8	3 (2)
TN9	North 8–9	4 (3)
TS6	South 5–6	3
TS7	South 6–7	2
TS8	South 7–8	5
TS9	South 8–9	3
TS10	South 9–10	1
TS11	South 10–11	3

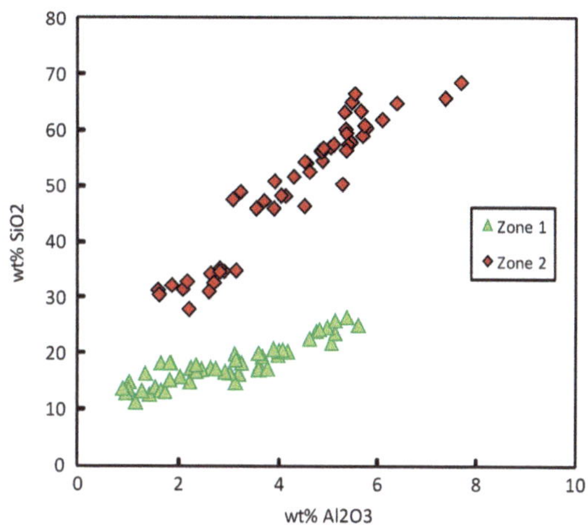

Fig. 14.5 NRC ratios for sample TN7H (zones 1 and 2 correspond to the squares in fig. 14.4)

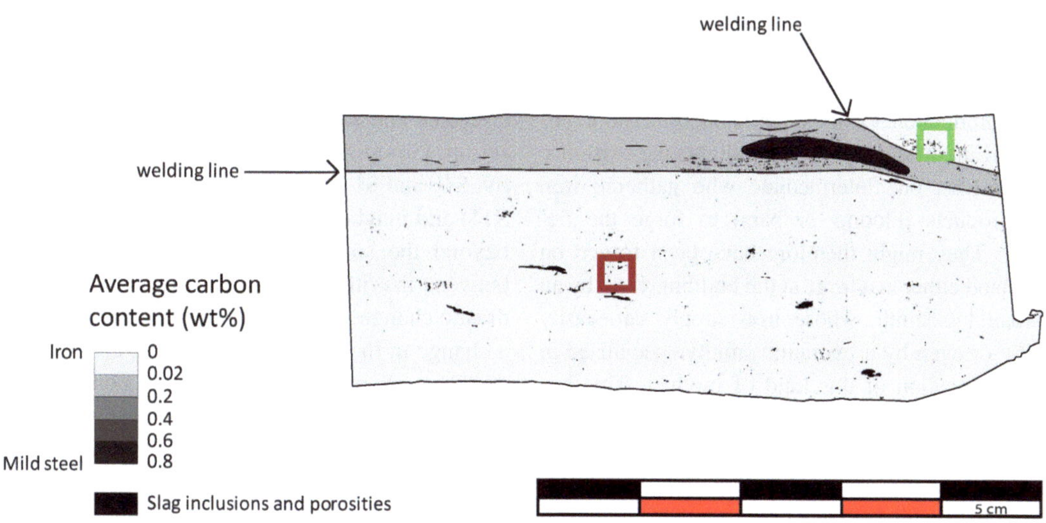

Fig. 14.4 Metallographic observations on sample TN7H with location of SI analyses by SEM-EDS (zone 1 *green*; zone 2 *red*)

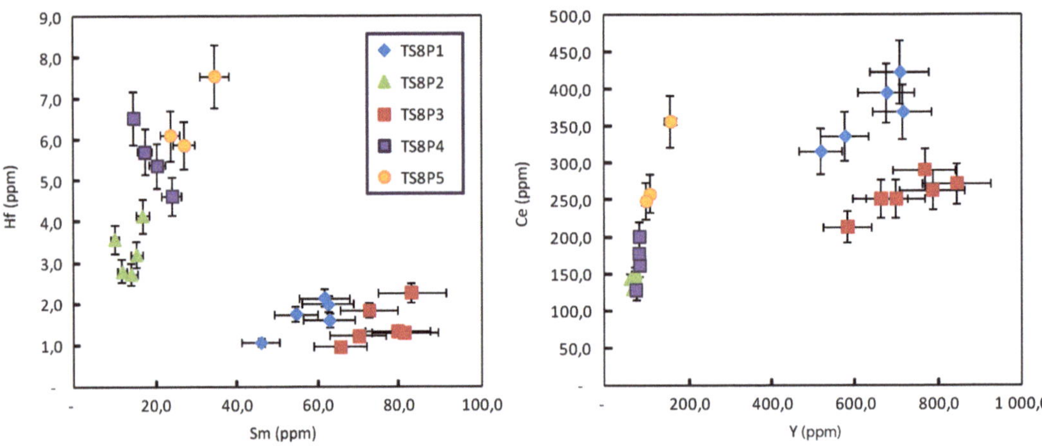

Fig. 14.6 Bivariate plots showing different ratios for samples of tie-rod TS8

correlation trends (Fig. 14.6). This is also probably true for TN5 and TS6, although chemical differences between the different samples are not as marked as for other tie-rods. The homogeneity of TS10 could not be checked as only one sample was taken from that tie-rod.

Thus, most bars were made by welding iron pieces from different origins together (Table 14.6), providing quite a different picture than was foreseen on the basis of major element analysis. Although there seems to be no more than two different origins per tie-rod, the repartition of samples of diverse provenance reveals that at least four blooms or bars were required to forge tie-rod TS8. This first result proves that these massive tie-rods were not imported as a whole from a single bloomery controlling all stages of iron production. On the other hand, total mixing of iron from different origins highlights the existence of at least one intermediate who gathered iron by-products (blooms or bars) to forge the tie-rods. They might therefore have been forged on demand either in a forge at the building yard, by an urban locksmith, whose iron supply can easily vary or even by a hydraulic smithy specialised in the fabrication of this kind of product. This last hypothesis might be most likely given the huge dimensions of these tie-rods, as neither a regular urban locksmith nor the building yard were likely to have the suitable equipment to easily forge such bars. Account books of Troyes cathedral from the early fifteenth century give an example of such orders for 150 lb iron bars "according to the dimensions given by the fabric master" to the distant "great forge" of Doulevant, 75 km away from Troyes, but also to a closer forge in the Pays d'Othe (L'Héritier et al. 2010).

Trace element analysis also allows comparisons between the tie-rods. A few bivariate plots (notably Nd/Sm, Y/La and U/Th), as well as the use of multivariate statistics in this case, Principal Component Analysis as proposed by Leroy et al (2011a) strongly discriminate a set of SI with a particular distinct composition, here labeled 'Group A' (Fig. 14.7). The specific Nd/Sm ratio of Group A is not compatible with local production areas like the Allogny forest (Bordeloup 1995) or the region around Noirlac abbey (Dunikowski 1987) in which slags were collected and analysed using the same analytical set-up. Group A includes tie-rods TN8, N9, S7, S9, S10 and S11 (although only partly for S9 and S11) and maybe also S6, which all happen to lie beyond the so-called "Branner break" located between the 5th and 6th bays (Table 14.6). This drastic change in SI composition might illustrate a change in the supplying smithy. It also reveals that these tie-rods match the known construction phases and were therefore probably installed at the beginning of the thirteenth century to help the side walls sustain the transverse thrust of the vault.

Trace element analysis on SI from tie-rods at Bourges cathedral permitted us to clarify several

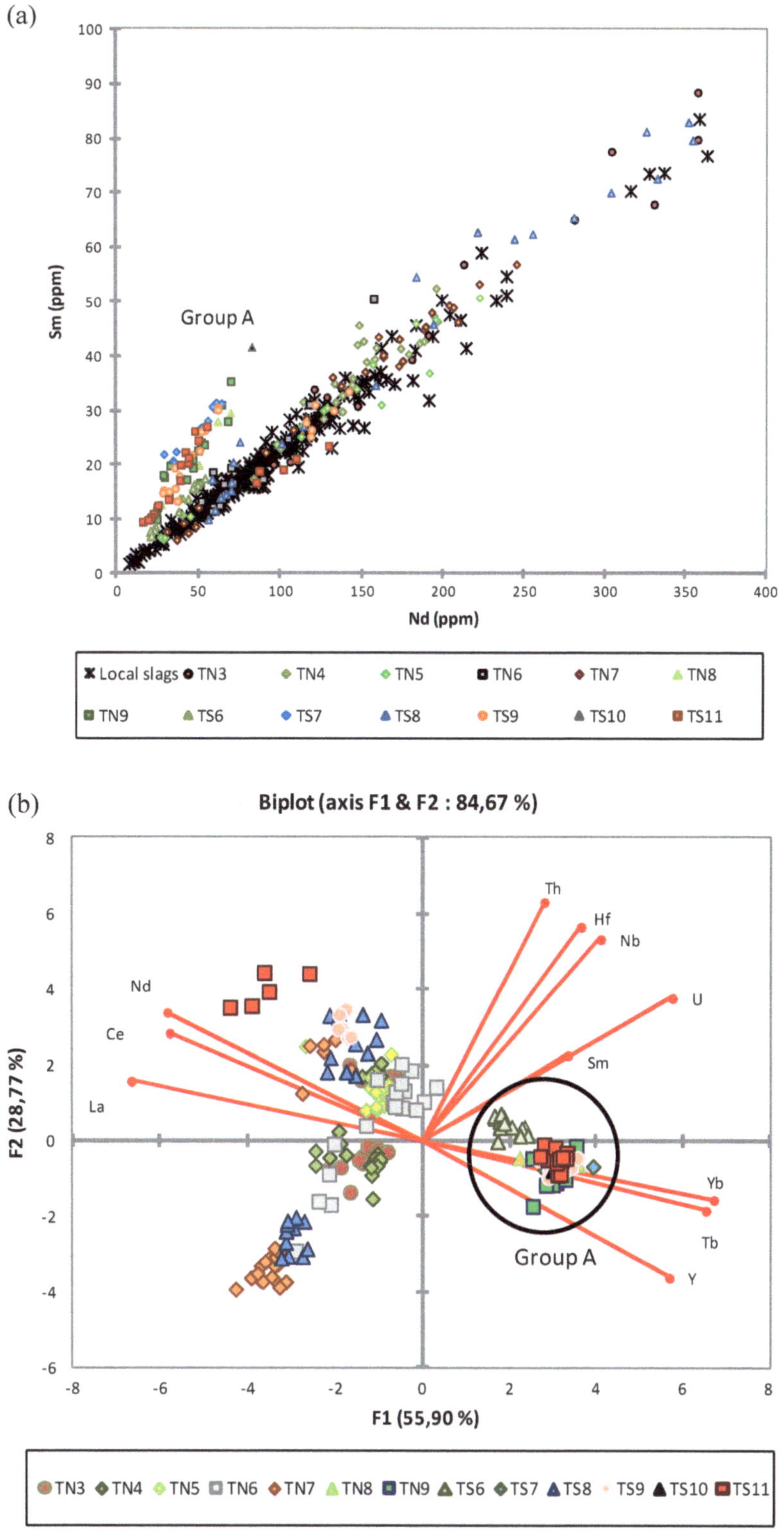

Fig. 14.7 Bivariate Sm/Nd plot (**a**) and Principal Component Analysis (**b**) on the tie-rods samples. The bivariate plot also shows a comparison with local slag coming from Allogny forest & Noirlac abbey surroundings

Table 14.6 Main results summary regarding major and trace element analyses

Tie-rod reference	Number of samples	Homogeneity of NRC ratios (MEB-EDS)	Groups of different origins (LA-ICP-MS)	Presence of "Group A"
TN3	4	Yes	2 (P1&P2/P3&P4)	
TN4	4	Yes	3 (P1&P4/P2/P3)	
TN5	3	?	2 (P1/P2&P3) ?	
TN6	4	No	3 (P1/P2/P3&P4)	
TN7	3	No	2 (F&G&Ha/Hb)	
TN8	3 (2)	Yes	1	X
TN9	4 (3)	Yes	1	X
TS6	3	Yes	2 (P1/P2&P3) ?	X (?)
TS7	2	Yes	1	X
TS8	5	No	3 (P1/P3/P2&P4&P5)	
TS9	3	Yes	2 (P1&P2/P3)	X (P1&P2)
TS10	1	–	–	X
TS11	3	Yes	2 (P1/P2&P3)	X (P2&P3)

"?" means uncertain

questions raised up in the 1990s about the structure of the cathedral and the role of metallic armatures. A further step would involve geographical provenancing of the iron, first with local production areas (except for Group A) and then with extra-regional areas. Leroy et al. (2011a) have demonstrated the possibility of determining the origin of iron artifacts by applying a methodology based, in particular, on a combination of LA-ICP-MS analysis and multivariate statistical methods. The multivariate methodology developed permits linking an elemental-based signature of a given ore extraction area with ore to artifacts of unidentified origin by comparison of chemical signatures. This approach was tested to characterize the medieval iron market of Ariège, France (Leroy et al. 2011a).

14.5 Conclusion

LA-ICP-MS is to date the best suited method to quantify trace element composition in SI entrapped in ancient iron objects as a means of determining provenance. Unlike other methods tested to date (confocal SR-μ-XRF, EPMA), LA-ICP-MS can quantify most lithophile elements below the ppm level. The only hindrance in the analytical set-up used so far is the ablation crater diameter, which limits the minimal analysis size to about 80 μm and therefore prevents the study of refined objects presenting only smaller SI, especially below 30 μm. However, decreasing the wavelength of the laser (e.g. from 266 nm to 193 nm), would improve ablation yield and increase signal strength, even for smaller SI, and thus allow analysis of smaller SI. Similarly, using an ICP-MS with better sensitivity than the one used in this study, it would also be possible to generate reliable results on a reduced quantity of ablated matter. Therefore, characterization by LA-ICP-MS of small inclusions should be possible by changing the wavelength of the laser or the increasing the sensitivity of the ICP-MS.

To obtain relevant results, a very rigorous protocol has to be followed, however, beginning with an initial complete metallographic analysis of the artifacts under study, including major element analysis by SEM-EDS to understand the genesis of SI. The application of this combined metallographic/LA-ICP-MS protocol, in conjunction with recent developments in data treatment such as the use of multivariate statistics, opens new prospects in the field of provenancing iron objects.

Acknowledgements The study of the cathedral of Bourges was funded by a Region Centre research project grant. We would like to thank Patrick Ponsot, Architecte en Chef des Monuments Historiques for Bourges cathedral, and Jean-Pierre Blin, former Curator of Historical Monuments of Région Centre for granting us the authorisation to work and sample on the cathedral.

References

Berranger M, Fluzin P (2007) Organisation de la chaîne opératoire en métallurgie du fer aux IIe-Ier siècle av. J.-C. sur l'oppidum d'Entremont (Aix-en-Provence, Bouches-du-Rhône): la circulation du métal. La Revue d'Archéométrie 31:7–22

Blakelock E, Martinon-Torres M, Veldhuijzen HA, Young T (2009) Slag inclusions in iron objects and the quest for provenance: an experiment and a case study. J Archaeol Sci 36:1745–1757

Bordeloup J-M (1995) Nouvelles observations sur la métallurgie ancienne du fer à Allogny (Cher). Bulletin du Groupe de Recherches Archéologiques et Historiques de Sologne 1:143–158

Branner R (1962) La Cathédrale de Bourges et sa place dans l'architecture gothique. Éditions Tardy, Paris

Brill RH (1999) Chemical analyses of early glasses. Corning Museum of Glass, New York

Buchwald VF (2005) Iron and steel in ancient times. Kgl Danske Videnskabernes Selskab, Copenhagen

Buchwald VF, Wivel H (1998) Slag analysis as a method for the characterization and provenancing of ancient iron objects. Mater Charact 40:73–96

Coustures M-P, Beziat D, Tollon F (2003) The use of trace element analysis of entrapped slag inclusions to establish ore-bar iron links: examples from two Galloroman iron-making sites in France (Les Martys, Montagne Noire, and Les Ferrys, Loiret). Archaeometry 45:599–613

Coustures M-P, Rico C, Beziat D, Djaoui D, Long L, Domergue C, Tollon F (2006) La provenance des barres de fer romaines des Saintes-Mariesde-la-Mer (Bouches-du-Rhône). Etude archéologique et archéométrique Gallia 63:243–261

Desaulty AM, Mariet C, Dillmann P, Joron JL, Fluzin P (2008) The study of provenance of iron objects by ICP-MS multi-elemental analysis. Spectrochim Acta B 63:1253–1262

Desaulty AM, Dillmann P, L'Héritier M, Mariet C, Gratuze B, Joron JL, Fluzin P (2009) Does it come from the Pays de Bray? Examination of an origin hypothesis for the ferrous reinforcements used in French medieval churches using major and trace element analyses. J Archaeol Sci 36:2445–2462

Dillmann P, L'Héritier M (2007) Slag inclusion analyses for studying ferrous alloys employed in French medieval buildings: supply of materials and diffusion of smelting processes. J Archaeol Sci 34:1810–1823

Disser A (2014) Production et circulation du fer en Lorraine (VIe siècle av. J.-C. –XVe siècle ap. J.-C.). Ph.D Dissertation, University Technologique de Belfort-Montbéliard

Dunikowski K (1987) Etude d'un complexe métallurgique à Farges Allichamps (Centre). Revue archéologique du centre de la France 26:85–86

Epaud F (2011) La charpente de la nef de la cathédrale de Bourges. Revue archéologique du Centre de la France 50:501–554

Férauge M, Mignerey P (1996) La cathédrale de Bourges et l'utilisation du fer dans l'architecture. B Monumental 154:129–148

Gimeno Adelantado JV, Ferrer Eres MA, Valle Algarra FM, Peris Vicente J, Bosch Reig F (2003) Analytical study by SEM/EDX and metallographic techniques of materials used in the iron production process during the Iberian period. Talanta 60:895–910

Gratuze B (1999) Obsidian characteristion by laser ablation ICP-MS and its application to prehistoric trade in the Mediterranean and the Near East: sources and distribution of obsidian within the Aegean and Anatolia. J Archaeol Sci 26:869–881

Gratuze B, Blet-Lemarquand M, Barrandon J-N (2001) Mass spectrometry with laser sampling: a new tool to characterize archaeological materials. J Radioanal Nuclear Chem 247:645–656

Hedges REM, Salter C (1979) Source determination of iron currency bars through the analysis of slag inclusions. Archaeometry 21:161–175

Hollocher K, Ruiz J (1995) Major and trace element determination on NIST glass standard reference material 611, 612, 614 and 1834 by inductively coupled plasma mass spectrometry. Geostand Newslett 19:27–34

Kronz A (2003) Ancient iron production compared to medieval techniques in Germany: Fayalitic slag and elemental mass balance. In: Proceedings of the conference Archaeometallurgy in Europe, September 24–26, 2003, Milano. Associazione Italiana di Metallurgia, Milano, pp 555–564

L'Héritier M, Dillmann P, Aumard S, Fluzin P (2013) Iron? Which iron? Methodologies for metallographic and slag inclusion studies applied to ferrous reinforcements of Auxerre cathedral. In: Humphris J, Rehren T (eds) The world of iron. Archetype Publications, London, pp 409–420

Leroy M (1997) La sidérurgie en Lorraine avant le haut fourneau: L'utilisation du minerai de fer oolithique en réduction directe. CRA-Monograph 18, Paris

Leroy S (2010) Circulation au Moyen Âge des matériaux ferreux issus des Pyrénéesariégeoises et de la Lombardie. Apport du couplage des analyses en éléments traces et multivariées. Ph.D Dissertation, University Technologique de Belfort-Montbéliard

Leroy S, Cohen SX, Verna C, Gratuze B, Téreygeol F, Fluzin P, Bertrand L, Dillmann P (2011a) The medieval iron market in Ariège (France). Multidisciplinary analytical approach and multivariate analyses. J Archaeol Sci 39:1080–1093

Leroy S, Simon R, Bertrand L, Foy E, Williams A, Dillmann P (2011b) First examination of slag inclusions in medieval armours by confocal SR–μXRF and LA-ICP-MS. J Anal Atom Spectrom 26:1078–1087

L'Héritier M, Dillmann P, Benoit P (2010) Iron in the building of gothic churches: its role, origins and production using evidence from Rouen and Troyes. Hist Metall 44:21–35

Lychatz B, Janke D (2000) Experimentelle Simulation der frühen Eisenverhüttung. Arbeits- und Forschungsberichte zur Sächsischen Bodendenkmalpflege 42:287–306

Mariet C, Belhadj O, Leroy S, Carrot F, Métrich N (2008) Relevance of NH4F in acid digestion before ICP-MS analysis. Talanta 77:445–450

Norman MD, Pearson NJ, Sharma A, Griffin WL (1996) Quantitative analysis of trace elements in geological materials by laser ablation ICPMS: instrumental operating conditions and calibration values of NIST glasses. Geostand Newslett 20:247–262

Pagès G, Dillmann P, Fluzin P, Long L (2011) A study of the Roman iron bars of Saintes-Maries-de-la-Mer (Bouches-du-Rhône, France). A proposal for a comprehensive metallographic approach. J Archaeol Sci 38:1234–1252

Paynter S (2006) Regional variations in bloomery smelting slag of the Iron Age and Romanobritish periods. Archaeometry 48:271–292

Pearce NJG, Perkins WT, Westgate JA, Gorton MP, Jackson SE, Neal CR, Chenery SP (1997) A compilation of new and published major and trace element data for NIST SRM 610 and NIST SRM 612 glass reference materials. Geostand Newslett 21:115–144

Ribault J-Y (1995) Un chef d'œuvre gothique: la cathédrale de Bourges. Anthèse éditions, Arcueil

Schwab R, Heger D, Höppner B, Pernicka E (2006) The provenance of iron artefacts from Manching: a multi-technique approach. Archaeometry 48:433–452

Tylecote RP (1970) The composition of metal artifacts: a guide to provenance? Antiquity 44:19–25

Verna C (2001) Le temps des Moulines. Fer, technique et société dans les Pyrénées centrales (XIIIe–XVIe siècles). Publications de la Sorbonne, Paris

Part IV

Expanding the Application of LA-ICP-MS in Archaeology

Expanded Applications of Laser Ablation-ICP-MS in Archaeology

15

Mark Golitko

Abstract

This chapter briefly reviews novel and expanded archaeological applications of LA-ICP-MS as presented in the chapters in Part IV of this volume.

15.1 Expanded Applications of LA-ICP-MS

The papers in this section detail new applications of LA-ICP-MS, either to materials not previously studied with this technique, or by using the particular strengths of the technique to expand study of material types previously analyzed by other means. An informal review of applications of LA-ICP-MS published in two leading archaeological sciences journals reveals an overwhelming use of the technique for analyzing ceramics, glass (both natural and synthetic), and metals and alloys. These materials of course each have a long history of study by other means of chemical characterization, and LA-ICP-MS has been utilized in some cases as an alternative means of generating essentially the same kinds of information, either because of small sample size, the need for low detection limits, or minimal destructiveness when compared to other available options, as detailed in prior chapters in this volume. For instance, applications of LA-ICP-MS for obsidian provenience studies (see Chap. 10) have largely sought to mimic bulk characterizations by INAA and other techniques, albeit with less damage to artifacts and the ability to avoid phenocrysts and other inclusions that might impact bulk composition. Analysis of ceramics, glass and metals have utilized unique properties of LA to analyze only particular phases or components of ceramic vessels (e.g., slip/paint, tempers, or paste), while analyses of glass and metals have used the time- and depth-resolved capabilities of LA to avoid corrosion layers and analyze multi-component objects.

The first two papers in this section deal with expanding the use of LA-ICP-MS to study rock types that are less frequently subject to archaeometrical study. Carter (Chap. 16) examines the provenance of garnet beads found throughout Southeast Asia, exploring the potential of trace elements to sort out provenance issues in a geological material that is characterized by end-members with highly

M. Golitko (✉)
Department of Anthropology, University of Notre Dame, 621 Flanner Hall, Notre Dame, IN 46556, USA
e-mail: mgolitko@nd.edu

L. Dussubieux et al. (eds.), *Recent Advances in Laser Ablation ICP-MS for Archaeology*, Natural Science in Archaeology, DOI 10.1007/978-3-662-49894-1_15

variable major element concentrations, moving beyond studies that have relied on SEM-EDS and other such techniques. Baron and Gratuze (Chap. 17) utilize LA-ICP-MS to measure trace element content in what they term "black rocks," a diverse group of rocks unified by high carbon content (jet, lignite, oil shale). While carbon cannot be effectively measured by ICP-MS directly, they produce concentrations of trace elements normalized to an assumed carbon content, and distinguish major sources of carbonaceous rocks in Europe utilized during the Iron Age.

These studies, as well as other published LA-ICP-MS studies on materials such as ochre (Scadding et al. 2015; Zipkin et al. 2015) and carnelian (Gliozzo et al. 2014; Insoll et al. 2004) highlight the difficulties inherent in analysis of highly heterogeneous rock categories, including the selection of appropriate means of quantification. The wider range of standards being produced specifically for LA-ICP-MS analysis currently, for instance USGS powder rock standards fused as glass discs, opens up new potential for quantifying concentrations in some of these types of materials, although some elements like carbon remain beyond the ability of LA-ICP-MS to quantify and studies using LA-ICP-MS would in many cases still benefit from complimentary measurement by other techniques such as SEM-EDS.

Dudgeon and colleagues (Chap. 18) follow such a joint approach in their study of human teeth from Bronze Age Armenian burials. After laser ablation, they use subsequent SEM-EDS quantification of major element content at the same locations, using a series of in-house apatite standards and USGS powder rock standards to quantify trace elements. Their study utilizes LA-ICP-MS raster mapping of tooth surfaces to explore impacts of elemental uptake from the burial environment in relation to elements that may have entered teeth and bones through life-activities (Arsenic-Bronze production in this case). While LA has to date been employed only sporadically to analyze biological materials in the field of archaeology, there is potential in the application of multi-collector ICP-MS with laser sampling to generate isotopic ratios at fine spatial scale across teeth and bones in order to explore subtle or short-term changes in residence, diet, and activity in prehistoric people, although interferences have been noted that impact accurate measurement of Sr isotopes (Copeland et al. 2010; Simonetti et al. 2008).

LA-MC-ICP-MS represents a promising approach for a number of materials, but has been only occasionally used in the past to study materials such as lead glazes on ceramics (e.g., Habicht-Mauche et al. 2002). Iñañez and his coauthors (Chap. 19) use LA-MC-ICP-MS to analyze such glazes on colonial ceramics to determine whether these were produced from European or Central/South American ore sources. While employed primarily for analysis of lead glazes, LA-MC-ICP-MS could hypothetically be used to examine isotope ratios in other components of ceramics such as temper grains as a means of estimating geological age of the rock deposits from which these materials came, or even to conduct isotopic studies on ceramic paste itself to provide complementary provenance analysis information with trace element chemistry. For instance, differing isotopic ratios between temper grains and paste might conclusively indicate different basic geological and geographical origins for the components of that ceramic matrix.

References

Copeland SR, Sponheimer M, Lee-Thorp JA, le Roux PJ, de Ruiter DJ, Richards MP (2010) Strontium isotope ratios in fossil teeth from South Africa: assessing laser ablation MC-ICP-MS analysis and the extent of diagenesis. J Archaeol Sci 37:1437–1446

Gliozzo E, Mattingly DJ, Cole F, Artioli G (2014) In the footsteps of Pliny: tracing the sources of Garamantian carnelian from Fazzan, sout-west Libya. J Archaeol Sci 52:218–241

Habicht-Mauche JA, Glenn ST, Schmidt MP, Franks R, Milford H, Flegal AR (2002) Stable lead isotope analysis of Rio Grande glaze paints and ores using ICP-MS: a comparison of acid dissolution and laser ablation techniques. J Archaeol Sci 29:1043–1053

Insoll T, Polya DA, Bhan K, Irving D, Jarvis K (2004) Towards an understanding of the carnelian bead trade from Western India to sub-Saharan Africa: the application of UV-LA-ICP-MS to carnelian from Gujarat, India, and West Africa. J Archaeol Sci 31:1161–1173

Scadding R, Winton V, Brown V (2015) An LA-ICP-MS trace element classification of ochres in the Weld Range environ, Mid West region, Western Australia. J Archaeol Sci 54:300–312

Simonetti A, Buzon MR, Creaser RA (2008) In-situ elemental and Sr isotope investigation of human tooth enamel by Laser Ablation-(MC)-ICP-MS: successes and pitfalls. Archaeometry 50(2):371–385

Zipkin AM, Hanchar JM, Brooks AS, Grabowski MW, Thompson JC, Gomani-Chindebvu E (2015) Ochre fingerprints: distinguishing among Malawian mineral pigment sources with Homogenized Ochre Chip LA-ICPMS. Archaeometry 57(2):297–317

Determining the Provenience of Garnet Beads Using LA-ICP-MS

16

Alison K. Carter

Abstract

Stone and glass beads have been found at Iron Age (500 BC–AD 500) sites across Southeast Asia and are often assumed to be indicators of contact with South Asia. However, recent research on glass, agate, and carnelian beads indicates there may have also been local production of these materials. In Cambodia, two different types of garnet beads have been identified at several Iron Age sites. The first type is spherical, well-polished, and appears to have been drilled with a diamond drill, a drilling technique that is strongly associated with Indian bead production. The second type of garnet bead is unpolished, unshaped, and drilled using a variety of different drilling methods. Based on these initial differences, it was hypothesized that there were two different bead-making traditions represented amongst the garnet beads, and that the second type of garnet bead may have been locally produced. To investigate this question more thoroughly the garnet beads were analyzed using laser ablation-inductively coupled plasma-mass spectrometry (LA-ICP-MS) in order to determine their chemical composition. Geological source samples from a variety of garnet sources across South and Southeast Asia were also analyzed using LA-ICP-MS. Results indicate that LA-ICP-MS is an excellent tool for differentiating between garnet sources and analyzing archaeological garnet artifacts with minimal damage. Furthermore, the results of the study confirm that the two types of beads were made from distinctly different garnet sources, although the locations of these sources remain unknown.

16.1 Introduction

The Iron Age period of mainland Southeast Asia dates from approximately 500 BC–AD 500 and is marked by the first appearance and use of iron,

A.K. Carter (✉)
Department of Anthropology, University of Illinois at Champaign-Urbana, 109 Davenport Hall, Urbana, IL 61801, USA
e-mail: alisonkyra@gmail.com

L. Dussubieux et al. (eds.), *Recent Advances in Laser Ablation ICP-MS for Archaeology*, Natural Science in Archaeology, DOI 10.1007/978-3-662-49894-1_16

increasing social stratification, and the first signs of contact with South Asia (Higham 2014). Long-distance trade and interaction with South Asia, primarily the modern countries of Bangladesh, India, Pakistan, and Sri Lanka, is assumed to have been an important factor in the processes leading to the development of the first complex societies in Southeast Asia (e.g. Glover 1996). Therefore, the study of trade and interaction between these two regions has important implications for understanding socio-political development in Southeast Asia during the Iron Age period. Stone and glass beads are often the first archaeological indicators of contact with South Asia and are ubiquitous at many Southeast Asian sites. Consequently, beads have been at the center of several studies examining their role in trade and socio-political development (e.g. Bellina 2003, 2007, 2014; Carter 2013, 2015; Francis 2002; Theunissen et al. 2000).

More recent research has focused specifically on evidence for possible stone and glass bead manufacture in Southeast Asia, in addition to the importation of these materials from South Asia (Bellina 2007, 2014; Lankton and Dussubieux 2006, 2013; Lankton et al. 2008; Theunissen et al. 2000). The possibility of local production of beads in Southeast Asia has several implications for our understanding of sociopolitical development in Southeast Asia. Vidale and Miller (2000) have argued that crafts that employ complex technologies, such as stone bead production, but made with local materials are a hallmark of increasing social stratification and hierarchy, most commonly associated with complex chiefdoms or state level societies. This allows elites to control the production of highly valuable prestige objects, whilst also having easy access and control of these raw materials. Theunissen et al. (2000) have argued that there may have been local production of agate beads in Thailand using a local central Thai agate source. At the peninsular Thai site of Khao Sam Kaeo there is evidence for stone and glass bead and ornament manufacture (Bellina 2007, 2014; Lankton et al. 2008). Bellina (2003, 2007, 2014) has argued that evidence from Khao Sam Kaeo also indicates that South Asian craftsmen may have been living and making stone beads under the patronage of Southeast Asian elites. Rather than controlling all aspects of production of beads, Bellina argues that, "patronizing foreign craftsmen was, in itself, a symbol of power and authority" (2003: 294). In order to further investigate organization of bead production in Southeast Asia, it is crucial to understand where and how beads were being made. This can be achieved in part through the use of compositional analysis and sourcing studies of archaeological beads.

The studies discussed above have focused solely on glass, agate, and carnelian beads. However, beads made of other materials have also been found, although in lesser quantities. This study focuses on two sets of garnet beads found at Iron Age sites in Cambodia. Previous examination of their morphology as well as their drill holes indicates that these two sets of beads are distinct from one another, and that beads found at sites in southeast Cambodia may represent a local bead-making tradition (Carter 2012). The current study investigates this question further by using laser ablation-inductively coupled plasma-mass spectrometry (LA-ICP-MS) analysis to determine the composition of these garnet bead artifacts and compare them with a variety of garnet geologic sources from South and Southeast Asia. This reflects a methodological advancement in that the measurement of major, minor, and trace element concentrations allows for both the identification of garnet types, and when compared to geologic sources samples, can be used to determine where the raw material was obtained. The results show that the two groups of garnet bead artifacts are compositionally distinct from one another, although the sources for both sets of artifacts remain unknown.

16.2 Garnet Samples

Garnet is a type of gemstone commonly found in metamorphic as well as igneous rocks and detrital sediments (Deer et al. 1982). Garnets can be found in a variety of colors, from the more common deep red and purple, to orange, yellow, and

Table 16.1 Chemical formulas for the different varieties of major garnet species (from Deer et al. 1982)

Garnet species	Chemical formula
Pyrope	$Mg_3Al_2Si_3O_{12}$
Almandine	$Fe_3Al_2Si_3O_{12}$
Spessartine	$Mn_3Al_2Si_3O_{12}$
Grossular	$Ca_3Al_2Si_3O_{12}$
Andradite	$Ca_3(Fe^{3+}, Ti)_2Si_3O_{12}$
Uvarovite	$Ca_3Cr_2Si_3O_{12}$
Hydrogrossular	$Ca_3Al_2Si_2O_8(SiO_4)_{1-m}\ (OH)_{4m}$

green. The garnet group contains several species, all of which share the same chemical formula: $X_3Y_2Z_3O_{12}$ (Table 16.1). These different species of garnet have been sub-divided into two series in which there is fairly continuous variation between the species within the series. The pyralspite series contains pyrope, almandine, and spessartine garnets and is the primary focus of the current study. The ugrandite series contains uvarovite, grossular, and andradite garnets. However, garnets rarely occur in their pure end-member forms and a single sample can often contain a mix of these species. Geochemical analysis of garnet is an ideal way to identify the different elements within a sample and recent research can assist with the conversion of this data into molar proportions of various end-member components (Locock 2008). It should also be noted that previous geochemical studies of garnets have identified instances of chemical zoning, in which certain elements are not evenly distributed throughout the sample (Deer et al. 1982: 546, discussed further below).

Due to the varying proportions of different garnet species in a single sample, geochemical analyses of garnets have been employed to assist in the identification of specific optical and physical characteristics that can be used to classify species of garnet using standard gem-testing equipment (Stockton and Manson 1985). Geologically, garnets have also been used in sediment provenance studies (Morton 1985). Different species of garnet are also known to be strongly associated with specific rock types (Wright 1938) and therefore have been used as indicator minerals for more valuable stones, such as diamonds. For the purposes of this study, it is significant that the composition of a garnet is associated with that of its host rock (Bimson et al. 1982: 51). While major and minor elements can assist with determining the type of garnet and perhaps identifying the type of host rock in which garnets were formed, it was expected that the additional minor and trace elements associated with the host rock should assist in providing a unique fingerprint for garnets from specific sources. LA-ICP-MS is an ideal technique for measuring the concentrations of these minor and trace elements.

16.2.1 Previous Geochemical Studies of Garnet Artifacts

There have been several recent archaeological studies attempting to provenience garnets using a variety of methods including PIXE (Calligaro et al. 2002; Farges 1998; Mathis et al. 2008; Perin et al. 2007) electron microprobe (Rösch et al. 1997; Quast and Schüssler 2000; Velde and Courtois 1983), X-ray fluorescence analysis (Bimson et al. 1982) and X-ray diffraction analysis (Schüssler et al. 2001). While these studies were able to successfully analyze garnets using non-destructive methods, they were only able to provide preliminary conclusions about the source of garnet artifacts because no geological source samples were analyzed. Several prior studies (e.g. Calligaro et al. 2002; Perin et al. 2007) relied on previously published data of non-specific sources from Europe and South Asia from Quast and Schüssler (2000). Other studies focused more specifically on classifying garnet end-member components and determining different groups, but not identifying their source, or else only compared analyzed garnets to other archaeological garnets (Bimson et al. 1982; Farges 1998; Mathis et al. 2008; Rösch et al. 1997; Schüssler et al. 2001). This study aims to rectify these problems and build on this previous research by examining both geologic sources and archaeological artifacts using a more robust method of chemical characterization, LA-ICP-MS.

LA-ICP-MS was chosen over other methods for several reasons. Firstly, LA-ICP-MS is virtually non-destructive and requires no sample preparation, making it ideal to use on archaeological artifacts. Secondly, LA-ICP-MS has low limits of detection and can determine concentrations of major, minor, and trace elements. Previous studies have focused solely on the major and minor elements in garnet samples, and one of the primary goals of the present study was to evaluate the possibility that trace element concentrations would vary systematically enough to distinguish between sources and allow for accurate assignment of archaeological artifacts to sources.

16.2.2 Garnet Artifacts Analyzed in This Study

A total of 15 garnet artifacts were analyzed using LA-ICP-MS (Table 16.2). Five garnet beads, which I call AB (Angkor Borei) Garnet Drill Type DD 1, were recovered from the site of Angkor Borei (Fig. 16.1), and all were compositionally characterized. Angkor Borei was an important center during the Iron Age and Early Historic period, and may have been the capital of the ancient polity known as Funan (Stark 2004). The beads were excavated from a densely packed cemetery that has been dated from 200 BC to AD 200 (Stark 2001). These beads are purple, round or spherical in shape, and range from approximately 3–4 mm in diameter, with one bead being slightly larger at about 5–6 mm in diameter (Fig. 16.2). In the context of this study, the classification of these beads as AB Garnet Drill Type DD 1 is not related to their specific shape or morphology, but is instead determined by their manufacturing method. Type DD 1 beads have been shaped and polished, and SEM (Scanning Electron Microscope) analysis of drill-hole impressions indicates they were drilled using a double-diamond drill (Fig. 16.3; Carter 2012). This drill uses two small diamond chips crimped at the tip of a metal drill. As a bead is drilled, the two chips create unique and identifiable concentric grooves in the bead hole (Gwinnett and Gorelick 1986). Diamond drilling techniques, especially double-diamond drills, are strongly associated with West Indian bead-making technology (Kenoyer and Vidale 1992).

A second set of garnet beads classified as SEA (Southeast Asia) Garnet Beads were excavated or collected at the Iron Age sites of Village 10.8, Prohear, and Bit Meas in southeast Cambodia (Fig. 16.1). A total of 18 beads were recorded from these sites, primarily from burial or possible burial contexts, and 8 were chosen for LA-ICP-MS analysis. These sites are considered to be roughly contemporaneous with one another and with Angkor Borei (Table 16.2). The SEA Garnet Beads are unusual in that they appear to be natural stones or pebbles, deep purple in color, and approximately 5–10 mm in length (Fig. 16.4). Unlike the AB Garnet Drill Type DD 1 beads, they show no evidence of shaping or polishing. SEM analysis of drill hole impressions confirms that their drilling method is different from those beads. The SEA Garnet Bead holes are conical, as opposed to the cylindrical drill holes found in diamond-drilled beads. Furthermore, the patterns on the sidewall of the bead are distinct from the regular concentric grooves found in diamond-drilled beads (Fig. 16.3; Carter 2012). Experimental studies are ongoing, however it appears the SEA Garnet Beads may have been drilled with a copper drill, or perhaps a bamboo or wood drill with abrasive.

In addition to the beads discussed above, two additional garnet artifacts were opportunistically included in the LA-ICP-MS analysis sample (Fig. 16.5). Both were part of a larger selection of glass artifacts analyzed by Laure Dussubieux at the Field Museum Elemental Analysis Facility (EAF). The first is a garnet bead from the site of Porunthal in south India (Fig. 16.6). This bead appears to have been a natural pebble that was polished and drilled. The exact archaeological context of this garnet bead is unknown, however a cultural deposit at Porunthal containing thousands of glass beads and other materials has been tentatively dated from the first century BC to third century AD (Rajan 2009). The second is a piece of raw undrilled garnet found in excavations at the site of Phu Khao Thong, henceforth PKT (Fig. 16.7). PKT is located on

Table 16.2 Garnet beads and artifacts analyzed using LA-ICP-MS

Database ID	Site	Context	Approximate dates	Sum-normalized Ca	Sum-normalized Fe	Sum-normalized Mg	Sum-normalized Mn
AB1767	Angkor Borei	Burial F16	200 BC–AD 200	1.20 %	87.40 %	6.70 %	4.70 %
AB1835	Angkor Borei	Burial F21	200 BC–AD 200	1.24 %	89.85 %	6.70 %	2.21 %
AB2801	Angkor Borei	Block 42, Layer 5	200 BC–AD 200	1.22 %	88.19 %	7.59 %	3.00 %
AB3284	Angkor Borei	Burial 42	200 BC–AD 200	1.36 %	88.87 %	7.60 %	2.17 %
AB3333	Angkor Borei	Block 4 N, Layer 5	200 BC–AD 200	1.42 %	89.74 %	6.78 %	2.06 %
AKC00310	Village 10.8	Burial 28	400 BC–AD 50	14.00 %	36.79 %	47.98 %	1.24 %
AKC00311	Village 10.8	Burial 34	400 BC–AD 50	14.05 %	39.71 %	44.95 %	1.28 %
AKC00338	Village 10.8	Burial 1	400 BC–AD 50	13.31 %	36.37 %	49.11 %	1.21 %
AKC00339	Village 10.8	Burial 1	400 BC–AD 50	13.08 %	34.74 %	51.10 %	1.08 %
AKC00590	Prohear	Burial 20	200 BC–AD 200	13.25 %	43.52 %	41.82 %	1.40 %
AKC00606	Prohear	Burial 11	200 BC–AD 200	13.28 %	43.47 %	41.88 %	1.38 %
AKC00651	Bit Meas	N/A	Contemporaneous with Prohear?	12.65 %	37.32 %	48.77 %	1.25 %
AKC00731	Bit Meas	N/A	Contemporaneous with Prohear?	13.41 %	39.36 %	45.94 %	1.29 %
Porunthal garnet	Porunthal, India	N/A	First century BC to third century AD	1.10 %	90.13 %	6.09 %	2.68 %
PKT garnet	Phu Khao Thong, Thailand	N/A	Iron Age	10.01 %	54.19 %	33.29 %	2.50 %

Fig. 16.1 Archaeological sites in Cambodia and southern Vietnam mentioned in the text

the Thai peninsula, close to the important trading and craft production site of Khao Sam Kaeo (Chaisuwan 2011).

It should be noted that no diamond-drilled beads have been identified at sites in southeast Cambodia and no SEA Garnet Beads have been found at Angkor Borei. However, similar SEA Garnet Beads have also been reported at other sites in the region, including the site of Krek 52/62, Cambodia (Albrecht et al. 2001) and Go O Chua, Vietnam (Reinecke personal communication) (Fig. 16.1). Garnet beads made from natural stones are not unique to Southeast Asia. In addition to the Porunthal sample discussed above, garnet beads made from natural stones similar to the SEA Garnet Beads have also been reported at other sites in India, including Arikamedu (Francis 2004: 493) and Kodumanal (Gwen Kelly personal communication) (Fig. 16.6). While it appears that many of these beads were not shaped, they may have been polished. As the drill holes on these beads have not been investigated, it

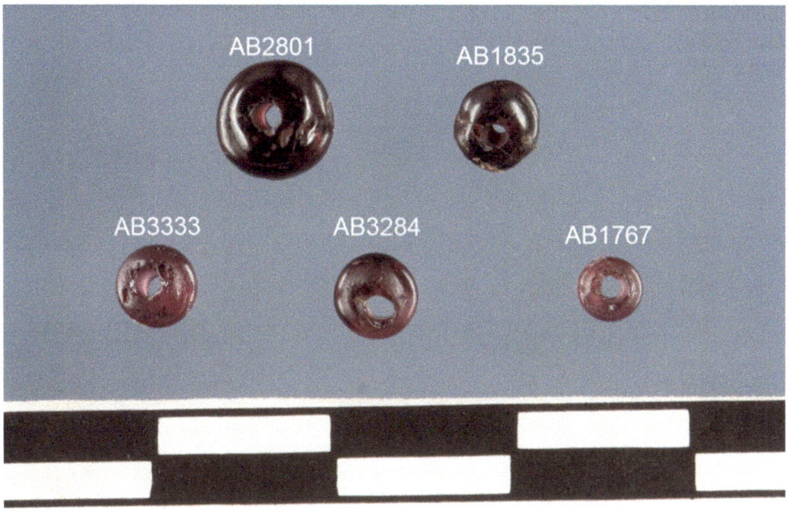

Fig. 16.2 AB (Angkor Borei) Garnet Drill Type DD1 beads from Angkor Borei. Scale: One bar is equal to 1 cm

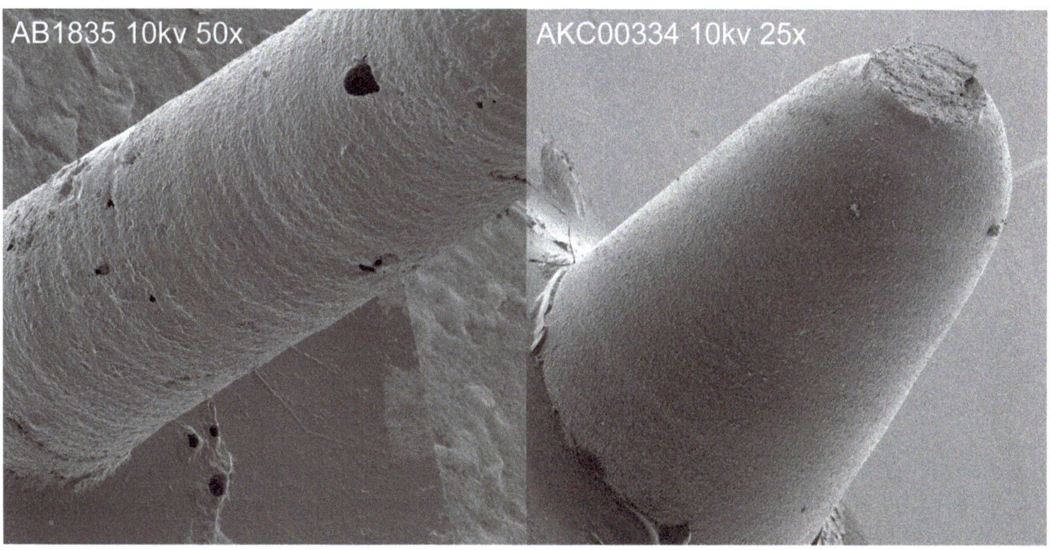

Fig. 16.3 Scanning Electron Microscope (SEM) images of drill-hole impressions taken from an AB Garnet Drill Type DD1 garnet bead (*left*) and a SEA garnet bead (*right*)

is however unclear whether or not they were drilled using diamond drills. However, visual inspection of the Porunthal sample shows the drill hole to be more similar to the AB Garnet Drill Type DD 1 than the SEA Garnet beads. Polished, spherical garnet beads similar to those found at Angkor Borei, as well as shaped and polished beads in a variety of other forms have been reported at the coastal site of Giong Ca Vo in southern Vietnam (Nguyen 2001) and at the nearby Mekong Delta site of Oc Eo, Vietnam (Malleret 1962: 221–222) (Fig. 16.1).

16.2.3 Garnet Source Samples Included in This Study

A total of 69 samples from nine geologic sources were analyzed as a part of this study (Table 16.3). Three geologic samples from India were acquired

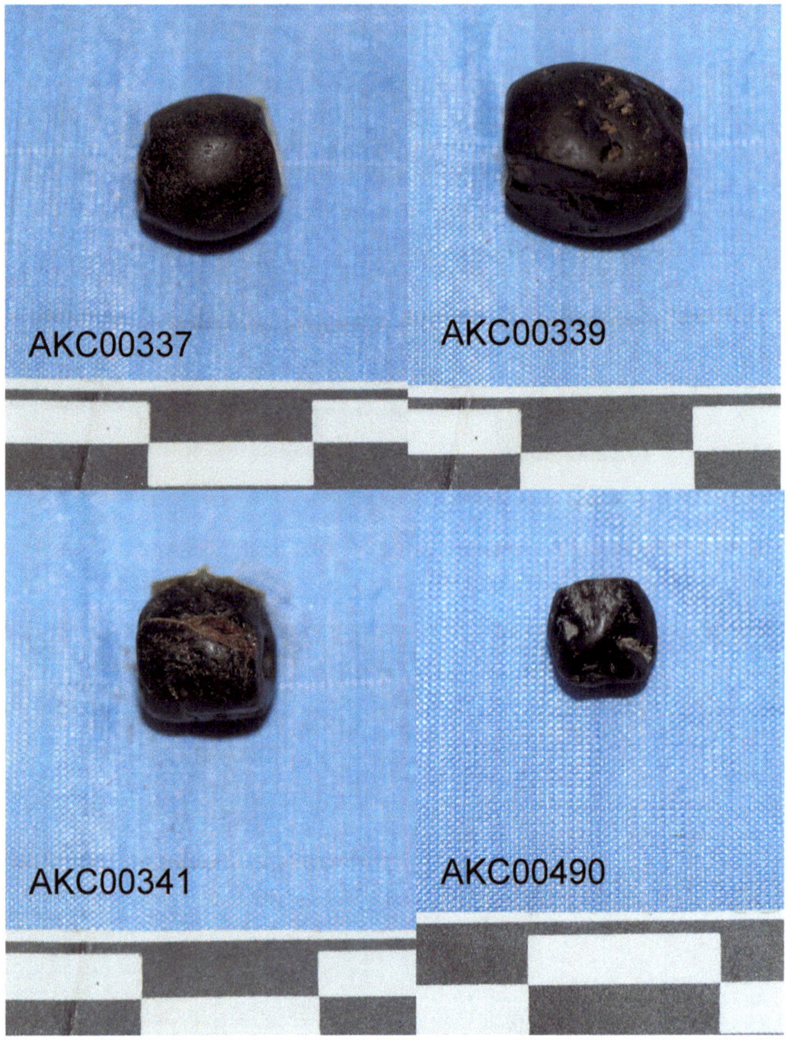

Fig. 16.4 Examples of "SEA garnet beads" from Southeast Cambodia. Scale: One bar is equal to 1 cm

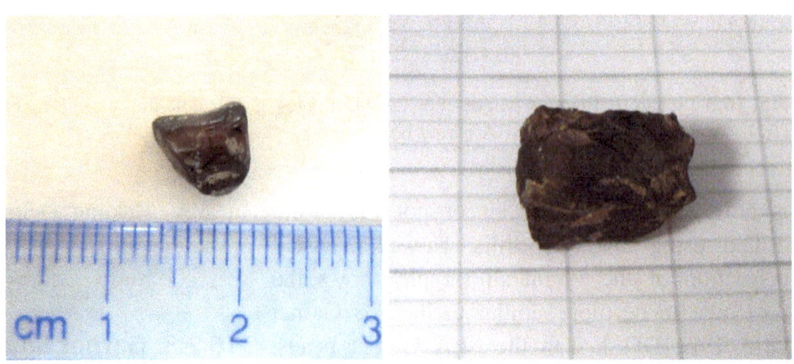

Fig. 16.5 Garnet artifacts from Porunthal, India (*left*) and Phu Khao Thong, Thailand (*right*)

from gem dealers, including samples from south India, Orissa, and an additional unknown source location in India (Fig. 16.6). The exact location of the south Indian source is unknown, however garnet is currently mined in Tamil Nadu and Andhra Pradesh (Shigley et al. 2010). Additionally, the

Fig. 16.6 Map of South Asian garnet sources and sites discussed in the text

south Indian village of Kondapalli, Andhra Pradesh is well known for its garnets and garnet gemstones (Bauer 1968: 354), as are Mahurjhari, Arikamedu, and Kodumanal (Francis 2002).

Sri Lanka is another important center for gemstones (Fig. 16.6), and may have been discussed by ancient travelers as early as the fourth century BC (Zwaan 1982). Garnet beads may have been produced at the two Early Historic period sites of Anuradhapura and Tissamahara (Coningham et al. 2006; Hannibal-Deraniyagala 2001, 2005) (Fig. 16.6). Currently, almandine and pyrope-almandine garnets are mined in central Sri Lanka, especially at the mines of Elahera (Gunawardene and Rupasinghe 1986) and in the southern part of Sri Lanka, including the famous mine of Ratnapura (Dissanayake et al. 2000: 14; Shigley

Fig. 16.7 Map of Vietnamese and Thai garnet sources and sites discussed in the text

et al. 2010) (Fig. 16.6). Two groups of samples were acquired from Sri Lanka, including one group from an unknown source and a second that may be from the mines at Ratnapura (Katie Lindstrom, personal communication).

Garnet is also quite widespread across mainland Southeast Asia, however it is not mined in the quantities seen in India and Sri Lanka. Gem quality garnets have been found in Myanmar (Shigley et al. 2010), Thailand (Aranyakanon Payome 1983), Cambodia (Chrea Vichett et al. 1999), and Vietnam (Van Long et al. 2004). Two groups of garnet geologic samples were obtained from sources in Vietnam (Fig. 16.7). The first are from a garnet mine in Phu Tho province in northern Vietnam. The second set of samples consists of two garnet inclusions in a piece of quartz from Lam Dong

Table 16.3 Geologic sources analyzed in the current study

Source name	Number of samples analyzed	Range of sum-normalized Ca	Range of sum-normalized Fe	Range of sum-normalized Mg	Range of sum-normalized Mn
Sri Lanka Group 1	10	2–4 %	73–80 %	10–12 %	4–15 %
Sri Lanka Group 2 (Ratnapura?)	6	2–4 %	80–85 %	9–17 %	<1–2 %
South India	10	3–4 %	80–84 %	10–14 %	1–6 %
Phu Tho province, Vietnam	10	4–8 %	52–67 %	22–41 %	1–7 %
"India"	9	1–4 %	67–75 %	22–29 %	1–2 %
Orissa, India	10	2–7 %	57–71 %	25–40 %	1–3 %
Anthill, Arizona	6	10–27 %	31–56 %	27–58 %	1 %
Mozambique, Africa	6	7–9 %	52–57 %	33–38 %	1–2 %
Lam Dong province, Vietnam	2	1 %	12–16 %	<1 %	83–87 %
Total	69				

province in southern Vietnam. This source was selected for analysis based on a suggestion that garnets from Lam Dong province may have been the source of the garnet beads found at the Early Historic site of Oc Eo, Vietnam (Francis 2002: 142).

Garnets from the Anthill garnet source in Arizona, and garnets from a garnet mine in Mozambique, Africa, well-known sources outside of Asia, were also sampled to determine if LA-ICP-MS could accurately distinguish between these samples and those collected from Asian sources. Although it is ideal to have all possible sources included in any sourcing study, the current analysis relies primarily on materials obtained through retailers and colleagues, and in several cases the exact location from which source materials were collected is unknown. It is anticipated that this ambiguity will be corrected with future studies that analyze a broader selection of accurately sourced geologic samples.

16.3 Method

The 15 garnet artifacts and 68 geological samples were analyzed by the author in the LA-ICP-MS laboratory at the Field Museum's Elemental Analysis Facility (EAF) in Chicago, Illinois. The equipment used includes an Analytik Jena (formerly Varian) quadrupole ICP-MS connected to a New Wave UP213 laser unit (for more details on the technique and its performance, see Dussubieux et al. 2009). To be able to determine elements with concentrations in the range of ppm and below without leaving a visible trace on the surface, we use the single point analysis mode with a laser beam diameter of 100 μm, operating at 70 % of the laser energy (0.2 mJ) and at a pulse frequency of 15 Hz. Signal acquisition begins after a 20 s pre-ablation period to avoid surface contamination as much as possible. A portion of the minute stone particles released is transported to the argon plasma torch using helium gas. The 8000 °C plasma rapidly dissociates the stone particles into constituent ions, and these ions pass through the quadrupole mass spectrometer. The quadrupole flight path through the spectrometer separates the ions by mass and charge before they are measured in the ion detector. Detection limits for most elements range from 10 ppb to 1 ppm, with an accuracy of 5–10 % depending on the elements and their concentrations.

The ICP-MS was set to scan the mass range three times and average the resulting signal. This process was repeated eight times, for a total of nine replicates. The nine replicates were then combined into a single average, and this process was repeated at five locations per garnet sample. If the signal for a particular element was less than three times the blank, it was noted as being below the limits of detection (<LOD) in our spreadsheet (Appendix). As many of the garnet pieces analyzed were quite small, the locations for sampling were randomly selected from the portion of the sample that was clearly visible within the chamber. The data from each of the five point ablations were then averaged together to calculate elemental concentrations in each sample.

In order to obtain quantitative data, four glass standards with known compositions were also analyzed. National Institute of Standards and Technology Standard Reference Materials (NIST SRM) 610 and 612 are silica glasses made with sodium and calcium and doped with varying levels of trace elements, approximately 500 ppm for NIST SRM 610 and 50 ppm for NIST SRM 612. Corning glass standards B and D were also used for the analysis of major elements. NIST SRM 610 and 612 were run before beginning analysis of the garnet samples. During analysis, NIST SRM 610 and another glass standard were run every 3–4 samples in order to correct for possible instrumental drift. Silica (^{29}Si) was used as an internal standard. As part of the calibration procedure, the Si concentration was measured for each analysis location along with all other elements. These were averaged to produce a value for the sample as a whole. The analytical protocol and calculation methods used were adapted from Gratuze (1999). Final measurements for all archaeological and geological samples are listed in Appendix; 54 elements are included with major and minor elements reported as percent oxide and trace elements as parts per million.

16.4 Identifying Compositional Differences Between Garnet Groups

As a garnet crystal grows, the distribution of elements can vary between the core and outer layers in a process called zoning. Identification of the zoning patterns of major elements (Fe, Ca, Mn, Mg) has assisted with understanding the temperature and pressure under which garnets were formed (Tracy et al. 1976). Trace element zoning of garnets has also been identified (see Hickmott et al. 1987; Hickmott 1988; Spear and Kohn 1996). In previous archaeological studies involving garnets, zoning has produced samples with a wide spread of values in certain elements, for example, yttrium values in an earlier study ranged from 100 to 1000 ppm (Calligaro et al. 2002). Heterogeneity in garnet compositions was common for most of the elements in the current study. LA-ICP-MS has successfully measured heterogeneous materials such as ceramics (Dussubieux et al. 2007) and agate (Carter 2013; Carter and Dussubieux 2016). Therefore, the inclusion of heterogeneously distributed elements was not expected to adversely affect results. However, it is important that multiple samples from a single geologic source are examined in order to measure the range of elemental variation within a source and effectively record this heterogeneity.

In the course of calculating concentrations, it was also discovered that several elements were not present in high enough quantities so as to be easily measured by the LA-ICP-MS. As noted above these elements were recorded as being below the limits of detection (LOD). The use of elements that were measured in all or nearly all of the samples assists in more accurately characterizing and differentiating the sources. Table 16.4 lists the elements and number of samples from each geologic source in which the sample was at or below the LOD for that element. Elements such as Ag, Au, Be, Cs, K, and Na, which were at the LOD in 85 % or more of the samples were not used in statistical analyses. However, the presence and absence of certain elements was notable and in some cases helpful in distinguishing between sources. For example, copper (Cu) was at the LOD for all of the samples from the India source, whereas the presence and varying concentrations of phosphorus (P) helped distinguish the Mozambique, India, and Orissa sources from others in the Almandine-Pyrope group.

Table 16.4 This table indicates for a given element and for each category the number of samples with concentrations below the LOD

	Artifacts N = 15	Anthill N = 6	India N = 9	Lam Dong N = 2	Mozambique N = 6	Orissa N = 10	PhuTho N = 10	South India N = 10	Sri Lanka 1 N = 10	Sri Lanka 2 N = 6	Total
Ag	13	4	9	0	5	8	9	9	10	4	71
Al	0	0	0	0	0	0	0	0	0	0	0
Au	13	6	9	1	6	10	9	10	10	6	80
B	12	2	9	0	2	8	10	5	10	6	64
Ba	5	0	5	0	2	2	1	9	7	6	37
Be	14	4	9	0	6	10	10	10	10	6	79
Bi	6	4	3	0	2	10	4	8	5	1	43
Ca	0	0	0	0	0	0	0	0	0	0	0
Ce	0	0	1	0	0	1	0	2	4	1	9
Co	0	0	0	2	0	0	0	0	0	0	2
Cr	0	0	0	2	0	0	0	0	0	1	3
Cs	9	3	8	1	4	10	9	10	9	6	69
Cu	2	0	9	0	0	0	0	0	10	0	21
Dy	0	0	0	0	0	0	0	0	0	0	0
Er	0	0	0	0	0	0	0	0	0	0	0
Eu	0	0	0	2	0	0	0	0	0	0	2
Fe	0	0	0	0	0	0	0	0	0	0	0
Gd	0	0	0	0	0	0	0	0	0	0	0
Hf	0	1	0	0	0	0	0	0	0	0	1
Ho	0	0	0	0	0	0	0	0	0	0	0
In	1	0	0	0	0	1	0	1	0	4	7
K	14	6	9	2	6	10	10	10	10	6	83
La	1	0	2	0	2	8	0	7	6	5	31
Li	8	2	0	0	0	1	0	0	0	0	11
Lu	0	0	0	0	0	0	0	0	0	0	0
Mg	0	0	0	0	0	0	0	0	0	0	0
Mn	0	0	0	0	0	0	0	0	0	0	0
Mo	7	5	8	1	2	6	0	3	10	6	48
Na	14	6	9	2	6	10	10	10	10	6	83
Nb	10	1	6	0	5	7	0	7	5	6	47
Nd	0	0	4	0	0	0	0	1	0	0	5

(continued)

Table 16.4 (continued)

	Artifacts N = 15	Anthill N = 6	India N = 9	Lam Dong N = 2	Mozambique N = 6	Orissa N = 10	PhuTho N = 10	South India N = 10	Sri Lanka 1 N = 10	Sri Lanka 2 N = 6	Total
Ni	0	0	9	2	0	0	8	6	10	6	41
P	9	5	1	2	0	2	10	8	10	0	47
Pb	0	0	8	1	1	4	0	1	4	4	23
Pr	1	0	3	0	0	1	0	4	8	1	18
Rb	7	4	8	0	6	10	7	10	4	6	62
Sb	4	5	9	0	3	10	0	10	10	4	55
Sc	0	0	0	0	0	0	0	0	0	0	0
Si	0	0	0	0	0	0	0	0	0	0	0
Sm	0	0	0	0	0	0	0	0	0	0	0
Sn	2	3	4	0	3	1	1	0	0	0	14
Sr	1	0	0	0	1	2	0	1	0	1	6
Ta	6	6	8	0	4	10	4	10	6	2	56
Tb	0	0	0	0	0	0	0	0	0	0	0
Th	5	1	8	0	4	10	1	4	7	5	45
Ti	0	0	0	0	0	0	0	0	0	0	0
Tm	0	0	0	0	0	0	0	0	0	0	0
U	7	2	5	0	4	6	7	2	5	1	39
V	0	0	0	0	0	0	0	0	0	0	0
W	11	5	8	0	5	10	0	9	5	6	59
Y	0	0	0	0	0	0	0	0	0	0	0
Yb	0	0	0	0	0	0	0	0	0	0	0
Zn	0	0	0	0	0	0	0	0	0	0	0
Zr	0	0	0	0	0	0	0	0	0	0	0

Elements that assisted with the discrimination of the different garnet sources were identified through exploratory bivariate plots and ternary diagrams, both of which have been used in earlier studies (e.g. Calligaro et al. 2002). In addition to these methods, principal components analysis (PCA), using IBM SPSS v. 20.0, assisted in the evaluation of a large number of elements measured by LA-ICP-MS and the identification of compositional differences between sources. Prior to performing all multivariate statistical analysis, all major element concentrations were converted from weight percent to parts per million (ppm) and all elements were $\log_{10}$ transformed. Elements that were below the LOD were replaced with a zero for the statistical analyses.

16.5 Results and Discussion

Following LA-ICP-MS analysis, the first step was to determine which garnet species were present. As noted earlier, garnets rarely occur in their pure end-member form, and are often a mixture of several sub-species. Recent research has provided a way to easily convert chemical analyses into molar proportions of various end-member components (Locock 2008). However, this method assumes a different stoichiometry than that assumed for converting from silica-normalized oxide weight percentages to elemental concentrations during calibration and LA-ICP-MS measurements, and therefore a different approach was needed. For this reason, the major elements of almandine (Fe), pyrope (Mg), spessartine (Mn), and ugrandite series garnets (Ca) were sum-normalized to 100 % as a proxy to determine the major sub-species present within a single garnet sample (see Tables 16.2 and 16.3). Based on the varying quantities of these major elements, the garnet sources were divided into three groups (Fig. 16.8).

The almandine-rich group contains samples from Sri Lanka Group 1, Sri Lanka Group 2, and South India. Per the sum-normalized table, garnets in this group are predominantly iron rich almandine (73–85 % Fe) with moderate amounts of pyrope (9–17 % Mg) and lesser amounts of spessartine and ugrandite-series garnet. Figure 16.9 presents these groups plotted by their first and second components using the elements Co, Dy, Tb, Y, and Zn, which account for 97 % of the variation. These sources are clearly compositionally different from one another. The Sri Lanka Group 1 source contains higher concentrations of Co (17–21 ppm), Dy (45–131 ppm), Tb (3–8 ppm) and Y (322–1148 ppm) than the other groups. Conversely, Sri Lanka Group 2 had lower concentrations of Co (3–5 ppm) and Zn (4–13 ppm). The South India group generally had moderate to low amounts of the five elements.

The second group, which I will refer to as almandine-pyrope, contains garnets with a high percentage of Fe (31–75 %), but higher levels of pyrope than the almandine group. These garnets also contain higher concentrations of Ca than Mn (Table 16.3). The sources in this group include Phu Tho, the unknown Indian source, Orissa, Anthill, and Mozambique. Bivariate plots identified elements that discriminated the groups and a PCA was performed using the elements Al, Ca, Cu, Fe, Li, Ni, P, Sc, Tb, Ti and V. The first three components are plotted in Fig. 16.10, which account for 80 % of the variation. There is fairly good separation between these five source groups. The Anthill source is notable for higher concentrations of CaO (approx. 5–13 wt%) and Ni (3–22 ppm). However, there is also variation in the major elements, as three samples from this source (Anthilll2, Anthill4, and Anthill5) had lower levels of iron than magnesium. The unknown "India" source was differentiated by low concentrations of CaO (approx. 1–2 wt%) and higher concentrations of Fe_2O_3 (25–29 wt %), as well as high levels of P (433–712 ppm). However, one India sample (IndiaGarnet2) consistently plotted with the Orissa samples. The Mozambique samples were notable for higher concentrations of Li (13–16 ppm) and lower Sc (41–49 ppm). The Phu Tho source contained elevated levels of Al_2O_3 (23–24 wt%) and CuO (0.001 wt%) compared to the other sources, as well as high Sc (69–264 ppm), Tb (0.3–8 ppm) but lower Ni, with several samples at the LOD.

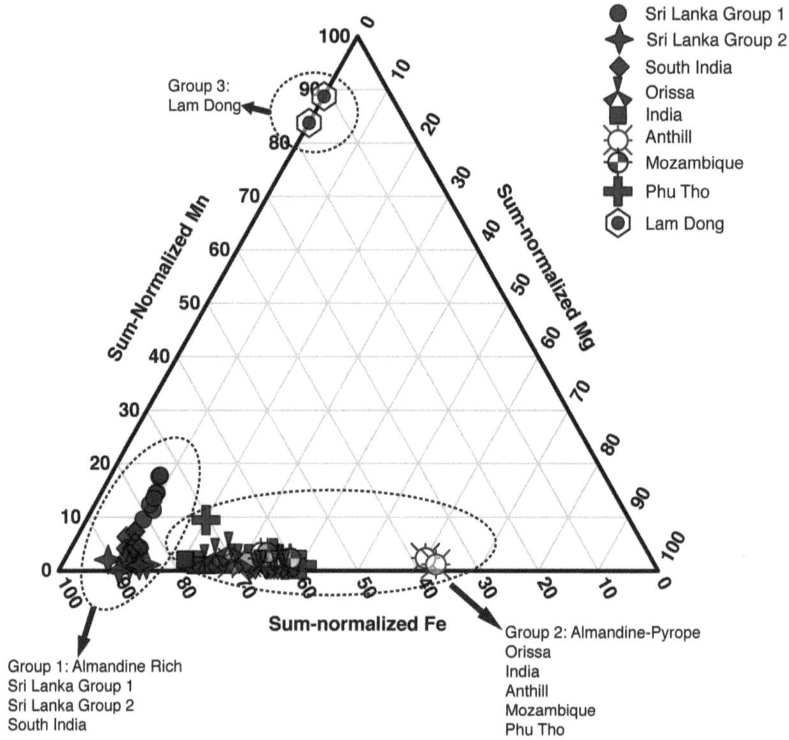

Fig. 16.8 Ternary diagram of sum-normalized Fe, Mn, and Mg showing three different groups of garnets: group 1 (almandine rich), group 2 (almandine-pyrope), and group 3 (Lam Dong garnet)

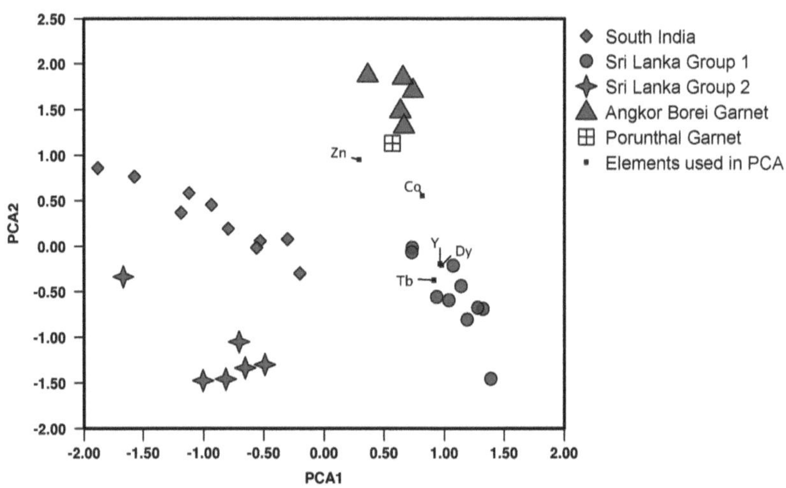

Fig. 16.9 Principal component analysis of group 1 almandine garnets and artifacts from Angkor Borei and Porunthal. The first component summarizes 69 % of the variance and the second component accounts for 28 %

The Orissa source was perhaps the most problematic source group to evaluate, as the samples are not tightly clustered. This highlights a drawback of not personally collecting geologic samples directly from a source, as we must consider that the variation could be due to the inclusion of samples from multiple distinct sources. It is hoped that future studies will address these issues with additional analyses of well provenienced source samples.

The final group consists of garnets from a single source: Lam Dong. Although only two garnets were analyzed, these samples are clearly different from the other sources and artifacts due

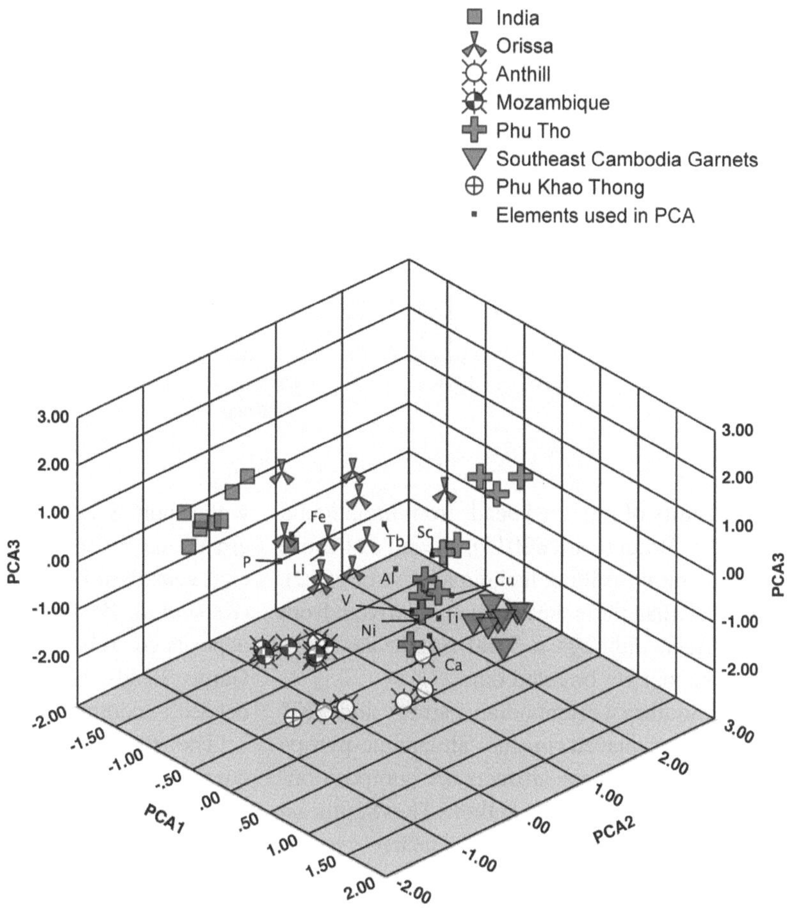

Fig. 16.10 Principal components analysis of group 2 almandine-pyrope garnets and artifacts from southeast Cambodia and Phu Khao Thong. The first component summarizes 50 % of the variance, the second component accounts for 20 %, and the third component accounts for 10 % of the variance

to their high levels of MnO (83–87 %), classifying them as spessartine garnets. These samples are also notable for their high concentrations of Ti (864–1102 ppm) and Zn (187–261 ppm), the highest concentration of these elements found in any of the analyzed sources.

In their recent study, Calligaro et al. (2002) identified three garnets that they believe originated from South Asia. Type 1, from an unknown Indian source, is described as quasi-pure almandine. Type II, believed to be from Rajasthan, India was classified as almandine with manganese. Type III, an intermediate "rhodolite" or almandine-pyrope, was believed to come from Sri Lanka. Although compositional data were only provided for the elements Fe, Mg, Ca, Mn, Cr, and Y (see Calligaro et al. 2002), it appears that the South Asian garnets in this study are distinct from the sources studied by Calligaro. This highlights the great diversity of garnet sources found across South Asia, and the need for analysis of well-provenienced garnet source samples.

The garnet artifacts can be similarly classified into different groups based on the sum-normalized totals of their major elements (Table 16.2). The Angkor Borei and Porunthal garnet artifacts belong to the almandine rich group. However, despite similarities in major elements, these artifacts are clearly distinct from the geologic source samples due to higher levels of Co (25–42 ppm) and Zn (83–138 ppm) (Fig. 16.9). It is also notable that the Angkor Borei garnet beads and Porunthal garnets share a similar composition. Although it is clear that the geologic source material for these artifacts has not been found, it appears that they may derive from the same source area. Due to the

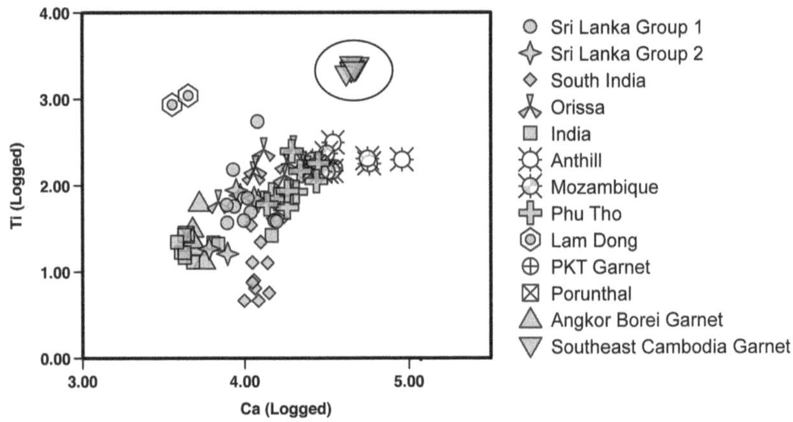

Fig. 16.11 Bivariate plot of logged Ca vs. Ti concentrations. Southeast Cambodia garnets (*circled*) have a high concentration of both elements in comparison with other sources and artifacts

high quantity of imported beads from India found at Angkor Borei (Carter 2015), as well as the use of the diamond-drilling technique (Carter 2012), I suggest that these beads likely derived from South Asia, although a local Southeast Asian source cannot yet be ruled out.

The Southeast Cambodian garnets and PKT garnet were placed in the almandine-pyrope group, although these artifacts are compositionally distinct from one another. The Southeast Cambodian garnets have nearly equal proportions of Fe_2O_3 (12–14 wt%) and MgO (14–17 wt%), high concentrations of CaO (5–7 wt%) and low concentrations of MnO (<1 wt%). The PKT garnet has slightly higher levels of Fe_2O_3 and lower levels of MgO. However, when included in PCA with the almandine-pyrope geologic source groups, these artifacts plot away from the other sources (Fig. 16.10). The Southeast Cambodian garnets are distinguished by their moderate levels of Ca and extremely high levels of Ti (approx. 2000 ppm), the highest of any garnet artifact or source sampled in this study (Fig. 16.11). This composition is similar to the group 3 "common eclogite group" garnets identified by Dawson and Stephens (1975: 603). Diamonds are also commonly found within the common eclogite group, and thus examining material from contemporary diamond mining operations in South and Southeast Asia may assist in identifying this garnet source. Kane et al. (1991: 153) have also noted the widespread presence of garnets associated with sapphire, zircon, and ruby deposits in southern Vietnam. Large garnets have also been discovered near Bien Ho in central Vietnam (Fig. 16.7) (Kane et al. 1991). It is possible that one of these deposits could be the source location for the SEA Garnet Beads, however further analyses will be needed to confirm this hypothesis.

The PKT garnet is distinguished by low concentrations of several elements, including Sc (34 ppm), V (22 ppm), and Y (17 ppm) (Fig. 16.12). Although the PKT garnet plots near the Anthill sources, it is highly unlikely that this garnet piece originated in Arizona. Rather, it is more likely that the PKT derives from an unknown source that shares a similar composition to the Anthill source. It is notable that the PKT sample is also compositionally distinct from the other garnet artifacts studied, and this suggests that complex bead production and trade networks were in play during the Iron Age period. As PKT was an important international trading center for artifacts from South Asia (Chaisuwan 2011), it is possible that this piece could derive from a source in either South or Southeast Asia.

16.6 Conclusion

Although the results of this study are still preliminary, they demonstrate the success of using LA-ICP-MS on garnet. The ability for LA-ICP-MS to measure minor and trace elements allows for a

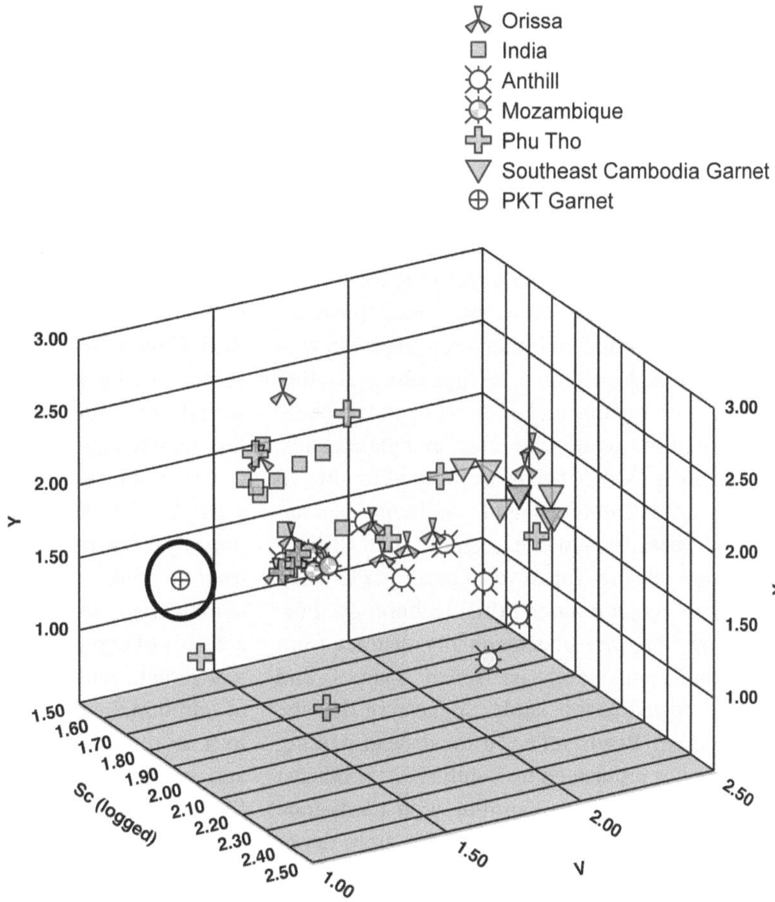

Fig. 16.12 A three-dimensional scatterplot of Sc, V, and Y (logged) showing the low concentrations of these elements in the PKT sample

more conclusive differentiation between various geologic sources and artifacts than when using major and minor elements alone. However, the inability to accurately characterize the proportion of garnet end-members using Locock's methodology (2008) is a drawback to this technique. In this case, a combination of LA-ICP-MS with techniques such as SEM-EDS to measure major elements may be the most effective way to characterize garnet samples. Additionally, in future studies it may be advisable to use more than five point ablations per sample in order to better account for the zoning of elements, as well as including additional samples in order to better characterize and understand the variation within a single source.

The current geochemical study, combined with previous studies of bead morphology and drill-holes, demonstrates that there are two distinct bead-making traditions represented amongst the garnet beads from Iron Age sites in Cambodia. Furthermore, analysis of additional artifacts from peninsular Thailand and southern India point toward at least three distinct garnet sources used during the Iron Age period. Compositional similarity between AB Garnet Drill Type DD1 beads and the analyzed bead from Porunthal, the diamond drilling technique used on the Angkor Borei beads, and the long history of bead production in South Asia strongly suggest an origin for these beads in South Asia. This connection with South Asia is not surprising as Angkor Borei and the nearby site of Oc Eo, Vietnam were believed to have been important trading centers. Excavations at Oc Eo have uncovered numerous artifacts derived from South Asia, including jewelry, coins, and pottery (Malleret 1962) and glass beads from Angkor Borei have been linked compositionally to a possible production center at Giribawa, Sri Lanka (Carter 2013; Dussubieux 2001).

However, we cannot yet rule out possible bead production in Southeast Asia. There is strong evidence for hard stone bead production, perhaps by Indian craftsmen, at the site of Khao Sam Kaeo in peninsular Thailand (Bellina 2014). Despite their unknown manufacturing location, it is interesting to note that similar beads have not yet been recorded at other Iron Age sites in Cambodia and only in small quantities at other sites in the region, and thus were not broadly distributed like agate, carnelian, and glass beads.

The SEA Garnet beads are limited to sites in southeastern Cambodia and southern Vietnam and link these sites to one another. Their drilling technique sets them apart from beads made using the South Asian diamond-drill technology. Furthermore, they are geochemically distinct from the South and Southeast Asian sources and artifacts tested in this study, suggesting that the SEA Garnet Beads reflect a local bead-making tradition not influenced by South Asia. Southeast Asian craftsmen were familiar with hard-stone jewelry production, as objects made from nephrite had been manufactured and widely distributed in Southeast Asia since the Neolithic period (e.g. Hung et al. 2007; Nguyen 1996). Nephrite ear ornaments have been found at sites in mainland Southeast Asia, as well as Taiwan, Sarawak, and the Philippines. In contrast, SEA Garnet Beads seem to have been manufactured in smaller quantities and had a more localized distribution. They reflect a different scale and type of craft manufacturing and distribution from local nephrite jewelry industries. Additionally, the distinctive SEA Garnet Beads may have also had unique and localized meanings and values, separate from imported carnelian and glass beads, whose value may have been related to their association with foreign craft producers (Bellina 2003; Carter 2012).

The identification of SEA Garnet Bead workshops may be difficult, as it appears natural stones were drilled without having been shaped or polished. Ethnoarchaeological studies of contemporary agate and carnelian bead-making workshops in India have highlighted the importance of debitage and flaking debris in identifying ancient bead workshops (see Kenoyer et al. 1991). While ongoing experimental studies will assist in the better understanding how the SEA Garnet Beads were drilled, drilling natural stones produces little material evidence that would be visible archaeologically, and it is for this reason that continued geochemical analysis is important, as it may confirm the presence of a localized Southeast Asian bead-making tradition by identifying a local geological source used to make these beads. The present study demonstrates the potentiality of LA-ICP-MS as a means of compositionally sourcing archaeological garnets, and lays the groundwork for analysis of additional garnet artifacts that will contribute to a more complete picture of bead production and exchange during the Iron Age period in Southeast Asia.

Acknowledgements I would like to thank the Cambodian government and the Ministry of Culture and Fine Arts for allowing me to work on and analyze the archaeological materials. Heng Sophady, Vuthy Voeun, Seng Sonetra, and Vin Laychour from the Memot Centre for Archaeology kindly provided materials from Village 10.8 and Bit Meas. Dr. Andreas Reinecke assisted with materials and information from Prohear. I am grateful to Dr. Miriam Stark for providing materials from Angkor Borei and ongoing advice on this project. Sokha Tep provided helpful advice on beads from Bit Meas. Drs. Rajan and Chaisuwan provided materials from Porunthal and Phu Khao Thong respectively. David Bernstein assisted with samples from Vietnam and Katie Lindstrom kindly provided samples Sri Lanka. I am grateful for the ongoing help of Laure Dussubieux at the LA-ICP-MS lab at the Field Museum. Thanks to Mark Golitko and two anonymous reviewers for their comments on a draft of this chapter, as well as my colleagues at UW-Madison, including Drs. James Burton, J. Mark Kenoyer, Gwen Kelly, Marc Kissel, and Gregg Jamison.

Appendix

Compositional results for garnet artifacts and geologic sources. Major and minor elements reported as percent oxide and trace elements as parts per million

Database ID	Location	SiO_2	Na_2O	MgO	Al_2O_3	P_2O_3	K_2O	CaO	MnO	Fe_2O_3	CuO	SnO_2	PbO_2	Li	Be	B	Sc	Ti	V
Anthill1	Anthill, Arizona	40.4 %	<LOD	9.28 %	22.6 %	0.07 %	<LOD	8.4%	0.35 %	19.3 %	0.0004 %	0.0001 %	0.0003 %	1.2	0.2	1.3	40	178	94
Anthill2	Anthill, Arizona	44.3 %	<LOD	18.5 %	21.8 %	<LOD	<LOD	4.84 %	0.43 %	10.1 %	0.001 %	0.0001 %	0.00008 %	0.3	0.4	1.3	93	313	170
Anthill3	Anthill, Arizona	40.0 %	<LOD	8.32 %	22.5 %	<LOD	<LOD	12.72 %	0.27 %	16.2 %	0.0002 %	0.00002 %	0.00003 %	<LOD	<LOD	<LOD	38	198	197
Anthill4	Anthill, Arizona	43.9 %	<LOD	18.6 %	22.0 %	<LOD	<LOD	4.77 %	0.47 %	10.3 %	0.0005 %	<LOD	0.00005 %	<LOD	<LOD	0.3	100	147	122
Anthill5	Anthill, Arizona	44.5 %	<LOD	18.8 %	21.8 %	<LOD	<LOD	4.43 %	0.38 %	10.1 %	0.0005 %	<LOD	0.00002 %	0.9	<LOD	0.5	111	234	108
Anthill6	Anthill, Arizona	41.3 %	<LOD	11.5 %	22.1 %	<LOD	<LOD	7.85 %	0.29 %	17.0 %	0.0003 %	<LOD	0.0002 %	1.3	<LOD	<LOD	32	205	154
IndiaGarnet1	Unknown source, India	38.5 %	<LOD	10.1 %	23.1 %	0.16 %	<LOD	0.57 %	0.46 %	27.0 %	<LOD	0.00001 %	<LOD	10	<LOD	<LOD	65	17	25
IndiaGarnet2	Unknown source, India	39.2 %	<LOD	10.7 %	22.9 %	<LOD	<LOD	2.05 %	0.21 %	25.0 %	<LOD	0.00001 %	<LOD	3.2	<LOD	<LOD	45	26	60
IndiaGarnet3	Unknown source, India	38.5 %	<LOD	9.99 %	22.9 %	0.16 %	<LOD	0.67 %	0.31 %	27.4 %	<LOD	<LOD	<LOD	12	<LOD	<LOD	47	13	29
IndiaGarnet4	Unknown source, India	38.6 %	<LOD	9.36 %	22.6 %	0.13 %	<LOD	0.60 %	0.49 %	28.2 %	<LOD	0.00001 %	0.00001 %	9.0	<LOD	<LOD	75	15	27
IndiaGarnet5	Unknown source, India	38.2 %	<LOD	9.51 %	23.0 %	0.14 %	<LOD	0.90 %	0.32 %	27.9 %	<LOD	<LOD	<LOD	10	<LOD	<LOD	63	22	25
IndiaGarnet6	Unknown source, India	38.4 %	<LOD	9.69 %	22.5 %	0.21 %	<LOD	0.59 %	0.46 %	28.2 %	<LOD	<LOD	<LOD	15	<LOD	<LOD	52	17	35
IndiaGarnet7	Unknown source, India	38.3 %	<LOD	9.40 %	23.2 %	0.16 %	<LOD	0.96 %	0.43 %	27.5 %	<LOD	0.000004 %	<LOD	10	<LOD	<LOD	65	21	26
IndiaGarnet8	Unknown source, India	38.1 %	<LOD	8.85 %	22.9 %	0.21 %	<LOD	0.59 %	0.62 %	28.6 %	<LOD	<LOD	<LOD	14	<LOD	<LOD	52	27	43
IndiaGarnet9	Unknown source, India	38.0 %	<LOD	8.65 %	23.0 %	0.14 %	<LOD	0.54 %	0.67 %	28.9 %	<LOD	0.00001 %	<LOD	10	<LOD	<LOD	83	22	41
LamDong1	Lam Dong, Vietnam	37.1 %	<LOD	0.055 %	21.3 %	<LOD	<LOD	0.50 %	34.2 %	6.73 %	0.001 %	0.08 %	0.00002 %	84	0.5	3.1	759	864	2.2
LamDong2	Lam Dong, Vietnam	37.2 %	<LOD	0.045 %	21.4 %	<LOD	<LOD	0.63 %	35.7 %	4.71 %	0.001 %	0.19 %	<LOD	47	1.8	8.8	926	1102	1.7
Mozambique1	Unknown source, Mozambique	41.7 %	<LOD	11.8 %	21.9 %	0.04 %	<LOD	4.16 %	0.60 %	19.9 %	0.0003 %	0.00001 %	0.000003 %	13	<LOD	0.2	41	183	61
Mozambique2	Unknown source, Mozambique	40.7 %	<LOD	11.9 %	22.8 %	0.05 %	<LOD	3.64 %	0.76 %	20.1 %	0.0003 %	<LOD	0.000004 %	13	<LOD	0.2	46	195	54
Mozambique3	Unknown source, Mozambique	40.4 %	<LOD	12.5 %	23.2 %	0.05 %	<LOD	4.45 %	0.64 %	18.8 %	0.0005 %	<LOD	0.000003 %	14	<LOD	<LOD	49	143	48
Mozambique4	Unknown source, Mozambique	40.9 %	<LOD	12.2 %	22.4 %	0.04 %	<LOD	3.79 %	0.77 %	19.9 %	0.0003 %	<LOD	<LOD	12	<LOD	<LOD	43	153	44
Mozambique5	Unknown source, Mozambique	40.7 %	<LOD	13.3 %	23.2 %	0.05 %	<LOD	4.09 %	0.45 %	18.3 %	0.0004 %	0.00001 %	0.00001 %	15	<LOD	0.2	47	143	53

(continued)

Database ID	Location	SiO_2	Na_2O	MgO	Al_2O_3	P_2O_3	K_2O	CaO	MnO	Fe_2O_3	CuO	SnO_2	PbO_2	Li	Be	B	Sc	Ti	V
Mozambique6	Unknown source, Mozambique	40.7 %	<LOD	13.4 %	23.2 %	0.05 %	<LOD	3.71 %	0.49 %	18.4 %	0.0004 %	0.00001 %	0.0001 %	16	<LOD	0.2	49	165	57
Orissa1	Unknown source, Orissa, India	41.7 %	<LOD	13.8 %	22.5 %	0.07 %	<LOD	1.59 %	0.40 %	20.0 %	0.0004 %	0.00005 %	0.00001 %	7.6	<LOD	<LOD	75	138	98
Orissa2	Unknown source, Orissa, India	41.1 %	<LOD	10.7 %	22.2 %	0.04 %	<LOD	2.47 %	0.63 %	22.9 %	0.0003 %	0.00005 %	0.00001 %	2.2	<LOD	0.1	119	130	154
Orissa3	Unknown source, Orissa, India	41.3 %	<LOD	11.0 %	21.9 %	0.05 %	<LOD	2.54 %	0.59 %	22.7 %	0.0004 %	0.00004 %	0.00001 %	<LOD	<LOD	<LOD	32	185	56
Orissa4	Unknown source, Orissa, India	41.8 %	<LOD	13.1 %	22.6 %	0.09 %	<LOD	1.84 %	0.33 %	20.2 %	0.001 %	0.00005 %	0.0002 %	8.3	<LOD	<LOD	79	261	28
Orissa5	Unknown source, Orissa, India	41.6 %	<LOD	12.0 %	22.4 %	0.05 %	<LOD	3.20 %	0.30 %	20.5 %	0.0003 %	0.00001 %	<LOD	1.1	<LOD	<LOD	74	180	59
Orissa6	Unknown source, Orissa, India	40.4 %	<LOD	10.1 %	22.5 %	<LOD	<LOD	3.40 %	1.16 %	22.4 %	0.001 %	0.00005 %	0.00001 %	2.4	<LOD	<LOD	125	161	139
Orissa7	Unknown source, Orissa, India	41.6 %	<LOD	14.1 %	22.5 %	0.03 %	<LOD	1.63 %	0.25 %	19.9 %	0.0004 %	0.00001 %	0.00002 %	4.4	<LOD	<LOD	58	163	32
Orissa8	Unknown source, Orissa, India	40.7 %	<LOD	9.35 %	22.2 %	0.09 %	<LOD	0.97 %	0.67 %	26.0 %	0.0004 %	0.00004 %	<LOD	2.5	<LOD	<LOD	72	64	23
Orissa9	Unknown source, Orissa, India	41.0 %	<LOD	9.62 %	22.2 %	<LOD	<LOD	2.23 %	0.95 %	24.0 %	0.0003 %	<LOD	<LOD	4.8	<LOD	<LOD	116	49	53
Orissa10	Unknown source, Orissa, India	41.3 %	<LOD	12.0 %	22.8 %	0.03 %	<LOD	2.80 %	0.38 %	20.6 %	0.0004 %	0.00004 %	<LOD	4.6	<LOD	0.3	80	274	60
PhuThoGarnet1	Phu Tho, Vietnam	40.3 %	<LOD	7.77 %	23.5 %	<LOD	<LOD	2.05 %	2.64 %	23.8 %	0.001 %	0.00002 %	0.0003 %	12	<LOD	<LOD	210	78	43
PhuThoGarnet2	Phu Tho, Vietnam	41.3 %	<LOD	12.2 %	24.2 %	<LOD	<LOD	2.20 %	0.87 %	19.2 %	0.001 %	0.00005 %	0.0002 %	22	<LOD	<LOD	264	69	80
PhuThoGarnet3	Phu Tho, Vietnam	41.1 %	<LOD	11.0 %	24.0 %	<LOD	<LOD	2.71 %	0.66 %	20.6 %	0.001 %	0.00008 %	0.001 %	10	<LOD	<LOD	106	248	35
PhuThoGarnet4	Phu Tho, Vietnam	41.4 %	<LOD	12.1 %	23.6 %	<LOD	<LOD	3.93 %	0.56 %	18.5 %	0.001 %	0.00001 %	0.0001 %	5.9	<LOD	<LOD	69	181	32
PhuThoGarnet5	Phu Tho, Vietnam	41.7 %	<LOD	13.0 %	23.8 %	<LOD	<LOD	3.83 %	0.33 %	17.4 %	0.001 %	0.00006 %	0.0002 %	6.8	<LOD	<LOD	80	112	29
PhuThoGarnet6	Phu Tho, Vietnam	42.0 %	<LOD	13.8 %	23.9 %	<LOD	<LOD	2.56 %	0.34 %	17.4 %	0.001 %	0.00006 %	0.0002 %	6.9	<LOD	<LOD	79	55	13
PhuThoGarnet7	Phu Tho, Vietnam	41.5 %	<LOD	12.7 %	24.0 %	<LOD	<LOD	1.92 %	0.43 %	19.4 %	0.001 %	0.00001 %	0.0001 %	8.3	<LOD	<LOD	104	60	20
PhuThoGarnet8	Phu Tho, Vietnam	41.0 %	<LOD	12.6 %	24.1 %	<LOD	<LOD	3.05 %	0.45 %	18.8 %	0.001 %	<LOD	0.00004 %	8.6	<LOD	<LOD	80	146	20
PhuThoGarnet9	Phu Tho, Vietnam	41.1 %	<LOD	13.0 %	24.0 %	<LOD	<LOD	2.88 %	0.35 %	18.7 %	0.001 %	0.00001 %	0.0002 %	7.1	<LOD	<LOD	77	84	38
PhuThoGarnet10	Phu Tho, Vietnam	43.2 %	<LOD	12.1 %	23.5 %	<LOD	<LOD	2.57 %	0.47 %	18.2 %	0.001 %	0.00002 %	0.0001 %	10	<LOD	<LOD	111	86	47
SouthIndia1	Unknown source, South India	39.8 %	<LOD	5.11 %	21.4 %	<LOD	<LOD	1.56 %	0.98 %	31.1 %	0.0001 %	0.00007 %	0.00001 %	2.6	<LOD	<LOD	108	13	18
SouthIndia2	Unknown source, South India	38.6 %	<LOD	3.81 %	22.7 %	<LOD	<LOD	1.62 %	2.17 %	31.2 %	0.0001 %	0.0003 %	0.00001 %	3.7	<LOD	0.4	122	6.5	15

SouthIndia3	Unknown source, South India	39.4 %	<LOD	3.94%	21.8 %	<LOD	<LOD	1.59%	2.05 %	31.2 %	0.0001 %	0.0001%	0.000004 %	2.7	<LOD	<LOD	93	8.0	17
SouthIndia4	Unknown source, South India	39.5 %	<LOD	4.08%	21.7 %	<LOD	<LOD	1.39%	1.24 %	32.1 %	0.0001 %	0.0005%	0.00001 %	2.7	<LOD	0.2	87	4.6	15
SouthIndia5	Unknown source, South India	38.8 %	<LOD	5.52%	22.1 %	<LOD	<LOD	1.93%	0.23 %	31.4 %	0.0001 %	0.003%	0.00001 %	1.6	<LOD	<LOD	46	13	12
SouthIndia6	Unknown source, South India	39.2 %	<LOD	4.33%	21.7 %	0.017 %	<LOD	1.55%	2.15 %	31.1 %	0.0001 %	0.00007 %	0.00002 %	3.5	<LOD	0.6	93	7.5	17
SouthIndia7	Unknown source, South India	40.1 %	<LOD	4.80%	21.0 %	<LOD	<LOD	1.53%	1.49 %	31.1 %	0.0001 %	0.0001%	0.00001 %	0.9	<LOD	6.0	68	35	25
SouthIndia8	Unknown source, South India	38.8 %	<LOD	4.42%	22.3 %	0.018 %	<LOD	1.70%	1.33 %	31.4 %	0.0001 %	0.00002 %	0.00001 %	2.8	<LOD	0.1	84	4.6	11
SouthIndia9	Unknown source, South India	39.4 %	<LOD	4.86%	21.9 %	<LOD	<LOD	1.97%	1.17 %	30.8 %	0.0001 %	0.0006%	0.00002 %	2.0	<LOD	<LOD	37	5.7	10
SouthIndia10	Unknown source, South India	39.3 %	<LOD	4.73%	22.5 %	<LOD	<LOD	1.75%	0.94 %	30.8 %	0.0001 %	0.00006 %	<LOD	2.6	<LOD	<LOD	71	22	26
SriLankaGroup1_1	Unknown source, Sri Lanka	44.0 %	<LOD	3.92%	19.7 %	<LOD	<LOD	2.12%	2.77 %	27.5 %	<LOD	0.00004 %	0.0003%	11	<LOD	<LOD	151	39	37
SriLankaGroup1_2	Unknown source, Sri Lanka	37.4 %	<LOD	4.00%	22.2 %	<LOD	<LOD	1.21%	5.99 %	29.2 %	<LOD	0.00003 %	0.00001 %	19	<LOD	<LOD	152	57	31
SriLankaGroup1_3	Unknown source, Sri Lanka	37.4 %	<LOD	4.20%	23.2 %	<LOD	<LOD	1.09%	4.94 %	29.2 %	<LOD	0.00003 %	0.00003 %	14	<LOD	<LOD	184	59	36
SriLankaGroup1_4	Unknown source, Sri Lanka	37.4 %	<LOD	4.32%	22.5 %	<LOD	<LOD	1.52%	4.71 %	29.5 %	<LOD	0.000005 %	<LOD	20	<LOD	<LOD	182	49	34
SriLankaGroup1_5	Unknown source, Sri Lanka	37.5 %	<LOD	4.55%	22.5 %	<LOD	<LOD	1.38%	3.69 %	30.4 %	<LOD	0.00002 %	0.00004 %	10	<LOD	<LOD	150	39	39
SriLankaGroup1_6	Unknown source, Sri Lanka	37.2 %	<LOD	4.05%	22.1 %	<LOD	<LOD	1.19%	6.12 %	29.3 %	<LOD	0.00008 %	0.00001 %	15	<LOD	<LOD	145	153	40
SriLankaGroup1_7	Unknown source, Sri Lanka	37.2 %	<LOD	4.90%	22.5 %	<LOD	<LOD	2.20%	1.64 %	31.5 %	<LOD	0.00001 %	0.00001 %	12	<LOD	<LOD	261	38	46
SriLankaGroup1_8	Unknown source, Sri Lanka	36.7 %	<LOD	4.25%	22.4 %	<LOD	<LOD	1.10%	4.33 %	31.2 %	<LOD	0.00002 %	<LOD	14	<LOD	<LOD	175	37	33
SriLankaGroup1_9	Unknown source, Sri Lanka	37.0 %	<LOD	4.47%	22.2 %	<LOD	<LOD	1.46%	4.79 %	30.1 %	<LOD	0.00004 %	<LOD	13	<LOD	<LOD	153	71	45
SriLankaGroup1_10	Unknown source, Sri Lanka	36.9 %	<LOD	4.32%	22.2 %	<LOD	<LOD	1.68%	4.50 %	30.4 %	<LOD	0.00006 %	<LOD	13	<LOD	<LOD	153	553	46
SriLankaGroup2_1	Unknown source, Sri Lanka	41.7 %	<LOD	4.06%	25.2 %	0.11%	<LOD	1.32%	0.70 %	26.9 %	0.0001 %	0.0003%	<LOD	31	<LOD	<LOD	15	76	6
SriLankaGroup2_2	Unknown source, Sri Lanka	41.3 %	<LOD	5.60%	25.6 %	0.11%	<LOD	1.10%	0.21 %	26.0 %	0.001%	0.0001%	<LOD	31	<LOD	<LOD	3.0	16	0.4
SriLankaGroup2_3	Unknown source, Sri Lanka	40.1 %	<LOD	4.60%	26.7 %	0.06%	<LOD	1.70%	0.18 %	26.6 %	0.0004 %	0.00008 %	<LOD	30	<LOD	<LOD	21	66	3.1
SriLankaGroup2_4	Unknown source, Sri Lanka	40.2 %	<LOD	5.39%	26.2 %	0.08%	<LOD	0.85%	0.16 %	27.2 %	0.001%	0.00006 %	<LOD	31	<LOD	<LOD	22	19	1.6
SriLankaGroup2_5	Unknown source, Sri Lanka	39.0 %	<LOD	4.33%	25.4 %	0.10%	<LOD	1.24%	0.12 %	29.8 %	0.001%	0.00005 %	0.00001 %	38	<LOD	<LOD	15	88	2.9

(continued)

Database ID	Location	SiO_2	Na_2O	MgO	Al_2O_3	P_2O_3	K_2O	CaO	MnO	Fe_2O_3	CuO	SnO_2	PbO_2	Li	Be	B	Sc	Ti	V
SriLankaGroup2_6	Unknown source, Sri Lanka	39.0 %	<LOD	3.41 %	24.5 %	0.09 %	<LOD	1.60 %	0.82 %	30.5 %	0.001 %	0.0003 %	<LOD	18	<LOD	<LOD	6.1	71	1.3
AB1767	Angkor Borei, Cambodia	36.0 %	<LOD	2.65 %	24.1 %	0.10 %	<LOD	0.67 %	1.86 %	34.6 %	<LOD	0.00007 %	0.0001 %	21	<LOD	0.8	91	32	24
AB1835	Angkor Borei, Cambodia	36.1 %	<LOD	2.69 %	23.5 %	0.04 %	<LOD	0.70 %	0.89 %	36.1 %	0.0001 %	0.00007 %	0.0001 %	20	<LOD	<LOD	97	19	25
AB2801	Angkor Borei, Cambodia	36.7 %	<LOD	2.97 %	23.9 %	0.05 %	<LOD	0.67 %	1.18 %	34.5 %	0.0001 %	0.00006 %	0.0001 %	19	<LOD	<LOD	109	23	32
AB3284	Angkor Borei, Cambodia	35.8 %	0.20 %	2.96 %	24.6 %	0.07 %	0.020 %	0.74 %	0.85 %	34.7 %	0.0004 %	0.00008 %	0.0002 %	21	<LOD	1.4	161	63	36
AB3333	Angkor Borei, Cambodia	36.4 %	<LOD	2.66 %	24.2 %	0.02 %	<LOD	0.78 %	0.81 %	35.2 %	0.0001 %	0.00003 %	0.0001 %	15	<LOD	0.4	107	13	19
AKC00310	Village 10.8, Cambodia	40.4 %	<LOD	16.6 %	23.1 %	<LOD	<LOD	6.78 %	0.43 %	12.7 %	0.001 %	0.00004 %	0.00001 %	<LOD	<LOD	<LOD	131	2206	163
AKC00311	Village 10.8, Cambodia	40.5 %	<LOD	15.8 %	22.5 %	<LOD	<LOD	6.89 %	0.45 %	13.9 %	0.001 %	0.00009 %	0.00002 %	<LOD	<LOD	<LOD	111	2394	190
AKC00338	Village 10.8, Cambodia	40.5 %	<LOD	17.0 %	23.0 %	<LOD	<LOD	6.44 %	0.42 %	12.6 %	0.001 %	0.00008 %	0.0001 %	<LOD	<LOD	<LOD	132	2227	168
AKC00339	Village 10.8, Cambodia	40.9 %	<LOD	17.4 %	23.3 %	<LOD	<LOD	6.23 %	0.37 %	11.8 %	0.001 %	0.00009 %	0.0001 %	<LOD	<LOD	<LOD	83	2118	159
AKC00590	Prohear, Cambodia	42.3 %	<LOD	14.2 %	22.0 %	<LOD	<LOD	6.29 %	0.48 %	14.7 %	0.001 %	0.00006 %	0.00004 %	<LOD	0.1	<LOD	64	2084	144
AKC00606	Prohear, Cambodia	42.3 %	<LOD	14.0 %	22.5 %	<LOD	<LOD	6.20 %	0.46 %	14.5 %	0.001 %	0.0001 %	0.0001 %	<LOD	<LOD	<LOD	72	2430	162
AKC00651	Bit Meas, Cambodia	42.2 %	<LOD	16.2 %	23.0 %	<LOD	<LOD	5.88 %	0.41 %	12.4 %	0.001 %	0.00004 %	0.0001 %	<LOD	<LOD	<LOD	96	1886	167
AKC00731	Bit Meas, Cambodia	40.4 %	<LOD	15.9 %	23.1 %	<LOD	<LOD	6.49 %	0.45 %	13.6 %	0.001 %	0.0001 %	0.0002 %	<LOD	<LOD	<LOD	99	2135	159
PKTgarnet	Phu Khao Thong, Thailand	39.9 %	<LOD	11.9 %	23.0 %	<LOD	<LOD	5.01 %	0.89 %	19.4 %	0.0002 %	<LOD	0.00001 %	1.4	<LOD	<LOD	34	160	22
PorunthalGarnet	Porunthal, South India	37.5 %	<LOD	2.45 %	22.0 %	0.06 %	<LOD	0.62 %	1.08 %	36.2 %	<LOD	<LOD	0.00001 %	20	<LOD	<LOD	111	27	23

Database ID	Location	Cr	Ni	Co	Zn	Rb	Sr	Zr	Nb	Ag	In	Sb	Cs	Ba	La	Ce	Pr	Ta	Au
Anthill1	Anthill, Arizona	262	2.6	51	76	0.05	2.4	5	0.01	0.02	0.05	<LOD	0.02	1.7	0.4	0.6	0.2	<LOD	<LOD
Anthill2	Anthill, Arizona	14770	19	40	18	0.4	0.5	3	0.24	0.2	0.04	<LOD	0.1	1.4	0.08	0.3	0.03	<LOD	<LOD
Anthill3	Anthill, Arizona	290	6.7	47	33	<LOD	1.0	8	<LOD	<LOD	0.05	<LOD	<LOD	0.9	0.05	0.4	0.3	<LOD	<LOD
Anthill4	Anthill, Arizona	12718	15	38	13	<LOD	0.08	1	0.16	<LOD	0.03	<LOD	<LOD	0.2	0.04	0.1	0.02	<LOD	<LOD
Anthill5	Anthill, Arizona	10088	22	41	18	<LOD	0.1	5	0.25	<LOD	0.04	0.5	<LOD	0.07	0.02	0.1	0.06	<LOD	<LOD
Anthill6	Anthill, Arizona	290	5.2	62	33	<LOD	0.6	8	0.07	<LOD	0.05	<LOD	0.01	1.4	2.1	5.3	0.84	<LOD	<LOD
IndiaGarnet1	Unknown source, India	78	<LOD	66	69	0.1	0.05	7	0.08	<LOD	0.1	<LOD	0.01	0.02	0.04	0.04	0.03	0.03	<LOD
IndiaGarnet2	Unknown source, India	69	<LOD	23	57	<LOD	0.03	15	<LOD	<LOD	0.1	<LOD	<LOD	<LOD	0.01	0.05	0.06	<LOD	<LOD
IndiaGarnet3	Unknown source, India	73	<LOD	33	83	<LOD	0.1	6	<LOD	<LOD	0.1	<LOD	<LOD	0.07	0.03	0.09	0.01	<LOD	<LOD
IndiaGarnet4	Unknown source, India	63	<LOD	59	77	<LOD	0.4	4	<LOD	<LOD	0.07	<LOD	<LOD	0.1	0.003	0.02	<LOD	<LOD	<LOD
IndiaGarnet5	Unknown source, India	45	<LOD	50	60	<LOD	0.09	126	<LOD	<LOD	0.04	<LOD	<LOD	0.05	<LOD	0.005	<LOD	<LOD	<LOD
IndiaGarnet6	Unknown source, India	63	<LOD	80	57	<LOD	0.03	13	0.02	<LOD	0.06	<LOD	<LOD	<LOD	0.003	0.01	0.02	<LOD	<LOD
IndiaGarnet7	Unknown source, India	67	<LOD	45	80	<LOD	0.05	31	0.01	<LOD	0.09	<LOD	<LOD	<LOD	0.01	0.01	0.01	<LOD	<LOD
IndiaGarnet8	Unknown source, India	75	<LOD	66	58	<LOD	0.1	12	<LOD	<LOD	0.06	<LOD	<LOD	<LOD	0.002	<LOD	<LOD	<LOD	<LOD
IndiaGarnet9	Unknown source, India	74	<LOD	46	66	<LOD	0.02	6	<LOD	<LOD	0.07	<LOD	<LOD	<LOD	<LOD	0.003	0.001	<LOD	<LOD
LamDong1	Lam Dong, Vietnam	<LOD	<LOD	<LOD	261	0.06	0.1	5	9	0.4	1.8	0.2	<LOD	0.10	0.07	1.1	0.1	2.0	0.02
LamDong2	Lam Dong, Vietnam	<LOD	<LOD	<LOD	187	0.9	0.09	20	103	0.05	8.4	0.3	0.2	0.05	0.07	0.7	0.5	11.8	<LOD
Mozambique1	Unknown source, Mozambique	128	3.6	60	51	<LOD	0.05	24	<LOD	<LOD	0.08	0.1	0.01	0.03	0.01	0.07	0.04	0.005	<LOD
Mozambique2	Unknown source, Mozambique	110	3.0	63	38	<LOD	0.1	41	<LOD	<LOD	0.04	<LOD	<LOD	0.04	0.005	0.05	0.02	<LOD	<LOD
Mozambique3	Unknown source, Mozambique	124	3.6	56	42	<LOD	0.02	29	<LOD	<LOD	0.03	<LOD	<LOD	<LOD	<LOD	0.05	0.03	<LOD	<LOD
Mozambique4	Unknown source, Mozambique	175	2.2	49	36	<LOD	<LOD	30	<LOD	<LOD	0.03	<LOD	<LOD	<LOD	<LOD	0.03	0.02	<LOD	<LOD
Mozambique5	Unknown source, Mozambique	139	4.1	59	54	<LOD	0.1	29	0.04	0.02	0.08	0.7	0.03	0.1	0.05	0.1	0.07	0.05	<LOD
Mozambique6	Unknown source, Mozambique	696	4.7	61	88	<LOD	0.08	30	<LOD	<LOD	0.03	0.3	<LOD	0.1	0.002	0.08	0.03	<LOD	<LOD
Orissa1	Unknown source, Orissa, India	554	7.7	67	100	<LOD	0.06	28	0.003	<LOD	0.08	<LOD	<LOD	<LOD	<LOD	0.03	0.04	<LOD	<LOD
Orissa2	Unknown source, Orissa, India	329	6.2	45	38	<LOD	0.02	15	<LOD	<LOD	0.1	<LOD	<LOD	0.02	<LOD	0.08	0.07	<LOD	<LOD
Orissa3	Unknown source, Orissa, India	13	2.9	37	59	<LOD	0.04	31	0.01	<LOD	0.1	<LOD	<LOD	0.2	0.02	0.4	0.2	<LOD	<LOD
Orissa4	Unknown source, Orissa, India	195	20	55	39	<LOD	0.6	80	<LOD	0.03	0.07	<LOD	<LOD	0.4	0.01	0.05	0.09	<LOD	<LOD
Orissa5	Unknown source, Orissa, India	69	2.3	30	19	<LOD	0.09	25	<LOD	<LOD	0.05	<LOD	<LOD	0.03	<LOD	0.07	0.04	<LOD	<LOD
Orissa6	Unknown source, Orissa, India	243	4.2	39	59	<LOD	0.06	20	<LOD	<LOD	0.07	<LOD	<LOD	0.1	<LOD	0.1	0.09	<LOD	<LOD
Orissa7	Unknown source, Orissa, India	160	18	59	54	<LOD	0.07	44	<LOD	0.01	0.08	<LOD	<LOD	0.3	<LOD	0.2	0.14	<LOD	<LOD
Orissa8	Unknown source, Orissa, India	22	0.5	16	63	<LOD	0.08	10	<LOD	<LOD	0.3	<LOD	<LOD	0.1	<LOD	0.04	0.04	<LOD	<LOD

(continued)

Database ID	Location	Cr	Ni	Co	Zn	Rb	Sr	Zr	Nb	Ag	In	Sb	Cs	Ba	La	Ce	Pr	Ta	Au
Orissa9	Unknown source, Orissa, India	126	0.7	32	44	<LOD	<LOD	12	0.03	<LOD	<LOD	<LOD	<LOD	<LOD	<LOD	0.00	<LOD	<LOD	<LOD
Orissa10	Unknown source, Orissa, India	44	7.1	51	52	<LOD	<LOD	53	<LOD	<LOD	0.07	<LOD	<LOD	0.2	<LOD	0.2	0.2	<LOD	<LOD
PhuThoGarnet1	Phu Tho, Vietnam	96	<LOD	33	129	0.6	1.0	6	0.25	<LOD	0.2	0.1	<LOD	4.7	1.6	3.1	0.3	<LOD	<LOD
PhuThoGarnet2	Phu Tho, Vietnam	708	<LOD	30	132	<LOD	0.2	6	0.14	<LOD	0.1	0.1	<LOD	0.7	0.06	0.1	0.03	0.002	<LOD
PhuThoGarnet3	Phu Tho, Vietnam	285	<LOD	20	42	0.2	0.5	34	0.61	0.05	0.2	0.9	0.1	1.2	0.3	0.5	0.2	0.1	0.06
PhuThoGarnet4	Phu Tho, Vietnam	291	<LOD	20	38	<LOD	0.1	8	0.67	<LOD	0.04	0.1	<LOD	0.1	0.06	0.2	0.05	0.01	<LOD
PhuThoGarnet5	Phu Tho, Vietnam	488	2.3	34	36	<LOD	0.2	17	0.32	<LOD	0.02	0.1	<LOD	0.3	0.09	0.2	0.04	0.07	<LOD
PhuThoGarnet6	Phu Tho, Vietnam	151	<LOD	21	37	<LOD	0.2	6	0.18	<LOD	0.01	0.1	<LOD	0.4	0.1	0.3	0.04	0.003	<LOD
PhuThoGarnet7	Phu Tho, Vietnam	104	<LOD	26	45	<LOD	0.07	4	0.07	<LOD	0.06	0.1	<LOD	0.4	0.03	0.06	0.01	<LOD	<LOD
PhuThoGarnet8	Phu Tho, Vietnam	37	<LOD	34	39	<LOD	0.1	9	0.18	<LOD	0.05	0.2	<LOD	<LOD	0.3	0.1	0.05	<LOD	<LOD
PhuThoGarnet9	Phu Tho, Vietnam	99	<LOD	40	68	<LOD	0.2	11	0.18	<LOD	0.05	0.1	<LOD	0.5	0.1	0.3	0.05	0.01	<LOD
PhuThoGarnet10	Phu Tho, Vietnam	1126	1.2	25	53	0.10	0.3	7	0.10	<LOD	0.05	0.2	<LOD	1.0	0.1	0.3	0.04	<LOD	<LOD
SouthIndia1	Unknown source, South India	57	0.80	7.7	28	<LOD	0.01	28	<LOD	<LOD	0.09	<LOD	<LOD	<LOD	<LOD	0.0	0.004	<LOD	<LOD
SouthIndia2	Unknown source, South India	74	<LOD	6.8	22	<LOD	<LOD	11	0.03	<LOD	0.1	<LOD	<LOD	<LOD	<LOD	0.04	0.04	<LOD	<LOD
SouthIndia3	Unknown source, South India	57	<LOD	7.1	26	<LOD	0.01	10	<LOD	<LOD	0.08	<LOD	<LOD	<LOD	0.004	0.01	<LOD	<LOD	<LOD
SouthIndia4	Unknown source, South India	65	<LOD	6.9	25	<LOD	0.03	45	<LOD	<LOD	0.08	<LOD	<LOD	<LOD	<LOD	<LOD	<LOD	<LOD	<LOD
SouthIndia5	Unknown source, South India	71	0.19	6.9	29	<LOD	0.02	32	<LOD	<LOD	0.08	<LOD	<LOD	0.01	<LOD	0.01	0.01	<LOD	<LOD
SouthIndia6	Unknown source, South India	60	<LOD	8.5	28	<LOD	0.08	135	<LOD	<LOD	0.03	<LOD	<LOD	<LOD	<LOD	0.2	<LOD	<LOD	<LOD
SouthIndia7	Unknown source, South India	58	0.3	6.5	25	<LOD	0.03	38	0.04	<LOD	<LOD	<LOD	<LOD	<LOD	<LOD	<LOD	<LOD	<LOD	<LOD
SouthIndia8	Unknown source, South India	59	<LOD	6.8	24	<LOD	0.4	6	0.02	<LOD	0.1	<LOD	<LOD	<LOD	0.3	0.9	0.1	<LOD	<LOD
SouthIndia9	Unknown source, South India	50	<LOD	6.1	28	<LOD	0.02	14	<LOD	0.005	0.08	<LOD	<LOD	<LOD	<LOD	<LOD	0.002	<LOD	<LOD
SouthIndia10	Unknown source, South India	66	1.5	8.8	26	<LOD	0.02	7	<LOD	<LOD	0.1	<LOD	<LOD	<LOD	0.002	0.01	0.01	<LOD	<LOD
SriLankaGroup1_1	Unknown source, Sri Lanka	112	<LOD	17	9	0.3	0.30	76	0.23	<LOD	0.08	<LOD	0.01	0.3	0.1	0.4	0.1	0.05	<LOD
SriLankaGroup1_2	Unknown source, Sri Lanka	53	<LOD	18	19	0.09	0.03	5	0.01	<LOD	0.05	<LOD	<LOD	0.01	0.02	0.04	0.01	<LOD	<LOD
SriLankaGroup1_3	Unknown source, Sri Lanka	149	<LOD	19	28	0.1	0.04	4	0.39	<LOD	0.05	<LOD	<LOD	<LOD	<LOD	0.01	<LOD	0.04	<LOD
SriLankaGroup1_4	Unknown source, Sri Lanka	83	<LOD	19	20	0.2	0.05	4	<LOD	<LOD	0.05	<LOD	<LOD	<LOD	<LOD	<LOD	<LOD	<LOD	<LOD
SriLankaGroup1_5	Unknown source, Sri Lanka	83	<LOD	20	29	0.05	0.02	5	<LOD	<LOD	0.06	<LOD	<LOD	<LOD	0.002	<LOD	<LOD	<LOD	<LOD

SriLankaGroup1_6	Unknown source, Sri Lanka	276	<LOD	19	25	0.1	0.05	6	7.37	<LOD	0.07	<LOD	<LOD	0.08	0.01	0.03	<LOD	0.9	<LOD
SriLankaGroup1_7	Unknown source, Sri Lanka	93	<LOD	22	16	<LOD	0.01	5	<LOD	<LOD	0.09	<LOD	<LOD	<LOD	<LOD	<LOD	<LOD	<LOD	<LOD
SriLankaGroup1_8	Unknown source, Sri Lanka	132	<LOD	18	28	<LOD	0.03	6	<LOD	<LOD	0.07	<LOD	<LOD	<LOD	<LOD	<LOD	<LOD	<LOD	<LOD
SriLankaGroup1_9	Unknown source, Sri Lanka	36	<LOD	21	15	<LOD	0.01	7	<LOD	<LOD	0.06	<LOD	<LOD	<LOD	<LOD	<LOD	<LOD	<LOD	<LOD
SriLankaGroup1_10	Unknown source, Sri Lanka	302	<LOD	19	21	<LOD	0.01	37	7.68	<LOD	0.06	<LOD	<LOD	<LOD	<LOD	<LOD	<LOD	0.6	<LOD
SriLankaGroup2_1	Unknown source, Sri Lanka	32	<LOD	3.7	4.4	<LOD	0.01	5	<LOD	<LOD	<LOD	<LOD	<LOD	<LOD	<LOD	0.01	0.01	0.05	<LOD
SriLankaGroup2_2	Unknown source, Sri Lanka	54	<LOD	3.0	13	<LOD	0.1	8	<LOD	0.1	<LOD	<LOD	<LOD	<LOD	0.06	0.2	0.04	0.01	<LOD
SriLankaGroup2_3	Unknown source, Sri Lanka	3.3	<LOD	4.7	5.5	<LOD	0.02	8	<LOD	<LOD	0.01	<LOD	<LOD	<LOD	<LOD	0.01	0.01	<LOD	<LOD
SriLankaGroup2_4	Unknown source, Sri Lanka	16	<LOD	3.6	8.4	<LOD	0.01	6	<LOD	<LOD	<LOD	<LOD	<LOD	<LOD	<LOD	<LOD	<LOD	<LOD	<LOD
SriLankaGroup2_5	Unknown source, Sri Lanka	4.8	<LOD	4.2	8.5	<LOD	0.05	8	<LOD	0.04	<LOD	0.3	<LOD	<LOD	<LOD	0.01	0.005	0.007	<LOD
SriLankaGroup2_6	Unknown source, Sri Lanka	<LOD	<LOD	3.5	5.7	<LOD	<LOD	6	<LOD	<LOD	0.005	0.4	<LOD	<LOD	<LOD	0.01	0.01	0.06	<LOD
AB1767	Angkor Borei, Cambodia	24	0.8	37	125	0.6	3.6	3	<LOD	<LOD	0.06	<LOD	0.03	2.00	0.07	0.2	0.02	<LOD	<LOD
AB1835	Angkor Borei, Cambodia	48	1.4	39	133	0.9	0.2	3	0.08	<LOD	0.1	0.07	0.02	0.7	0.08	0.2	0.03	<LOD	<LOD
AB2801	Angkor Borei, Cambodia	43	1.3	42	138	0.3	0.3	5	0.01	<LOD	0.1	0.2	0.05	0.4	0.06	0.1	0.06	0.05	<LOD
AB3284	Angkor Borei, Cambodia	20	0.5	29	124	0.6	2.1	13	0.11	<LOD	0.1	<LOD	<LOD	7.2	0.3	0.5	0.06	0.004	<LOD
AB3333	Angkor Borei, Cambodia	39	0.1	25	114	0.3	0.3	2	0.01	<LOD	0.1	<LOD	0.005	0.4	0.05	0.1	0.02	0.001	<LOD
AKC00310	Village 10.8, Cambodia	21	43	64	37	<LOD	0.3	52	<LOD	<LOD	0.1	0.1	<LOD	<LOD	0.2	0.4	0.1	0.003	<LOD
AKC00311	Village 10.8, Cambodia	45	38	67	42	<LOD	0.5	51	<LOD	<LOD	0.1	0.1	<LOD	<LOD	0.15	0.4	0.1	0.009	<LOD
AKC00338	Village 10.8, Cambodia	27	42	64	37	<LOD	0.4	54	<LOD	<LOD	0.1	0.1	<LOD	<LOD	0.06	0.3	0.1	<LOD	<LOD
AKC00339	Village 10.8, Cambodia	61	49	63	35	<LOD	0.5	57	<LOD	0.07	0.1	0.2	<LOD	0.3	0.1	0.3	0.1	0.004	0.12
AKC00590	Prohear, Cambodia	11	25	69	43	0.7	0.6	39	<LOD	<LOD	0.1	0.2	0.09	1.2	0.2	0.5	0.2	0.04	<LOD
AKC00606	Prohear, Cambodia	12	23	65	45	2.3	1.2	51	0.3	0.09	0.1	0.2	0.2	2.7	0.5	0.9	0.3	0.03	<LOD
AKC00651	Bit Meas, Cambodia	17	32	62	34	<LOD	0.3	45	<LOD	<LOD	0.1	0.1	<LOD	0.1	0.06	0.3	0.09	0.002	0.01
AKC00731	Bit Meas, Cambodia	42	38	66	42	<LOD	0.3	41	<LOD	<LOD	0.1	0.2	<LOD	<LOD	0.05	0.3	0.09	<LOD	<LOD
PKTgarnet	Phu Khao Thong, Thailand	176	3.4	62	27	<LOD	<LOD	19	<LOD	<LOD	<LOD	0.2	<LOD	<LOD	0.002	0.05	<LOD	<LOD	<LOD
PorunthalGarnet	Porunthal, South India	29	0.9	25	83	0.6	0.8	3.2	<LOD	<LOD	0.06	<LOD	<LOD	3.2	<LOD	0.1	0.01	<LOD	<LOD

(continued)

Database ID	Location	Y	Bi	U	W	Mo	Nd	Sm	Eu	Gd	Tb	Dy	Ho	Er	Tm	Yb	Lu	Hf	Th
Anthill1	Anthill, Arizona	28	0.01	0.008	<LOD	0.05	2.0	2.1	1.2	3.6	0.7	4.9	1.1	3.2	0.5	3.6	0.5	0.1	0.31
Anthill2	Anthill, Arizona	13	<LOD	0.003	<LOD	<LOD	0.2	0.2	0.1	0.6	0.2	1.8	0.5	1.7	0.3	2.1	0.3	0.1	0.03
Anthill3	Anthill, Arizona	14	<LOD	0.02	<LOD	<LOD	3.1	2.3	1.1	2.1	0.4	2.4	0.5	1.5	0.2	1.4	0.2	0.1	0.05
Anthill4	Anthill, Arizona	7.8	<LOD	<LOD	<LOD	<LOD	0.1	0.2	0.1	0.4	0.1	1.2	0.3	0.9	0.1	1.1	0.2	<LOD	0.03
Anthill5	Anthill, Arizona	31	<LOD	<LOD	<LOD	<LOD	0.4	0.3	0.2	0.9	0.3	3.5	1.2	4.3	0.7	6.1	1.0	0.1	<LOD
Anthill6	Anthill, Arizona	7.4	0.02	0.009	0.06	<LOD	3.9	1.5	0.9	1.6	0.2	1.5	0.3	0.7	0.1	0.8	0.1	0.1	0.36
IndiaGarnet1	Unknown source, India	129	0.04	0.02	<LOD	<LOD	0.0	0.1	0.2	1.3	0.9	14	4.5	14.0	2.1	13	1.9	0.2	0.01
IndiaGarnet2	Unknown source, India	116	0.01	0.03	<LOD	<LOD	0.5	2.1	0.5	11.2	2.8	19	4.2	11.1	1.6	10	1.4	0.4	<LOD
IndiaGarnet3	Unknown source, India	110	0.01	<LOD	<LOD	0.02	0.05	0.2	0.1	1.5	0.8	11	3.5	10.7	1.5	9.1	1.5	0.2	<LOD
IndiaGarnet4	Unknown source, India	84	<LOD	<LOD	<LOD	<LOD	<LOD	0.1	0.0	0.9	0.6	7.9	2.5	8.8	1.4	10	1.6	0.1	<LOD
IndiaGarnet5	Unknown source, India	142	<LOD	0.2	<LOD	<LOD	0.07	0.6	0.2	6.9	2.5	22	5.0	13.7	2.1	14	2.0	3	<LOD
IndiaGarnet6	Unknown source, India	108	0.03	0.01	0.01	<LOD	<LOD	0.2	0.1	2.3	1.2	14	3.7	10.9	1.7	11	1.6	0.3	<LOD
IndiaGarnet7	Unknown source, India	281	0.01	<LOD	<LOD	<LOD	0.06	0.6	0.3	8.6	4.0	40	8.9	22.6	3.0	19	2.3	0.6	<LOD
IndiaGarnet8	Unknown source, India	131	<LOD	<LOD	<LOD	<LOD	<LOD	0.1	0.07	1.6	1.1	15.2	5.2	18.6	3.1	22	3.4	0.3	<LOD
IndiaGarnet9	Unknown source, India	81	0.00	<LOD	<LOD	<LOD	<LOD	0.1	0.05	0.9	0.6	8.8	3.1	10.6	1.7	12	1.6	0.1	<LOD
LamDong1	Lam Dong, Vietnam	508	0.03	0.13	0.05	<LOD	1.7	6.5	<LOD	20	7.4	59	13	41	8.3	74	11	0.6	0.01
LamDong2	Lam Dong, Vietnam	632	0.07	0.72	2.9	0.3	5.5	15.3	<LOD	37	11.5	81	17	51	10	96	15	4.5	0.10
Mozambique1	Unknown source, Mozambique	21	0.04	0.01	<LOD	0.01	0.3	0.4	0.3	1.2	0.3	3.1	0.8	2.4	0.4	2.9	0.5	0.4	<LOD
Mozambique2	Unknown source, Mozambique	24	0.005	<LOD	<LOD	<LOD	0.3	0.4	0.3	1.2	0.4	3.3	0.9	2.8	0.5	3.5	0.6	0.5	0.01
Mozambique3	Unknown source, Mozambique	29	0.01	<LOD	<LOD	0.01	0.3	0.6	0.3	1.6	0.4	4.2	1.1	3.5	0.5	4.2	0.7	0.4	<LOD
Mozambique4	Unknown source, Mozambique	22	<LOD	<LOD	<LOD	<LOD	0.2	0.3	0.3	1.1	0.3	2.9	0.8	2.7	0.4	3.3	0.5	0.4	<LOD
Mozambique5	Unknown source, Mozambique	20	0.05	0.05	0.04	0.02	0.3	0.4	0.3	1.2	0.4	3.0	0.8	2.4	0.4	2.7	0.5	0.5	0.04
Mozambique6	Unknown source, Mozambique	22	<LOD	<LOD	<LOD	0.01	0.2	0.3	0.2	1.3	0.3	3.3	0.8	2.4	0.4	2.6	0.4	0.5	<LOD
Orissa1	Unknown source, Orissa, India	47	<LOD	0.002	<LOD	<LOD	0.7	2.3	1.1	7.5	1.5	9.4	1.9	4.9	0.7	4.8	0.7	0.6	<LOD
Orissa2	Unknown source, Orissa, India	253	<LOD	<LOD	<LOD	0.03	1.0	2.8	0.6	12	3.9	38	9.7	29.3	4.2	28	3.9	0.3	<LOD
Orissa3	Unknown source, Orissa, India	220	<LOD	0.05	<LOD	0.01	3.2	6.2	1.3	17	4.6	38	9.0	26.8	3.9	26	3.6	0.6	<LOD
Orissa4	Unknown source, Orissa, India	80	<LOD	<LOD	<LOD	0.04	1.7	4.1	1.9	9.3	1.8	13	2.7	7.2	1.0	6.3	0.9	1.4	<LOD
Orissa5	Unknown source, Orissa, India	68	<LOD	0.03	<LOD	<LOD	0.5	0.7	0.5	3.0	1.0	9.2	2.5	8.2	1.3	9.6	1.6	0.5	<LOD
Orissa6	Unknown source, Orissa, India	204	<LOD	0.03	<LOD	<LOD	1.5	3.9	1.0	15	4.3	36	8.3	24.7	3.7	27	4.0	0.4	<LOD
Orissa7	Unknown source, Orissa, India	31	<LOD	<LOD	<LOD	<LOD	2.5	3.5	1.5	4.9	0.9	5.9	1.2	3.3	0.5	3.1	0.5	1.0	<LOD
Orissa8	Unknown source, Orissa, India	268	<LOD	<LOD	<LOD	0.06	0.5	1.5	0.3	8.3	3.5	34	9.0	29.8	5.0	37	5.8	0.2	<LOD
Orissa9	Unknown source, Orissa, India	80	<LOD	<LOD	<LOD	<LOD	0.5	2.0	0.6	10	2.5	16	3.1	8.3	1.2	8.9	1.3	0.2	<LOD
Orissa10	Unknown source, Orissa, India	44	<LOD	<LOD	<LOD	<LOD	2.6	3.9	1.9	5.5	1.1	7.9	1.9	5.8	0.9	7.0	1	1.0	<LOD
PhuThoGarnet1	Phu Tho, Vietnam	508	0.08	0.01	1.8	0.06	1.6	0.9	0.2	4.4	2.7	46	20	99	21	188	35	0.2	0.51
PhuThoGarnet2	Phu Tho, Vietnam	193	0.02	<LOD	2.2	0.03	0.3	1.0	0.1	7.2	3.2	30	6.2	15	1.9	12	2	0.2	0.01
PhuThoGarnet3	Phu Tho, Vietnam	703	0.2	0.2	0.7	0.14	0.5	2.2	0.9	18	8.0	89	24	66	8.7	51	6	0.8	0.20
PhuThoGarnet4	Phu Tho, Vietnam	42	0.05	<LOD	1.1	0.04	0.7	3.7	1.3	9.3	1.5	7.8	1.3	2.7	0.3	1.8	0.2	0.3	0.11
PhuThoGarnet5	Phu Tho, Vietnam	60	<LOD	<LOD	0.4	0.01	0.5	1.5	0.7	6.3	1.5	11	2.0	4.5	0.5	3.0	0.3	0.4	0.17
PhuThoGarnet6	Phu Tho, Vietnam	16	<LOD	<LOD	0.6	0.01	0.3	1.0	0.6	3.9	0.8	3.8	0.6	1.2	0.1	1.3	0.2	0.2	0.07
PhuThoGarnet7	Phu Tho, Vietnam	70	0.01	<LOD	0.2	0.02	0.2	1.4	0.6	7.7	2.0	13	2.1	4.2	0.5	2.5	0.3	0.1	0.02
PhuThoGarnet8	Phu Tho, Vietnam	339	<LOD	<LOD	1.7	0.01	0.5	2.6	1.3	12	3.5	37	13	56	11	95	14	0.2	<LOD

PhuThoGarnet9	Phu Tho, Vietnam	4.4	0.07	0.06	1.1	0.02	0.5	1.2	0.4	1.4	0.3	0.8	0.2	0.6	0.1	1.2	0.2	0.3	0.11
PhuThoGarnet10	Phu Tho, Vietnam	89	<LOD	<LOD	1.3	0.04	0.4	1.1	0.7	2.6	0.9	10	3.1	11	1.8	14	2	0.2	0.002
SouthIndia1	Unknown source, South India	27	<LOD	0.2	<LOD	0.06	0.2	0.4	0.3	1.9	0.6	5.3	1.1	2.7	0.4	2.4	0.4	1.0	0.01
SouthIndia2	Unknown source, South India	160	<LOD	0.04	<LOD	<LOD	<LOD	0.3	0.3	2.9	1.5	20	7.0	28	5.1	41.2	6.9	0.2	0.04
SouthIndia3	Unknown source, South India	64	<LOD	<LOD	<LOD	0.09	0.1	0.4	0.3	3.3	1.4	12	2.2	5.5	0.7	4.8	0.7	0.2	<LOD
SouthIndia4	Unknown source, South India	52	<LOD	0.3	<LOD	0.07	0.02	0.2	0.2	2.2	0.9	8.5	2.0	5.5	0.7	4.3	0.7	1.2	<LOD
SouthIndia5	Unknown source, South India	9	0.01	0.1	<LOD	0.03	0.2	0.8	0.6	3.7	0.6	2.4	0.4	1.0	0.1	1.1	0.2	0.9	0.02
SouthIndia6	Unknown source, South India	78	<LOD	0.9	<LOD	<LOD	0.1	0.5	0.3	4.1	1.7	14	3.1	9.0	1.4	11.0	1.7	4.2	0.49
SouthIndia7	Unknown source, South India	60	<LOD	0.5	<LOD	<LOD	0.1	1.0	0.5	4.9	1.4	12	2.4	6.5	0.9	6.6	1.0	1.1	0.71
SouthIndia8	Unknown source, South India	18	0.02	0.09	0.01	0.09	0.7	0.7	0.4	3.3	0.9	4.6	0.8	1.7	0.2	1.6	0.3	0.2	0.05
SouthIndia9	Unknown source, South India	4.7	<LOD	0.02	<LOD	0.04	0.1	0.6	0.6	2.8	0.5	1.7	0.2	0.4	0.1	0.4	0.1	0.4	<LOD
SouthIndia10	Unknown source, South India	29	<LOD	<LOD	<LOD	0.06	0.2	0.7	0.5	4.0	0.9	5.6	1.1	2.6	0.3	2.3	0.3	0.1	<LOD
SriLankaGroup1_1	Unknown source, Sri Lanka	1148	0.09	0.7	0.1	<LOD	0.5	0.8	0.4	12	7.9	131	52	181	27	162	20	2.2	0.24
SriLankaGroup1_2	Unknown source, Sri Lanka	622	0.004	0.01	0.02	<LOD	0.1	0.6	0.3	8.3	4.9	68	22	74	12	85	12	0.1	0.04
SriLankaGroup1_3	Unknown source, Sri Lanka	459	0.002	0.01	0.2	<LOD	0.01	0.3	0.2	4.5	2.8	45	18	75	15	116	19	0.1	<LOD
SriLankaGroup1_4	Unknown source, Sri Lanka	993	<LOD	<LOD	<LOD	<LOD	0.05	0.5	0.3	9.0	6.0	103	42	171	30	221	33	0.1	<LOD
SriLankaGroup1_5	Unknown source, Sri Lanka	525	<LOD	<LOD	<LOD	<LOD	0.03	0.4	0.2	8.0	4.9	67	19	53	7.3	44	5	0.1	<LOD
SriLankaGroup1_6	Unknown source, Sri Lanka	617	0.01	0.03	0.2	<LOD	0.05	0.5	0.3	9.0	5.5	78	21	56	7.3	42	4	0.1	0.004
SriLankaGroup1_7	Unknown source, Sri Lanka	322	0.004	<LOD	<LOD	<LOD	0.06	0.6	0.3	9.9	5.4	57	12	25	3.2	20	3	0.1	<LOD
SriLankaGroup1_8	Unknown source, Sri Lanka	428	<LOD	<LOD	<LOD	<LOD	0.01	0.2	0.1	4.3	2.9	48	19	72	12	89	14	0.1	<LOD
SriLankaGroup1_9	Unknown source, Sri Lanka	589	<LOD	<LOD	<LOD	<LOD	0.06	0.8	0.4	13	6.7	80	22	68	10	67	9	0.2	<LOD
SriLankaGroup1_10	Unknown source, Sri Lanka	658	<LOD	0.3	3.4	<LOD	0.06	0.8	0.3	13	7.4	91	22	54	7.1	41	4.4	1.1	<LOD
SriLankaGroup2_1	Unknown source, Sri Lanka	65	0.01	0.1	<LOD	<LOD	0.2	1.0	0.9	4.4	1.2	8.5	1.7	4.5	0.7	4.6	0.6	0.1	<LOD
SriLankaGroup2_2	Unknown source, Sri Lanka	19	0.02	0.07	<LOD	<LOD	0.3	1.0	0.5	3.2	0.7	3.1	0.4	0.9	0.1	1.0	0.1	0.2	0.01
SriLankaGroup2_3	Unknown source, Sri Lanka	116	0.01	0.03	<LOD	<LOD	0.2	1.1	0.7	3.9	1.4	13	3.3	10	1.8	14	1.9	0.2	<LOD
SriLankaGroup2_4	Unknown source, Sri Lanka	172	<LOD	0.008	<LOD	<LOD	0.05	0.5	0.2	3.7	1.6	17	4.6	16	3.0	24	3.6	0.2	<LOD
SriLankaGroup2_5	Unknown source, Sri Lanka	109	0.01	<LOD	<LOD	<LOD	0.06	0.4	0.2	3.4	1.3	11	2.8	8.5	1.4	11	1.6	0.2	<LOD
SriLankaGroup2_6	Unknown source, Sri Lanka	72	0.005	0.2	<LOD	<LOD	0.3	2.5	1.7	7.4	1.7	10	1.9	4.6	0.7	4.8	0.6	0.2	<LOD
AB1767	Angkor Borei, Cambodia	184	0.02	0.04	<LOD	0.08	0.08	0.1	0.1	1.7	1.2	18	6.9	25	5.1	39	5.5	0.1	<LOD
AB1835	Angkor Borei, Cambodia	273	0.02	0.01	<LOD	0.10	0.1	0.1	0.1	2.1	1.8	30	12	32	4.5	25	3.2	0.0	0.06
AB2801	Angkor Borei, Cambodia	258	0.05	0.06	0.03	0.08	0.06	0.2	0.1	1.9	1.5	24	11	36	6.1	40	6.2	0.1	0.05
AB3284	Angkor Borei, Cambodia	232	<LOD	0.2	0.01	0.05	0.2	0.2	0.1	3.3	2.1	28	9.1	30	5.7	42	6.4	0.3	0.16
AB3333	Angkor Borei, Cambodia	318	0.04	0.03	<LOD	0.04	0.05	0.1	0.1	2.5	2.0	30	11	29	4.1	24	2.8	0.03	0.003
AKC00310	Village 10.8, Cambodia	89	<LOD	<LOD	<LOD	<LOD	1.3	1.5	0.8	4.0	1.1	10	2.8	9.5	1.6	12	1.9	1.1	0.04
AKC00311	Village 10.8, Cambodia	101	<LOD	<LOD	<LOD	<LOD	1.1	1.4	0.8	4.4	1.2	12	3.2	11	1.7	12	1.9	1.0	0.01
AKC00338	Village 10.8, Cambodia	86	0.005	<LOD	<LOD	<LOD	1.2	1.5	0.8	4.0	1.0	10	2.8	9.3	1.5	11	1.8	1.1	<LOD
AKC00339	Village 10.8, Cambodia	62	<LOD	0.01	<LOD	<LOD	1.2	1.2	0.7	3.5	0.9	8	2.1	6.4	1.0	6.7	1.0	1.2	0.03
AKC00590	Prohear, Cambodia	94	0.05	0.03	0.03	0.07	1.1	1.3	0.8	4.2	1.2	11	3.0	9.2	1.4	9.1	1.3	0.8	0.07
AKC00606	Prohear, Cambodia	99	0.02	0.05	0.04	0.06	1.4	1.5	0.9	4.7	1.3	12	3.1	10	1.5	10	1.4	1.0	0.22
AKC00651	Bit Meas, Cambodia	87	<LOD	<LOD	<LOD	<LOD	0.9	1.1	0.7	3.6	1.0	10	2.7	8.9	1.4	10	1.5	0.9	0.01
AKC00731	Bit Meas, Cambodia	96	0.01	<LOD	<LOD	0.03	1.0	1.3	0.7	4.2	1.2	11	3.1	11	1.7	12	1.9	0.9	<LOD
PKTgarnet	Phu Khao Thong, Thailand	17	<LOD	<LOD	<LOD	<LOD	0.06	0.1	0.2	0.7	0.2	1.9	0.5	1.8	0.3	2.4	0.4	0.2	<LOD
PorunthalGarnet	Porunthal, South India	267	0.004	<LOD	<LOD	<LOD	0.02	0.1	0.1	2.4	1.9	29	11	36	6.3	40	5.8	0.01	<LOD

References

Albrecht G, Haidle M, Sivleng C, Hong H, Sophady H, Than H, Someaphyvath M, Kada S, Sophal S, Chanthourn T, Laychour V (2001) Circular earthwork Krek 52/62: recent research on the prehistory of Cambodia. Asian Perspect 39:20–46

Aranyakanon P (1983) The gem deposits of Thailand. Department of Mineral Resources, Bangkok

Bauer M (1968) Precious stones. Dover Publications, New York, NY

Bellina B (2003) Beads, social change and interaction between India and South-east Asia. Antiquity 77:285–297

Bellina B (2007) Cultural exchange between India and Southeast Asia: production and distribution of hard stone ornaments (VI c. BC–VI c. AD). Éditions de la Maison des Sciences de l'Homme, Paris

Bellina B (2014) Maritime silk roads' ornament industries: socio-political practices and cultural transfers in the South China Sea. Camb Archaeol J 24:345–377

Bimson M, La Neice S, Leese M (1982) The characterisation of mounted garnets. Archaeometry 24:51–58

Calligaro T, Colinart S, Poirot JP, Sudres C (2002) Combined external-beam PIXE and μ-Raman characterisation of garnets used in Merovingian jewellery. Nucl Instrum Meth B 189:320–327

Carter A (2012) Garnet beads in Southeast Asia: evidence for local production? In: Bonatz D, Reinecke A, Tjoa-Bonatz ML (eds) Crossing borders in Southeast Asian archaeology. Selected papers from the 13th international conference of the European association of Southeast Asian archaeologists, Berlin, 2010. NUS, Singapore, p 296–306

Carter AK (2013) Trade, exchange, and, sociopolitical development in Iron Age (500 BC – AD 500) mainland Southeast Asia: an examination of stone and glass beads from Cambodia and Thailand. Ph.D. Dissertation, University of Wisconsin-Madison

Carter AK (2015) Beads, exchange networks and emerging complexity: a case study from Cambodia and Thailand (500 BCE–CE 500). Camb Archaeol J 25(4):733–757

Carter AK, Dussubieux L (2016) Geologic provenience analysis of agate and carnelian beads using laser ablation-inductively coupled plasma-mass spectrometry (LA-ICP-MS): a case study from Iron Age Cambodia and Thailand. J Archaeol Sci Rep 6:321–331

Chaisuwan B (2011) Early contacts between India and the Andaman Coast in Thailand from the second century BCE to eleventh century CE. In: Manguin PY, Mani A, Wade G (eds) Early interactions Between South and Southeast Asia. Reflections on cross-cultural exchange. ISEAS Publishing, Singapore, pp 82–112

Chrea V, Sieng S, Mak B, Tep V, Phok S, Mom S (1999) Mineral deposits in Cambodia. General Department of Mineral Resources, Ministry of Industry, Mines, and Energy, Phnom Penh

Coningham R, Burroni D, Donahue R, Ford L (2006) Stone objects. In: Coningham R (ed) Anuradhapura: the British-Sri Lankan excavations at Anuradhapura Salgaha Watta 2. Archaeopress, Oxford, pp 377–413

Dawson JB, Stephens WE (1975) Statistical classification of garnets from kimberlite and associated xenoliths. J Geol 83:589–607

Deer WA, Howie RA, Zussman J (1982) Rock forming minerals, vol 1A: orthosilicates. Longman Group, London

Dissanayake CB, Chandrajith R, Tobschall HJ (2000) The geology, mineralogy and rare element geochemistry of the gem deposits of Sri Lanka. Bull Geol Soc Finl 72:5–20

Dussubieux L (2001) L'Apport de l'ablation laser couplée a l'ICP-MS à la caractérisation des verres: Application a l'étude du verre de l'océan Indien. Ph.D. Dissertation, Université d'Orléans

Dussubieux L, Golitko M, Williams PR, Speakman R (2007) Laser ablation-inductively coupled plasma-mass spectrometry analysis applied to the characterization of Peruvian Wari ceramics. In: Glascock M, Speakman R, Popelka-Filcoff R (eds) Archaeological chemistry: analytical techniques and archaeological interpretation. American Chemical Society, Washington, DC, pp 349–363

Dussubieux L, Robertshaw P, Glascock M (2009) LA-ICP-MS analysis of African glass beads: laboratory inter-comparison with an emphasis on the impact of corrosion on data interpretation. Int J Mass Spectrom 284:152–161

Farges F (1998) Mineralogy of the Louvres Merovingian garnet cloisonné jewelry; origins of the gems of the first kings of France. Am Mineral 83:323

Francis P Jr (2002) Asia's maritime bead trade: 300 B.C. to the present. University of Hawaii Press, Honolulu, HI

Francis P Jr (2004) Beads and selected small finds from the 1989–1992 excavations. In: Begley V (ed) The ancient port of Arikamedu: new excavations and researches 1989–1992, vol 2. Ecole française d'Extrême-Orient, Paris, pp 447–604

Glover I (1996) The southern silk road: archaeological evidence for early trade between India and Southeast Asia. In: Chutiwongs N (ed) Ancient trades and cultural contacts in Southeast Asia. The Office of the National Culture Commission, Bangkok, pp 57–94

Gratuze B (1999) Obsidian characterization by laser ablation ICP-MS and its application to prehistoric trade in the Mediterranean and the Near East: sources and distribution of obsidian within the Aegean and Anatolia. J Archaeol Sci 26:869–881

Gunawardene M, Rupasinghe M (1986) The Elahera gem field in central Sri Lanka. Gems Gemol 22:80–95

Gwinnett AJ, Gorelick L (1986) Evidence for the use of a diamond drill for bead making in Sri-Lanka c. 700–1000 A.D. Scan Electron Micros II:473–477

Hannibal-Deraniyagala A (2001) Beads from Tissamaharama: a typology of Sri Lankan glass and semi-precious stone beads. In: Weisshaar H-J, Roth H, Wijeyapala W (eds) Ancient Ruhuna. Sri Lankan-German archaeological project in the southern province, vol 1. Von Zabern, Mainz, pp 203–226

Hannibal-Deraniyagala A (2005) Beads from Anuradhapura and Tissmaharama, Sri Lanka: trade contacts in the early historic period. J Indian Ocean Archaeol 2:21–24

Hickmott DD (1988) Trace element zoning in garnets: implications for metamorphic processes. Ph.D. Dissertation, Massachusetts Institute for Technology

Hickmott DD, Shimizu N, Spear FS, Selverstone J (1987) Trace-element zoning in a metamorphic garnet. Geology 15(6):573–576

Higham C (2014) Early mainland Southeast Asia: from first humans to Angkor. River Books, Bangkok

Hung H, Lizuka Y, Bellwood P, Nguyen KD, Bellina B, Silapanth P, Dizon E, Santiago R, Datan I, Manton J (2007) Ancient jades map 3000 years of prehistoric exchange in Southeast Asia. Proc Natl Acad Sci USA 104:19745–19750

Kane R, McClure S, Kammerling R, Nguyen D, Mora C, Repetto S, Nguyen D, Koivula J (1991) Rubies and fancy sapphires from Vietnam. Gem Gemol 2001 (Fall):136–155

Kenoyer J, Vidale M (1992) A new look at stone drills of the Indus Valley Tradition. In: Vandiver P, Druzick JR, Wheeler GS, Freestone I (eds) Materials issues in art and archaeology, vol III. Materials Research Society, Pittsburgh, PA, pp 495–518

Kenoyer J, Vidale M, Bhan KK (1991) Contemporary stone bead making in Khambhat India: patterns of craft specialization and organization of production as reflected in the archaeological record. World Archaeol 23:44–63

Lankton J, Dussubieux L (2006) Early glass in Asian maritime trade: a review and an interpretation of compositional analysis. J Glass Stud 48:121–144

Lankton J, Dussubieux L (2013) Early glass in Southeast Asia. In: Janssens K (ed) Modern methods for analysing archaeological and historic glass. Wiley, Chichester, pp 415–457

Lankton J, Dussubieux L, Gratuze B (2008) Glass from Khao Sam Kaeo: transferred technology for an early Southeast Asian exchange network. Bull l'École FR d'Extrême-Orient 93:317–351

Locock AJ (2008) An excel spreadsheet to recast analyses of garnet into end-member components, and a synopsis of the crystal chemistry of natural silicate garnets. Comput Geosci 34(12):1769–1780

Malleret L (1962) L'Archéologie du Delta du Mékong, Part 3, La Culture du Fou-Nan. École Française d'Extrême-Orient, Paris

Mathis F, Vrielynck O, Laclavetine K, Chene G, Strivay D (2008) Study of the provenance of Belgian Merovingian garnets by PIXE at IPNAS cyclotron. Nucl Instrum Meth B 266:2348–2352

Morton A (1985) A new approach to provenance studies: electron microprobe analysis of detrital garnets from Middle Jurassic sandstones of the northern North Sea. Sedimentology 32:553–556

Nguyen KD (1996) The Trang Kenh jewellery workshop site: an experimental and microwear study. Bull Indo Pac Pre Hist 14:161–165

Nguyen KD (2001) Jewellery from late prehistoric sites recently excavated in South Vietnam. Bull Indo Pac Pre Hist 21:107–113

Perin P, Calligaro T, Vallet F, Poirot JP, Bagault D (2007) Provenancing Merovingian garnets by PIXE and μ-Raman spectrometry. In: Henning J (ed) Post-Roman towns, trade and settlement in Europe and Byzantium, vol 1, The heirs of the Roman West. Walter de Gruyter, Berlin, pp 69–76

Quast V, Schüssler U (2000) Mineralogische Untersuchungen zur Herkunft der Granate merowingerzeitlicher Cloisonnéarbeiten. Germania 78:75–96

Rajan K (2009) Archaeological excavations at Porunthal 2009. Pondicherry University, Pondicherry

Rösch C, Hock R, Schüssler U, Yule P, Hannibal A (1997) Electron microprobe analysis and X-ray diffraction methods in archaeometry: investigations on ancient beads from the Sultanate of Oman and from Sri Lanka European. J Mineral 9:763–783

Schüssler U, Rösch C, Hock R (2001) Beads from ancient Sri Lanka—first results of a systematic material analysis. In: Weisshaar H-J, Roth H, Wijeyapala W (eds) Ancient Ruhuna. Sri Lankan-German archaeological project in the southern province, vol 1. Von Zabern, Mainz, pp 227–242

Shigley J, Laurs B, Janse AJA, Elen S, Dirlam D (2010) Gem localities of the 2000s. Gem Gemol 2010 (Fall):188–216

Spear FS, Kohn MJ (1996) Trace element zoning in garnet as a monitor of crustal melting. Geology 24 (12):1099–1102

Stark M (2001) Some preliminary results of the 1999–2000 archaeological field investigations at Angkor Borei, Takeo Province. Udaya, J Khmer Stud 2:19–35

Stark M (2004) Pre-Angkorian and Angkorian Cambodia. In: Glover I, Bellwood P (eds) Southeast Asia: from prehistory to history. Routledge Curzon, London, pp 89–119

Stockton C, Manson V (1985) A proposed new classification for gem-quality garnets. Gem Gemol 1985:205–218

Theunissen R, Grave P, Bailey G (2000) Doubts on diffusion: challenging the assumed Indian origin of Iron Age agate and carnelian beads in Southeast Asia. World Archaeol 32:84–105

Tracy RJ, Robinson P, Thompson AB (1976) Garnet composition and zoning in the determination of temperature and pressure of metamorphism, Central Massachusetts. Am Mineral 61:762–775

Van Long P, Giuliani G, Garnier V, Ohnenstetter D (2004) Gemstones in Vietnam: a review. Austral Gemmol 22:162–168

Velde B, Courtois L (1983) Yellow garnets in roman amphorae: a possible tracer of ancient commerce. J Archaeol Sci 10:531–539

Vidale M, Miller H (2000) On the development of Indus technical virtuosity and its relation to social structure. In: Taddei M, De Marco G (eds) South Asian archaeology 1997. IsIAO, Rome, pp 115–132

Wright WI (1938) The composition and occurrence of garnets. Am Mineral 23:436–449

Zwaan PC (1982) Sri Lanka: the gem Island. Gem Gemol 18:62–71

Application of LA-ICP-MS to Black Stone Objects Used During the Iron Age in Celtic Europe

17

Anne Baron and Bernard Gratuze

Abstract

Black stones—"mixed rocks" composed of both an organic and a mineral fraction—were widely used by ancient populations to make ornamental objects. During the Iron Age in Europe, the Celts made rings, bracelets, beads, and carved vessels out of these rocks. The identification of the geological resources used for the production of these objects is essential for evaluating the distance traveled by these raw materials and for reconstructing ancient exchange networks. The archaeological objects are difficult to characterize due to their heterogeneity, and their characterization is not as advanced as that of other stones used by ancient populations. Characterization of geological samples using LA-ICP-MS has created a preliminary reference database of geological materials, which allowed for the analysis of more than 200 objects from different archaeological sites in Europe. It has thus been possible to demonstrate the great heterogeneity of archaeological materials used and to document a shift in raw material usage between the Hallstatt and La Tène periods. Two different acquisition patterns are proposed: in the first, raw material is transported to manufacturing sites which can be located far from the extraction site, while in the second, production occurs directly at the extraction site. These results allow us to present the distribution network of black stones in Europe. This new approach using LA-ICP-MS is a first step for characterizing black stone artifacts in order to give a first estimation of the chemical composition of these stones. Due to its non-destructiveness, as well as its potential for analyzing large objects

A. Baron (✉)
Chercheur associé, Institut de Recherche sur les Archéomatériaux, Centre Ernest Babelon, UMR 5060 CNRS/Université d'Orléans, 3D rue de la Férollerie, 45071 Orléans Cedex 2, France
e-mail: anne.baron@netcourrier.com

B. Gratuze
Institut de Recherche sur les Archéomatériaux, Centre Ernest Babelon, CNRS/Université d'Orléans, 3D rue de la Férollerie, 45071 Orléans Cedex 2, France
e-mail: gratuze@cnrs-orleans.fr

L. Dussubieux et al. (eds.), *Recent Advances in Laser Ablation ICP-MS for Archaeology*, Natural Science in Archaeology, DOI 10.1007/978-3-662-49894-1_17

such as bracelet or vessel fragments, LA-ICP-MS constitutes a useful characterization tool for black stone artifacts.

17.1 Introduction

During the Iron Age in Europe, black stones were used by Celtic peoples to make ornamental objects such as bracelets and rings, beads, and also carved vessels (Fig. 17.1). The large number of such artifacts at Iron Age sites in Europe makes understanding the modes of raw material acquisition, manufacturing, and reconstruction of ancient exchange networks of great interest. Relating archaeological objects to their geological sources is essential for understanding the distribution of the raw material and the production system of these artifacts, and for reconstructing the socio-economic system through which black stones were transported.

The terminology relating to these materials is complex because ancient people used different black lithic materials (Baron et al. 2007). After manufacturing and polishing, these materials have a similar appearance. Previously, archaeologists have referred to these materials as jet, lignite, or sometimes fossilized wood, because these materials are composed of among others things, decaying wood structures in organic clays which have been subjected to carbonization. The actual material utilized cannot be identified by naked eye alone. As a result, most of the objects

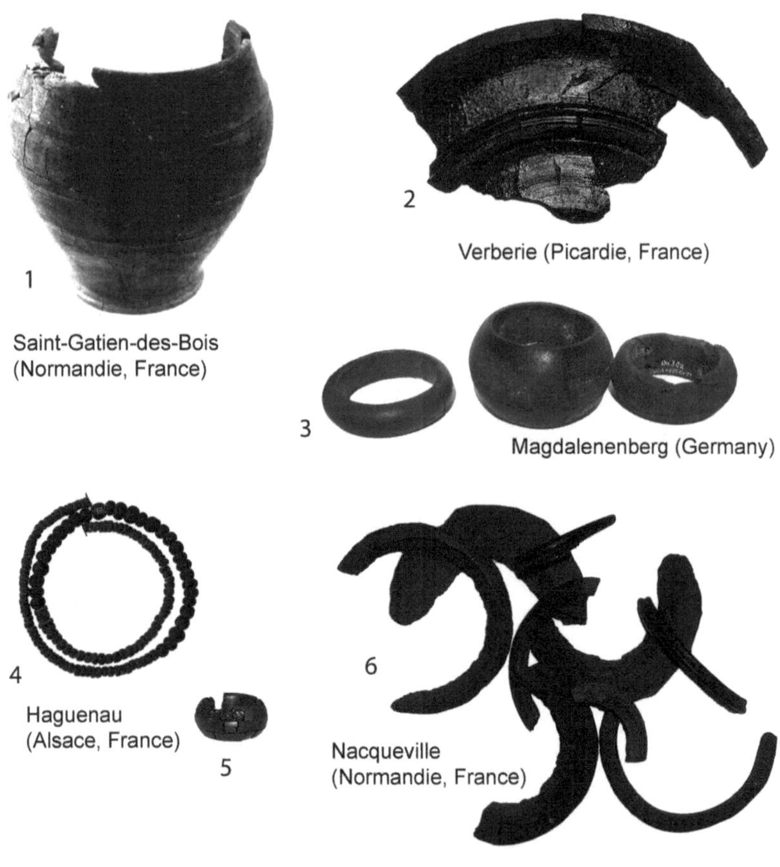

Fig. 17.1 Examples of black stone objects from Iron Age sites in Celtic Europe. (1–2) Carved vessel, La Tène period; (3) bracelets, Hallstatt period; (4) beads, Hallstatt period; (5) pendant; Hallstatt period; (6) raw material, half-fabricated and finished bracelets, La Tène period. (Photos: A. Baron, except no. 1 Musée de Normandie ©)

are often inventoried with several terms such as jet, lignite, sapropelit or shale in the archaeological literature and in museum collections. This problem of designation results from the variety of materials used by ancient people, but also through the lack of characterization studies of these archaeomaterials. Consequently, studies on these kinds of artifacts have often grouped several kinds of black lithic materials. To simplify the terminology and to allow for the inclusion of several types of sedimentary rocks, the general term *black stones* is proposed for all of these materials. By this, we mean black sedimentary rocks with variable organic and mineral fractions. Although their geological sources are unknown, multiple sources could exist throughout Europe (Alpern 1981; Barrabé and Feys 1965). Therefore, it is extremely difficult to characterize these objects and to identify the resources exploited.

Only a few studies have been dedicated to this topic and most present limitations. From an archaeological point of view, these studies are carried out on a very local scale and hence lack general perspective. From an analytical point of view, the use of destructive techniques on archaeological objects from museum collection constitutes a limiting factor. Additionally, data on geological reference materials are generally lacking. This geological component is important for linking archaeological objects with the geological resources exploited to produce them. Recent work has partly resolved this problem (Baron 2009).

In this paper, we propose a new methodology for characterizing these black stones by an elemental chemical approach using Laser Ablation-Inductively Coupled Plasma-Mass Spectrometry (LA-ICP-MS). This application on archaeological and geological materials is an exploratory study, and we present here the results obtained in order to reconstruct the relations between artifacts and sources. Firstly, we show how a change of material between the earlier (Hallstatt) and the later (La Tène) Iron Age resulted in modification of the techniques used for manufacturing objects from these stones. Secondly, we present different case studies of archaeological sites for which we were able to reconstruct the relations between artifacts and sources. For example, geological sources located in southern Germany were exploited to make objects recovered at Eckbolsheim (Alsace, France), while some objects discovered on the Isle of Yoc'h (Brittany, France) come from southern England, where a black stone source was exploited during the Iron Age. These different results allow us to suggest the structure of the diffusion network of black stones during the Iron Age in Europe.

17.2 Prior Characterization Studies of Black Stones

Research on the geological origin of black stone artifacts is not well developed, in contrast to sourcing studies carried out on other rock types present in archaeological contexts (Gratuze et al. 2001; Janssens and Van Grieken 2004). Only a few studies have been dedicated to this topic. The first attempts to define the origin of those artifacts were carried out in Germany between the 1960s and 1990s (Grasselt and Volkmann 1991; Ligouis 2000; Rochna 1961, 1962; Rochna and Mädler 1974; Teichmüller 1992). Since 1995, English researchers have conducted localized studies within Great Britain, specifically on Kimmeridge shale, an important raw material source exploited beginning in the Neolithic period. Between 1995 and 2000, a research program dealing with the La Tène occupation of the Lodenice region of the Czech Republic led by N. Venclová demonstrated that black stones there were imported from more than 300 km away (Venclová 2001). Different methods have been utilized to characterize and distinguish black lithic materials, especially in England where several studies have been conducted. Several techniques were used in concert including: i) characterization of the inorganic fraction using techniques such as X-Ray

Fluorescence and Scanning Electron Microscopy (Bussell et al. 1982; Davis and Sheridan 1993; Hunter et al. 1993; Pollard et al. 1981; Sheridan and Davis 2002), ii) examination of organic petrology and palynology (Allason-Jones 1996; Allason-Jones and Jones 1994, 2001; Ligouis 2000, 2006; Teichmüller 1992; Watts 1996; Zakova 2001), and iii) characterization of organic matter using techniques such as Fourier Transformed Infrared Spectroscopy (FTIR), Electron Spin Resonance (ESR), and Pyrolysis Gas Chromatography Mass Spectrometry (Py-GC/MS) and isotopy (Thierry 1999; Watts and Pollard 1997, 1998, 1999).

In her PhD thesis, Watts (1996) tested most of these methods in order to define an adapted protocol for black lithic materials: X-radiography and XRF to study their inorganic component, and FTIR, Py-GC/MS, and extractable biological markers to study their organic part. These methods were carried out on geological and archaeological samples from York and Verulamium, except Py-GC/MS, which was only used on geological samples. X-radiography and XRF allowed for differentiation between six categories of black stones: 1) Jet, 2) Jet/lignite, 3) Lignite/cannel-coal, 4) Cannel-coal, 5) Cannel-coal/oil shale, and 6) Oil shale.

The identification of such broad categories quickly reaches its limits in terms of source identification for archaeological artifacts because it is impossible to identify a unique chemical signature for different sources of the same category of black stone. In addition to using particular elements measured by XRF such as Fe or Zr to distinguish between different sources of jet and black stones materials, Watts (1996) also tried to interpret the XRF data by application of principal components analysis (PCA). This last approach was unsuccessful and compositional groups identified by PCA were not relatable to different sources. FTIR appears also to be a good technique for differentiating between jet and jet-like material, but within the same type of material, interpretation of results is more complex because spectrums are very similar. Watts also found that even if some compositional trends could be observed, not all of them applied to all kinds of materials.

Comparison of the results obtained by XRF and FTIR demonstrated the complexity of classifying these different types of black stones. Using both methods, some artifacts have been identified as a different type of black stone, thus revealing the limitations of these methods, which only provide broad categories. When categories were too difficult to assign, discriminants observed in geological samples such as high levels of Fe or Zr were applied to archaeological samples. For this reason, these techniques have been coupled with organic chemical methods in order to improve results. The Py-GC/MS and identification of extractable biomarkers have given promising results on geological specimens, but have not been applied to archaeological artifacts to date (Watts 1996; Watts and Pollard 1999). Furthermore, only a few geological samples were analyzed, meaning that a good understanding of inter-source variability on a broader geographical scale. Nevertheless, the initial results of this new approach are promising, although these techniques are destructive.

In an unpublished study, Thierry (1999) tested characterization of different black stones using carbon and nitrogen isotopy. His results have shown that $\delta^{13}C$ and $\delta^{15}N$ are good tracers for differentiating between black stones recovered from archaeological sites and those from geological sources, as well as between material from different archaeological sites. However, he found that variations in $\delta^{13}C$ and $\delta^{15}N$ exist within individual outcrops, which are difficult to estimate if the sampling is not representative. This method also requires samples measuring 5 mm or more in diameter.

These studies have revealed the problems linked to the characterization of black stones, which constitute a barrier to their compositional measurement and identification of potential geological sources:

Heterogeneity is the main problem in characterizing these materials, because the variability and distribution of chemical elements are different in different types of black stones and depend on the mode of burial and geological context. Migrations of chemical elements can also occur. Composition can be different between the beginning and the end of the same seam within a deposit. Moreover, the contribution of other inorganic elements can modify the original matrix. It is difficult to understand the origin and distribution of inorganic elements in black stone deposits.

Alteration of the Materials after their extraction, archaeological materials are affected by different climatic factors (temperature, air, oxidation) but also those linked due to the "life history" of the object (wear, burial conditions, etc.). These modifications can lead to mineral formations, alteration, or migrations of chemical elements.

Conservation some products used during artifact conservation, such as resins, for example, can alter the initial composition of black stone objects.

Destructive Methods most of the techniques used in previous studies were destructive and do not allow for the analysis of the archaeological artifacts conserved in museums.

Geological Sampling Representativeness - geological outcrops of black stones are not always sampled in their stratigraphic entirety, which does not allow for adequate determination of the compositional variability possible within rocks extracted from that outcrop. In some cases, only a few samples from each outcrop were analyzed, making it difficult to confirm how distinct they are compositionally from one another. Because concentrations can vary dramatically between the beginning and end of a particular seam, composition observed in only one sample of an outcrop may be unrepresentative.

Archaeological Sampling Representativeness archaeological burial conditions, conservation problems, and the use of destructive techniques (see above) do not allow for the analysis of all black stone objects. If all artifacts from a particular site are not analyzed, the presence of certain black stone resources may be missed.

Consequently, considerable work on the representativeness of different kinds of samples (archaeological and geological) must be undertaken before starting any research on black lithic materials in order to better understand the materials themselves.

From a geological point of view, reference data is lacking because raw material collections do not exist. Therefore, we have created and expanded a new referential database of potential European black stone sources (see Sect. 17.3.1). This database is essential for establishing connections between archaeological objects and geological sources. From an archaeological standpoint, other limitations can also be observed in these prior studies. From a geographical and chronological point of view, these studies included analysis of black stones on only a local scale or site-specific scale (except for the Lodenice project in Bohemia) and did not examine exploitation patterns over the *longue durée*. Consequently, it is difficult to reconstruct cultural or material usage over time. From an analytical point of view, the use of destructive techniques on archaeological objects from museum collections constitutes a limiting factor. For example, petrographic or palynological studies require crushing a portion of each sample. Consequently, it has not been possible to analyze a large number of artifacts, thus resulting in potential bias in the representativeness of the sampling for archaeological artifacts. Furthermore, due to the use of different analytical techniques, it has not been possible to compare the results of different studies.

These limitations require the development of an effective methodology using a minimally-destructive analytical method. Laser Ablation-Inductively Coupled Plasma-Mass Spectrometry (LA-ICP-MS) was chosen for several reasons. Firstly, its minimally destructive sampling capability (nearly non-visible) is ideal for analyzing museum collections and expanding archaeological sampling to increase the number of analyses. Secondly, a large number of elements can be analyzed, making it possible to produce a near complete chemical characterization. Thirdly, samples do not require any preparation prior to analysis, allowing for substantial time savings. Finally, use of a series of quartz cells of different sizes allows analysis of large artifacts such as bracelets or vessel fragments. LA-ICP-MS allows us to perform systematic characterization of both archaeological and geological materials, and consequently to have comparable data. The present characterization of black stones using LA-ICP-MS is an exploratory study, as this technique has never been tested on this material type before. Preliminary results are promising, however (Baron 2009, 2011, 2012; Baron et al. 2007). Its application provides an estimation of the chemical composition of black stones from geological and archaeological contexts.

17.3 Materials and Methods

17.3.1 Sampling

Samples were collected from several archaeological sites and geological sources throughout Europe (Fig. 17.2), with particular focus on archaeological sites where traces of fabrication have been identified (raw material, partially-fabricated objects, etc.). However, sites containing only finished black stone objects were also considered (Fig. 17.2). Additionally, artifact types specific to the La Tène period–vessels, plates, and

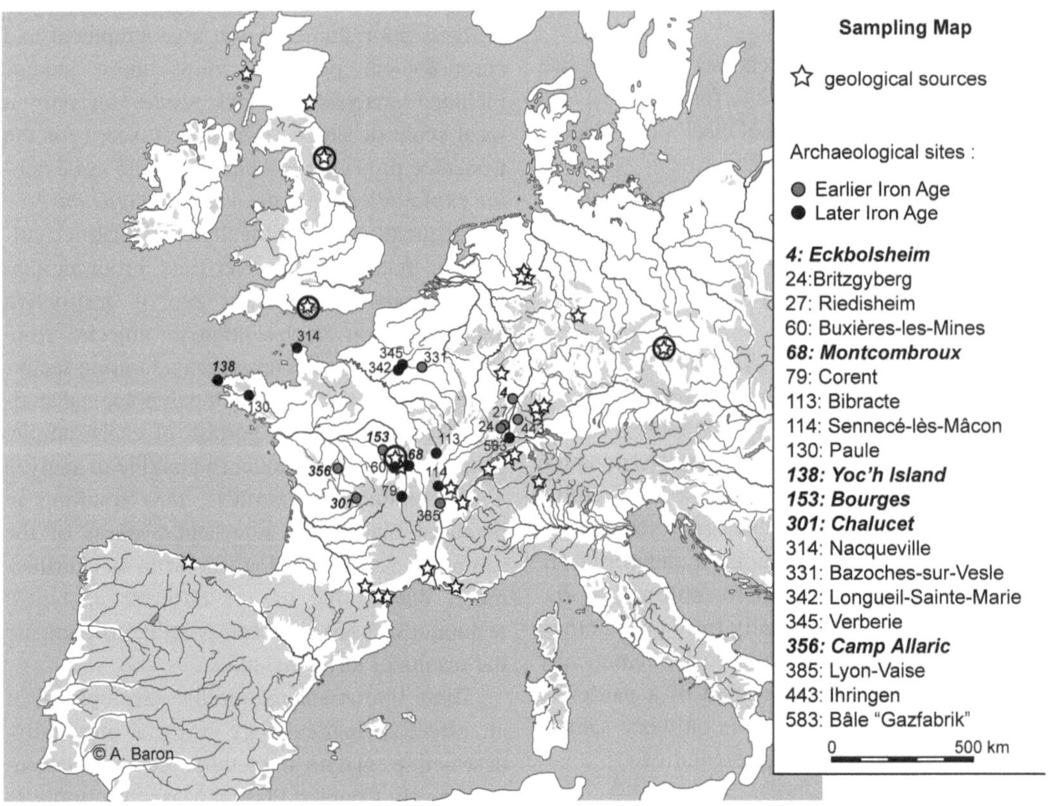

Fig. 17.2 Map of archaeological sites and geological sources analyzed using LA-ICP-MS (CAD A. Baron, Baron 2009: 30, Fig. 7). *Stars* in the *circles* indicate the sources known to have been exploited during the Iron Age. The names of the archaeological sites in bold italics are case studies presented in this paper

lids–have also been sampled. Altogether, black stones objects from 20 archaeological sites (ten sites dating to the Hallstatt period and ten sites to the La Tène period) were analyzed, representing more than 200 objects in total (Baron 2009, 2012).

Samples were collected from several different European geological sources (Fig. 17.2 and Table 17.1). The creation of this geological reference database has taken a considerable amount of time, and it has been to date impossible to sample all known potential geological sources of black stones. Therefore, collection of material from geological sources, known to have been exploited in the past and identified in prior analytical studies (Allason-Jones and Jones 1994, 2001; Bussell et al. 1982; Hunter et al. 1993; Venclová 2001) was first prioritized (see Fig. 17.2 and geological sources in bold italics in Table 17.1). Secondly, we have focused on

Table 17.1 List of geological samples analyzed using LA-ICP-MS

Country	Region	Locale	Material	Geochronology
Germany	Baden-Würtemberg	Boll	Posidonia shale	Toarcian
Germany	Baden-Würtemberg	Ohmden	Posidonia shale	Toarcian
Germany	Baden-Würtemberg	Aichelberg	Posidonia shale	Toarcian
Germany	Baden-Würtemberg	Holzmaden	Posidonia shale	Toarcian
Germany	Baden-Würtemberg	Erzingen	Posidonia shale	Toarcian
Germany	Baden-Würtemberg	Schlierbach-Bolzhaüser	Posidonia shale	Toarcian
Germany	Baden-Würtemberg	Dotternhausen	Posidonia shale	Toarcian
Germany	Baden-Würtemberg	Geislingen	Posidonia shale	Toarcian
Germany	Nordrhein-Wesphalen	Wistinghausen bei Oerlinghausen	Posidonia shale	Toarcian
Germany	Nordrhein-Wesphalen	Vehrte	Black coal	Toarcian
Germany	Saar	Saarbruck	Black coal	Unknown
Germany	Thuringen	Wandersleben	Posidonia shale	Toarcien
England	***Dorset***	***Kimmeridge***	***Blackstone/ blackshales***	***Kimmeridgian***
England	***Yorkshire***	***Whitby***	***Hard jet***	***Low Jurassic***
Scotland	Hybrides	Isle of Mull	Black shale	Unknown
Scotland	West coast	Wemyss, Bay Fife	Cannel-coal	Unknown
Spain	ASTURIES	Oles, mine of los Molinos	Jet	Kimmeridgian
France	Ain	Corbonod, mine of Orbagnoux	Bituminous shale	Kimmeridgian
France	***Allier***	***Buxières-les-Mines***	***Bituminous shale***	***Permo-carboniferous***
France	Allier	Mine of Saint-Hilaire, near of Moulins	Bituminous shale	Permian
France	Ariège	Near Foix	Vitrain	Unknown
France	Aude	Durban	Coal	Unknown
France	Aude	Rennes-les-Bains	Vitrain	Unknown
France	Bourgogne	Autun	Bituminous shale	Autunian
France	Gard	Barjac	Lignite	Eocene
France	Provence-Alpes-Côte d'Azur	Gardanne	Lignite	Unknown
France	Savoie	Voglans, Chambéry	Lignite	Unknown
Czech Republic	***Bohème***	***Mšecke Zehrovice***	***Kounov-coal***	***Westphalian***
Switzerland	Vaud	Belmont, Lutry	Lignite or jet	Unknown
Switzerland	Bern	Blapbach, Emmental	Lignite?	Unknown
Switzerland	Bern	Gondiswil	Sapropelit	Unknown

potential sources mentioned in the archaeological literature that have not been demonstrated analytically. Lastly, various other geological sources were sampled in order to confirm or disconfirm possible exploitation. Some of the geological sources were sampled during new fieldwork, while samples from others were obtained through collaboration with various geological institutions (Landesamt für Geologie in Germany, Bureau des Ressources Géologiques et Minières in France, geologic associations in France, University of Geology of Oviedo in Spain and the Scotland Museum). In total, it has been possible to collect black stones from more than 20 geological sources (Table 17.1). All samples were analyzed by LA-ICP-MS using the same protocol to produce comparable data, thus constituting the most comprehensive archaeological, geological, and analytical referential database for black stones available to date.

17.3.2 Analytical Protocol

Our method is based on the assumption that every raw material source has a specific chemical signature and can be characterized by the differential concentrations of chemical elements. For the principle of this assumption (the so-called "provenance postulate"), see previous chapters in this volume and Baron et al. (2007). Previous studies—specifically Watts's work—have shown that some black stone outcrops have a characteristic concentration of chemical elements according to their sedimentation environment. XRF analyses have shown that jet is characterized by a high zirconium and low iron content. An opposite trend can be observed in cannel-coals, where low zirconium and high iron contents are observed (Watts 1996; Watts and Pollard 1997). It is of interest to see whether using LA-ICP-MS—capable of analyzing more elements to lower detection limits than XRF—can add further useful chemical trends to our suite of tools for distinguishing different occurrences of black stones.

The instrument used is a VG Plasma Quad PQXS ICP-MS with a VG UV laser probe ablation sampling device operating at 266 nm. The Nd: YAG laser is operated at 7 Hz, with laser spot size was varied between 20 and 100 μm to obtain adequate sensitivity, but spot size also depended on object size, and was varied to avoid detector saturation for some elements such as manganese. The ablation depth was typically about 200 μm. There is 20 s of uptake time prior to 50 s SI analysis. Ablations were performed on object surfaces or in section when possible in order to estimate the compositional heterogeneity within each sample. Ablation produced no marks visible to the naked eye (Fig. 17.3).

The compositions of black stones are generally relatively similar to those of sedimentary, clay-rich rocks that are mostly carbonated or carbonaceous. However, it is the mineral fraction that is important for distinguishing between sources, and we have accordingly measured chemical elements present in this fraction. Silicon is an abundant element, and we can assume that its distribution is relatively homogeneous within black stones. Consequently, silicon was chosen as our internal standard during analysis. International standard glasses (NIST 610 and CORNING B) were used to calibrate our instrument and monitor instrumental precision (Brill 1999; Bronk and Freestone 2001; Pearce et al. 1997; Popelka et al. 2005; Verità et al. 1994).

However, using LA-ICP-MS to measure carbon is problematic because no certified carbon standards exist. The carbon concentration in black stone samples was consequently measured semi-quantitatively using a graphite pencil tip as an "internal laboratory relative standard." This material appears to have a composition similar to our studied rock samples (pencil tips are made by mixing graphite with a calcareous clay binder) and was the only carbon-rich material available which produces a fairly stable and reproducible signal at the scale of laser ablation (μms). The composition of the pencil tip for the main oxides (SiO_2, Al_2O_3, CaO, K_2O, Na_2O, MgO, Fe_2O_3,

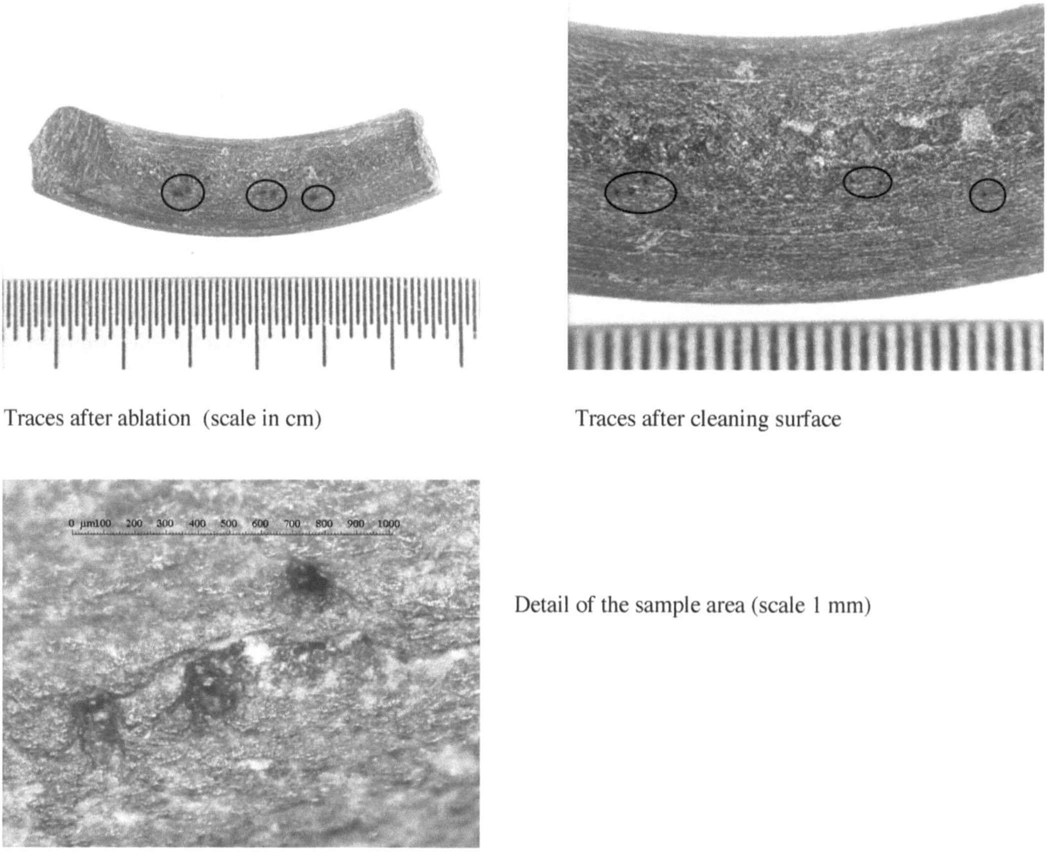

Traces after ablation (scale in cm)

Traces after cleaning surface

Detail of the sample area (scale 1 mm)

Fig. 17.3 Details of laser ablation on archaeological black stone objects (photos: B. Gratuze)

and TiO_2) and the main trace elements (Li, P, V, Cr, Mn, Cu, Rb, Zr, Sr, Ba, Ce) was determined first by using standard reference materials (SRMs) NIST 610, Corning B and D. However, as none of these SRMs contain carbon, and as our calculation protocol assumes that the sum of the contents of the determined elements (in weight percent) is equal to 100 %, it was not possible to determine the carbon concentration of the tip either directly or by difference with the sum of other elements. We have thus decided to arbitrarily set the weight percent concentration of carbon in the pencil tip at 12 %. This value was chosen in order to be low enough to produce relatively high concentrations for other elements. Major elements are linked with oxygen as silicates, oxides, or carbonates, and form the main constituents of rocks. If we include volatile elements such as oxygen (introduced by stoichiometry), hydrogen, sulfur, or nitrogen (we consider that these elements are present in very low concentrations)—even if these elements are not measured by LA-ICP-MS—their total concentration constitutes close to 100 % of each sample. Using this approach it is obviously not possible to measure real absolute concentrations for carbon in the studied shale samples. However, we were able to obtain relative carbon concentration values allowing comparison and classification of the different measured samples within our study. Comparison with other studies is unfortunately impossible without obtaining the real concentration of carbon within our samples. Although caution must be exercised, the carbon measurements obtained allow us to distinguish between carbonaceous materials and carbonated rocks.

17.3.3 Reliability of the Results and Errors

When enough samples from the same geological outcrop are available, the compositional variability present within that outcrop can be estimated. However, several sources of error or imprecision must be considered, which can further add to the variability observed within a geological source. These include error and imprecision resulting from instrumentation and calibration (matrix effect, electronic drifts etc...); error resulting from the assumption that the major elements measured constitute 100 % of each sample's composition (leading to potential enrichment or dilution of measured elements if this assumption is not correct); and error and imprecision resulting from using point sampling—although several ablations were performed on each sample, black stone samples are not homogeneous, as previously noted. Compositional variability within the sample volume volatilized during different ablations was observed, as a result of compositional heterogeneity within each sample (i.e., the presence of different mineral phases of various sizes). Ablation yields for these different mineral phases (compositions and sizes) are not equal, and thus some elements may be artificially enriched or depleted in the plasma.

Consequently, each sample was ablated twice to test its compositional variability. The standard deviation for each element and for all samples of an outcrop allow for an estimation of the global variability in each source. Detection limits depend of the sensibility of the device, the value of the background noise and the coefficient of sensibility for each element. The order of range is included between the fraction ppm to few tens of ppm for the elements where $Z < 30$ and of the order of the tens ppb to the fraction of ppm where $Z > 30$. These detection limits depends also of the quantity sent in the torch, the size of the ablation hole and the frequency of the laser shooting. They are calculating by taking three times the root of the background noise. The average value is calculated on two or three analytical sessions. Therefore, the relative errors for major, minor, and traces elements are presented in Table 17.2. Giving an estimate of the error for each chemical element is difficult because black stones are very heterogeneous, and errors could be different from one type of black stone to another.

17.4 Results

Complete compositional measurements for each analyzed sample can be found in Baron (2009, 2012). Only a selection of data for measured black stones is shown in Appendix 1. All geological samples are shown in Table 17.1, but only

Table 17.2 Relative errors for major, minor, and trace elements for the elements measured by LA-ICP-MS in black stones

Elements	Concentration	Maximal relative error
Major elements (SiO_2, Al_2O_3, CaO, K_2O, P_2O_5, Na_2O, MgO, MnO, Fe_2O_3, TiO_2)	<~2.5 %	20 %
	between 2.5 and 10 %	10 %
	>~10 %	5 %
Minor and trace elements (Li, Be, V, Cr, Cu, Zn, Ga, As, Rb, Sr, Y, Zr, Nb, Mo, Sb, Cs, Ba, La, Ce, Pr, Tb, Lu, W, Th, Pb, Th, U)	>~500 ppm	10 %
	Between 50 and 500 ppm	15 %
	Between 5 and 50 ppm	20 %
	<~5 ppm	Detection limits

some of the analyzed archaeological samples are discussed in the following Sects. 17.4.2 and 17.4.3. We have chosen to present only the archaeological sites for which we are able to suggest robust relationships with geological sources, but also in some cases connections between black stone objects from different archaeological sites. Therefore, only some characteristic case studies are presented (see Fig. 17.2). Other archaeological sites listed on the sampling map are therefore not discussed in the following sections. We deliberately represent geological samples and/or archaeological samples as shaded regions on many graphs instead of as individual sample symbols because inclusion of all samples in every case would be visually uninterpretable.

Black stones are very difficult to characterize due to their heterogeneity. Furthermore, some modification of their composition is possible due to the use-life of the object or its burial in the ground: for instance, the fixation of compounds and elements such as carbonates, phosphates, sulfates, or iron or by the leaching of humic acids or partially soluble components of the rock. For example, an anomaly in P_2O_5 contents can be observed—archaeological samples have higher concentrations (about 4–5 %), while geological samples have lower concentrations (about 2 %). This difference in concentrations can be explained by contamination and/or anthropogenic alteration, or may have resulted from a specific sedimentary facies (dependent on geological context: continental or marine deposit). Although our geological sampling is not exhaustive, the significantly lower P_2O_5 content in geologic samples may be due to a contamination of an anthropogenic nature for the archaeological objects.

Several different factors contribute to the formation of sedimentary rocks, including variable formation environment and the occurrence of different minerals within their matrices. Particular chemical associations result from the occurrence of these minerals in different black stones. Samples can, for instance, be more clayey, carbonated, siliceous, or carbonaceous depending on the minerals present. De La Roche's triangular diagrams, which display particular mineral chemical characteristics in the form of parametric poles, allow for the identification of groups or sub-groups characteristic of different sedimentary facies (De La Roche 1968, 1972; cf. Fig. 17.4). These parameters are defined by a quartz axis (parameter Q), clayey axis (parameter A) or calcareous axis (calcium). Other parameters, Al/3-K, Al/3-Na, which relate concentrations of Al_2O_3, Na_2O, and K_2O, relate to the composition of the clay component of the samples (illite vs. kaolinite).

Calculation of these parameters involves a conversion of the oxide weight of chemical mineral compositions to milliatom-grams for every 100 g of rock. This conversion allows us to situate minerals according to their quantitative proportions in relation to their mass. According to these parameters and the distribution of the data points, it is possible to define the sedimentary facies of a sample according to its place in the diagram in relation to the other known minerals. These parameters include minerals like feldspaths, which are essential and characteristic components of metamorphic rocks. During the sedimentary process, feldspaths are transformed or deteriorate due to sedimentary contribution or erosion of the rocks in which they are contained. Some are transformed into clays or carbonates, or become partially deteriorated. According to their degree of alteration, the connections between the different mineralogic poles (quartz, clays, feldspaths) demonstrate compositional variation.

17.4.1 Characterization of Geological Sources

The variability of the chemical elements contained in black stones results from the formation environment of the different rocks and their affiliation to certain sedimentary categories. Accordingly, this variability suggests which chemical elements should provide the most important distinguishing markers between

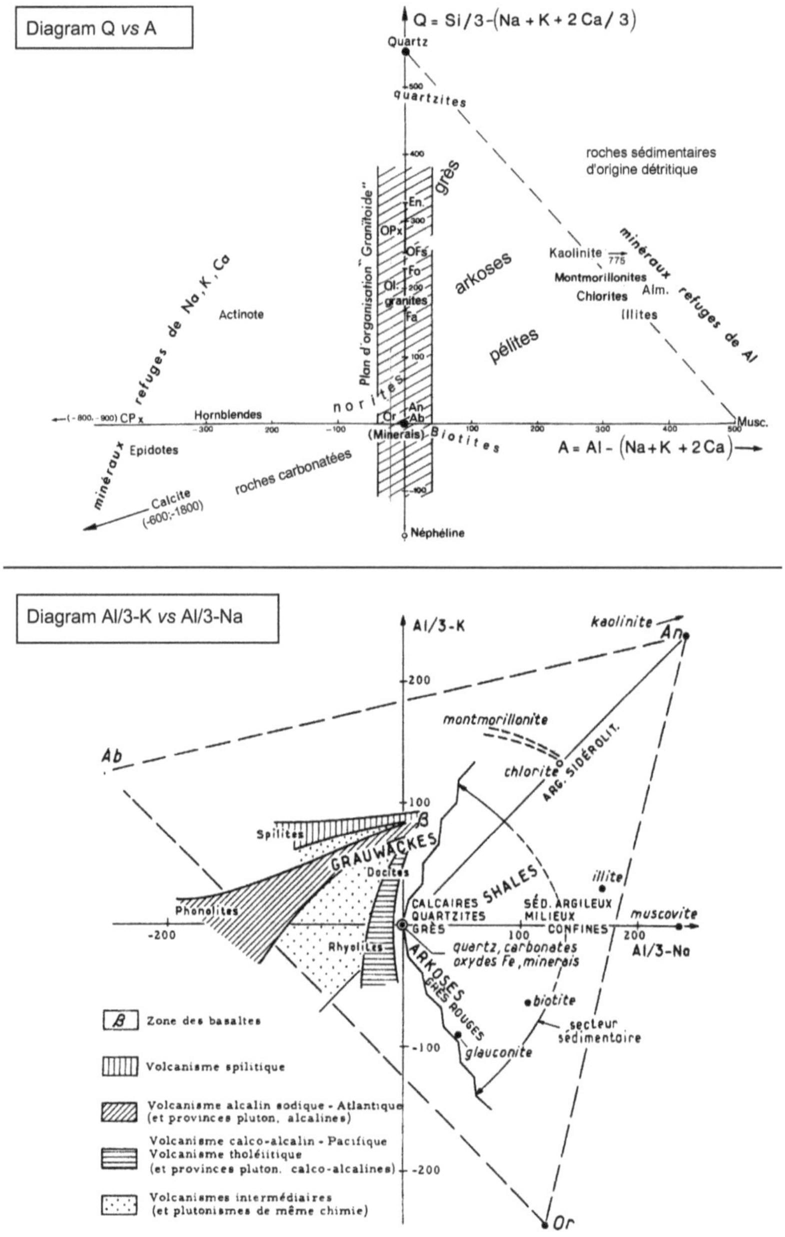

Fig. 17.4 De la Roche triangular diagrams (after Ploquin 1975: planche I; Guillot 1987: 121; De la Roche 1972: Fig. 3). Diagrams displaying the Q parameter (quartz axis) *vs.* A parameter (clay axis) distinguish between detritic sedimentary rocks and those that are more carbonated (Fig. 17.5 and Table 17.4). More clay-rich sedimentary rocks are localized in a zone with A > 50 while carbonated rocks occupy a zone with A < 50. Diagrams displaying Al/3-K *vs.* Al/3-Na distinguish between advanced sedimentary rocks and metamorphic rocks (igneous rocks). The alkalines Na and K are divided according to their rock types. For igneous rocks, Na, K, and Al exhibit the similar behavior, while in sedimentary rocks, Na deviates from K

different sources of black stones. As black stones include different kinds of materials—some of which are more organic or mineral—the resulting distributions of chemical compositions are extremely variable (Fig. 17.5). The use of different parameters allows us to characterize geographic supply areas according to the lithology of the materials available there. Sample groupings have been formed according to the distribution of points in the different resulting graphs (Figs. 17.6 and 17.7). On the Q *vs.* A diagram (Fig. 17.6), geological samples vary between pelitic sandstone, sedimentary rocks, and carbonated rocks. All of these are fairly

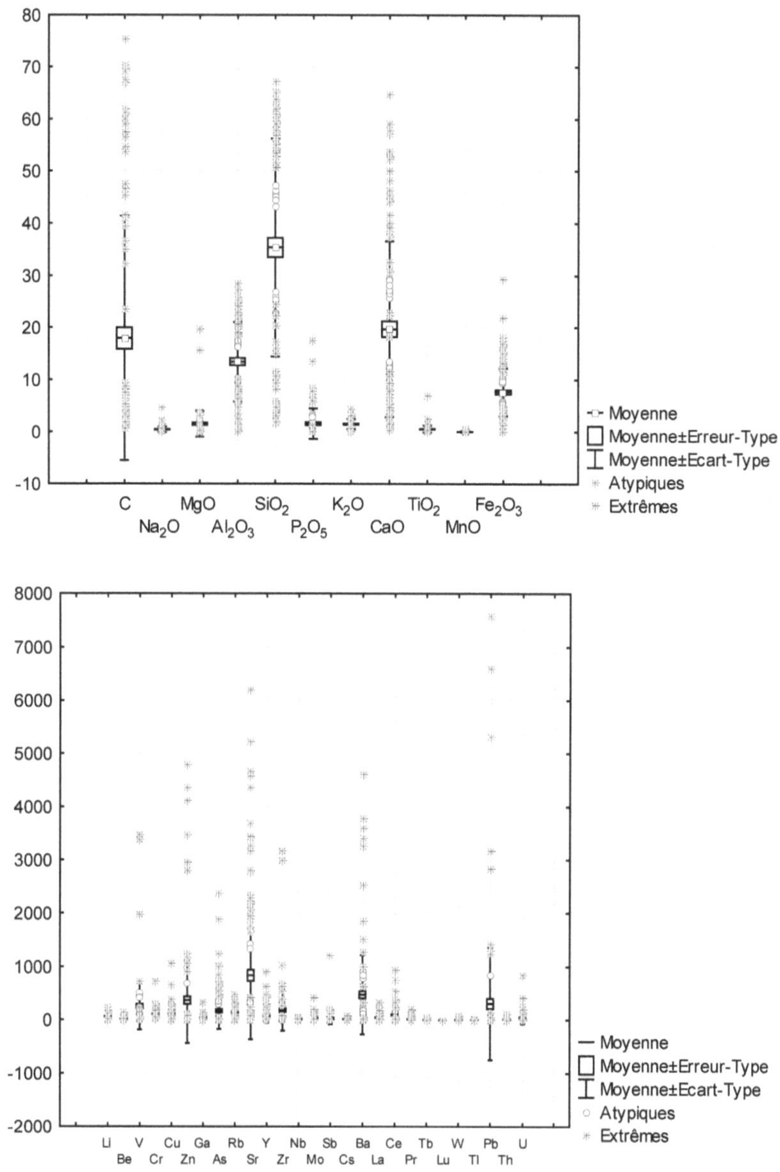

Fig. 17.5 Variability observed for major, minor, and trace elements for geological samples, which show significant heterogeneity

rich in calcium and different types of clays (a marly trend).

On the diagram Al/3-Na *vs.* Al/3-K (Fig. 17.7), carbonated materials are situated close to the axes, whereas clayey materials deviate from the axes. Some samples fall nearer to the kaolinite or illite regions on this graph. The Al/3-Na *vs.* Al/3-K graph shows that the geological materials sampled in this study include several different kinds of sedimentary rocks: 1) rocks with a pelitic sandstone trend that are fairly rich in quartz and also display a clayey trend (mainly kaolinite); 2) rocks with a pelitic trend that are finely detritic and fairly clayey (kaolinite or illite mainly); 3) rocks with a marly trend consisting of a mix of calcareous rocks and clays; 4) rocks with a calcareous trend and a strong carbonated component that is fairly rich in magnesium; 5) Carbonaceous rocks, rich in carbon and characterized by carbonaceous materials.

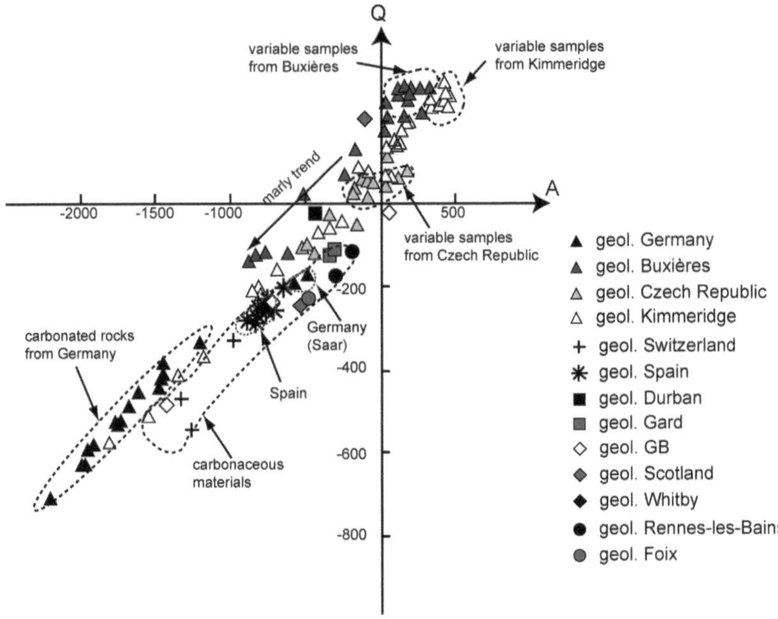

Fig. 17.6 Q *vs.* A (in milliatom-grams) for geological samples. When Q and A >50, rocks are pelitic; when A is between 50 and −400, rocks present a marly trend; when A > −600, rocks are carbonated. Samples from the Czech Republic, Kimmeridge, and Buxières-les-Mines show significant variability between marly and clay trends. Only a few groups can be identified

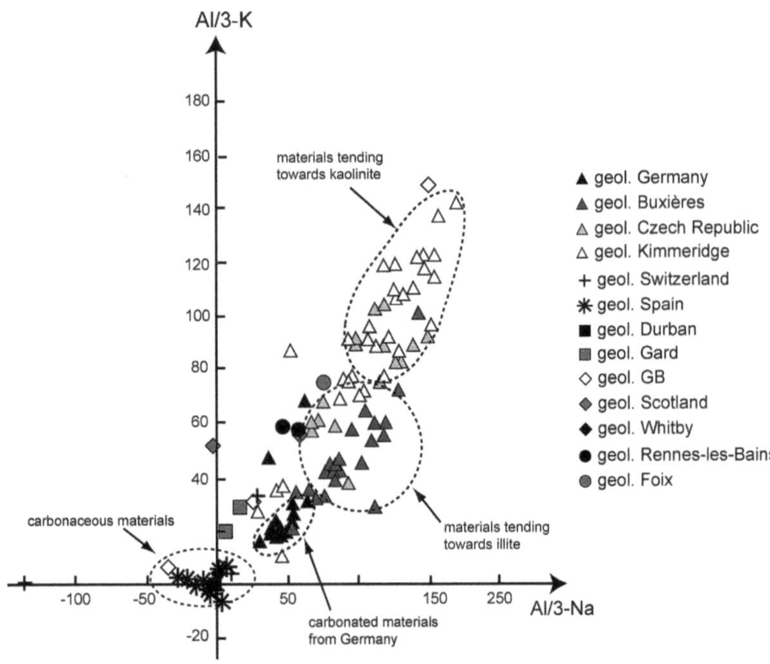

Fig. 17.7 Diagram of Al/3-K *vs.* Al/3-Na (in milliatom-grams) for geological samples. Carbonated materials are localized near the axis while the other samples are localized according to the amount of clays in their mineral composition. When Al/3-Na is between 50 and 150 and Al/3-K <50 materials tend towards illite and those above 50 tend towards kaolinite

These different geological materials correspond (more or less) to geological facies localized in well-defined geographical regions of Europe (Table 17.3). The following five geological sources can be distinguish from one another using their chemical characteristics: 1) The geological sources of the samples studied from Germany are composed of carbonated rocks that are rich in calcium, with siliceous contents lower than 30 % (unlike the other

Table 17.3 Main sedimentary facies represented in the geological samples

Geological sources	Rock types Greso-pelitic (quartz and kaolinite)	Pelitic	Marley pelites (calcareous and clays)	Carbonated	"Dolomitic" (Ca/Mg)	Carboneous
Germany: G-Aich 021, G-Aich-S 027, G-Geis 065, G-Geis 065bis, G-Dott 028, G-Ohm 019, G-Ohm-B 020, G-Ohm-L 018, G-Holz 022, G-Holz 025, G-Schl-B 026, G-Bol 017, G-Erz 024, G-Wan 066				X		
G-Wis 015			X			
G-Saar 078 (Saar)						X
Buxières-les-Mines (France): G-Bux 006, 007, 009, 011, 043, 044, 046, 049, 050, 052, 054, 056, 057	X					
G-Bux 008, 010, 041, 042, 045, 047, 048, 051, 053, 055		X				
Saint-Hilaire (France): G-Hil 038		X				
Chambéry (France): G-Vog 060				X		X
Czech Republic (Bohemia): A-Mkz 013, A-Mkz 121			X			
A-Srb 122	X					
A-Mkz 246, 247, 248		X				
Kimmeridge (England): G-Kim 012, 085, 086, 093, 107, 110, 101			X			
G-Kim 084, 087, 088, 090, 091, 092, 094, 095, 103, 104, 105	X					
G-Kim 097, 098, 099, 100, 102, 106, 108, 109, 111		X				
G-Kim 096, 089				X	X	
G-Kim 033				X		
Switzerland: G-Bel 079, G-Blap 080, G-Gond 081						X
Asturias (Spain): G-Ole 072, 073, 074, 075, 076, G-Ole-Mol 067, 068, 069, 070, 071						X
France various: G-Dur 058, G-Gar 061, G-Foi 004, G-Rlb 062						X
Great Britain various: G-Gb 077, G-Gb 082, G-Whi 030, G-Gb031						X
Scotland: G-Mull 029, G-Wem 032						X

studied materials); 2) The Kimmeridge source samples, which are pelites, but with a kaolinic trend; 3) The Buxières-les-Mines source samples, which are characterized by both a sandstone group and a pelitic group, both of which are rich in quartz and clays—a small number of samples have a slight carbonated trend; 4) Samples from the Czech Republic, the sources of which are relatively pelitic, however two groups can be observed: one with an illitic trend and the other with a kaolinic trend. One sample is more calcic. They are also richer in carbon than the other studied sources samples; 5) Carbonaceous materials from various localities, which are characterized by high carbon content and a trend towards potassium or sodic feldspaths. Jets from Asturias are better characterized than the other samples in this group.

The characterization of the geological materials is also based on discrimination using minor and trace elements, and characterization using minor and trace elements can be used to refine the relationship between archaeological and geological samples, and allow us to confirm that different samples belong to the same sedimentary facies. Using a matrix correlations study (i.e. evaluating the associations between the different chemical elements using PCA), some minor chemical elements function as discriminants between different studied sources of black stone materials—particularly lithium (Li), rubidium (Rb), niobium (Nb), cesium (Cs) and tungsten (W). The significance of varying contents of these elements could be the result of the particular sedimentation conditions during geological formation of these various black stone sources. During these sedimentation processes and according to the erosion of the parent rocks, these elements can concentrate in the form of clay minerals, or oxides, but are also present in the ashes of marine plants. The presence of these different elements can be favored by sedimentation in a lagoonal environment. Some sources have a distinctive chemical trace element signature, whereas others do not. For example, W content is significant in discriminating sources from Allier (Buxières-les-Mines and Montcombroux) from others. It is also possible to distinguish between sources using Rb and Cs content as different concentrations of these elements could be the result of different behavior during sedimentary evolution; this is particularly so with clay-rich black stones. Cs and Rb are more typically enriched during lakeside organic sedimentation in humid environments. These elements can also be found in some potassic feldspaths, however.

17.4.2 Characterization of Archaeological Materials

Table 17.4 presents all the results obtained on archaeological samples and also shows the diversity of materials used by Iron Age Europeans. Generally, the composition of the archaeological objects reflects a pelitic-marly trend, i.e. a mix of calcareous matrix and clays in variable proportions. The affinity with specific kinds of clay (kaolinite or illite, for example) might indicate that the source used to produce these objects was heterogeneous, or that these objects were made using material from different sources.

17.4.3 Relationships Between Artifacts and Sources

The correlations between sources and artifacts can be assessed by use of De la Roche triangular diagrams or by using major elements such as iron or potassium, for example. The distribution of points can differ for geological and archaeological samples depending on the data projection chosen. For example, an apparent relationship between a source and an archaeological site on a Q *vs.* A diagram could appear completely different when viewed on an Al/3-K *vs.* Al/3-Na diagram. If the materials are similar, they should display the same chemical characteristics on all projections. Any differences in compositional trends or concentrations between geological and archaeological samples means that a particular geological source (or at least the samples taken from there) was not the supply source for archaeological materials. If there is no correlation between the two main parameters (Q *vs.* A, Al/3-K *vs.* Al/3-Na), further analysis of the data is not necessary. In order to strengthen the hypothetical relationship between artifacts and

Table 17.4 Main sedimentary facies represented in the archaeological samples

	Rock types					
Archaeological sites	Greso-pelitic (quartz and kaolinite)	Pelitic	Marley pelites (calcareous and clays)	Carbonated	"Dolomitic" (Ca/Mg)	Carboneous
Illfurth "Britzgyberg" (Alsace, France): A-Btz 126, 130		X				
A-Btz 127, 128, 129				X		
Riedisheim "Glaisière Hartmann" (Alsace, France): A-Rid 131, 133, 134				X		
A-Rid 132			X with calcic trend			
Eckbolsheim "Zénith" (Alsace, France): A-Eck 191, 192, 193, 194, 195, 196, 197, 198, 199, 200, 201, 202, 203, 204, 205, 206, 207, 208, 209, 210				X		
Ihringen "Hinter Ehlen" (Baden-Wurttembergn, Germany): A-Ihr 112, 113, 114, 115, 117, 120			X with marly trend			
A-Ihr 116, 118, 119			X with pelitic trend			
Aslonnes "Camp Allaric" (Poitou-Charentes, France): A-Alc 161, 162, 169		X with clayey trend				
A-Alc 163, 164, 165, 166, 167, 168, 170, 171, 172		X				
St-Jean-Ligoure "Chalucet" (Poitou-Charentes, France): A-Chal 173, 174, 175		X with clayey trend				
Lyon "Dr. Horand" (Rhône-Alpes, France): A Lyo 032, 033	X					
Lyon "Gorge de Loup" (Rhône-Alpes, France): A Lyo 034, 035, 037, 038	X					
Lyon "Sgt. Berthet II" (Rhône-Alpes, France): A Lyo 039, 041, 043, 045	X					
A Lyo 040, 042, 044			X			
Bourges "Port-Sec" (Centre, France): A Brg 001, 006, 007		X				
A Brg 002, 003, 004, 005		X more calcic				
A Brg 008						X more calcic
A Brg 009						X
Bourges "St-Martin-des-Champs" (Centre, France): A Brg 010		X more calcic				
A Brg 011, 012, 013		X				
Bourges "Nation" (Centre, France): A-Brg 014, 015, 016		X				
Bourges "Littré" (Centre, France): A Brg 017, 018, 019, 020, 021, 022, 023, 024		X				
Bourges "hôtel-Dieu" (Centre, France): A Brg 025, 026		X				
A Brg 027		X more calcic				
Bourges-Lazenay "chemin de Gionne" (Centre, France): A Brg 028		X more calcic				

(continued)

Table 17.4 (continued)

Archaeological sites	Rock types Greso-pelitic (quartz and kaolinite)	Pelitic	Marley pelites (calcareous and clays)	Carbonated	"Dolomitic" (Ca/Mg)	Carboneous
A Brg 029		X				
Bourges-Lazenay "La Rottée" (Centre, France): A Brg 030, 030Bis						X more calcic
A Brg 031		X				
Bazoches/Vesle "les Chantraines" (Picardie, France): A-Baz176, 179,180,181, 183			X			
A-Baz 177, 178, 182				X		
Paule "St-Symphorien" (Bretagne France): A-Pau 097, 098, 102, 104		X with clayey trend				
A-Pau 099, 100, 101		X with illitic trend				
Nacqueville "la Batterie Basse" (Normandie, France): A-Nac 135, 136, 137, 225, 226, 227, 228, 230, 232, 233, 234, 235, 236, 238, 240, 241, 242, 244		X with clayey trend				
A-Nac 224, 229, 231, 237, 239, 243, 245				X		
île d'Yoc'h (Bretagne, France): A Yoc 123, 124, 125	X					
Longueil-Ste-Marie "l'Ormeon" (Picardie, France): A Lsm 108			X			
A Lsm 109				X		
Longueil-Ste-Marie "Vivier des grès" (Picardie, France): A Lsm 110, 111			X			
Verberie "Plaine d'Herneuse" (Picardie, France): A Vbr 105			X			
A Vbr 106		X				
Les Martres-de-Veyre "Corent" (Auvergne, France): A-Cor 184, 186, 187, 189			X with marly trend			
A-Cor 185, 188		X				
Montcombroux (Auvergne, France): A-Mtc 046, 050, 055, 056, 057, 058, 066, 067,082		X with clayey trend				
A-Mtc 054, 059, 062, 063, 068, 071, 072, 073, 074, 075, 078, 080, 081, 084, 085, 086, 088, 090		X				
Buxières-les-Mines "la Chassagne" (Auvergne, France): A-Bux 138, 139, 140, 141, 142, 143, 144,145, 146,147, 148,149, 150,151, 152, 153, 154,155, 157, 160		X				
A-Bux 156, 159		X with clayey trend				
Bâle "Gasfabrik" (Switzerland): A-Bal 212		X				
A-Bal 213	X					
A-Bal 214, 216, 220, 221, 222			X with marly trend			

sources, the workshops of the Allier region (France) are also represented on the graphs. A link between half-fabricated artifacts and finished products can also help in assessing the distribution range of material from a particular production place.

First inspection of these plots (Q *vs.* A and Al/3-K *vs.* Al/3-Na) reveals an area of non-overlap between archaeological and geological samples included in the study (Figs. 17.8 and 17.9). This indicates that not all sampled geological materials were used to produce archaeological artifacts, and that some of the archaeological artifacts included in the study cannot be linked to any analyzed geological materials. However, the archaeological sites discussed in the following sections were chosen as examples because they are assignable to one of the geological sources analyzed in this study. Although the black stones analyzed in this study were found to be very heterogeneous, a general change in raw material usage has been identified between the earlier (Hallstatt) and later (La Tène) Iron Age (Fig. 17.10): materials of the fifth and sixth centuries BC were made of carbonated rocks while materials from the La Tène period were made of more clayey rocks. This change could be connected to a change in the manufacturing techniques identified for the two periods. During the earlier Iron Age, the technique used for

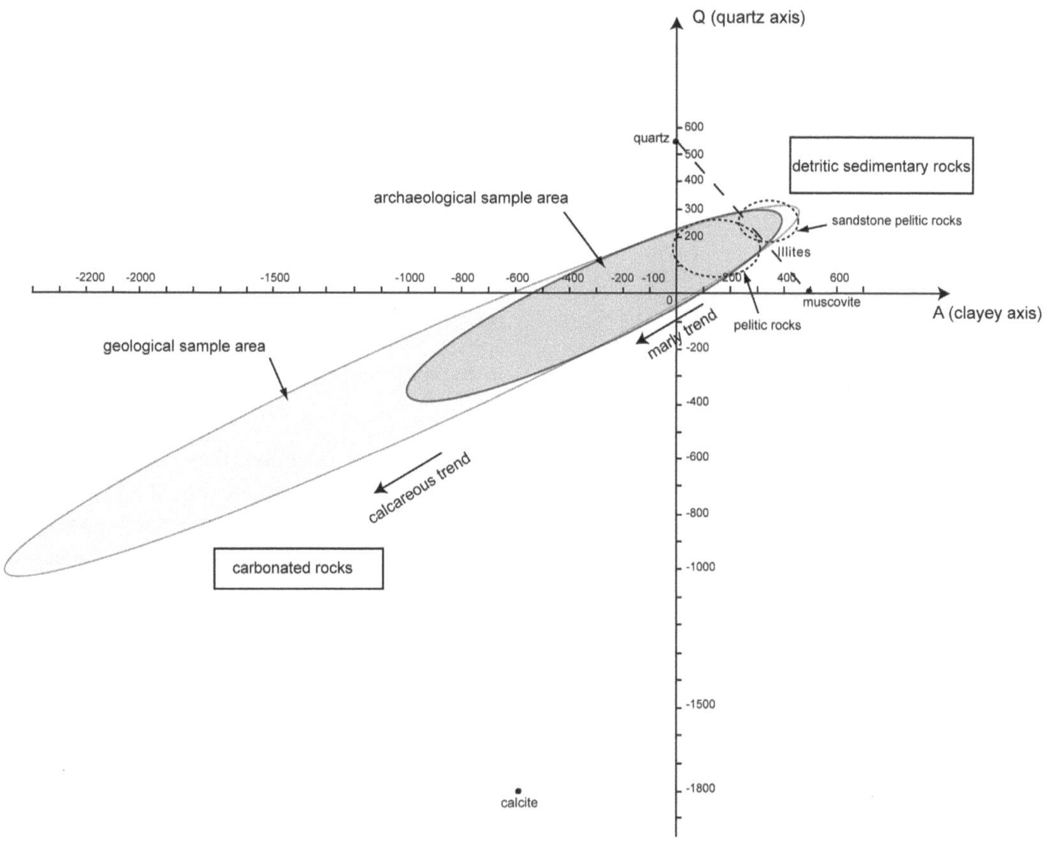

Fig. 17.8 Binary diagram of Q *vs.* A (in milliatoms-gram) for archaeological and geological samples. This graph shows the diversity of black stones, varying from pelitic sandstone to material with high carbonate content. The area of overlap between geological and archaeological samples is localized in a zone representing variably marly pelitic rocks and pelitic rocks. Good correlation can be observed between archaeological and geological samples within this region of the graph

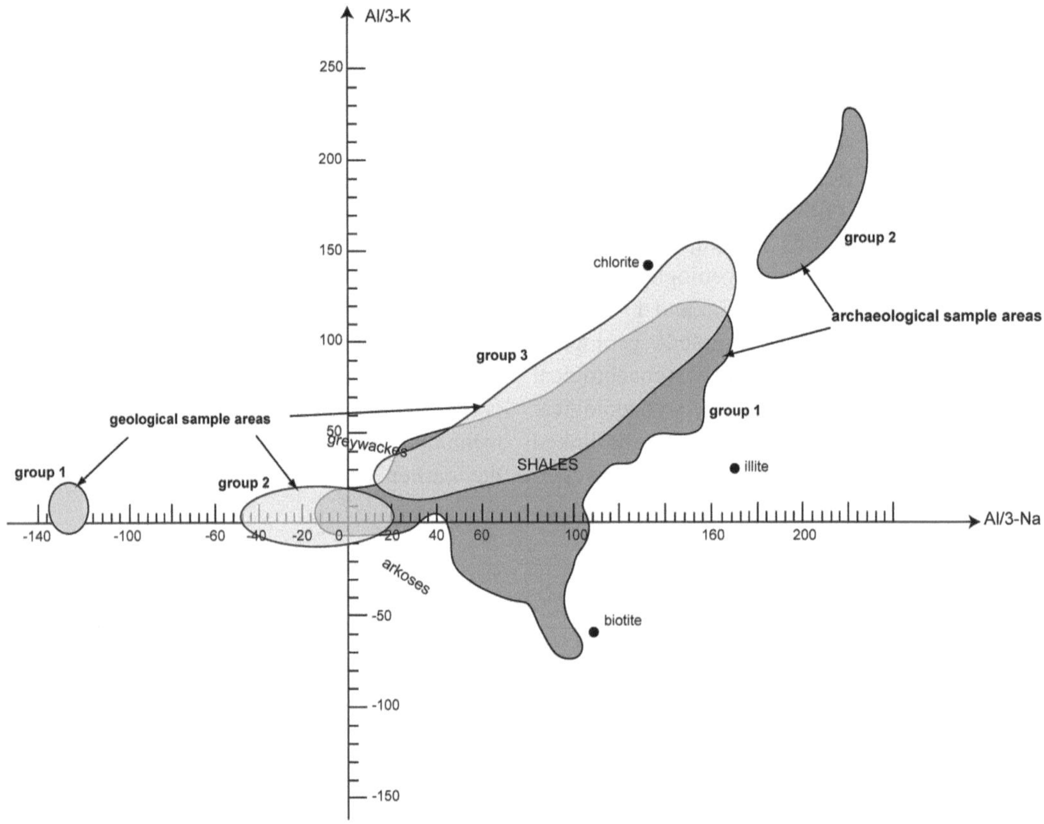

Fig. 17.9 Binary diagram of Al/3-K *vs.* Al/3-Na (in - milliatoms-gram) for archaeological and geological samples. Although there are areas that do not no overlap between some geological samples and archaeological artifacts, these sources may have been exploited, but not in the production of the studied archaeological samples

manufacturing bracelets was a central perforation that is enlarged during carving, whereas during the later Iron Age, bifacial carving was used, with the optional use of a rotary system in the final steps of fabrication (Fig. 17.10). Thus, there seems to be a relationship between materials utilized and techniques employed, which resulted, during the later Iron Age, in artisans choosing raw materials that were easier to carve and work. Some connections between artifacts and specific sources can be proposed for two sites included in the study.

Eckbolsheim (Alsace, France) the method employed in the analysis of black stone objects from Eckbolsheim (Alsace) has been recently published in Baron (2011). The objects from Eckbolsheim are marls with significant clayey and calcareous fractions; these objects are also linked to carbonated rocks (Fig. 17.11). A comparison between these artifacts and geological samples demonstrates that they are made of *Posidonia shales* from southern Germany (Fig. 17.12). The material analyzed from Eckbolsheim demonstrates that raw material was sometimes imported from distant sources, thus implying the presence of exchange networks (Baron 2011).

Montcombroux a complex of "archaeological sites (workshops)-sources," localized in Allier (France), was identified during excavations conducting during the 1990s (Pouenat and Vernet 2002). Analytic studies have shown that the artifacts from there and the local sources are linked (Thierry 1999; Thomas 2003). In

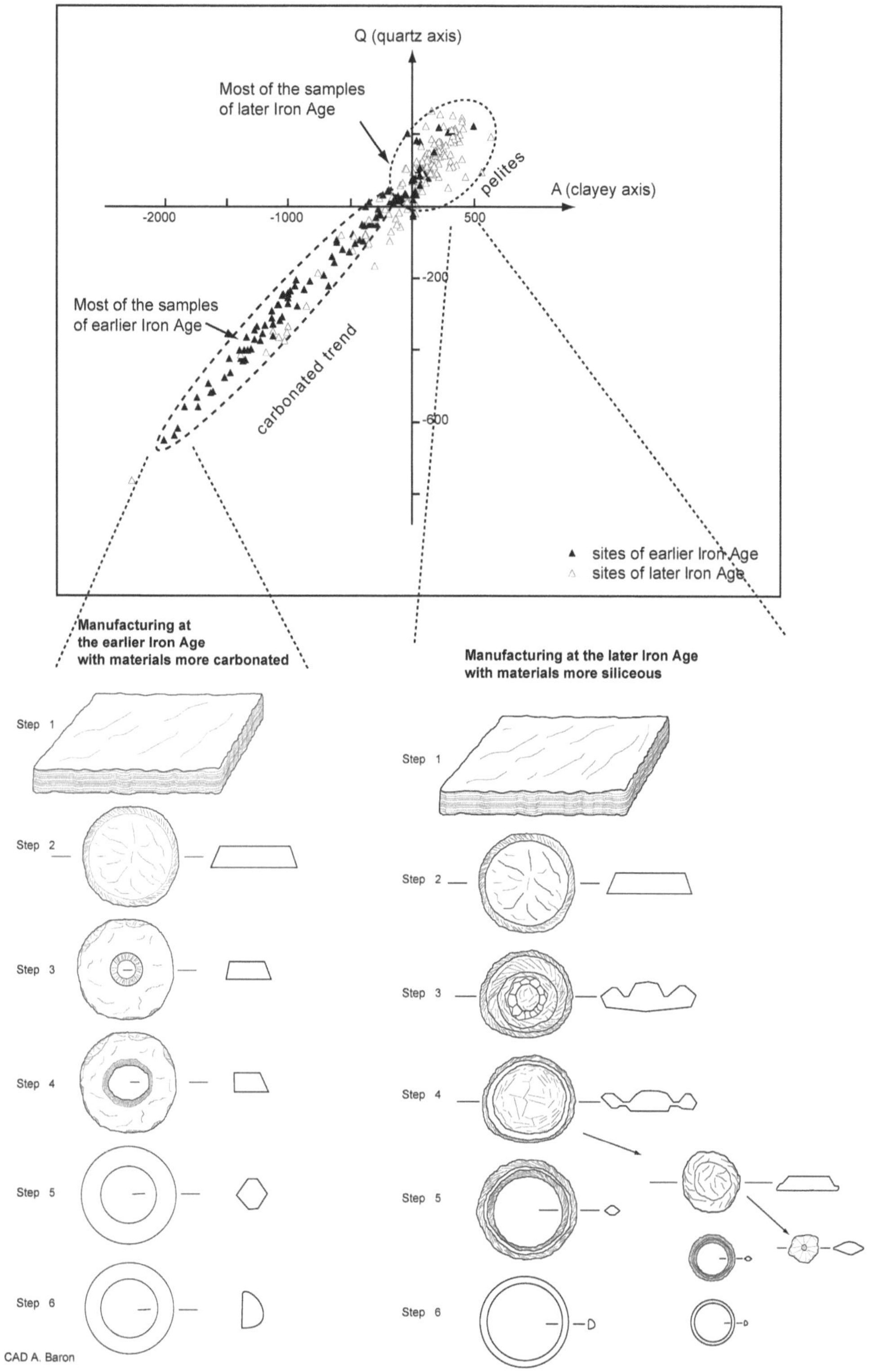

Fig. 17.10 Binary diagram of Q *vs.* A (in milliatoms-gram) for analyzed archaeological objects. A significant difference can be observed between objects of the early Iron Age (carbonated rocks) and those of the later Iron Age (pelitic rocks). This change of material also relates to a change in manufacturing techniques between the two

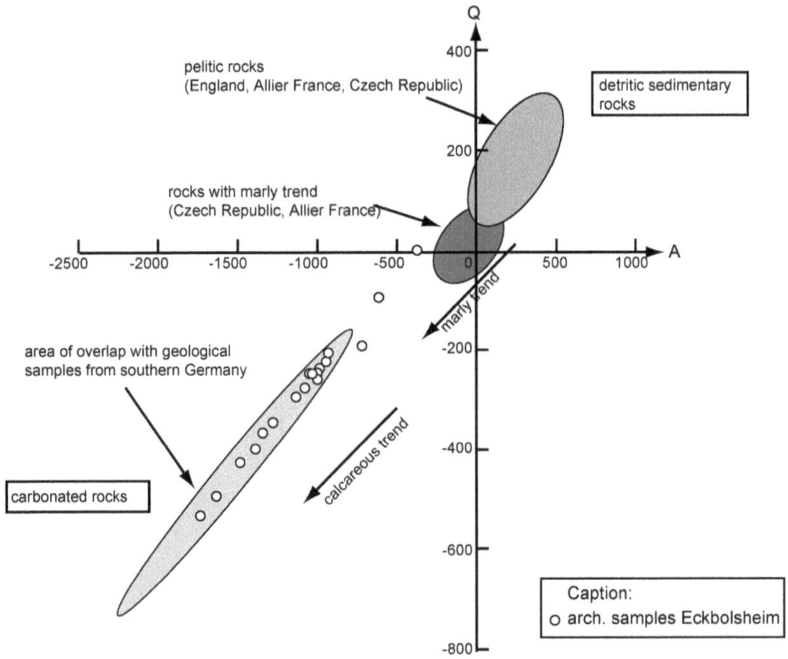

Fig. 17.11 Q *vs.* A diagram for archaeological samples from Eckbolsheim and geological sources. The plot shows some relations between archaeological samples and the Posidonia shales of southern Germany. Only part of the Eckbolsheim sample shows the same geological characteristics as this German source

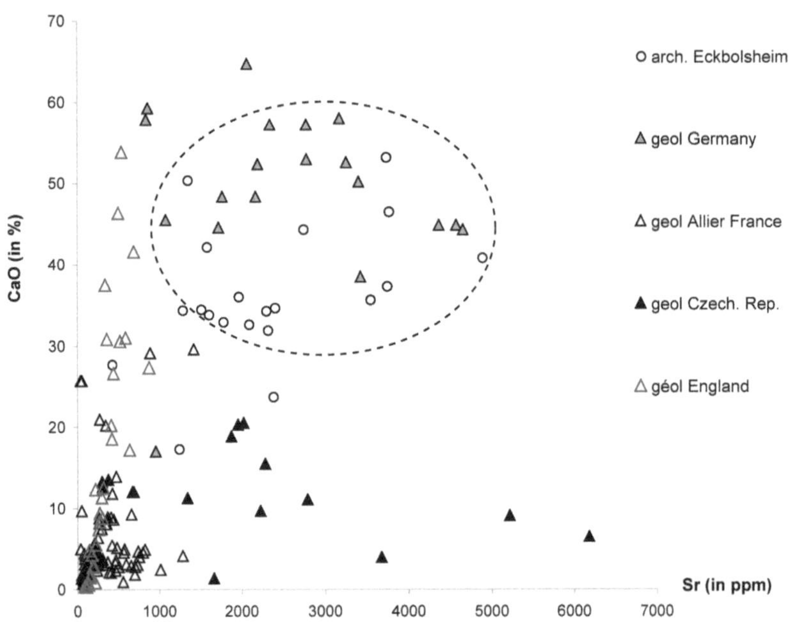

Fig. 17.12 Binary diagram of strontium and calcium (CaO in % and Sr in ppm) for archaeological samples from Eckbolsheim and geological sources. The archaeological samples and the geological samples from Germany are characterized by high concentrations of Sr and CaO (*black circle* in *dotted line*)

Fig. 17.10 (continued) periods, particularly the carving method used to shape bracelets. When carving a central perforation, the risk of breakage is higher than when using bifacial carving because the material is more malleable. Moreover, the waste material of the later Iron Age technique could have been reused to make other objects of smaller size the same region, another archaeological site—Montcombroux-les-Mines—is also known to have been a workshop for black stones. The results obtained in this study demonstrate that the materials of this region (Buxières-les-Mines and Montcombroux-les-Mines) are similar compositionally and that only one geological source was exploited at the two archaeological sites (Fig. 17.13; Baron 2009, 2012). These materials are characterized by pelitic rocks rich in iron and potassium, but with low calcium content. Another chemical characteristic of artifacts from the two sites is high content of Cs and W (Fig. 17.14). Even if several parts of the same seam might have been exploited, generally, these materials present the same geological facies and belong to the same family of rocks. Therefore, the study of these two archaeological sites constitutes a geographical reference point for production of other archaeological objects from other sites (Baron 2012).

Yoc'h Island (Brittany, France) the use of De La Roche's geochemical parameters allows for the demonstration of a connection between archaeological artifacts from Yoc'h Island (Brittany) and the geological source at Kimmeridge (Dorset) in England. In contrast to many of the studied geologic outcrops, particularly those from Czech and German sources, the samples from Yoc'h Island have a pelitic sandstone trend. Marly, carbonated, and carbonaceous rocks are all represented among the Yoc'h Island samples (Fig. 17.15). In Fig. 17.16, it can be seen that the Yoc'h archaeological samples are relatively similar to the Kimmeridge source, but also overlap with bituminous shale from Montcombroux (Allier, France). Discrimination using minor and trace elements is necessary in order to refine this distinction.

The Allier source is relatively rich in tungsten (W), while the objects from Yoc'h Island are not, suggesting that the Yoc'h Island objects did not originate from Allier (Appendix 1 of Ch. 17). As noted, other elements that discriminate between

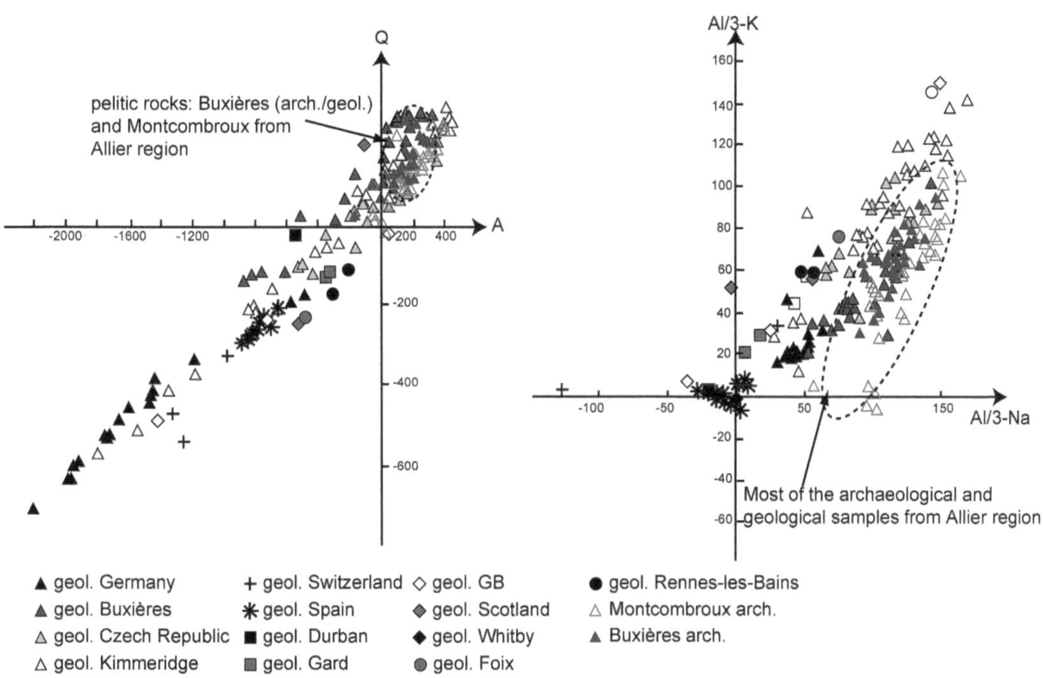

Fig. 17.13 Diagram of Q *vs.* A and Al/3-K *vs.* Al/3-Na (in milliatoms-gram) for archaeological samples from the Allier region and geological sources. The archaeological and geological samples from Buxières-les-Mines and the archaeological samples from Montcombroux have the same pelitic nature

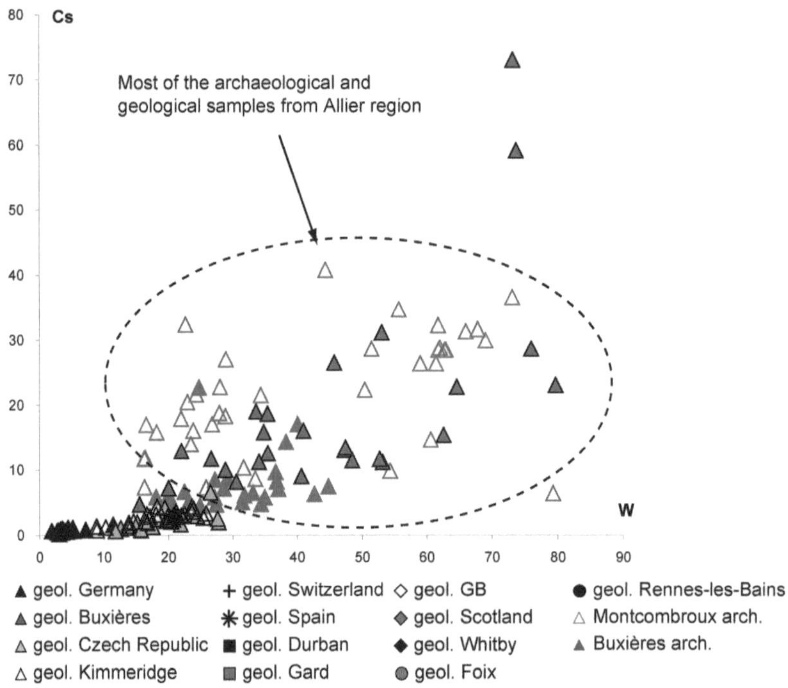

Fig. 17.14 Binary diagram of cesium and tungsten (in ppm) for archaeological samples from the Allier region and geological sources. Most of the samples from the Allier region have high contents of W and Cs in comparison to the rest of the analyzed samples

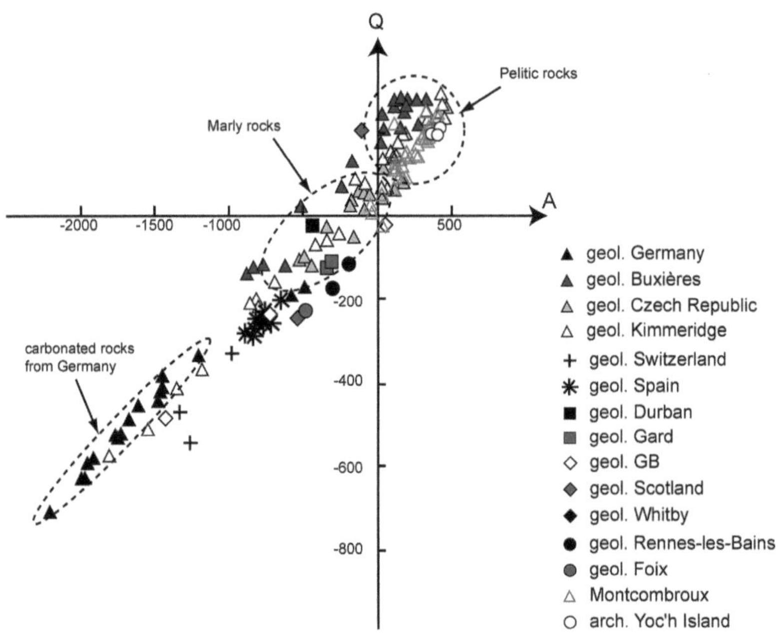

Fig. 17.15 Binary diagram of Q *vs.* A (in milliatoms-gram). The objects from Yoc'h Island belong to the group of pelitic rocks (*black circles* within the *dotted ellipse*). Some outcrops, such as the German sources, for example, are clearly distinct from the Yoc'h archaeological materials. However, a general similarity can be observed between Yoc'h objects and geological samples from Kimmeridge, but also with the Allier source (Buxières-les-Mines and Montcombroux) in France

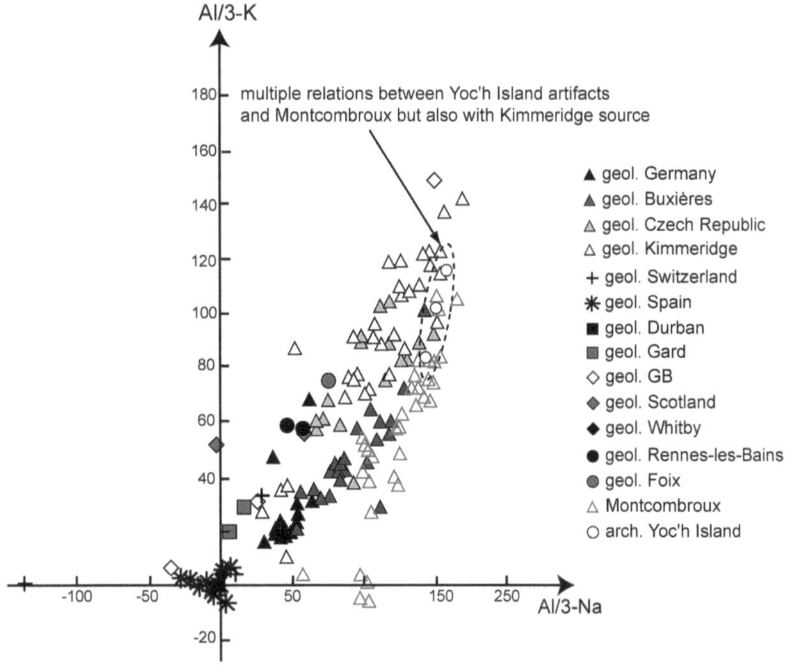

Fig. 17.16 Binary diagram of Al/3-K *vs*. Al/3-Na (in milliatoms-gram). Correlations can be observed between the geological samples from Kimmeridge and archaeological materials from Montcombroux (*black circles* within the *dotted ellipse*)

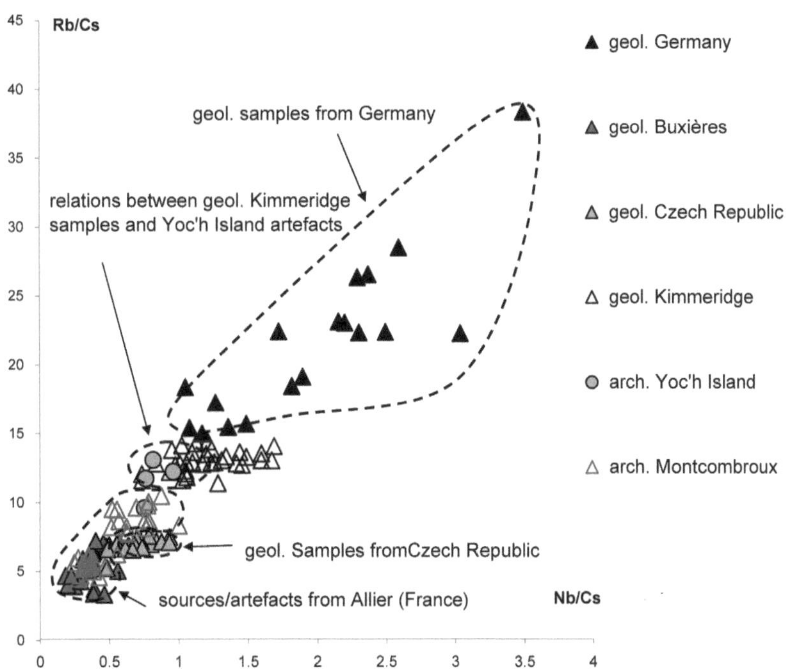

Fig. 17.17 Binary diagram of Nb/Cs *vs*. Rb/Cs. A clear connection between the objects from Yoc'h and Kimmeridge shale can be observed (*grey circles* within the *dotted line*). Other sources have distinct elemental ratios

geological sources of black stones are rubidium (Rb), cesium (Cs), and niobium (Nb). A further overlap between the objects from Yoc'h and the Kimmeridge geological samples can be observed in Fig. 17.17, a plot of Rb and Nb normalized using Cs concentration. Using this elemental projection, it is possible to isolate each geological source and propose relationships to the Yoc'h

Island samples. The German, Czech, and Allier sources are all well differentiated from the archaeological materials. Therefore, the artifacts from Yoc'h Island appear to have originated from Kimmeridge, where raw material is available on the coast.

Black Stones from Other Archaeological Sites Comparisons were also made between the archaeological materials from other archaeological sites included in this study. The results of this comparison demonstrate that some artifacts from several different archaeological sites display a similar composition indicative of a common source of supply, even if its location is presently unknown. For instance, the analysis of decorated bracelets discovered at the Hallstatt-period sites of Chalucet and Allaric's Camp (Poitou-Charentes, France) indicates that only one resource was used for production at both sites (Baron 2009; Appendix 1). Additionally, some bracelets from the site of Bourges (Centre, France) have a similar composition to another group of undecorated bracelets found at Allaric's Camp. These interconnections between sites allow us to suggest a common supply source, or at least trade and exchange networks operating at different geographical scales (Baron 2009, 2012).

17.5 The Procurement of Black Stones During the Iron Age in Europe

The results of this research in combination with prior studies allow us to map the distribution system of black stones across Iron Age Europe (Fig. 17.18). During the early Iron Age period (Hallstatt), the locations of utilized

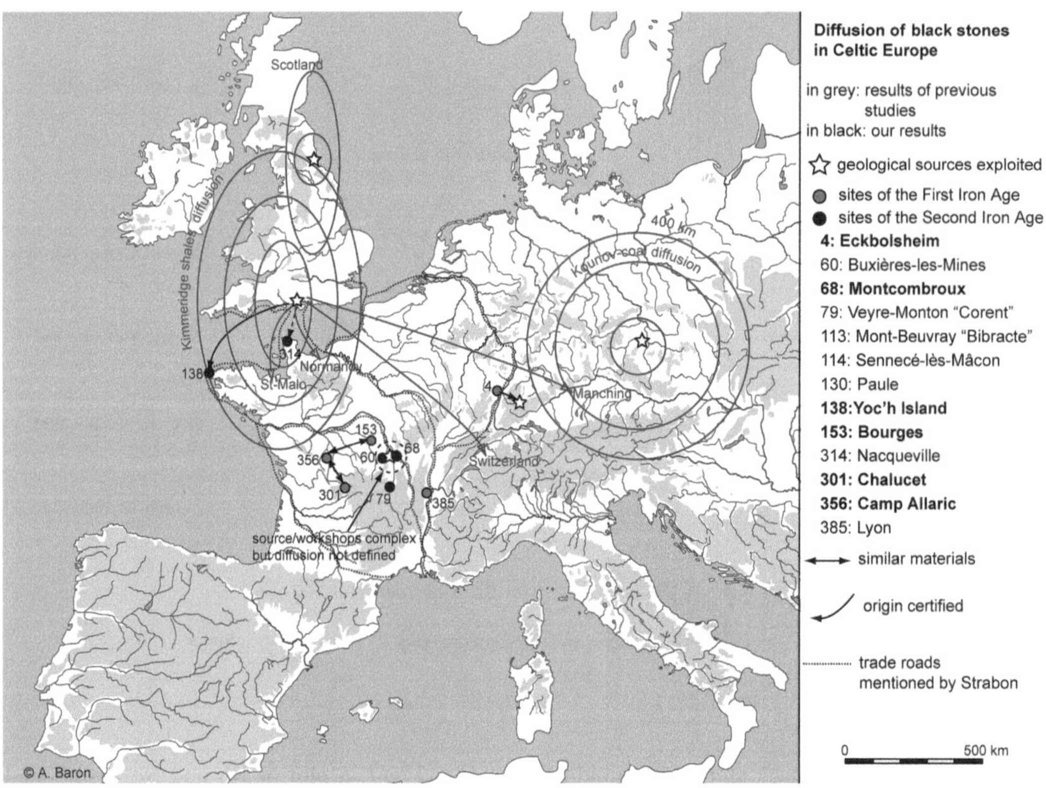

Fig. 17.18 Distribution map of black stones in the Iron Age in Celtic Europe (Adapted from Baron (2009, 2012) and previous studies; CAD A. Baron)

geological resources are presently mostly unknown, except for the *Posidonia shales* in Germany and Kimmeridge shale in England. In the later Iron Age (La Tène), reconstructed distribution systems are more complex. Two supply sources have been identified in previous studies: one at Kimmeridge, which was exploited during both the early and later Iron Age, the material from which was primarily distributed throughout Western Europe; and the other in the Czech Republic, material from which is present only in Central Europe. Analyzed artifacts dating to the La Tène period demonstrate that this source was not exported to Gaul during the later Iron Age. No connections between La Tène period artifacts and the Czech source have been identified to date. In addition, a third source of supply has been identified in the central part of France in the Allier region, but its distribution in Europe is not yet well defined. However, the *Posidonia shales* exploited during the early Iron Age appear not to have been used during the later Iron Age. Obviously, this preliminary reconstruction requires further research and the characterization of additional sources of black stones potentially utilized during the Iron Age in Europe.

17.6 Conclusions

This exploratory research using LA-ICP-MS on black stones is a first step towards characterizing these kinds of materials to provide a preliminary estimation of their chemical compositions. The use of geochemical compositional measurements allows us to suggest connections between artifacts and sources, but also between different archaeological sites even when sources of some objects remain unknown. A change of the supply source between the early (Hallstatt) and the later (La Tène) Iron Age corresponded to a transformation in the mode of manufacture. This evidence allows us to demonstrate the circulation and diffusion of black stones objects across Iron Age Europe.

We have adapted the approach typically employed for LA-ICP-MS analysis of other types of archaeomaterials (e.g., ceramics, metals, volcanic rocks) in order to characterize the inorganic constituents of black stones. In the future, this methodology will be coupled with other analytical techniques to characterize the organic fraction of these stones in order to produce a complete characterization of geological and archaeological materials. Studying the organic fraction of geological and archaeological black stones will allow identification of additional compositional characteristics of these archaeomaterials, and can give us information about the environment of their deposition, and therefore direct us to a possible geographical localization of their source of extraction. For instance, future research will explore the method utilized by Watts Py-GC-MS in order to provide a first estimation of chemical markers for archaeological materials, even if this method is partially destructive. Characterization of the organic component of black stones by this method, when augmented with characterization of the elemental content of the inorganic fraction by LA-ICP-MS, will allow a complete characterization of archaeological and geological black stone materials.

Acknowledgements The authors would like to thank all involved archaeological and geological institutions for the loan of samples: Institut National d'Archéologie Préventive in France; the Regierungprasidium of Freiburg-im-Breigsau in Germany; Museum of Poitiers; the Municipal service of archaeology of Bourges and the Berry's museum; the museums of Mulhouse, Poitiers, Lyon, Cherbourg, and Autun in France; the Archäologische Bodenforschung of Bâle in Switzerland; the regional services of archaeology in France; and the Institute of Archaeology of Prague in the Czech Republic. They would also like to thank the following geologic institutions: the Landesamt für Geologie in Germany; Bureau des Ressources Géologiques et Minières in France; the geologic associations in France; the University of Geology of Oviedo in Spain; and the Scotland Museum.

Appendix 1

Major Results

	C	Na_2O	MgO	Al_2O_3	SiO_2	P_2O_5	K_2O	CaO	TiO_2	MnO	Fe_2O_3	Q	A	Al/3-K	Al/3-Na
Geological samples															
G-AICH 021	2.35	0.16	2.00	8.55	25.93	0.15	1.69	52.88	0.34	0.16	5.46	−526.34	−1759.27	19.98	50.61
G-AICH-S 027	3.21	0.31	2.20	8.12	27.04	0.25	1.53	50.09	0.32	0.19	6.44	−488.43	−1669.71	20.67	42.92
G-GEIS 065	3.57	0.39	1.20	6.51	17.33	6.50	1.26	57.79	0.22	0.18	4.85	−630.27	−1972.38	15.77	29.91
G-GEIS 065	3.05	0.30	1.39	9.63	29.62	0.00	1.54	44.80	0.35	0.13	8.45	−411.24	−1451.07	30.22	53.16
G-DOTT 028	7.25	0.30	1.44	9.55	25.39	0.24	1.67	45.43	0.25	0.19	8.18	−444.75	−1477.74	26.98	52.74
G-DOTT 028	3.49	0.36	1.45	8.19	20.31	0.00	1.46	59.19	0.23	0.22	4.96	−633.75	−1992.56	22.49	41.96
G-OHM 019	1.83	0.25	1.74	9.51	33.58	0.47	1.72	44.48	0.31	0.00	5.76	−387.54	−1444.01	25.73	54.09
G-OHM-B 020	2.32	0.16	1.91	7.60	22.66	0.23	1.46	57.15	0.32	0.16	5.69	−590.10	−1925.07	18.71	44.49
G-OHM-L 018	3.31	0.34	2.03	8.44	24.63	0.21	1.61	52.46	0.32	0.15	5.96	−532.16	−1749.99	21.09	44.38
G-HOLZ 022	2.58	0.23	1.88	9.27	25.29	0.18	1.74	52.26	0.41	0.14	5.49	−525.48	−1726.03	23.61	53.35
G-HOLZ 025	3.65	0.24	1.79	7.92	28.60	0.28	1.51	48.27	0.37	0.16	6.86	−455.47	−1605.85	19.66	44.09
G-SCHL-B 026	4.41	0.36	2.32	11.38	31.36	0.18	2.01	38.41	0.34	0.15	8.66	−337.47	−1200.79	31.65	62.95
G-BOL 017	2.82	0.38	2.00	7.71	22.65	0.17	1.38	57.91	0.29	0.16	4.17	−604.31	−1955.27	21.21	38.31
G-ERZ 024	4.00	0.45	1.58	8.45	28.05	0.12	1.75	44.22	0.33	0.15	10.39	−422.05	−1462.67	18.10	40.85
G-WAN 066	3.75	0.21	1.72	6.62	16.03	0.15	1.16	64.66	0.24	0.33	4.76	−710.93	−2207.11	18.67	36.68
G-WIS 015	1.28	0.29	1.45	19.60	50.64	0.51	2.77	16.95	0.66	0.00	5.57	10.05	−287.87	69.33	118.92
G-BUX 006	15.53	0.34	0.75	13.36	63.60	0.20	2.17	1.17	0.76	0.00	2.06	280.06	163.29	41.31	76.30
G-BUX 007	3.52	0.28	1.46	17.21	63.60	1.50	2.24	2.67	0.70	0.13	6.63	262.67	185.76	64.92	103.49
G-BUX 008	1.82	0.11	1.47	11.08	46.22	0.77	1.91	29.52	0.41	0.34	6.23	−139.44	−879.20	31.96	69.05
G-BUX 009	1.71	0.31	1.37	21.12	57.32	1.37	3.05	1.90	0.60	0.00	11.14	219.08	271.93	73.34	128.17
G-BUX 010	1.67	0.00	1.10	12.33	54.32	1.23	1.70	20.11	0.53	0.61	6.33	24.85	−511.09	44.53	80.67
G-BUX 011	3.07	0.23	0.98	14.35	62.12	1.56	2.23	1.15	0.43	0.21	13.48	274.26	185.66	46.46	86.34
G-BUX 041	4.42	0.24	1.14	13.89	58.45	2.92	2.48	4.60	0.44	0.15	11.20	207.76	48.21	38.29	83.24
G-BUX 042	4.22	0.13	1.61	16.25	53.66	1.09	2.88	13.83	0.52	0.27	5.44	66.45	−239.69	45.11	101.94
G-BUX 043	4.15	0.29	1.53	18.46	59.39	1.68	4.30	3.17	0.72	0.00	6.16	189.61	148.55	29.37	111.50
G-BUX 044	4.30	0.22	1.35	15.48	64.99	1.63	2.05	4.08	0.43	0.00	5.36	259.84	107.77	57.69	94.28
G-BUX 045	5.39	0.28	1.17	13.01	64.88	2.09	2.41	4.54	0.35	0.17	5.66	244.04	33.09	33.97	76.04
G-BUX 046	2.96	0.00	1.32	18.13	62.85	0.80	2.79	1.07	0.55	0.00	9.42	275.04	258.54	59.43	118.60
G-BUX 047	1.86	0.00	1.58	12.51	47.20	0.55	1.88	29.04	0.38	0.35	4.48	−124.46	−830.17	41.78	81.79

G-BUX 048	2.61	0.00	0.86	8.64	37.39	17.45	1.03	25.66	0.32	0.67	5.25	−120.45	−767.73	34.53	56.48
G-BUX 049	3.75	0.17	1.72	17.72	56.22	1.68	2.62	3.70	0.44	0.22	11.73	205.30	154.51	60.23	110.51
G-BUX 050	2.59	0.00	1.23	21.93	61.35	0.21	1.98	1.69	0.48	0.00	8.34	276.49	327.95	101.37	143.43
G-BUX 051	14.92	0.37	0.70	11.69	53.79	2.96	1.95	9.62	0.44	0.00	3.48	129.15	−167.21	34.97	64.41
G-BUX 052	3.01	0.00	1.48	16.53	65.47	1.72	2.58	2.87	0.47	0.00	5.74	272.58	167.22	53.40	108.10
G-BUX 053	4.26	0.00	1.15	13.27	51.67	3.31	2.09	5.61	0.37	0.11	18.09	174.12	15.75	42.32	86.77
G-BUX 054	4.98	0.00	1.27	17.80	62.09	1.45	2.83	3.01	0.66	0.00	5.81	246.93	181.80	56.36	116.41
G-BUX 055	3.35	0.00	0.80	8.06	29.30	13.46	1.48	20.84	0.22	0.34	22.03	−117.30	−616.58	21.25	52.68
G-BUX 056	4.46	0.30	1.26	14.22	67.07	1.75	2.31	3.27	0.42	0.13	4.76	272.79	103.66	44.02	83.39
G-BUX 057	6.36	0.00	1.16	18.13	63.67	0.19	2.71	1.29	0.73	0.00	5.68	278.64	252.28	61.13	118.60
G-HIL 038	3.45	0.00	1.91	20.58	55.49	1.97	3.53	4.02	0.70	0.00	8.12	183.53	185.41	59.59	134.61
G-VOG 060	16.76	0.00	1.85	15.36	21.97	0.78	0.40	38.98	0.38	0.00	3.25	−350.40	−1097.13	91.88	100.47
G-VOG 060	13.92	0.00	1.72	17.20	26.34	0.45	0.50	36.55	0.45	0.00	2.64	−299.29	−976.34	101.92	112.48
G-VOG 060	18.79	0.00	1.94	14.05	18.84	1.01	0.34	40.72	0.34	0.00	3.68	−386.99	−1183.67	84.71	91.90
G-MKZ 013	20.29	0.40	0.63	15.96	28.52	2.80	3.16	11.02	0.52	0.00	16.22	−53.32	−159.59	37.31	91.54
A-MKZ 121	15.29	0.36	0.55	14.47	26.82	8.19	1.68	18.76	0.37	0.00	12.43	−122.02	−432.36	59.05	82.99
A-MKZ 121	14.91	0.34	0.54	11.65	30.37	5.76	0.86	20.41	0.47	0.12	13.97	−103.98	−528.30	57.94	65.32
A-MKZ 121	13.36	0.31	0.54	12.36	31.19	6.13	0.93	20.21	0.43	0.18	13.52	−97.60	−507.81	61.20	70.83
A-MKZ 121	17.20	0.49	0.70	13.91	37.89	2.74	1.09	9.60	0.52	0.00	15.32	56.15	−108.29	67.82	75.18
A-MKZ 121	17.78	0.37	0.50	11.90	33.82	5.85	0.79	15.41	0.40	0.00	12.71	−25.14	−344.78	61.01	65.93
A-SRB 122	11.77	0.24	0.83	19.08	35.23	6.39	1.68	11.95	0.45	0.00	12.17	9.01	−95.24	89.10	116.89
A-SRB 122	8.44	0.37	0.82	18.71	43.20	7.32	0.96	12.53	0.46	0.00	6.99	57.15	−112.32	101.95	110.35
A-SRB 122	7.00	0.36	0.74	16.63	38.64	8.41	0.90	13.42	0.46	0.10	13.17	23.27	−182.87	89.68	97.18
A-SRB 122	10.04	0.41	0.87	16.92	40.49	7.92	0.90	13.18	0.69	0.00	8.32	34.66	−170.36	91.54	97.55
A-SRB 122	8.10	0.38	0.95	19.66	45.76	6.04	1.17	8.86	0.55	0.00	8.30	110.00	32.40	103.74	116.18
A-MKZ 246	8.02	0.15	0.68	23.38	44.94	0.99	2.84	1.29	0.57	0.14	16.69	167.49	347.44	92.61	147.95
A-MKZ 247	9.54	0.12	0.52	18.19	31.78	3.04	2.04	3.85	0.55	0.12	29.28	82.25	171.99	75.55	114.97
A-MKZ 247bis	15.42	0.17	0.79	22.11	39.18	2.55	2.62	9.00	0.69	0.11	6.02	48.17	51.75	89.05	139.03
A-MKZ 247	12.48	0.15	0.66	20.15	35.48	2.79	2.33	6.43	0.62	0.12	17.65	65.21	111.87	82.30	127.00
A-MKZ 248	10.88	0.11	0.68	20.46	42.55	1.34	2.40	11.18	0.72	0.00	9.17	47.51	−51.84	82.77	130.39
G-KIM 012	12.62	0.33	1.10	17.49	28.72	1.58	2.00	26.54	0.51	0.00	8.97	−209.80	−656.28	71.89	103.74
G KIM 033	7.65	0.20	0.96	7.90	20.45	1.46	1.94	53.75	0.22	0.00	5.31	−573.38	−1809.43	10.49	45.24
G-KIM 084	0.90	0.73	1.93	25.87	60.39	0.24	2.20	2.69	0.74	0.00	4.24	231.40	341.66	122.55	145.76

(continued)

	C	Na_2O	MgO	Al_2O_3	SiO_2	P_2O_5	K_2O	CaO	TiO_2	MnO	Fe_2O_3	Q	A	Al/3-K	Al/3-Na
G-KIM 085	2.47	0.66	1.73	16.81	38.21	0.55	1.55	30.49	0.73	0.00	6.65	−205.58	−811.95	76.94	88.65
G-KIM 085	2.33	0.72	1.85	17.38	41.00	0.60	1.75	27.28	0.55	0.00	6.43	−158.14	−692.14	76.56	90.32
G-KIM 086	2.91	0.61	1.62	16.16	38.21	0.68	1.70	30.91	0.59	0.00	6.47	−211.92	−840.87	69.64	86.10
G-KIM 087	1.00	1.55	2.24	25.74	59.48	0.31	2.34	1.41	0.80	0.00	5.03	211.82	354.80	118.57	118.33
G-KIM 088	1.18	0.56	1.60	28.60	57.88	0.27	2.15	1.67	0.82	0.00	5.20	236.05	437.90	141.41	169.06
G-KIM 089	1.35	0.37	19.67	6.11	10.86	0.18	0.60	46.29	0.21	0.13	14.15	−514.70	−1555.47	27.20	28.05
G-KIM 089	1.62	0.22	15.66	7.48	17.28	0.26	0.63	41.52	0.22	0.13	14.81	−418.50	−1354.59	35.54	41.65
G-KIM 090	0.94	0.62	1.52	24.60	61.96	0.27	1.84	2.78	0.95	0.00	4.44	249.68	324.00	121.69	140.77
G-KIM 091	1.11	0.65	1.48	27.19	60.62	0.25	1.92	0.77	0.82	0.00	5.09	263.73	444.21	137.11	156.83
G-KIM 092	4.84	0.70	1.29	21.91	53.82	1.09	2.44	0.72	0.85	0.00	12.27	214.24	330.00	91.57	120.75
G-KIM 093	5.98	0.65	1.42	22.47	40.76	0.92	1.32	18.45	0.82	0.00	7.10	−43.18	−266.16	119.01	125.93
G-KIM 094	5.35	0.68	1.27	24.60	57.62	0.37	2.41	0.52	1.00	0.00	6.10	238.82	390.97	109.66	139.04
G-KIM 095	1.57	0.49	1.12	24.94	64.89	0.21	2.15	0.28	0.82	0.00	3.48	293.60	418.18	117.50	147.40
G-KIM 096	1.68	0.62	15.65	10.33	22.20	0.34	1.44	37.41	0.31	0.11	9.83	−372.48	−1181.91	37.03	47.49
G-KIM 097	3.29	0.39	1.25	21.97	54.02	0.34	1.72	8.27	0.65	0.00	7.99	150.75	86.87	107.18	131.00
G-KIM 098	0.76	0.99	1.45	19.10	56.99	1.05	2.29	12.46	0.58	0.00	4.25	86.16	−149.92	76.39	93.09
G-KIM 099	0.91	0.47	1.11	21.91	55.41	0.41	2.64	11.27	0.71	0.00	5.10	100.98	−42.99	87.25	128.22
G-KIM 100	1.67	0.89	1.08	20.99	52.81	0.83	1.96	12.24	0.70	0.00	6.75	75.93	−94.91	95.64	108.67
G-KIM 101	1.28	0.87	2.42	19.73	46.36	0.67	2.75	20.12	0.63	0.00	5.05	−69.80	−417.09	70.53	100.82
G-KIM 102	1.27	0.48	3.37	18.66	53.55	0.82	1.46	4.19	0.92	0.00	15.12	199.28	170.07	91.10	106.45
G-KIM 103	1.05	0.36	1.63	25.48	60.71	0.31	2.44	0.29	0.87	0.00	6.77	268.22	426.27	114.77	155.04
G-KIM 104	1.40	0.50	1.50	26.03	60.88	0.24	2.28	0.34	0.80	0.00	5.95	267.34	433.88	121.91	153.97
G-KIM 105	1.58	0.44	1.14	25.31	61.76	0.31	3.29	0.29	0.85	0.00	4.95	253.48	402.12	95.75	151.32
G-KIM 106	0.54	0.81	1.51	21.08	57.81	0.74	2.34	8.71	0.81	0.00	5.56	139.92	27.37	88.17	111.90
G-KIM 107	1.70	2.24	2.36	19.02	47.06	0.82	1.76	17.10	0.71	0.00	7.09	−53.03	−346.33	86.99	52.02
G-KIM 108	8.62	0.73	1.27	22.95	44.32	0.92	2.08	9.30	0.70	0.00	9.00	66.48	51.06	106.04	126.43
G-KIM 109	1.42	0.84	1.38	23.32	54.33	0.36	2.07	7.34	1.01	0.00	7.79	141.63	124.78	108.69	125.28
G-KIM 109	1.44	0.76	1.12	21.64	59.75	0.42	3.05	5.43	0.72	0.00	5.59	176.17	141.87	76.90	117.04
G-KIM 110	2.11	0.70	1.19	18.15	38.19	0.88	1.94	30.76	0.77	0.00	5.19	−218.48	−804.66	77.42	96.03
G-KIM 111	1.88	1.27	1.07	20.66	59.06	1.41	2.07	4.70	0.65	0.00	7.14	185.26	152.72	91.04	94.20
G-BEL 079	57.56	0.11	1.05	0.74	1.80	0.27	0.00	27.99	0.00	0.00	9.24	−325.98	−986.88	4.86	1.40

G-BLAP 080	40.84	4.63	3.47	3.43	2.23	0.14	0.92	32.50	0.00	0.00	10.89	−542.75	−1260.60	2.81	−126.92
G-GOND 081	46.72	0.13	1.18	5.20	1.71	0.35	0.00	40.05	0.31	0.00	4.11	−470.39	−1330.22	33.98	29.91
G-SAAR 078	41.58	0.49	0.19	11.71	15.03	0.15	0.36	19.68	0.70	0.00	9.53	−174.10	−495.34	69.05	60.94
G-SAAR 078	47.61	0.46	0.16	8.02	11.59	0.15	0.27	20.08	0.84	0.00	10.25	−195.14	−579.31	46.77	37.61
G-OLE 072	70.44	0.44	0.00	0.21	3.65	0.00	0.00	24.44	0.00	0.00	0.58	−284.42	−881.50	1.40	−12.87
G-OLE 073	61.93	0.34	0.19	1.16	5.68	0.19	0.47	18.13	2.50	0.00	9.34	−205.22	−645.05	−2.47	−3.50
G-OLE 074	62.04	1.11	0.14	1.09	3.94	0.15	0.19	19.57	0.00	0.00	11.67	−250.53	−716.08	3.11	−28.62
G-OLE 075	67.65	0.00	0.14	1.44	5.94	0.41	0.25	22.57	0.00	0.00	1.47	−240.75	−782.01	4.03	9.42
G-OLE 076	53.88	0.33	0.38	2.21	11.10	0.12	0.95	21.54	6.87	0.00	2.56	−225.45	−755.46	−5.62	3.76
G-OLE-MOL 067	66.85	0.63	0.18	0.88	5.33	0.42	0.28	21.59	0.00	0.00	3.71	−253.35	−778.80	−0.24	−14.58
G-OLE-MOL 067	69.77	0.73	0.00	0.40	3.53	0.28	0.00	23.37	0.00	0.00	1.68	−281.61	−849.03	2.60	−20.88
G-OLE-MOL 068	69.47	0.38	0.11	0.66	4.91	0.00	0.26	22.34	0.00	0.00	1.70	−256.03	−801.38	−1.11	−7.99
G-OLE-MOL 069	69.73	0.14	0.00	0.83	3.75	0.38	0.00	23.51	0.00	0.00	1.44	−263.25	−826.77	5.46	0.85
G-OLE-MOL 070	67.53	0.47	0.11	0.86	4.73	0.14	0.19	23.18	0.00	0.00	2.69	−268.41	−828.93	1.67	−9.48
G-OLE-MOL 071	69.21	0.00	0.12	1.06	4.07	0.32	0.00	18.67	0.00	0.00	6.35	−199.40	−645.13	6.93	6.93
G-DUR 058	39.24	1.19	0.45	2.52	39.29	0.18	0.62	15.13	0.13	0.00	1.13	−14.40	−541.57	3.37	−21.95
G-GARx 061	60.10	0.62	0.49	5.70	9.37	0.12	0.35	11.37	0.53	0.00	11.21	−110.89	−321.16	29.90	17.24
G-GARx 061	61.35	0.74	0.45	4.60	8.13	0.13	0.44	11.34	0.50	0.00	12.18	−122.84	−346.94	20.81	6.39
G-FOI 004	61.02	0.00	1.44	11.58	1.48	0.14	0.00	19.98	0.00	0.00	3.87	−229.14	−485.24	75.70	75.70
G-RLB 062	61.18	0.55	1.39	9.97	3.06	0.00	0.27	13.43	0.00	0.00	9.74	−166.20	−306.84	59.43	47.51
G-RLB 062	58.95	0.00	1.27	8.89	3.07	0.00	0.00	10.50	0.00	0.33	16.78	−107.75	−199.94	58.14	58.14
G-XXX 077	34.97	0.25	0.36	24.19	24.46	0.00	0.44	11.55	1.51	0.00	2.06	−19.58	45.34	148.99	150.09
G-XXX 082	54.61	1.28	0.00	0.93	3.19	0.10	0.00	39.05	0.22	0.00	0.49	−487.60	−1415.66	6.07	−35.20
G MULL 029	23.53	0.00	0.78	8.76	54.94	0.00	0.00	7.88	0.67	0.00	3.16	209.60	−109.28	57.29	57.29
G WEM 032	56.43	2.12	0.76	10.08	10.35	0.00	0.64	18.04	0.25	0.00	1.19	−239.28	−527.62	52.34	−2.60
G WHI 030	75.36	0.00	0.00	0.00	1.89	0.15	0.00	22.08	0.11	0.00	0.11	−251.85	−787.26	0.00	0.00
G XXX 031	45.45	0.50	0.40	6.37	10.78	1.38	0.49	22.69	0.43	0.00	11.33	−236.79	−710.91	31.20	25.36
Archaeological samples															
A-ALC 161	5.62	0.00	0.81	12.88	37.52	4.44	4.74	5.76	0.49	0.14	27.07	37.97	−53.32	−16.47	84.26
A-ALC 162	4.12	0.60	1.30	16.05	39.67	4.15	6.61	5.14	0.51	0.70	20.65	−1.96	−28.32	−35.40	85.52
A-ALC 163	5.33	0.36	1.63	13.60	39.97	7.28	2.58	17.06	0.43	0.42	11.20	−48.24	−407.84	34.22	77.40
A-ALC 164	9.32	0.15	1.34	18.77	42.05	1.26	2.52	16.14	0.64	0.18	7.55	−17.72	−265.27	69.36	118.00
A-ALC 165	5.53	0.28	1.61	15.21	44.14	4.73	2.17	12.28	0.58	0.12	13.13	42.52	−194.72	53.43	90.31

(continued)

	C	Na_2O	MgO	Al_2O_3	SiO_2	P_2O_5	K_2O	CaO	TiO_2	MnO	Fe_2O_3	Q	A	Al/3-K	Al/3-Na
A-ALC 166	5.51	0.24	1.58	13.77	45.70	4.82	1.81	16.06	0.53	0.18	9.62	15.03	−349.12	51.57	82.14
A-ALC 167	10.02	0.12	1.23	10.03	21.02	3.47	1.47	43.00	0.33	0.15	9.00	−430.01	−1371.85	34.42	61.61
A-ALC 168	5.40	0.25	0.94	18.01	47.72	2.80	3.75	5.44	0.46	0.00	14.81	111.14	71.71	38.24	109.64
A-ALC 169	4.67	0.39	0.91	16.18	46.55	2.40	4.05	4.72	0.48	0.00	19.18	102.40	50.48	19.83	93.28
A-ALC 170	12.47	0.15	1.37	12.26	25.09	4.58	1.66	31.54	0.39	0.00	10.29	−276.27	−924.33	44.95	75.40
A-ALC 171	6.63	0.23	1.61	20.69	41.55	0.57	2.91	17.48	0.55	0.00	7.61	−47.62	−286.72	73.55	127.83
A-ALC 172	7.35	0.41	1.21	15.27	34.30	1.55	2.36	18.47	0.47	0.10	18.32	−93.27	−422.35	49.82	86.78
A-CHAL 173	4.11	0.24	1.11	16.60	36.10	5.20	8.16	3.84	0.49	0.23	22.62	−27.43	7.76	−64.73	100.76
A-CHAL 174	5.44	0.00	0.62	13.36	41.99	2.36	6.07	2.25	0.42	0.00	26.34	76.21	53.22	−41.48	87.40
A-CHAL 174	5.33	0.00	0.66	14.95	43.17	2.16	5.23	3.29	0.37	0.00	23.67	88.08	64.76	−13.35	97.75
A-CHAL 175	3.22	0.33	1.28	16.89	45.35	3.14	3.78	2.45	0.54	0.00	22.27	130.26	152.93	30.18	99.77
A YOC 123	2.96	0.48	1.29	24.30	53.95	0.49	3.52	0.55	0.88	0.00	11.46	201.07	366.83	84.23	143.46
A-YOC 124	2.56	0.48	1.35	26.55	54.75	0.98	2.71	0.94	0.79	0.00	8.81	218.18	414.52	116.17	158.17
A-YOC 125	4.39	0.41	1.69	25.10	52.31	0.24	2.94	0.94	0.84	0.00	11.06	201.77	383.15	101.76	150.87
A-MTC 046	3.05	0.00	1.97	21.80	61.14	0.72	3.44	0.84	0.93	0.00	6.00	254.35	324.47	69.43	142.55
A-MTC 050	1.78	0.23	2.38	23.73	50.50	2.76	3.81	1.05	0.90	0.12	12.65	178.11	340.01	74.34	147.91
A-MTC 054	2.92	0.43	2.28	23.22	52.81	1.24	3.26	2.94	0.56	0.00	10.10	173.59	267.75	82.63	138.07
A-MTC 055	1.16	0.00	2.03	23.53	54.78	1.16	3.27	1.52	0.62	0.18	11.66	214.93	338.02	84.45	153.87
A-MTC 056	0.23	0.00	1.95	22.93	54.22	1.03	3.20	1.08	0.57	0.10	14.56	218.66	343.62	82.08	149.95
A-MTC 057	1.11	0.00	2.30	25.44	57.84	0.19	2.90	0.93	0.81	0.16	8.21	246.63	404.26	104.85	166.37
A-MTC 058	0.92	0.42	2.12	25.31	57.34	0.59	2.76	0.65	0.63	0.11	9.09	236.64	401.21	106.84	152.03
A-MTC 059	0.91	1.08	1.99	21.06	48.06	2.57	4.15	3.90	0.80	0.29	14.81	95.87	151.03	49.71	102.78
A-MTC 062	0.34	0.23	1.67	20.12	47.64	1.90	3.43	4.32	0.60	0.61	18.94	131.43	160.53	58.73	124.18
A-MTC 063	1.71	0.42	2.07	20.70	49.46	1.32	4.51	4.49	1.09	0.18	13.87	110.22	136.67	39.54	121.78
A-MTC 066	5.38	0.48	1.89	23.61	53.63	0.81	3.85	0.88	0.57	0.00	8.78	188.37	334.63	72.65	138.93
A-MTC 067	1.98	0.42	1.99	22.50	51.87	1.30	3.52	1.28	0.63	0.00	14.30	183.02	307.71	72.48	133.77
A-MTC 068	1.04	0.60	1.74	21.68	50.11	1.06	3.96	2.89	0.67	1.69	14.44	138.84	218.99	57.71	122.48
A-MTC 071	4.25	0.27	1.34	17.35	42.78	8.72	3.10	4.47	0.50	0.17	16.62	108.66	106.68	47.72	104.79
A-MTC 072	5.00	0.23	1.37	16.40	34.46	12.51	2.50	8.71	0.45	0.53	17.51	26.38	−49.25	54.26	99.91
A-MTC 073	1.91	0.19	0.77	15.60	47.42	6.12	5.02	1.98	0.46	0.28	19.67	125.30	122.53	−4.61	95.77
A-MTC 074	5.77	0.21	0.85	16.86	47.47	3.23	5.48	2.35	0.53	0.21	15.61	111.15	124.06	−6.05	103.64

A-MTC 075	6.92	0.42	1.00	18.14	39.89	3.58	4.30	4.05	0.49	0.00	20.95	67.16	106.55	27.40	105.03
A-MTC 078	3.29	0.41	1.05	17.14	35.62	2.36	3.27	1.75	0.43	0.00	34.35	93.24	191.11	42.61	98.98
A-MTC 080	4.93	0.23	0.99	20.01	50.25	2.57	4.42	2.92	0.62	0.00	12.82	141.28	187.08	36.90	123.34
A-MTC 080	6.83	0.23	0.86	16.06	39.44	5.02	4.75	4.85	0.46	0.00	21.18	51.85	34.03	4.29	97.50
A-MTC 081	3.83	0.44	0.87	17.55	64.22	0.19	2.99	4.44	0.95	0.00	4.37	224.19	108.34	51.33	100.68
A-MTC 082	5.54	0.20	1.81	21.67	46.53	2.22	3.05	5.52	0.57	1.80	10.71	120.19	157.30	77.01	135.32
A-MTC 082	2.91	0.00	1.89	23.41	50.00	0.97	2.45	2.23	0.48	2.10	13.18	197.43	327.58	101.06	153.10
A-MTC 084	4.44	0.15	1.09	23.20	49.70	3.56	3.96	3.14	0.83	0.00	9.76	148.19	254.33	67.66	146.94
A-MTC 085	12.92	0.00	1.10	22.36	52.64	0.45	3.01	4.05	0.47	0.00	2.79	178.51	230.40	82.26	146.23
A-MTC 086	5.52	0.27	1.58	20.53	41.62	5.88	4.03	3.91	0.63	0.35	15.14	89.04	169.07	48.59	125.69
A-MTC 086	4.44	0.32	1.72	17.19	36.03	11.00	5.23	5.95	0.80	0.40	15.98	6.89	3.79	1.37	102.10
A-MTC 090	5.84	0.22	1.05	20.54	49.17	2.74	3.35	3.53	0.54	0.00	12.86	151.10	198.70	63.21	127.12
A-BUX 138	8.08	0.33	1.76	22.30	46.59	2.19	3.93	3.69	0.77	0.00	10.20	119.32	212.01	62.47	135.17
A-BUX 139	6.56	0.33	1.77	22.90	47.49	2.80	3.50	5.36	0.70	0.00	8.26	113.32	173.09	75.46	139.01
A-BUX 140	3.98	0.12	1.20	14.49	64.46	0.82	3.04	1.71	0.42	1.18	8.46	267.05	154.98	30.20	90.82
A-BUX 141	5.99	0.33	1.45	20.24	52.49	3.38	2.84	3.00	0.62	0.00	9.48	183.12	219.27	72.01	121.77
A-BUX 142	7.95	0.29	1.19	16.68	48.31	5.81	2.03	9.19	0.51	0.00	7.79	105.04	−52.86	65.97	99.67
A-BUX 143	5.16	0.29	1.52	16.00	39.90	6.13	2.16	11.74	0.44	0.33	15.84	25.65	−159.78	58.78	95.32
A-BUX 143	10.57	0.23	1.32	15.52	45.62	3.44	1.78	8.54	0.46	0.23	12.07	105.08	−45.50	63.76	93.96
A-BUX 144	8.45	0.36	1.08	19.44	44.77	4.60	3.15	6.28	0.65	0.12	10.81	94.15	78.92	60.30	115.62
A-BUX 145	6.24	0.36	1.74	19.75	50.93	2.13	2.98	4.81	0.70	0.00	10.12	149.09	141.02	65.81	117.52
A-BUX 146	4.15	0.18	1.66	17.66	61.15	1.22	2.45	1.91	0.49	0.00	8.06	256.83	220.37	63.40	109.58
A-BUX 147	6.20	0.35	1.81	20.72	54.31	1.62	3.22	3.53	0.81	0.00	7.20	178.07	200.72	67.18	124.08
A-BUX 148	7.18	0.38	1.55	20.83	47.87	4.47	2.56	7.41	0.72	0.00	6.81	109.50	77.59	81.80	123.88
A-BUX 149	6.68	0.21	1.48	17.05	54.54	1.78	3.40	2.70	0.53	0.00	11.40	190.14	159.56	39.41	104.77
A-BUX 150	4.58	0.31	1.96	21.32	53.86	1.06	2.82	3.75	0.64	0.29	9.25	182.86	214.62	79.50	129.45
A-BUX 151	6.76	0.38	1.21	21.92	51.03	1.56	3.22	3.76	0.85	0.00	8.39	156.44	215.50	75.06	131.06
A-BUX 152	7.00	0.16	1.35	18.04	46.26	3.85	2.36	7.98	0.44	0.00	12.20	105.31	14.17	67.82	112.90
A-BUX 153	5.11	0.24	1.51	18.39	52.84	2.59	3.42	3.03	0.72	0.00	11.77	175.35	172.45	47.73	112.57
A-BUX 154	8.05	0.39	1.82	18.91	47.16	2.64	2.81	5.08	0.61	0.00	11.92	128.00	117.85	64.09	111.19
A-BUX 155	4.83	0.40	1.76	17.45	49.42	4.41	3.68	4.64	0.71	0.43	12.03	126.78	86.03	36.10	101.21
A-BUX 156	3.98	0.23	1.53	23.60	53.66	0.51	2.82	1.35	0.82	0.10	11.31	212.79	347.49	94.54	146.88
A-BUX 157	7.01	0.38	1.60	19.68	44.45	5.13	2.75	8.81	0.63	0.00	9.12	70.19	1.42	70.41	116.59

(continued)

	C	Na_2O	MgO	Al_2O_3	SiO_2	P_2O_5	K_2O	CaO	TiO_2	MnO	Fe_2O_3	Q	A	Al/3-K	Al/3-Na
A-BUX 159	4.14	0.25	1.37	21.91	53.73	1.22	2.45	2.08	0.62	0.00	11.35	211.79	295.63	91.27	135.14
A-BUX 160	6.94	0.39	1.59	19.40	49.17	1.60	3.53	2.81	0.67	0.00	13.32	150.54	192.95	51.89	114.44
A BRG 001	6.25	0.27	1.32	18.40	43.17	0.75	2.61	19.17	0.79	0.00	7.14	−53.45	−386.53	64.99	111.76
A BRG 002	5.35	0.23	1.21	14.95	35.79	0.37	2.08	34.94	0.55	0.00	4.40	−269.20	−1004.35	53.57	90.25
A BRG 003	11.32	0.23	1.27	10.02	24.19	2.58	1.86	40.59	0.40	0.00	7.32	−395.43	−1297.70	26.02	58.24
A-BRG 004	17.54	0.22	0.72	6.98	16.17	5.29	0.67	44.51	0.21	0.15	7.42	−460.78	−1471.62	31.33	38.66
A-BRG 005	2.48	0.17	0.88	12.69	32.44	0.52	1.73	37.70	0.48	0.00	10.76	−310.85	−1137.50	46.33	77.50
A BRG 006	5.17	0.19	1.56	16.39	37.31	2.76	1.88	23.41	0.45	0.70	10.02	−118.33	−559.37	67.29	100.98
A BRG 007	5.55	0.38	1.72	16.05	39.53	5.85	2.63	16.65	0.58	0.78	10.11	−47.58	−346.83	49.12	92.87
A BRG 008	54.11	0.14	0.28	2.18	7.97	0.23	0.65	32.28	0.98	0.00	1.11	−357.95	−1126.80	0.44	9.58
A BRG 009	28.44	0.24	0.85	19.07	38.63	0.14	3.06	6.67	0.88	0.00	1.94	61.33	63.61	59.67	117.05
A BRG 010	10.73	0.28	1.62	7.07	17.76	4.27	1.71	47.85	0.29	0.37	7.80	−515.72	−1612.81	10.03	37.11
A BRG 011	4.52	0.31	1.06	16.16	40.72	3.98	6.30	3.98	0.61	0.18	20.67	33.75	31.38	−28.11	95.76
A BRG 012	6.53	0.35	1.46	17.82	42.18	1.38	2.83	15.99	0.67	0.00	10.57	−28.42	−291.79	56.44	105.32
A-BRG 013	0.77	0.12	1.70	15.13	39.75	6.15	2.88	24.73	0.56	0.40	7.59	−139.51	−650.10	37.82	94.96
A-BRG 014	1.83	0.14	1.52	20.26	60.50	1.77	3.47	2.84	0.56	0.00	6.94	221.85	217.80	58.73	127.93
A-BRG 015	8.91	0.37	1.64	22.59	44.09	1.26	3.81	9.06	0.76	0.00	7.41	42.77	27.16	66.85	135.74
A BRG 016	2.53	0.63	2.10	23.55	59.29	0.46	3.84	1.16	0.89	0.00	5.44	211.78	318.85	72.48	133.76
A BRG 017	5.96	0.60	1.04	12.60	43.25	4.65	5.44	6.27	0.38	0.32	19.12	29.42	−111.18	−33.02	62.99
A-BRG 018	8.44	0.69	0.73	10.86	42.95	4.51	2.76	10.39	0.92	0.00	17.47	32.82	−238.34	12.47	48.73
A BRG 019	4.45	0.28	0.97	19.79	45.80	4.75	3.72	6.52	0.55	0.00	12.84	87.35	67.86	50.49	120.33
A BRG 020	4.97	0.44	1.58	15.07	35.11	7.16	3.02	20.37	0.62	0.59	9.78	−126.55	−508.98	34.39	84.32
A BRG 021	3.71	0.34	1.44	15.69	42.47	4.46	3.57	6.30	0.48	2.66	18.63	72.85	−3.62	26.78	91.67
A BRG 022	5.81	0.40	1.26	13.27	36.39	4.38	3.91	6.90	0.50	2.25	24.70	22.92	−81.68	3.72	73.83
A BRG 023	4.78	0.22	1.45	16.45	58.42	2.10	3.40	5.11	0.65	0.12	7.18	182.64	61.31	35.42	100.61
A BRG 024	5.22	0.35	0.79	9.19	61.03	3.31	3.57	3.70	0.41	0.00	12.19	205.98	−38.55	−15.69	48.89
A BRG 025	7.31	0.30	0.72	10.07	43.66	6.40	4.16	8.21	0.37	0.15	18.33	45.74	−192.91	−22.45	56.31
A BRG 026	6.52	0.41	1.31	13.65	38.10	6.16	4.30	7.15	0.45	0.32	21.25	20.75	−91.79	−2.08	75.90
A BRG 027	11.64	0.25	0.93	5.90	14.59	2.84	1.54	55.31	0.26	0.00	6.46	−617.32	−1897.51	5.92	30.51
A BRG 028	12.87	0.22	1.02	10.38	21.75	2.37	1.61	42.51	0.37	0.00	6.71	−426.11	−1353.34	33.67	60.85
A BRG 029	2.25	0.22	2.01	23.31	57.50	1.25	3.27	2.59	0.97	0.00	6.50	210.18	288.61	83.03	145.48

A BRG 030	40.46	0.00	0.41	0.45	1.43	0.20	0.00	54.34	2.26	0.00	0.23	−637.68	−1928.97	2.95	2.95
A BRG 030bis	77.70	0.21	0.00	0.00	1.54	0.26	0.00	18.69	1.31	0.00	0.14	−220.12	−672.99	0.00	−6.66
A BRG 031	4.59	0.37	2.11	21.66	52.88	1.91	3.36	4.65	0.81	0.00	7.53	153.45	175.94	70.26	129.79
A-ECK 191	0.97	0.34	2.25	11.56	38.96	1.31	2.02	34.27	0.42	0.00	7.39	−245.91	−1049.11	32.63	64.75
A-ECK 192	1.37	0.37	1.78	13.20	39.48	0.52	2.09	33.83	0.45	0.00	6.58	−240.04	−1003.46	42.02	74.53
A-ECK 193	1.53	0.27	1.91	13.80	39.96	0.96	2.13	32.60	0.47	0.00	6.06	−220.44	−945.69	45.09	81.66
A-ECK 194	0.78	0.17	1.58	10.78	30.59	0.79	1.55	46.48	0.39	0.00	6.09	−421.75	−1484.30	37.64	64.88
A-ECK 195	1.24	0.32	1.80	12.50	39.33	1.44	2.02	34.45	0.45	0.00	6.16	−245.23	−1036.42	38.93	71.47
A-ECK 196	1.11	0.37	1.79	12.83	37.44	0.42	1.95	37.31	0.45	0.00	5.52	−290.14	−1132.32	42.40	71.84
A-ECK 197	3.19	0.49	1.38	17.44	37.93	1.64	2.57	27.67	0.54	0.00	6.97	−189.71	−714.96	59.50	98.30
A-ECK 198	1.12	0.34	1.64	11.81	34.06	0.90	1.63	40.83	0.61	0.11	6.36	−342.80	−1270.02	42.52	66.17
A-ECK 199	1.35	0.47	1.81	15.29	45.76	1.02	2.28	23.75	0.53	0.00	7.40	−93.10	−610.45	51.60	84.88
A-ECK 200	1.40	0.00	1.32	11.59	30.61	2.19	1.89	44.32	0.37	0.00	5.80	−397.61	−1393.08	35.67	75.78
A-ECK 201	1.25	0.30	1.55	14.42	38.32	1.26	2.00	34.68	0.50	0.00	5.34	−252.51	−1005.81	51.88	84.67
A-ECK 202	1.54	0.40	1.58	12.91	38.03	0.65	2.02	36.01	0.49	0.12	6.03	−273.46	−1086.26	41.64	71.67
A-ECK 203	1.29	0.37	1.59	13.00	38.66	0.43	2.39	35.63	0.47	0.00	5.37	−272.44	−1077.96	34.32	73.24
A-ECK 204	1.23	0.23	1.80	9.76	26.37	1.86	1.50	50.37	0.35	0.12	6.24	−491.95	−1643.80	31.96	56.54
A-ECK 205	1.79	0.37	1.65	12.80	39.72	1.29	2.37	32.93	0.45	0.00	6.33	−234.17	−985.29	33.49	71.74
A-ECK 206	0.85	0.36	1.81	15.76	49.28	4.30	2.59	17.25	0.50	0.00	7.05	0.42	−372.66	48.11	91.34
A-ECK 207	1.24	0.20	1.99	13.92	38.62	1.06	2.04	34.36	0.46	0.00	5.86	−245.03	−1002.24	47.60	84.44
A-ECK 208	1.40	0.27	1.56	13.13	41.41	1.52	2.10	31.89	0.47	0.00	5.92	−203.58	−932.94	41.20	77.17
A-ECK 209	1.71	0.19	1.46	9.58	24.95	1.24	1.40	53.16	0.32	0.11	5.51	−529.70	−1743.72	32.90	56.52
A-ECK 210	1.73	0.30	1.42	10.83	33.67	1.62	1.86	42.15	0.40	0.00	5.69	−364.01	−1339.52	31.27	61.25

Minor Results

	Li	Be	V	Cr	Cu	Zn	Ga	As	Rb	Sr	Y	Zr	Nb	Mo	Sb
Geological samples															
G-AICH 021	23	2	499	74	202	431	20	26	79	2780	54	96	7	43	3
G-AICH-S 027	21	1	427	58	163	231	18	24	71	3405	41	75	8	42	3
G-GEIS 065	9	1	94	38	77	1142	13	36	56	847	8	29	5	39	2
G-GEIS 065	0	1	202	63	112	242	19	91	82	4576	23	60	7	46	5
G-GEIS 065	0	0	255	66	83	190	19	32	89	4367	22	20	8	42	6
G-DOTT 028	27	1	188	86	49	56	21	122	99	1084	9	31	6	62	2
G-DOTT 028	21	1	151	69	40	45	17	97	79	867	7	25	5	49	2
G-OHM 019	19	2	238	78	116	373	27	41	109	1722	42	75	8	86	2
G-OHM-B 020	16	2	423	60	147	300	20	16	76	2776	264	102	7	47	2
G-OHM-B 020	14	1	423	60	197	215	21	24	75	2342	51	83	8	49	3
G-OHM-L 018	12	2	408	71	188	222	22	27	77	3256	42	93	7	59	3
G-HOLZ 022	15	2	450	81	153	348	25	32	100	2197	35	113	10	29	3
G-HOLZ 025	6	1	337	54	125	218	14	28	56	2173	23	51	4	33	1
G-HOLZ 025	0	2	437	65	184	708	18	37	64	1765	53	77	5	43	3
G-SCHL-B 026	28	2	502	82	153	249	26	28	101	3432	37	115	14	45	3
G-BOL 017	14	1	371	54	179	322	19	22	71	3173	45	143	5	38	3
G-ERZ 024	8	2	177	56	68	319	17	70	84	4654	32	120	8	41	3
G-WAN 066	22	1	131	68	73	1232	14	50	56	2067	16	44	5	13	1
G-WIS 015	62	3	262	153	94	218	52	39	181	962	33	119	17	31	3
G-BUX 006	76	32	124	115	231	105	30	28	271	80	61	341	29	12	167
G-BUX 007	97	18	309	121	77	107	35	221	272	126	12	138	18	25	58
G-BUX 008	66	2	106	60	53	105	21	150	142	1420	20	62	9	23	30
G-BUX 009	125	10	172	111	87	67	38	536	325	123	26	116	20	22	71
G-BUX 010	81	3	136	80	104	228	30	306	219	352	25	93	13	39	56
G-BUX 011	81	13	186	88	68	235	37	158	246	73	14	134	16	16	78
G-BUX 041	73	11	202	91	65	124	41	258	206	749	36	117	23	27	94
G-BUX 042	66	8	153	75	34	115	30	82	207	480	15	690	17	199	52
G-BUX 043	83	22	331	125	71	123	42	117	285	162	20	137	15	40	71
G-BUX 044	80	7	194	94	58	199	30	167	155	97	24	118	10	19	59

G-BUX 045	79	4	164	85	15	129	38	264	260	212	41	76	16	46	67
G-BUX 046	45	5	134	76	79	67	28	174	182	89	10	109	13	11	70
G-BUX 047	62	2	125	63	52	101	31	112	195	895	19	84	13	52	43
G-BUX 048	54	2	147	67	44	33	23	50	210	44	3	46	11	15	17
G-BUX 048	56	3	145	66	30	44	23	29	212	61	4	51	11	18	18
G-BUX 049	116	11	141	93	79	74	35	301	373	310	152	150	15	12	56
G-BUX 049	113	8	144	92	43	56	35	368	277	95	23	121	19	16	50
G-BUX 050	85	7	128	71	128	65	30	127	179	55	14	183	21	13	1215
G-BUX 050	32	3	65	29	36	125	61	17	493	48	0	10	0	0	6
G-BUX 051	41	34	96	72	4	17	23	18	166	58	38	194	8	8	151
G-BUX 052	73	18	203	90	104	157	39	467	207	199	43	129	14	30	119
G-BUX 053	68	7	155	73	65	97	28	694	159	213	28	85	12	48	127
G-BUX 054	43	5	124	65	50	55	23	100	142	90	5	76	9	12	42
G-BUX 054	60	6	188	95	37	79	37	150	247	604	7	106	14	18	38
G-BUX 055	38	6	83	40	5	68	15	46	108	277	6	48	8	38	17
G-BUX 056	99	9	247	109	62	126	40	79	279	375	73	133	17	22	43
G-BUX 057	92	5	97	70	48	49	37	99	244	53	10	110	34	5	215
G-HIL 038	37	5	90	45	50	44	20	61	117	109	14	47	7	21	28
G-VOG 060	42	2	130	115	94	24	21	20	30	651	39	41	10	1	23
G-VOG 060	54	3	148	143	117	29	24	18	31	786	30	69	8	3	24
G-VOG 060	36	2	111	99	80	21	18	17	25	558	33	35	9	1	20
G-MKZ 013	35	73	197	146	142	551	46	1020	95	2792	87	119	12	64	134
A-MKZ 121	33	60	125	105	127	470	62	711	91	1876	89	104	11	30	106
A-SRB 122	66	60	148	100	101	66	53	541	123	694	916	133	14	27	14
A-SRB 122	63	88	191	114	117	105	89	617	111	680	627	139	14	33	14
A-MKZ 121	44	64	152	139	256	4127	123	993	107	2027	231	104	8	28	68
A-MKZ 121	49	70	179	138	231	3461	126	1237	118	1957	208	116	11	34	87
A-MKZ 121	61	38	219	175	292	4785	63	1873	120	2229	226	117	12	83	164
A-MKZ 121	54	26	211	156	191	4348	62	2369	113	2286	37	97	12	68	149
A-SRB 122	84	23	181	115	139	267	45	60	169	308	280	92	16	32	7
A-SRB 122	67	29	194	106	229	131	51	619	135	384	420	108	13	26	11
A-SRB 122	72	28	197	107	137	173	44	465	144	304	193	97	14	20	9
A-SRB 122	83	38	232	119	257	238	57	432	177	373	216	111	16	26	8

(continued)

	Li	Be	V	Cr	Cu	Zn	Ga	As	Rb	Sr	Y	Zr	Nb	Mo	Sb
A-MKZ 246	66	34	245	175	1056	1151	73	850	184	1665	69	195	14	18	97
A-MKZ 247	55	102	144	130	229	630	116	643	120	3681	429	139	16	27	35
A-MKZ 247	73	109	158	150	301	488	86	451	152	5223	439	166	15	16	28
A-MKZ 247	96	139	223	181	222	462	117	659	186	6184	464	208	20	35	25
A-MKZ 248	42	35	148	105	187	537	44	540	78	1343	26	104	8	37	113
G-KIM 012	88	3	203	132	189	84	26	29	144	448	83	267	13	406	1
G KIM 033	50	2	106	68	106	166	19	15	109	551	59	97	7	15	1
G-KIM 084	141	4	198	164	44	72	43	9	257	193	22	143	23	4	1
G-KIM 085	97	4	207	141	178	157	34	46	201	528	47	134	17	138	1
G-KIM 085	94	3	241	141	154	100	33	43	198	882	53	108	16	175	1
G-KIM 086	92	3	211	134	122	97	31	31	202	596	44	114	14	146	1
G-KIM 087	128	4	212	171	71	320	51	13	312	157	35	168	24	15	1
G-KIM 088	162	4	193	167	61	225	44	22	278	166	23	145	28	12	0
G-KIM 089	21	2	105	62	104	184	10	72	44	508	9	62	4	1	1
G-KIM 089	33	3	154	79	103	133	12	78	63	698	12	67	7	14	1
G-KIM 090	138	4	192	171	53	113	44	13	271	203	23	165	25	4	1
G-KIM 091	160	4	191	172	58	98	44	19	275	222	24	204	26	7	0
G-KIM 092	179	5	287	258	244	122	48	51	284	135	27	185	29	99	2
G-KIM 093	140	4	293	213	261	204	36	50	235	426	51	150	19	193	2
G-KIM 094	225	5	323	288	654	197	58	22	335	140	26	197	40	90	1
G-KIM 095	173	4	254	231	40	94	56	25	306	114	21	203	35	24	1
G-KIM 096	38	2	89	69	51	93	16	26	104	344	15	55	7	5	0
G-KIM 097	237	5	278	235	170	117	45	31	268	276	41	163	30	83	1
G-KIM 098	92	3	171	130	66	90	39	11	264	322	32	110	17	7	2
G-KIM 099	128	3	284	156	120	79	44	22	271	299	25	149	33	65	2
G-KIM 100	74	3	194	135	128	119	41	28	255	222	31	161	25	51	1
G-KIM 101	76	1	136	91	0	21	31	0	234	419	25	101	18	0	0
G-KIM 102	114	3	165	134	39	84	39	23	214	181	115	134	18	8	1
G-KIM 103	147	2	189	140	20	49	34	0	242	93	18	117	21	21	1
G-KIM 104	173	4	227	189	104	107	43	23	265	108	23	155	23	24	1
G-KIM 105	226	4	251	206	41	75	45	15	296	101	15	161	33	48	1
G-KIM 106	147	4	208	171	59	67	49	26	285	311	32	141	24	6	1

G-KIM 107	129	5	223	163	88	124	41	31	244	647	116	197	27	42	1
G-KIM 108	149	4	280	206	205	123	38	38	228	278	54	175	24	209	1
G-KIM 109	125	4	243	166	121	81	43	33	288	266	61	157	21	58	2
G-KIM 109	148	4	292	200	121	382	48	33	320	229	34	149	28	57	2
G-KIM 110	101	2	166	136	141	161	29	35	178	365	32	97	14	31	1
G-KIM 111	126	3	172	159	138	153	43	53	237	152	36	198	30	30	1
G-BEL 079	1	2	2	6	4	18	17	8	0	978	27	1	0	1	0
G-BLAP 080	0	6	46	21	19	78	14	5	1	1647	25	8	5	42	5
G-GOND 081	3	0	39	17	6	6	2	26	4	54	4	3	1	1	1
G-SAAR 078	35	11	33	57	247	1188	55	14	11	83	56	53	7	5	2
G-SAAR 078	27	9	25	44	190	916	42	11	9	64	43	41	5	4	1
G-OLE 072	0	3	166	112	11	12	5	5	1	109	5	621	37	2	1
G-OLE 073	8	6	716	188	44	20	12	61	21	134	11	1031	56	9	4
G-OLE 073	1	8	140	138	21	26	18	16	6	97	8	632	51	5	1
G-OLE 074	27	8	4	22	14	37	64	18	4	126	11	6	0	8	1
G-OLE 075	11	14	164	94	35	29	50	16	4	154	13	253	33	4	2
G-OLE 076	0	11	151	100	55	68	34	19	13	364	22	553	12	2	0
G-OLE-MOL 067	3	3	425	112	49	82	7	16	4	226	40	278	46	6	1
G-OLE-MOL 067	2	2	281	74	33	54	5	11	2	150	27	184	31	4	1
G-OLE-MOL 068	0	7	26	38	18	31	18	2	2	86	9	16	4	2	1
G-OLE-MOL 069	13	8	127	44	0	15	32	0	0	110	26	76	7	3	5
G-OLE-MOL 070	0	4	311	70	16	32	4	14	2	126	8	144	30	2	4
G-OLE-MOL 071	14	15	10	42	5	49	52	5	3	198	19	217	4	2	6
G-DUR 058	52	2	16	12	5	49	3	4	11	50	15	11	2	1	2
G-GARx 061	3	31	22	45	73	164	57	192	8	263	77	42	13	4	17
G-GARx 061	4	37	27	53	86	193	67	227	9	311	91	50	15	5	20
G-FOI 004	0	11	46	16	3	13	41	2	1	556	150	17	2	1	1
G-RLB 062	0	51	12	19	13	112	327	4	0	429	233	7	0	26	3
G-RLB 062	0	51	12	19	13	2	329	4	0	431	234	7	0	26	3
G-XXX 077	106	14	191	144	90	112	30	20	17	178	56	381	28	4	3
G-XXX 082	1	5	52	25	43	32	4	3	0	3	10	28	5	3	5
G MULL 029	10	3	1964	715	150	19	24	15	6	157	56	257	15	35	3
G WEM 032	16	60	239	72	18	21	148	9	24	169	112	42	5	6	6
G WHI 030	1	3	3387	157	71	2804	10	1	0	64	6	2973	58	1	0

(continued)

	Li	Be	V	Cr	Cu	Zn	Ga	As	Rb	Sr	Y	Zr	Nb	Mo	Sb
G WHI 030	1	3	3476	173	5	70	2	1	0	66	6	3156	57	1	0
G XXX 031	13	18	28	63	103	1066	63	15	17	245	19	65	20	9	5
Archaeological samples															
A-ALC 161	70	7	192	111	89	315	149	1804	166	616	23	54	13	31	11
A-ALC 161	71	7	203	131	120	443	223	2081	184	766	48	110	21	33	11
A-ALC 161	77	8	197	120	93	308	157	1580	194	502	32	76	17	24	9
A-ALC 162	99	14	274	126	88	698	118	1943	237	586	96	80	19	12	7
A-ALC 162	99	14	204	139	107	410	71	1374	201	323	66	73	15	9	5
A-ALC 163	110	6	142	106	118	554	65	611	227	477	21	45	14	1	2
A-ALC 164	132	4	280	621	222	337	56	40	182	145	49	306	16	22	4
A-ALC 164	135	4	270	820	191	293	44	22	199	143	49	91	16	14	4
A-ALC 165	79	10	127	129	72	482	60	507	184	503	42	79	15	4	2
A-ALC 165	73	30	100	108	89	456	74	349	173	1808	329	108	14	4	2
A-ALC 166	83	9	100	112	106	368	56	444	168	651	38	67	13	3	2
A-ALC 166	78	9	110	130	81	350	84	487	177	888	25	77	17	4	3
A-ALC 167	58	4	178	195	238	232	32	32	108	406	136	78	9	22	3
A-ALC 167	64	5	184	195	240	253	32	29	111	335	137	99	10	22	3
A-ALC 168	73	14	107	115	79	357	120	966	191	735	43	104	17	4	2
A-ALC 168	89	16	112	115	108	390	150	1420	228	738	46	82	17	6	5
A-ALC 169	70	6	128	141	128	360	148	965	184	637	30	81	13	5	3
A-ALC 169	79	7	136	153	105	390	139	792	218	522	36	110	16	6	2
A-ALC 170	68	4	206	203	296	509	43	36	129	477	126	127	10	44	3
A-ALC 171	104	7	251	191	191	185	47	31	189	264	140	130	19	15	3
A-ALC 172	78	5	330	522	449	849	65	117	154	278	37	99	12	25	4
A-CHAL 173	93	9	268	134	544	424	148	555	173	966	43	59	16	12	8
A-CHAL 173	82	9	289	138	770	584	287	1046	191	1377	32	68	15	20	8
A-CHAL 174	77	9	168	103	368	699	428	844	149	1050	21	52	11	10	2
A-CHAL 174	123	13	171	163	354	389	466	793	281	1188	37	98	19	9	2
A-CHAL 175	108	7	177	261	504	414	171	390	254	738	17	138	16	7	5
A-CHAL 175	83	6	146	92	497	400	161	430	183	911	16	94	11	6	4
A-CHAL 175	78	6	155	87	415	356	145	258	155	834	27	83	11	6	4
A-CHAL 175	84	6	157	110	579	344	155	325	197	895	22	73	13	5	4

A YOC 123	127	4	356	200	38	66	57	75	371	133	16	142	23	23	1
A-YOC 124	126	3	198	152	52	133	72	186	246	124	12	103	16	30	2
A-YOC 124	197	4	300	232	59	139	87	207	365	186	13	127	29	22	2
A-YOC 125	149	3	269	199	37	97	59	37	269	119	7	103	21	51	1
A-MTC 046	132	10	168	136	103	381	125	655	365	565	18	129	28	9	30
A-MTC 050	134	12	178	130	96	383	69	211	267	113	14	69	18	8	17
A-MTC 054	166	12	212	138	115	426	67	281	336	119	16	61	20	6	16
A-MTC 055	171	12	280	140	98	352	64	324	350	89	18	109	31	11	20
A-MTC 055	182	14	300	180	141	773	82	268	384	130	16	96	25	12	15
A-MTC 056	142	12	299	144	144	514	62	539	324	83	21	85	19	16	18
A-MTC 056	156	13	355	181	189	688	68	432	367	120	50	87	24	15	18
A-MTC 057	150	9	161	137	78	241	57	260	318	92	11	93	20	2	18
A-MTC 057	158	11	178	143	53	234	60	205	362	113	17	528	22	3	6
A-MTC 058	159	12	153	108	45	149	43	109	276	69	9	62	25	5	3
A-MTC 059	98	16	262	147	88	368	135	167	276	244	27	110	21	31	73
A-MTC 062	97	19	258	147	189	335	81	156	275	85	42	98	18	11	64
A-MTC 063	91	22	192	113	95	214	107	126	246	166	45	96	14	17	55
A-MTC 063	60	6	74	51	35	137	26	51	135	47	8	31	8	8	21
A-MTC 066	141	13	164	136	96	273	54	259	277	93	17	119	17	5	8
A-MTC 066	123	12	160	241	106	304	51	325	282	81	14	77	26	6	8
A-MTC 067	161	16	234	151	86	291	60	514	292	104	17	69	18	6	12
A-MTC 068	173	14	204	144	66	379	131	644	277	264	9	46	17	8	22
A-MTC 068	191	11	178	134	103	403	84	350	330	109	7	49	20	4	18
A-MTC 071	89	22	100	105	47	523	300	781	190	575	31	66	23	6	3
A-MTC 072	64	25	79	80	56	251	83	1053	141	267	24	62	13	5	4
A-MTC 072	65	26	80	80	56	253	83	1062	143	270	24	62	13	5	4
A-MTC 073	71	9	111	101	58	430	242	1051	160	417	42	44	12	7	2
A-MTC 074	63	22	178	187	77	273	261	2527	276	1017	49	91	15	16	8
A-MTC 075	63	9	109	105	55	76	72	2319	204	1288	19	78	18	12	7
A-MTC 078	57	24	128	119	148	206	100	1582	192	707	31	83	16	18	8
A-MTC 080	72	17	172	125	58	80	83	1336	246	740	32	124	21	43	7
A-MTC 080	56	13	135	98	46	63	65	1049	193	581	25	97	17	34	5
A-MTC 081	92	17	209	105	151	88	60	27	474	251	38	135	22	4	6
A-MTC 082	117	37	145	117	85	378	115	269	211	228	66	134	17	7	8

(continued)

	Li	Be	V	Cr	Cu	Zn	Ga	As	Rb	Sr	Y	Zr	Nb	Mo	Sb
A-MTC 082	126	39	156	126	91	406	124	289	227	245	71	144	18	8	8
A-MTC 084	91	10	121	142	45	58	67	1418	266	469	30	115	22	8	4
A-MTC 085	279	20	182	110	230	513	80	92	374	194	18	79	21	3	18
A-MTC 085	198	14	127	69	105	107	60	49	306	145	17	106	13	0	6
A-MTC 086	101	19	121	135	58	265	122	454	226	760	31	95	18	4	2
A-MTC 086	80	18	105	110	49	224	134	568	195	431	33	77	16	4	2
A-MTC 086	106	20	128	143	62	280	130	481	239	805	33	100	19	4	2
A-MTC 088	91	10	124	317	69	158	66	1260	236	473	30	107	18	5	4
A-MTC 090	72	14	154	129	39	57	56	3241	229	478	40	104	22	32	11
A-BUX 138	63	12	212	109	38	115	43	116	195	254	30	183	16	18	64
A-BUX 139	63	10	208	114	88	346	91	229	227	429	46	141	17	40	83
A-BUX 140	77	3	172	89	176	387	46	179	184	96	21	56	14	59	66
A-BUX 141	72	9	293	137	64	100	60	268	262	312	23	168	25	57	121
A-BUX 142	66	23	154	94	51	142	68	211	167	364	30	95	16	27	187
A-BUX 142	84	47	200	96	178	174	99	521	187	667	15	85	18	57	323
A-BUX 143	73	22	145	97	84	2852	157	419	178	431	67	112	14	17	193
A-BUX 143	84	37	161	110	56	579	262	282	186	449	97	159	18	15	192
A-BUX 144	59	11	190	103	94	108	86	249	186	255	16	99	16	42	106
A-BUX 145	73	21	258	125	115	131	109	366	238	830	26	136	19	62	107
A-BUX 146	103	14	170	98	76	162	83	207	196	93	31	135	12	14	60
A-BUX 147	77	12	219	111	76	96	54	317	234	307	24	153	18	35	106
A-BUX 147	116	37	481	171	203	167	82	77	306	391	29	231	24	51	100
A-BUX 148	65	12	234	114	64	199	215	219	221	300	37	141	16	30	85
A-BUX 149	83	11	326	111	119	287	238	404	225	713	24	121	17	44	115
A-BUX 149	79	9	334	125	111	246	77	367	251	512	20	132	21	58	122
A-BUX 150	84	27	196	130	72	137	108	139	205	255	32	119	15	26	99
A-BUX 151	104	25	246	148	97	274	153	301	327	266	40	156	32	29	71
A-BUX 151	105	17	223	152	115	416	345	320	333	223	42	99	22	24	65
A-BUX 152	84	34	177	130	161	156	123	283	211	345	256	134	24	17	148
A-BUX 153	95	16	308	148	122	348	120	253	310	237	41	144	20	19	92
A-BUX 154	72	20	300	127	137	258	176	234	211	486	59	167	19	44	173
A-BUX 155	66	15	185	117	67	136	68	210	249	213	29	128	18	22	91
A-BUX 156	89	10	240	120	54	109	62	129	281	77	19	134	23	13	54

A-BUX 157	67	8	176	104	75	113	96	154	205	416	15	120	16	29	79
A-BUX 159	81	15	219	119	104	565	130	336	237	94	49	152	17	14	100
A-BUX 160	73	11	329	130	106	637	236	353	240	663	29	689	19	34	113
A-BRG 013	74	3	158	175	2486	570	57	46	213	730	13	44	13	2	2
A-BRG 014	79	9	113	124	101	183	50	108	272	143	38	469	18	0	5
A-BRG 015	134	3	285	269	136	138	43	28	242	162	34	113	21	16	2
A-BRG 018	51	3	102	125	101	111	39	153	114	123	10	50	8	3	1
A BRG 001	111	5	301	292	158	123	40	15	177	242	35	133	18	8	2
A BRG 002	113	4	173	187	160	84	35	12	163	561	37	92	14	9	2
A BRG 003	72	18	192	233	263	101	36	19	101	492	57	109	7	11	2
A-BRG 004	40	4	221	330	186	134	19	26	61	504	86	105	6	51	2
A-BRG 005	61	3	253	209	206	185	27	43	134	262	32	173	9	10	1
A BRG 006	81	9	146	118	57	146	59	909	155	756	34	160	16	5	2
A BRG 007	84	13	179	144	99	194	52	1012	208	494	43	99	25	3	6
A BRG 008	4	32	435	54	44	64	18	4	13	26	35	5	23	12	5
A BRG 009	54	22	276	130	56	96	53	33	299	179	35	136	53	15	70
A BRG 010	47	2	220	600	619	165	31	31	95	533	47	71	7	28	6
A BRG 011	104	11	141	105	181	348	109	672	268	675	47	60	17	4	2
A BRG 012	109	4	234	525	527	167	39	22	211	197	25	111	15	8	4
A BRG 016	179	10	174	143	131	411	64	190	492	154	37	110	29	6	11
A BRG 016	253	14	258	224	161	747	99	236	864	204	29	105	41	7	12
A BRG 017	93	13	204	152	139	875	64	652	223	579	42	67	10	11	2
A BRG 019	114	31	165	118	86	740	63	339	287	610	79	83	20	4	2
A BRG 020	69	11	145	129	88	1036	446	1877	219	1353	28	87	16	5	3
A BRG 021	86	8	237	128	166	522	41	2900	301	429	35	83	14	19	4
A BRG 022	63	10	280	138	102	512	31	1305	180	524	28	74	16	14	5
A BRG 023	67	13	302	140	119	243	38	321	220	135	17	93	15	15	57
A BRG 024	91	12	174	175	149	272	31	444	168	293	34	61	8	7	3
A BRG 025	100	5	246	202	210	411	109	1649	212	764	30	59	10	9	4
A BRG 025	135	5	223	222	232	431	122	1637	291	1173	25	87	14	15	4
A BRG 025	120	5	219	200	187	360	104	1041	235	762	65	211	11	10	3
A BRG 026	84	24	238	132	93	780	126	645	393	939	99	85	17	9	5
A BRG 027	40	14	129	337	258	336	54	32	103	514	13	69	6	25	4
A BRG 028	61	19	225	460	293	336	39	25	109	620	66	93	8	21	5

(continued)

	Li	Be	V	Cr	Cu	Zn	Ga	As	Rb	Sr	Y	Zr	Nb	Mo	Sb
A BRG 029	126	6	154	226	99	405	47	261	251	118	21	81	19	6	11
A BRG 030	0	5	565	166	14	196	7	1	1	123	18	1412	13	0	3
A BRG 030bis	0	2	1269	110	4	30	1	1	0	66	5	1110	22	1	0
A BRG 031	77	6	231	124	2269	186	39	199	232	207	30	1094	22	16	54
A BRG 031	18	2	74	26	1096	98	9	60	55	105	14	48	4	8	18
A-ECK 191	47	3	167	205	177	191	30	103	135	2294	33	114	11	22	3
A-ECK 192	43	2	403	128	163	488	31	52	136	1600	22	142	10	18	4
A-ECK 193	50	2	428	163	195	350	33	64	142	2090	23	102	11	13	2
A-ECK 194	28	1	337	149	158	209	23	37	100	3765	17	92	10	20	2
A-ECK 195	46	2	425	130	111	121	29	34	127	1504	57	88	10	17	2
A-ECK 196	37	2	388	188	234	515	30	75	127	3746	21	117	12	29	4
A-ECK 197	91	2	219	329	168	158	34	14	187	429	41	157	15	7	3
A-ECK 198	29	3	325	166	233	245	25	100	108	4889	33	107	10	14	2
A-ECK 199	41	3	411	183	101	179	31	26	158	2381	31	136	14	24	2
A-ECK 200	33	1	271	130	234	335	26	62	120	2737	31	82	8	9	1
A-ECK 201	59	2	221	133	153	176	29	33	133	2400	28	97	9	30	1
A-ECK 202	33	2	366	196	271	240	28	51	137	1960	39	153	13	25	3
A-ECK 203	25	2	381	97	179	293	28	57	136	3546	30	113	12	19	3
A-ECK 204	28	1	168	146	220	203	20	117	85	1344	32	57	6	3	1
A-ECK 205	45	2	428	180	117	242	33	28	139	1773	32	138	11	43	4
A-ECK 206	72	3	449	199	58	206	37	17	170	1236	45	133	15	19	2
A-ECK 207	51	3	368	158	200	316	29	62	131	1284	25	101	14	42	2
A-ECK 208	46	2	517	183	228	441	35	115	138	2316	23	214	13	21	3
A-ECK 209	25	4	157	63	232	342	15	161	68	3733	32	73	7	11	0
A-ECK 210	35	2	418	176	194	276	29	92	114	1577	22	104	9	25	3

Continuation of the Results for the Minor, Traces Elements and Correlations Between Elements

	Cs	Ba	La	Ce	Pr	Tb	Lu	W	Tl	Pb	Th	U	Rb/Cs	Nb/Cs
Geological samples														
G-AICH 021	3	168	45	78	10	1	1	1	11	70	5	11	26	2
G-AICH-S 027	3	182	36	70	8	1	0	0	10	71	4	8	22	2
G-GEIS 065	4	88	17	31	3	0	0	1	8	30	2	5	15	1
G-GEIS 065	3	188	34	59	6	1	0	0	24	23	5	8	29	3
G-GEIS 065	3	130	33	60	5	1	0	1	17	18	3	10	27	2
G-DOTT 028	5	241	24	47	4	0	0	1	4	22	3	10	18	1
G-DOTT 028	4	193	20	38	3	0	0	0	4	17	3	8	18	1
G-OHM 019	7	308	36	89	11	1	0	1	15	37	11	12	15	1
G-OHM-B 020	3	226	141	352	57	8	2	0	9	57	25	28	23	2
G-OHM-B 020	3	300	40	74	10	1	1	1	12	77	5	9	22	2
G-OHM-L 018	2	274	38	70	9	1	1	1	14	79	5	9	38	3
G-HOLZ 022	5	191	47	87	9	1	0	1	8	73	5	12	19	2
G-HOLZ 025	4	155	29	53	5	1	0	0	8	48	3	8	15	1
G-HOLZ 025	3	192	40	91	11	2	1	1	10	57	6	13	22	2
G-SCHL-B 026	5	211	56	97	11	1	0	1	9	70	7	13	22	3
G-BOL 017	4	202	38	74	9	1	1	1	10	79	5	9	17	1
G-ERZ 024	4	187	37	63	8	1	0	1	20	28	6	10	23	2
G-WAN 066	3	151	26	41	4	0	0	0	1	86	2	7	18	2
G-WIS 015	12	894	53	108	11	1	0	2	7	61	11	6	16	1
G-BUX 006	73	356	60	78	9	1	1	73	1	7	18	13	4	0
G-BUX 007	47	422	29	45	5	0	0	13	4	48	17	56	6	0
G-BUX 008	22	350	30	51	6	1	0	13	1	34	12	21	6	0
G-BUX 009	76	386	59	75	8	1	0	29	3	69	34	27	4	0
G-BUX 010	35	355	37	66	8	1	0	16	3	52	14	19	6	0
G-BUX 011	63	442	27	36	3	0	0	15	3	33	16	27	4	0
G-BUX 041	41	661	52	83	10	1	0	16	3	27	28	44	5	1
G-BUX 042	29	344	27	49	6	0	0	10	3	25	16	18	7	1
G-BUX 043	49	404	37	57	6	0	0	11	4	34	18	92	6	0
G-BUX 044	34	258	32	72	8	0	0	19	1	70	9	21	5	0

(continued)

	Cs	Ba	La	Ce	Pr	Tb	Lu	W	Tl	Pb	Th	U	Rb/Cs	Nb/Cs
G-BUX 045	46	317	58	127	14	1	1	26	2	98	21	33	6	0
G-BUX 046	27	294	22	37	4	0	0	12	2	37	30	37	7	0
G-BUX 047	36	386	34	64	8	1	0	19	2	53	21	24	5	0
G-BUX 048	53	205	16	19	2	0	0	11	2	18	7	15	4	0
G-BUX 048	53	226	20	26	3	0	0	12	1	22	13	24	4	0
G-BUX 049	80	409	158	395	47	5	1	23	4	40	99	76	5	0
G-BUX 049	65	332	43	70	7	1	0	23	3	60	20	24	4	0
G-BUX 050	53	185	37	53	4	0	0	31	4	74	20	12	3	0
G-BUX 050	29	795	1	2	0	0	0	2	5	25	1	0	17	0
G-BUX 051	36	218	50	82	11	1	0	13	1	11	17	6	5	0
G-BUX 052	41	436	59	80	9	1	0	9	5	97	25	42	5	0
G-BUX 053	31	356	40	60	7	1	0	8	5	95	18	23	5	0
G-BUX 054	20	218	13	22	2	0	0	7	1	14	7	15	7	0
G-BUX 054	34	440	25	34	3	0	0	11	2	41	12	34	7	0
G-BUX 055	16	197	14	31	3	0	0	5	1	12	5	28	7	0
G-BUX 056	47	457	96	169	21	2	1	13	4	31	26	130	6	0
G-BUX 057	74	231	27	50	5	0	0	59	2	110	15	9	3	0
G-HIL 038	13	362	25	39	5	1	0	4	2	32	15	13	9	1
G-VOG 060	3	232	107	181	21	2	0	1	0	23	19	68	11	4
G-VOG 060	3	281	28	53	6	1	0	1	0	23	11	90	10	3
G-VOG 060	2	199	92	155	18	1	0	1	0	20	16	58	11	4
G-MKZ 013	13	553	46	71	8	1	1	1	6	491	26	17	7	1
A-MKZ 121	13	914	48	86	11	2	1	1	5	1313	20	54	7	1
A-SRB 122	18	437	247	746	159	44	4	1	5	115	18	30	7	1
A-SRB 122	16	373	169	526	105	28	3	1	6	79	17	18	7	1
A-MKZ 121	15	4592	102	249	35	6	2	2	9	5302	24	92	7	1
A-MKZ 121	23	3776	95	218	28	5	1	3	10	6591	28	103	5	0
A-MKZ 121	16	1270	79	202	22	4	2	1	12	7570	33	59	8	1
A-MKZ 121	16	1287	35	63	8	1	0	1	10	2824	25	30	7	1
A-SRB 122	25	436	125	313	54	11	2	3	2	523	21	81	7	1
A-SRB 122	20	548	141	392	73	17	2	2	2	156	16	251	7	1
A-SRB 122	20	536	81	205	36	7	2	4	1	239	25	72	7	1
A-SRB 122	27	853	100	236	39	9	2	6	2	410	28	65	7	1

A-MKZ 246	28	1504	90	127	17	2	1	2	3	1228	56	10	7	0
A-MKZ 247	17	3600	323	938	198	20	2	2	19	641	63	146	7	1
A-MKZ 247	22	1864	130	378	80	17	3	2	4	322	32	141	7	1
A-MKZ 247	28	3402	170	321	52	20	2	2	4	375	42	116	7	1
A-MKZ 248	12	832	25	59	9	1	0	1	3	1433	21	56	7	1
G-KIM 012	10	250	81	108	14	2	1	1	9	54	14	22	14	1
G KIM 033	9	281	79	165	19	2	1	1	2	84	10	15	12	1
G-KIM 084	22	379	48	87	10	1	0	3	2	32	15	4	12	1
G-KIM 085	14	281	55	84	10	1	1	2	7	56	12	15	14	1
G-KIM 085	14	276	70	113	13	1	0	1	6	52	12	20	14	1
G-KIM 086	15	252	73	123	14	1	1	2	7	49	11	16	14	1
G-KIM 087	26	413	59	104	12	1	1	3	2	43	15	4	12	1
G-KIM 088	21	371	50	90	10	1	0	3	2	42	15	4	13	1
G-KIM 089	4	99	13	18	2	0	0	1	1	18	2	1	12	1
G-KIM 089	5	142	18	27	3	0	0	1	1	28	3	2	14	2
G-KIM 090	21	395	51	92	10	1	0	3	2	44	15	4	13	1
G-KIM 091	21	350	50	95	10	1	0	3	2	38	15	3	13	1
G-KIM 092	21	414	55	99	13	1	1	4	4	89	31	24	13	1
G-KIM 093	17	317	75	102	12	1	1	3	9	95	16	13	14	1
G-KIM 094	24	416	53	85	10	1	0	4	6	62	27	11	14	2
G-KIM 095	24	421	47	73	8	1	0	4	3	20	15	6	13	1
G-KIM 096	9	180	41	62	7	0	0	1	1	13	8	2	12	1
G-KIM 097	21	424	60	99	11	1	0	3	4	76	20	8	13	1
G-KIM 098	21	360	64	107	12	1	0	2	3	54	14	4	13	1
G-KIM 099	21	379	52	95	11	1	0	2	6	89	14	6	13	2
G-KIM 100	20	352	62	94	11	1	0	2	5	78	13	8	13	1
G-KIM 101	18	266	47	82	9	1	0	2	1	18	10	2	13	1
G-KIM 102	17	261	63	102	12	3	1	2	2	44	15	4	13	1
G-KIM 103	20	352	46	91	9	1	0	3	2	30	14	5	12	1
G-KIM 104	22	410	50	94	10	1	0	3	3	78	15	5	12	1
G-KIM 105	26	418	40	75	8	0	0	3	2	27	21	6	11	1
G-KIM 106	22	430	62	111	12	1	0	3	2	60	16	4	13	1
G-KIM 107	18	287	95	158	18	4	1	4	4	56	21	8	13	1
G-KIM 108	17	323	75	125	16	2	1	3	8	86	19	15	14	1

(continued)

	Cs	Ba	La	Ce	Pr	Tb	Lu	W	Tl	Pb	Th	U	Rb/Cs	Nb/Cs
G-KIM 109	20	399	91	169	21	2	1	3	5	65	40	14	14	1
G-KIM 109	24	461	56	101	11	1	0	4	4	55	17	8	13	1
G-KIM 110	14	249	52	83	10	1	0	1	2	50	10	4	13	1
G-KIM 111	18	312	71	128	15	1	1	3	5	84	21	7	13	2
G-BEL 079	0	1013	9	12	2	1	0	3	4	0	0	0	0	0
G-BLAP 080	0	37	3	6	1	0	0	11	1	2	1	418	0	0
G-GOND 081	1	50	3	4	1	0	0	0	0	3	2	1	8	2
G-SAAR 078	1	3271	19	22	3	1	0	1	3	59	7	1	13	8
G-SAAR 078	1	2523	14	17	2	1	0	1	2	46	5	1	13	8
G-OLE 072	1	4	1	3	1	0	0	1	0	12	3	13	0	31
G-OLE 073	3	28	6	12	2	0	0	1	1	134	11	11	7	18
G-OLE 073	0	21	4	7	1	0	0	0	1	98	4	19	0	0
G-OLE 074	0	7	4	6	1	0	0	17	1	143	1	17	35	4
G-OLE 075	0	4	4	9	1	0	0	5	1	86	2	41	28	226
G-OLE 076	2	32	6	17	2	0	0	0	2	73	18	21	6	5
G-OLE-MOL 067	1	27	22	40	6	1	0	0	1	168	2	8	3	41
G-OLE-MOL 067	1	18	14	27	4	1	0	0	1	111	1	5	3	41
G-OLE-MOL 068	1	14	4	8	2	0	0	0	1	41	2	29	4	8
G-OLE-MOL 069	0	9	13	32	5	1	0	6	0	53	2	61	0	0
G-OLE-MOL 070	1	40	3	7	1	0	0	1	1	122	6	16	3	45
G-OLE-MOL 071	0	15	7	17	3	1	0	12	1	58	0	67	0	0
G-DUR 058	1	77	7	6	1	0	0	0	1	291	1	2	13	2
G-GARx 061	1	57	7	13	2	1	1	0	12	202	4	1	14	23
G-GARx 061	1	68	9	15	2	2	1	0	14	238	4	1	14	23
G-FOI 004	0	114	6	13	2	3	2	23	0	0	5	6	5	17
G-RLB 062	0	51	4	10	2	4	3	5	3	6	3	845	0	12
G-RLB 062	0	52	4	10	2	4	3	5	3	6	3	849	0	12
G-XXX 077	1	628	27	41	7	1	1	1	1	132	13	4	25	42
G-XXX 082	0	13	2	3	0	0	0	1	0	9	2	2	0	0
G MULL 029	0	5	30	43	7	1	1	6	0	3	3	30	16	39
G WEM 032	2	57	17	23	3	2	1	3	3	11	9	4	11	2
G WHI 030	0	10	2	3	0	0	0	0	0	0	0	2	0	0

G WHI 030	0	12	2	4	0	0	0	0	0	0	0	2	0	0
G XXX 031	2	315	18	21	3	0	0	1	3	43	4	2	9	11
Archaeological samples														
A-ALC 161	22	4263	55	108	11	1	0	12	8	414	16	18	8	1
A-ALC 161	21	5091	87	151	16	1	0	13	7	369	20	21	9	1
A-ALC 161	25	4094	69	116	12	1	0	16	5	356	19	20	8	1
A-ALC 162	24	4387	108	140	20	2	1	18	17	302	19	7	10	1
A-ALC 162	18	2266	102	110	18	2	1	10	4	184	20	6	11	1
A-ALC 163	25	977	34	73	7	1	0	10	7	217	14	5	9	1
A-ALC 164	12	653	63	103	13	1	0	1	2	72	12	10	16	1
A-ALC 164	12	413	68	87	12	1	0	1	1	59	13	9	16	1
A-ALC 165	21	1379	57	65	9	1	0	14	6	329	22	4	9	1
A-ALC 165	20	2072	284	545	59	7	2	17	3	216	16	17	9	1
A-ALC 166	19	1298	60	80	9	1	0	14	4	285	19	4	9	1
A-ALC 166	20	2351	43	58	7	1	0	12	4	285	23	4	9	1
A-ALC 167	8	553	149	101	25	3	1	1	2	78	7	11	14	1
A-ALC 167	7	546	143	92	23	2	1	1	2	79	9	11	15	1
A-ALC 168	24	3583	44	79	9	1	0	25	6	189	25	5	8	1
A-ALC 168	32	4635	89	154	18	2	0	27	12	402	31	5	7	1
A-ALC 169	19	4226	59	74	9	1	0	21	5	456	23	5	10	1
A-ALC 169	23	3707	62	74	9	1	0	22	5	372	22	6	9	1
A-ALC 170	8	788	155	139	21	2	1	1	3	87	10	15	16	1
A-ALC 171	17	605	160	140	28	3	1	2	2	174	18	10	11	1
A-ALC 172	11	1504	64	94	13	1	0	1	4	467	14	10	15	1
A-CHAL 173	21	5667	101	142	17	1	1	13	3	196	21	11	8	1
A-CHAL 173	22	9465	153	179	20	1	0	18	4	234	21	11	9	1
A-CHAL 174	23	8999	104	114	15	1	0	13	5	206	20	6	7	0
A-CHAL 174	41	9168	99	144	17	1	1	22	6	317	32	6	7	0
A-CHAL 175	24	4014	63	72	9	1	0	24	2	254	14	7	11	1
A-CHAL 175	18	4241	61	78	9	1	0	16	2	264	14	5	10	1
A-CHAL 175	15	3854	101	154	17	1	0	19	2	185	14	6	10	1
A-CHAL 175	18	3849	75	102	12	1	0	22	2	200	15	5	11	1
A YOC 123	28	305	40	66	8	0	0	3	2	85	20	5	13	1
A-YOC 124	21	200	22	39	4	0	0	2	3	189	12	3	12	1

(continued)

	Cs	Ba	La	Ce	Pr	Tb	Lu	W	Tl	Pb	Th	U	Rb/Cs	Nb/Cs
A-YOC 124	38	310	37	66	6	0	0	3	3	160	15	4	10	1
A-YOC 125	22	303	22	40	4	0	0	3	2	58	13	5	12	1
A-MTC 046	73	3957	53	89	8	1	0	36	5	398	43	28	5	0
A-MTC 050	51	746	47	105	9	0	0	22	2	186	30	23	5	0
A-MTC 054	63	923	48	107	10	1	0	28	3	359	35	27	5	0
A-MTC 055	66	570	53	120	8	1	0	31	3	250	35	46	5	0
A-MTC 055	68	906	46	107	9	1	0	32	4	500	34	44	6	0
A-MTC 056	59	482	50	125	10	1	0	26	3	415	49	33	5	0
A-MTC 056	63	727	64	140	13	1	1	28	3	432	33	33	6	0
A-MTC 057	61	653	34	67	6	0	0	26	3	373	26	14	5	0
A-MTC 057	69	745	42	71	7	1	1	30	3	204	36	17	5	0
A-MTC 058	44	428	30	64	5	0	0	41	2	150	33	24	6	1
A-MTC 059	34	3603	58	90	12	1	0	9	2	239	18	17	8	1
A-MTC 062	32	1618	70	107	16	2	0	10	2	344	15	12	9	1
A-MTC 063	26	2914	89	153	22	2	1	7	2	395	23	30	9	1
A-MTC 063	16	430	17	28	4	0	0	7	1	107	7	8	8	1
A-MTC 066	62	444	53	97	9	1	0	28	2	156	35	21	4	0
A-MTC 066	62	382	48	89	8	1	0	32	2	156	31	16	5	0
A-MTC 067	62	612	49	118	10	1	0	29	2	312	42	38	5	0
A-MTC 068	52	2248	32	63	6	0	0	29	2	398	41	23	5	0
A-MTC 068	56	1048	36	65	7	0	0	35	3	664	23	21	6	0
A-MTC 071	23	5958	68	129	14	1	1	32	4	182	22	9	8	1
A-MTC 072	16	1538	49	94	10	1	0	12	7	174	18	10	9	1
A-MTC 072	16	1551	50	95	10	1	0	12	7	176	18	11	9	1
A-MTC 073	17	4956	127	174	20	2	0	17	7	461	23	9	10	1
A-MTC 074	29	12422	322	439	50	3	0	27	20	1502	29	38	10	1
A-MTC 075	24	2128	63	83	10	1	0	16	18	329	24	6	9	1
A-MTC 078	18	4298	119	154	18	1	0	16	8	742	22	12	10	1
A-MTC 080	28	2252	123	194	21	1	0	23	8	493	31	15	9	1
A-MTC 080	22	1768	97	152	16	1	0	18	6	387	24	12	9	1
A-MTC 081	79	1107	88	152	18	1	0	6	2	93	20	18	6	0
A-MTC 082	27	2108	112	231	24	3	1	17	3	605	33	24	8	1

A-MTC 082	29	2266	121	248	26	3	1	18	4	650	35	26	8	1
A-MTC 084	34	1662	59	98	12	1	0	21	6	300	36	6	8	1
A-MTC 085	61	918	39	72	8	1	0	15	3	389	29	26	6	0
A-MTC 085	54	541	38	58	6	0	0	10	2	54	23	11	6	0
A-MTC 086	23	2273	66	102	13	1	0	20	4	290	27	14	10	1
A-MTC 086	24	2758	75	101	13	1	0	15	5	184	27	7	8	1
A-MTC 086	24	2407	70	108	13	1	0	22	5	307	29	14	10	1
A-MTC 088	24	1802	57	92	10	1	0	14	5	377	28	7	10	1
A-MTC 090	28	1027	67	102	11	1	0	19	10	466	37	9	8	1
A-BUX 138	23	694	45	67	9	1	1	7	2	53	30	20	9	1
A-BUX 139	27	2968	82	93	15	1	1	8	2	139	23	15	8	1
A-BUX 140	25	856	36	48	6	0	0	23	2	106	13	24	7	1
A-BUX 141	33	756	48	72	7	1	0	6	6	139	18	12	8	1
A-BUX 142	21	992	49	73	9	1	0	3	1	41	18	19	8	1
A-BUX 142	23	1530	49	52	5	0	0	3	2	87	7	27	8	1
A-BUX 143	23	3219	74	131	18	2	1	5	1	131	19	16	8	1
A-BUX 143	20	5524	96	170	25	3	1	6	1	83	22	19	9	1
A-BUX 144	18	1526	32	52	7	1	0	6	2	46	15	17	10	1
A-BUX 145	29	2374	67	95	11	1	0	9	2	81	22	26	8	1
A-BUX 146	28	1529	50	95	10	1	0	5	1	65	11	10	7	0
A-BUX 147	32	659	39	60	8	1	0	6	2	99	18	18	7	1
A-BUX 147	37	810	50	93	10	1	1	10	3	102	19	36	8	1
A-BUX 148	29	4156	55	92	11	1	0	7	1	67	33	21	8	1
A-BUX 149	32	5560	81	118	13	1	0	5	2	220	19	14	7	1
A-BUX 149	35	1300	71	96	10	1	0	6	2	102	14	12	7	1
A-BUX 150	20	1956	54	78	12	1	0	5	1	92	14	23	10	1
A-BUX 151	40	3170	80	128	15	2	1	17	4	311	18	9	8	1
A-BUX 151	45	7254	64	143	16	2	0	7	3	245	23	9	7	0
A-BUX 152	25	2771	227	339	67	12	2	4	2	231	17	29	8	1
A-BUX 153	37	2590	74	92	15	2	1	8	2	597	20	10	8	1
A-BUX 154	20	4062	109	194	21	2	1	5	2	294	21	19	10	1
A-BUX 155	35	1095	56	85	9	1	0	5	2	44	21	14	7	1
A-BUX 156	38	824	31	42	5	1	0	14	1	95	21	8	7	1

(continued)

	Cs	Ba	La	Ce	Pr	Tb	Lu	W	Tl	Pb	Th	U	Rb/Cs	Nb/Cs
A-BUX 157	27	1800	33	48	6	1	0	6	2	35	17	20	8	1
A-BUX 159	37	2647	53	91	13	2	1	7	2	459	21	6	6	0
A-BUX 160	43	5682	59	77	9	1	1	6	3	141	18	9	6	0
A-BRG 013	16	1368	30	84	9	0	0	2	2	97	7	3	14	1
A-BRG 014	37	773	54	98	11	1	1	16	2	60	19	7	7	0
A-BRG 015	15	469	47	90	10	1	0	2	1	45	14	7	16	1
A-BRG 018	12	930	21	28	4	0	0	14	1	116	9	2	10	1
A BRG 001	13	352	57	72	10	1	0	2	1	68	12	7	13	1
A BRG 002	12	272	55	85	11	1	0	1	1	58	9	6	13	1
A BRG 003	8	378	70	108	13	1	1	1	2	59	7	11	12	1
A-BRG 004	6	485	94	133	17	2	1	1	4	71	6	20	11	1
A-BRG 005	10	368	43	68	9	1	1	1	1	237	10	6	13	1
A BRG 006	19	1119	47	69	9	1	0	11	6	157	21	5	8	1
A BRG 007	26	1125	62	88	11	1	1	11	3	290	27	4	8	1
A BRG 008	1	37	2	4	1	1	1	1	0	23	2	12	10	17
A BRG 009	39	394	58	96	11	1	1	26	2	49	20	12	8	1
A BRG 010	7	929	58	98	10	1	0	1	4	82	6	15	14	1
A BRG 011	35	4263	40	82	10	1	0	34	6	136	30	5	8	0
A BRG 012	12	523	43	66	8	1	0	1	2	59	13	6	18	1
A BRG 016	73	805	40	94	10	1	1	52	3	342	56	17	7	0
A BRG 016	133	912	56	129	14	1	0	70	6	676	69	23	7	0
A BRG 017	27	2276	33	60	7	1	1	23	8	277	17	6	8	0
A BRG 019	42	1799	40	84	10	1	1	49	5	119	30	7	7	0
A BRG 020	24	20398	43	78	9	1	0	11	5	170	23	4	9	1
A BRG 021	32	2106	40	78	8	1	0	21	15	283	22	4	9	0
A BRG 022	20	1981	31	58	6	1	0	15	6	231	16	5	9	1
A BRG 023	26	868	30	56	6	1	0	13	2	231	22	51	9	1
A BRG 024	16	1608	40	78	8	1	0	28	3	206	12	3	10	0
A BRG 025	21	2998	50	91	10	1	0	18	12	555	17	7	10	0
A BRG 025	33	3361	84	134	13	1	0	33	13	561	24	10	9	0
A BRG 025	22	2952	84	142	15	2	1	26	14	635	24	17	11	1

A BRG 026	33	3151	105	146	13	2	1	18	5	275	17	7	12	1
A BRG 027	7	1400	23	44	4	0	0	1	1	133	5	22	15	1
A BRG 028	8	576	76	86	11	1	1	1	2	88	8	12	13	1
A BRG 029	54	342	38	82	10	1	0	46	2	276	36	13	5	0
A BRG 030	0	173	7	8	1	0	0	0	0	6	1	2	27	435
A BRG 030bis	0	8	2	6	1	0	0	0	0	0	1	1	0	0
A BRG 031	22	469	40	76	9	1	1	22	3	922	18	37	10	1
A BRG 031	5	141	12	24	3	0	0	6	1	734	10	7	11	1
A-ECK 191	8	459	39	61	7	1	0	2	1	25	6	12	18	1
A-ECK 192	8	383	33	53	6	1	0	1	4	58	7	10	17	1
A-ECK 193	9	408	34	57	6	1	0	1	2	41	7	14	16	1
A-ECK 194	5	325	32	45	5	0	0	1	4	46	4	9	21	2
A-ECK 195	8	451	51	109	13	2	1	1	2	40	15	12	16	1
A-ECK 196	6	365	43	64	7	0	0	2	4	53	6	9	20	2
A-ECK 197	16	542	56	94	10	1	1	2	1	108	12	8	12	1
A-ECK 198	6	451	39	67	8	1	0	1	7	35	10	10	17	2
A-ECK 199	9	409	59	91	10	1	0	2	3	44	11	14	17	2
A-ECK 200	6	351	40	66	8	1	0	1	3	38	6	7	19	1
A-ECK 201	9	422	44	74	8	1	0	2	2	32	8	9	15	1
A-ECK 202	7	378	49	64	8	1	1	1	2	42	7	9	20	2
A-ECK 203	6	341	41	69	8	1	0	2	1	40	7	10	21	2
A-ECK 204	4	431	217	326	51	3	0	1	1	27	12	4	19	1
A-ECK 205	8	444	39	69	8	1	0	1	3	40	9	11	17	1
A-ECK 206	11	744	60	119	13	2	0	2	2	44	21	12	16	1
A-ECK 207	8	434	36	59	7	1	0	2	2	40	8	8	16	2
A-ECK 208	7	541	41	60	7	1	0	2	3	46	7	12	19	2
A-ECK 209	3	638	36	64	7	1	0	1	6	26	4	11	22	2
AnnECK 210	6	395	34	50	5	1	0	2	3	34	5	11	19	2

References

Allason-Jones L (1996) Roman jet in the Yorkshire Museum. Yorkshire Museum, York

Allason-Jones L, Jones DM (1994) Jet and other materials in Roman artefact studies. Museum Notes. *Archaeologia Aeliana*, 5é série, vol XXII. The Society of Antiquaries of Newcastle-Upon-Tyne, Newcastle-Upon-Type, p 263–272

Allason-Jones L, Jones DM (2001) Identification of jet artifacts by reflected light microscopy. Eur J Archaeol 4(2):233–251

Alpern B (1981) Les schistes bitumineux : constitution, réserves, valorisations. Bulletin des Centres de Recherches exploration-production Elf-Aquitaine, vol 5, 2. Elf-Aquitaine, Pau, p 319–352

Baron A (2009) Les objets en roches noires (« lignite ») à l'âge du Fer: recherches de provenance, mise en œuvre et diffusion en Europe celtique du VIIIème au Ier s. av. J.-C. Ph.D. Dissertation, University of Strasbourg

Baron A (2011) Provenance et circulation des objets de parure en roches noires : l'exemple du site hallstattien d'Eckbolsheim (Bas-Rhin, Alsace). Bull Soc Prehist FR 108(2):345–358

Baron A (2012) Provenance et circulation des objets en roches noires (« lignite ») à l'âge du Fer en Europe celtique (VIIIème-Ier s. av. J.-C.), British Archaeological Report, BAR. International series 2453. Archaeopress, Oxford, 589 p

Baron A, Gratuze B, Querré G (2007) Les objets de parure en *black shales* à l'Âge du Fer en Europe celtique: recherche de provenance par l'analyse élémentaire (LA-ICP-MS). Archéosciences 31:87–96

Barrabé L, Feys R (1965) Géologie du charbon et des bassins houillers. éd Masson et Cie, Paris

Brill RH (1999) Chemical analyses of early glasses. The Corning Museum of Glass, New York, NY

Bronk H, Freestone I-C (2001) A quasi non-destructive microsampling technique for the analysis of intact glass objects by SEM/EDXA. Archaeometry 43 (4):517–527

Bussell GD, Pollard AM, Baird DC (1982) The characterization of early Bronze Age jet and jet-like material by X-ray fluorescence. Wiltshire Archaeol Mag 76:27–32

Davis M, Sheridan A (1993) Scottish prehistoric 'jet' jewellery: some new work. Proceedings of the Society of Antiquaries of Scotland Lecture Summaries 123:455–456

De La Roche H (1968) Comportement géochimique différentiel de Na, K et Al dans les formations volcaniques et sédimentaires: un guide pour l'étude des formations métamorphiques et plutoniques. C R Acad Sci Paris 267, série D:39–42

De La Roche H (1972) Revue sommaire de quelques diagrammes chimico-minéralogiques pour l'étude des associations ignées ou sédimentaires et de leurs dérivés métamorphiques. Sciences de la Terre (Nancy) 17:33–46

Grasselt T, Volkmann N (1991) Latènezeitlicher Ringschmuck aus Thüringen. Alt-Thüringen, Jahresschrift des Museums für Ur- und Frügeschichte Thüringens 26:179–196

Gratuze B, Blet-Lemarquand M, Barrandon J-N (2001) Mass spectrometry with laser sampling: a new tool to characterize archaeological materials. J Radioanal Nucl Chem 247:645–656

Guillot F (1987) Géologie de l'antepermien de Vanoise septentrionale (zone briançonaise interne, Alpes occidentales, Savoie, France), Ph.D. Dissertation, University of Lille

Hunter FJ, McDonnell JG, Pollard AM, Morris CR, Rowlands CC (1993) The scientific identification of archaeological jet-like artifacts. Archaeometry 35 (1):69–89

Janssens K, Van Grieken R (eds) (2004) Non-destructive microanalysis of cultural heritage materials, vol XLII, Comprehensive analytical chemistry. Elsevier, Amsterdam

Ligouis B (2000) Kohlenpetrographische Untersuchungen an Funden aus kohlenstoffreichen organogenen sedimenten aus der Heuneburg-Aussiedlung bei Hundersingen an der Donau. In: Kurz S (ed) Die Heuneburg Außensiedlung. Befunde and Funde, Forsch Ber Vor u Fruhgesch Baden-Württemberg 72, p 179–185

Ligouis B (2006) Jais, lignite, charbon et autres matières organiques fossiles: application de la pétrologie organique à l'étude des éléments de parure et des fragments bruts. In: Bullinger J, Leesch D, Plumettaz N (eds) Le site magdalénien de Monruz 1. Premiers éléments pour l'analyse d'un habitat de plein air, Service et musée cantonal d'archéologie, Archéologie neuchâteloise 33, Neuchâtel, p 197–216

Pearce NJG, Perkins WT, Westgate JA, Gorton MT, Jackson SE, Neal CR, Chenery SP (1997) A compilation of new and published major and trace element data for NIST SRM 610 and SRM 612 glass reference materials. Geostand Newsl XXI:114–115

Ploquin A (1975) Étude géochimique et pétrographique du Complexe de gneiss, migmatites et granites du Telemark - Aust Agder (Précambrien de Norvège méridionale). Sa plaçe dans l'ensemble épizonal à catazonal profond, du Haut-Telemark au Bamble. Sciences de la Terre, série Mémoires, no 38. Fondation scientifique de la géologie et des ses applications, Nancy

Pollard AM, Bussell GD, Baird DC (1981) The analytical investigation of Early Bronze Age jet and jet-like material from the Devizes Museum. Archaeometry 23(2):139–167

Popelka RS, Glascock MD, Robertshaw P, Wood M (2005) Laser ablation ICP-MS of African glass trade beads. In: Speakman RJ, Neff H (eds) Laser ablation ICP-MS in archaeological research. University of New Mexico Press, Albuquerque, NM, pp 84–93

Pouenat P, Vernet G (2002) Un atelier de fabrication d'anneaux en schiste bitumineux à Buxières-les-Mines (Allier). In: Maranski D, Guichard V (eds) Les Âges du Fer en Nivernais, Bourbonnais et Berry oriental, Regards européens sur les Âges du Fer en France, Actes du XVIIème colloque de l'A.F.E.A.F., Nevers mai 1993, Collection Bibracte 6. centre archéologique européen du Mont-Beuvray, Glux-en-Glenne, p 151–157

Rochna O (1961) Zur Herkunft der Manchinger Sapropelit-Ringe. Germania 39(3/4):329–354

Rochna O (1962) Hallstattzeitlicher Lignit-und Gagat-Schmuck Zur Verbreitung, Zeitstellung und Herkunft. Fundberichte aus Schwaben, Neue Folge 16:44–83

Rochna O, Mädler K (1974) Die Sapropelit-und Gagatfunde von Dürrnberg. In: Moosleitner F, Pauli L, Penninger E (eds) Der Dürrnberg bei Hallein II. Katalog der Grabfunde aus der Hallstatt-und Latènezeit, Munich, p 153–157

Sheridan A, Davis M (2002) Investigating jet and jet-like artifacts from prehistoric Scotland: The National Museums of Scotland project. Antiquity 76 (293):812–825

Teichmüller M (1992) Organic petrology in the service of archaeology. Int J Coal Archaeol 20:1–21

Thierry J (1999) Etude isotopique d'objets archéologiques en lignite (site de Buxières-les-Mines). Mémoire de géochimie isotopique, Université Pierre et Marie Curie

Thomas S (2003) Etude pétrographique et palynologique de mobilier en « lignite » protohistorique. Caractérisation et étude de provenance, Mémoire de DESS, Université de Bourgogne, Dijon

Venclová N (ed) (2001) Výroba a sídla v Době Laténské, Projekt Loděnice (Production and settlement in the La Tène period, the Lodenice project). Prague Institute of Archaeology, Prague

Verità M, Basso R, Wypyski MT, Koestler RJ (1994) X-ray microanalysis of ancient glassy materials: a comparative study of wavelength dispersive and energy dispersive techniques. Archaeometry 36(2):241–251

Watts S (1996) The geochemical characterization of jet and jet-like objects in archaeology. A discussion of analytical methods for the identification of archaeological black lithic materials, with a case study of objects from Roman York and Verulamium. Ph.D. Dissertation, University of Bradford

Watts S, Pollard AM (1997) Kimmeridge jet, a potential new source for British jet. Archaeometry 39:125–143

Watts S, Pollard AM (1998) Identifying archaeological jet and jet-like artifacts using FTIR. Résumé des communications du colloque de Infrared User's Group, September 1995, p 37–52

Watts S, Pollard AM (1999) The organic geochemistry of jet: pyrolysis-gas chromatography/mass spectrometry (Py-GCMS) applied to identifying jet and similar black lithic materials – preliminary results. J Archaeol Sci 26:923–933

Zakova B (2001) Petrologický výzkum švartny a příbuzných materiálu. In: Venclová N (ed) Výroba a sídla v době Laténské. Prague Institute of Archaeology, Prague, pp 227–266

18 Investigating Biogenic Versus Diagenetic Trace Element Incorporation in Archaeological Mineralized Tissues with LA-ICP-MS

John V. Dudgeon, Monica Tromp, Bryan K. Hanks, and Andrei V. Epimakhov

Abstract

In this chapter, we develop a method employing combined SEM-EDS and LA-ICP-MS analysis of mineralized tissue (bone apatite) to quantify trace element concentrations. We apply this method to bone and tooth samples from the Kammenyi Ambar 5 cemetery (Russia) to determine whether a trace element signal resulting from working of Arsenical Bronze can be distinguished from post-burial uptake of trace elements. Using raster mapping of bone and tooth cross-sections, we demonstrate that while some elements show a pattern consistent with post-burial uptake, other elements including arsenic appear to reflect uptake related to life activities including Bronze working. We review mechanisms by which trace elements may find their way into bone and tooth structure to explain these differences in incorporation pathway.

J.V. Dudgeon (✉)
Department of Anthropology, Center for Archaeology, Materials and Applied Spectroscopy, Idaho State University, 921 S. 8th Avenue, Stop 8005, Pocatello, ID 83209-8005, USA
e-mail: dudgeon@isu.edu

M. Tromp
Department of Anatomy, University of Otago, P.O. Box 913, Dunedin 9054, New Zealand
e-mail: monica.tromp@anatomy.otago.ac.nz

B.K. Hanks
Department of Anthropology, University of Pittsburgh, #3114 WWPH, 230 S. Bouquet St., Pittsburgh, PA 15260, USA
e-mail: bkh5@pitt.edu

A.V. Epimakhov
Institute of History and Archaeology RAS (South Ural Department), South Ural State University, Chelyabinsk 454080, Russia
e-mail: eav@susu.ac.ru

18.1 Introduction

Recent developments in the application of laser ablation-inductively coupled plasma-mass spectrometry (LA-ICP-MS) have spurred interest in its use to probe the inorganic biochemistry of mineralized tissues (Castro et al. 2010; Cucina et al. 2011; Durrant and Ward 2005; Scharlotta et al. 2013). Many questions have been addressed, such as elemental signatures associated with locale of origin (Arai et al. 2007; Beard and Johnson 2000; Castro et al. 2010; Cucina et al. 2005, 2007, 2011; Dudgeon 2008), consequences of epidemiological exposure to toxic or non-essential elements during growth and development (Abdullah et al. 2012; Budd et al. 1998,

L. Dussubieux et al. (eds.), *Recent Advances in Laser Ablation ICP-MS for Archaeology*, Natural Science in Archaeology, DOI 10.1007/978-3-662-49894-1_18

2000; Dolphin et al. 2005; Zaichick et al. 2011), and studying the process of post-depositional alteration, including fossilization, in the sedimentary context (Koenig et al. 2009; Kohn 2008; Hinz and Kohn 2010). These studies exploit the spatial resolving power of modern laser ablation sampling devices to characterize elemental variation over micrometer-level scales, and across heterogeneous and biologically-important domains associated with tissue growth and differentiation. LA-ICP-MS is capable of generating quantitative multi-isotope elemental data over several orders of magnitude concentration with low limits of detection (LOD) for 60+ elements nearly simultaneously, all with minimal sample preparation. Coupled with the ability to perform these analyses with minimal visible damage to the specimen, it is no wonder that LA-ICP-MS has been adopted enthusiastically by archaeologists to answer fundamental questions of artifact composition, environmental variability and diagenetic alteration. In this chapter, we use LA-ICP-MS and SEM-EDS to examine the occurrence and potential causes of elevated levels of trace elements in archaeological bone samples from central Eurasia to test the hypothesis that biogenic incorporation is distinguishable from diagenetic uptake infiltration after death and burial.

18.2 Mineralized Tissues

The utilization of LA-ICP-MS to study archaeological mineralized tissues has developed slowly over the past several years. This is probably due to a growing appreciation of the complex sets of physiological variables mediating dietary incorporation of minor and trace elements in bony tissues (Burton et al. 1999; Zaichick et al. 2011), often obscured by the unpredictable and rate-variable exchange of ions in hydrodynamic sedimentary contexts (Trueman and Tuross 2002). Unlike most other classes of archaeological material (but see Golitko et al. 2012), biominerals interact extensively with their environment both during and after the life of the individual, creating difficulties for separating the *biogenic* signature of life, versus *diagenetic* alteration after entry into the depositional environment. The basis for this high level of interaction is the composition of the hydroxyapatite crystal [$Ca_{10}(CO_3, PO_4)_6(OH)_2$] itself. The apatite molecule is capable of accommodating cation and anion substitutions: the hydroxyl (OH^-) ion can be readily substituted by anions Cl^- and F^-; the phosphate (PO_4^{3}) molecule by AsO_4^{3-}, SO_4^{2-}, CO_3^{2-}, SiO_4^{4-}; the Ca^{2+} ion by a variety of cations, including K^+, Na^+, Mn^{2+}, Ni^{2+}, Cu^{2+}, Co^{2+}, Zn^{2+}, Sr^{2+}, Ba^{2+}, Pb^{2+}, Cd^{2+}, Y^{3+}, UO^{2+} (Wopenka and Pasteris 2005). For this reason, identifying components of incorporation in both biogenic and diagenetic contexts requires careful assessment of the overall preservation of the element under consideration, the bio- or geochemical pathway of incorporation, a set of general rules for the expected concentration of elements in mineralized tissue, and provisional explanations for the departure of actual from expected element concentrations.

Previous research in LA-ICP-MS of archaeological trace element biomineralization have attempted to measure biogenic incorporation of trace elements for the purpose of determining diet (Rasmussen et al. 2009), resolving migration and residence locale (Cucina et al. 2005, 2007, 2011; Dudgeon 2008; Horstwood et al. 2008), and epidemiological exposure in past human populations (Uryu et al. 2003). More recently, LA-ICP-MS has been applied to the question of post-depositional alteration and estimating the extent of diagenetic alteration in both archaeological and paleontological contexts (Koenig et al. 2009; Kohn 2008). LA-ICP-MS has proven to be particularly useful under these conditions because it is capable of gathering discrete spatial information for bio-available elements when micro-scale structures can be sampled at high precision. This approach enables sampling across or around histologically-relevant morphological features such as dentin-enamel interfaces, seasonal annuli or compositional structures associated with osteogenesis. These features can contain chemical information that informs on growth and development, and the rate or frequency of elemental exposure, as well as the timing and mechanism of post-depositional or diagenetic alteration.

Although a variety of quantification strategies have been employed for laser ablation analysis of solids (Speakman and Neff 2005), precise and accurate data calibration is a subject of debate. A host of alternative calibration strategies have been offered, including semi-quantitative standardless approaches (Leach and Hieftje 2001; Yu et al. 2009), desolvated liquid standard additions (Aeschliman et al. 2003; Horn et al. 2000) and certified reference material (CRM) standardization approaches with varying levels of calibration assumptions (Craig et al. 2000; Castro et al. 2010; Gratuze 1999; Gratuze et al. 2001; Neff 2003. Also see Speakman and Neff (2005) for a concise summary). By far the most common calibration approach for archaeologists is the use of graded standards whose concentration brackets the expected concentration and whose matrix composition is similar to the sample unknowns (Bertini et al. 2011; Cochrane and Neff 2006; Eckert and James 2011; Golitko et al. 2012; van Elteren et al. 2009). This approach is similar to the method introduced by Gratuze (Gratuze 1999; Gratuze et al. 2001), which accounts for instrumental drift and matrix effects by using an internal standard naturally present at a known concentration in the unknown sample to calculate ablation efficiency, and by analyzing external reference standards between every few samples. To produce more accurate results, Gratuze suggests the use of several different element "menus" or groups of analytes at similar ranges of concentration. The first menu includes all major elements, which are then summed to 100 %, and the element chosen for an internal standard for the rest of the "menus" is then taken from this analysis.

Unfortunately, most non-silicate archaeomaterials are marked by a general lack of effective analytical calibrating standards spanning the range of typical occurrence in biological systems. We have arrived at a methodological solution to the problem of calibrating archaeological mineralized tissues which employs the internal standard correction method of Gratuze (Gratuze 1999; Gratuze et al. 2001), but augments it further by utilizing high-precision, semi-quantitative major and minor element measurements (proximate to the ablation site) of weight percent concentration using SEM-EDS for spatially discrete estimation of the internal standard concentration. This is an alternative to the full spectrum method, which employs assumption of major (Gratuze 1999) or all element oxide states (Neff 2003) for the calculation of final concentration. Even though full spectrum or 'dual mode' high–low concentration approaches based on oxide calculation produce data in reasonable agreement with internally-standardized measured values, internal standardization, where possible, is preferred (Speakman and Neff 2005).

Because biological mineralized tissues are rarely in stoichiometric equilibrium, significant and varying fractions of carbon, nitrogen and oxygen in biomolecules such as collagen, non-collagenous proteins and low-molecular weight peptides (Hedges and van Klinken 1992) comprise the total ablation volume. The non-mineralized, organic portion of unaltered biological apatites varies from around 2–5 % in dental enamel, 20–30 % in dentin (Gage et al. 1989; Nanci 2008) and up to 30+ % in cementum and bone (Hillson 1996). The percentage of these organic components is highly variable within each tissue type, depending on the amount of diagenetic matrix alteration. Ionic substitutions replace calcium and phosphates in the burial environment, creating spatially heterogeneous major, minor and trace element concentration gradients (Hinz and Kohn 2010; Koenig et al. 2009; Kohn 2008). To ascertain the degree of post-depositional alteration and the magnitude of a biogenic or living signature of trace elements at a few tens of μg/g to sub-μg/g levels, a robust, spatially-sensitive internal standardization procedure—coupled with a matrix-specific external calibration—is required.

Here, we adopted a dual mode strategy for overcoming instrumental sensitivity drift of the ICP-MS, the suppression of signal caused by high matrix loads in the plasma (space charge) and the spatial heterogeneity of major elements in the material ablated. These uncertainties can be divided into three primary standardization/correction strategies: 1) dual inlet GCMS interface for simultaneous liquid standard and ablated material measurement; 2) frequent external standard calibration during the ablation experiment; and 3) estimation of spatially-specific major element concentrations for internal standard correction of analyte signals, post laser ablation. The basic approach to data calibration is a slight modification of the approach described previously (Neff and Dudgeon 2006; Speakman and Neff 2005). ICP-MS data calibration was performed by fitting standardized concentrations (ratios to one isotope of calcium; in this case, ^{43}Ca) to normalized counts (ratios of raw analyte counts to raw ^{43}Ca counts in the standard). The standardized signal (*SS*) for analyte x (background subtracted and liquid internal standard-corrected) is defined as:

$$SS_x = ICPS_x / ICPS_{43Ca} \tag{18.1}$$

where ICPS are measured intensity counts per second for each analyte or isotope. The standardized concentration (*SC*) for analyte x is:

$$SC_x = Conc_x / Conc_{43Ca} \tag{18.2}$$

Using the external standards, regression of SC_x on SS_x for each element yields the slope parameter for calculating the standardized concentrations (SC_{unk}) in the unknowns. In principle, forcing the regression line through zero (as in the Excel function *linest*) minimizes the effect on the slope parameter of residual background signal on the unknowns, and experience has demonstrated that forcing the regression through zero produces the most consistent calibration curve at low signal intensities. With the intercept forced through zero, the slope (K) can now be used to calculate SC_x in the unknowns by:

$$SC_x = K(SS_x) \tag{18.3}$$

Because SC_x is defined as the actual concentration divided by the internal standard concentration, the concentration of each analyte x can be determined by multiplying the slope parameter and standardized signal by the internal standard concentration of calcium ($Conc_{43Ca}$), as follows:

$$Conc_x = K_y(SS_x)Conc_{43Ca} \tag{18.4}$$

The stoichiometric concentration of calcium in biological hydroxyapatite [$Ca_{10}(PO_4)_6(OH)_2$] is around 39.5 % wt/wt. In living systems however, calcium concentration by weight varies from this theoretical value. Because calcium is used to standardize the laser ablation signal intensities for each element acquisition, deviations from the expected stoichiometric concentration affect the accuracy of the concentration determination. We adopted a procedure to externally measure the actual calcium concentration at each ablation site, using a Bruker Quantax 200 SDD-EDS system on an FEI Quanta 200F SEM. A minimum of ten replicate EDS measures of calcium (normalized weight percent) were averaged and the mean value used in the internal standard correction ($Conc_{43Ca}$) of the unknowns (see Table 18.1 for SEM-EDS instrument parameters). SEM-EDS measures at each ablation site on the unknowns were validated by frequent reference to our synthetic hydroxyapatite blank standard (Ca = 39.5 % wt/wt) to verify EDS stability and by comparing unknown Ca/P ratios adjacent to the ablation site with experimental values for Ca and P from our ICP-MS acid digests of enamel and dentin in both modern and archaeological samples (enamel $\bar{x} = 2.15$, sd = 0.011; dentin $\bar{x} = 2.06$, sd = 0.005, determined by mass ratio) to assess deviations from expected Ca/P ratios.

18.3 Understanding the Biological Implications of Archaeometallurgy

We used our LA-ICP-MS—SEM-EDS analysis strategy to study the epidemiological implications for archaeometallurgy in a collection of skeletal material from the Kammenyi Ambar 5 cemetery (Middle Bronze Age [2035 to 1735 cal. BC]) in the Karagaily-Ayat River Valley of the southern Ural Mountains in Russia (Hanks 2010; Hanks and

Table 18.1 Typical instrument operational parameters for LA-ICP-MS and SEM-EDS analyses discussed in the text

Instrumental parameters for LA-ICP-MS analysis	
Laser ablation device	New Wave UP213 Nd:YAG
Wavelength	213 nm
Energy	0.6 mJ
Spot size	100 μm (imaged aperture)
Fluence	7.35 J cm^{-2}
Rep rate	20 Hz
Passes	1
Depth/pass	5 μm
Laser output power	60 %
Laser on time	40 s, spot mode
Sweep gas (He)	0.313 min^{-1}, constant
ICP-MS	Thermo X-series II
Operation mode	Dual-mode liquid/laser, GCMS interface
RF power	1403.92 W
Nebulizer flow rate (Ar, dual inlet system)	0.7 l min^{-1}
Sample gas flow rate (He)	0.3 l min^{-1}
Scan mode	Peak jumping
Number of mass sweeps per scan	35
Dwell time at standard resolution	10 ms
Dwell time at high resolution	2 ms; (Na, P)
Channels per mass	1
Channel spacing	0.02
Instrumental parameters for SEM-EDS analysis	
SEM	FEI Quanta 200F ESEM
Accelerating voltage	20 kV
Imaging detector	Solid-state backscattered electron detector (BSED)
Imaging mode	Low or high vacuum
Spot size	5.0
Working distance	Variable; 10.0–12.5 mm
Magnification	Variable
EDS detector	Bruker Quantax 200 SDD-EDS
Detector	Silicon drift detector (129 eV)
Analysis mode	Objects—multipoint
Spectra correction	Escape, shift, shelf and tail
Element identification	Automatic—preset list
Preset element list	C, N, O, Na, Mg, P, S, K, Ca and Sr
Spectral deconvolution	Series deconvolution
Quantification	PB-ZAF (standardless)
Results report format	Mass percent (normalized)

Doonan 2009). This project was spurred by previous research suggesting elevated levels of bone arsenic in archaeological skeletons were a consequence of long-term exposure to gaseous arsenic during the production of arsenical bronze (Oakberg et al. 2000). A dissenting opinion was provided by Pike and Richards (2002), who argued that substantial post-depositional incorporation resulted in the measured arsenic in the bone matrix. Due to the disagreement on the likelihood of preservation of biogenic signatures of metallurgy under complex soil/bone interaction in the post-depositional environment this question has engendered a lively debate over the proximate mechanism for arsenic incorporation in the skeleton (Martíneaz-García et al. 2005; Millard 2006;

Oakberg et al. 2000; Özdemir et al. 2010; Pike and Richards 2002). Because arsenic may substitute for phosphorus in the apatite matrix of bone and teeth as an $AsO^{3-}{}_4 \rightarrow PO^{3-}{}_4$ substitution, incorporation is possible from inhalation and drinking water (Kabata-Pendias and Mukherjee 2007) in the biogenic context and infiltration from soil pore water (Mahoney et al. 2005) in the diagenetic context.

To determine the concentration and mode of uptake of arsenic in apatite matrix, it is necessary to profile the distribution of the element in both histological and archaeological sedimentary orientation. When this is coupled with evidence from mobile elements present in the burial soil, a record of pre and post-depositional incorporation can be compiled to trace the source of arsenic contamination. Previous work on arsenic uptake in bone was performed using acid digestion ICP-MS (Oakberg et al. 2000; Özdemir et al. 2010). While careful precautions were taken to remove contamination from the cortical surfaces of bone, averaging effects from using whole bone samples prevent the interpretation of the mode of uptake. Our laser ablation approach to this problem allows us to visualize the spatially-discrete occurrence of arsenic and to test the alternate hypotheses of diagenetic diffusion uptake, versus incorporation during biogenic formation and remodeling.

18.4 Materials and Methods

Human skeletal samples consisted of 66 bone fragments (47 femora, 10 tibia, 4 humera, 2 radii, 1 ulna, 1 rib, 1 unknown element) and 15 permanent teeth (5 lower first or second molars, 9 upper first or second molars and 1 upper premolar). Animal skeletal samples consisted of 18 individual bone fragments from 7 genera (*Bos* sp. = 3, *Canus* sp. = 2, *Equus* sp. = 3, *Sus* sp. = 2, *Ovis* sp. = 3, *Marmota* sp. = 4, *Allactaga* sp. = 1 [Great Jerboa]). Small samples (±7.0 mm × 7.0 mm × 3.5 mm) were cut from the bones using a low speed saw (Buehler Isomet). The cut bone samples and whole teeth were mounted in uniform orientation in polystyrene petri dishes using cyanoacrylate glue and embedded in low viscosity epoxy (Buehler EpoThin). The samples were placed in a vacuum chamber to draw epoxy into voids in the bone matrix and to remove surface bubbles obscuring histological features. The epoxy discs were sectioned to expose the inner bone or tooth surface, ground using commercial silica and alumina wet/dry sanding paper and polished on a Buehler Ecomet 2 grinder/polisher with colloidal silica. Discs were cleaned in an ultrasonic cleaning bath using 18 MΩ water to remove grinding contamination and photographed prior to laser ablation so they could be used as sample maps when setting up the laser patterns (Fig. 18.1).

18.4.1 LA-ICP-MS Analysis

Laser ablation analyses were performed on the flattened and polished discs, and the sampling and instrument parameters are shown in Table 18.1. Ablation of each sample was performed using a variable grid of 100 μm spots spaced each 250 μm, after pre-ablating the surface with a lower powered 200 μm spot to assure removal of surface contaminants from grinding,

Fig. 18.1 Cut and polished epoxy disc with embedded archaeological human bone samples

polishing and handling. The circular 100 μm spot was used to provide a compromise between discreteness of ablation target and sufficient ablation volume to achieve quantification above minimum detection limits for all of the elements analyzed (Table 18.2). The sampling protocol used a pre-experiment delay of 5 min to monitor background signal intensity. Each sample was run with an uptake time of 15 s, a run time of 15 s and a washout time of 60 s. Calibration standards were run every 0.5 h, and were bracketed by three washouts (equivalent to a liquid aspiration reagent blank) for a total of 270 s on either side to allow the backgrounds to reach stability. A combination of calibration standards (GSC-1G, GSD-1G, GSE-1G, BHVO-2G, BCR-2G, BIR-1G and six synthetic apatites) were employed to calculate the concentration of 35 analytes of interest for evaluating biogenic incorporation versus diagenetic alteration. Calibration and calculation of element concentration was performed according to the procedure outlined above.

We performed a second experiment to compare paired cortical bone and tooth samples taken from the same individual from the KA-5 cemetery, which gives an empirical perspective on the process of diffusion and incorporation of trace elements across different skeletal elements in histological context (Fig. 18.2). In this second experiment, grids of laser spots were placed over the entire crown of the sectioned tooth (1333 analyzed spots) and bone (440 analyzed spots) surface (Fig. 18.3). After the laser ablation data collection was completed, an SEM visual inspection was made of the ablation surface to remove spots from the grid matrix that had missed the bone or tooth and ablated only epoxy.

18.4.2 SEM-EDS Analysis

Immediately after completion of laser ablation data collection, sample discs were carbon sputter coated (Structure Probe, Inc.) and major element concentration by normalized weight percent was obtained using a Bruker Quantax 200 SDD-EDS mounted to an FEI Quanta 200 FE ESEM (see Table 18.1 for instrument parameters). Three replicate EDS acquisitions were made immediately adjacent to individual ablation craters, to provide an estimate of calcium concentration to incorporate into the internal standard correction calculation. EDS measurements were performed in high vacuum mode using the backscattered electron detector (BSED) to pinpoint areas adjacent to ablation craters and to visualize density variation across heterogeneous domains.

Compact bone is neither solid nor homogeneous. It is composed of many morphological structures which supply blood, mineral transport and innervation (haversian system, Volkmann's canals, canaliculi and lacunae), perform bone remodeling (osteocytes, osteoblasts and osteoclasts, and chondrocytes) and define the cortical margins (periosteum and endosteum). Because bone histology varies from the outer cortical aspect toward the trabeculae or bone marrow at the inner cortical margins of the medullary cavity, it cannot be assumed to be three-dimensionally homogeneous or of uniform density. Laser ablations penetrating thin surfaces of bone may open voids in the matrix that are either filled with epoxy or are empty, resulting in a much lower time-averaged intensity signal on the ICP-MS. To address this, post-EDS internally standardized values were examined for normalized phosphorus concentration, to further exclude partial bone or tooth matrix ablations. Phosphorus concentrations at each ablation site were compared to the overall specimen average for phosphorus and to the average at adjacent ablation sites. Deviations from overall or localized average phosphorus concentrations were subsequently removed from the ablation matrix and replaced with a blanked node in the surface trace metal concentration maps plotted using Surfer (Golden Software, v. 8.01). This data validation procedure allows us to create surface plots that are generally free from artifacts related to laser ablation performance and increases the representativeness of the trace element profiles in histological context. By conducting our laser spot mode analysis over the entire crown region of tooth and the cortical to inner trabecular margin of the bone,

Table 18.2 LA-ICP-MS average concentration of archaeological samples, calibrating standard consensus values and limit of detection (LOD; $3*\sigma_b$) for selected elements (all values in ppm, parts per million; μg/g)

	KA-5 bone	KA-5 enamel	KA-5 dentin	GSC-1G	GSD-1G	Apatite 1	Apatite 2	Apatite 3	Apatite 4	LOD
^{7}Li	1.93	1.07	1.28	5.90	43.00	0.70	0.79	1.33	4.20	0.14
^{11}B	7.42	6.71	23.48	26.00	50.00	14.35	20.09	41.38	129.10	14.47
^{39}K	46.63	104.74	50.99	25,700.00	25,300.00	9.94	31.15	93.23	257.00	331.55
^{43}Ca	248,213.70	339,829.01	342,603.08	50,743.04	51,457.74	359,975.23	359,456.65	358,938.07	357,032.92	532.53
^{45}Sc	0.15	0.22	0.24	5.40	52.00	0.37	0.50	0.97	2.79	0.20
^{47}Ti	149.93	263.13	262.79	8211.00	7431.85	2904.00	2574.00	2412.00	2303.00	3.46
^{51}V	6.41	0.28	2.86	5.40	44.00	0.05	0.22	0.81	2.52	0.95
^{52}Cr	9.15	0.61	2.78	10.30	42.00	3.35	4.22	7.82	16.64	1.32
^{55}Mn	225.27	2.53	4.32	176.00	220.00	2.41	3.25	5.48	15.42	1.30
^{57}Fe	473.15	1964.75	1909.70	106,490.76	103,381.54	3615.00	2319.00	2473.00	2436.00	37.66
^{59}Co	0.94	0.54	0.53	5.90	40.00	1.40	1.65	2.55	6.39	0.02
^{60}Ni	19.30	25.25	26.40	21.00	58.00	47.51	43.47	44.12	45.84	4.46
^{65}Cu	11.17	2.90	3.62	16.00	42.00	1.38	2.21	4.52	14.79	0.52
^{66}Zn	137.51	48.78	103.71	12.70	54.00	1.66	41.75	145.10	577.40	1.69
^{69}Ga	5.42	0.18	0.77	10.00	54.00	0.57	0.71	1.12	2.54	0.03
^{72}Ge	0.03	0.04	0.03	4.00	32.00	0.19	0.20	0.18	0.20	0.17
^{75}As	1.13	0.59	0.63	3.20	27.00	0.32	0.54	0.97	3.08	0.42
^{82}Se	9.93	4.86	2.89	0.20	2.00	0.08	0.53	1.58	5.84	1.15
^{88}Sr	859.06	203.22	577.87	32.30	69.40	198.30	234.80	326.00	787.60	3.38
^{89}Y	0.84	0.07	0.01	4.80	42.00	0.00	0.12	0.42	1.67	0.01
^{111}Cd	0.12	0.02	0.02	1.90	18.00	0.03	0.89	3.10	12.20	0.02
^{118}Sn	0.13	0.21	0.16	5.30	29.00	0.29	1.06	3.02	10.29	0.21
^{121}Sb	21.38	1.40	26.70	5.30	43.00	0.75	1.00	1.68	4.45	0.02
^{137}Ba	226.54	9.11	40.73	34.80	67.00	27.16	33.53	49.37	121.50	0.55
^{139}La	0.97	0.06	0.02	4.36	39.10	0.02	0.19	0.68	2.61	0.00
^{140}Ce	4.44	0.09	0.02	4.62	41.40	0.03	0.22	0.84	2.74	0.01
^{202}Hg	0.51	0.41	0.37	–	–	0.01	0.08	0.23	0.75	0.14
^{208}Pb	0.64	0.12	0.21	14.00	50.00	0.45	5.75	20.16	79.19	0.57
^{232}Th	0.02	0.00	0.00	4.20	41.00	0.03	0.15	0.48	2.00	0.00
^{238}U	71.84	0.34	1.47	4.70	41.00	0.01	0.34	1.14	4.59	0.11

Experimental limit of detection calculated as 3× standard deviation of experiment reagent blanks. Underlined LOD values indicate detection limits at or below one or more unknown specimens

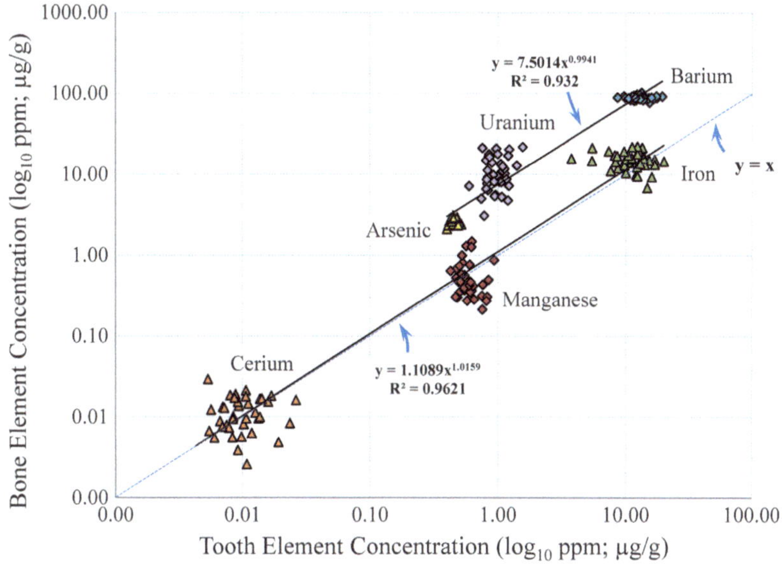

Fig. 18.2 Paired sample (single individual) comparison for six trace element concentrations in human cortical bone (y-axis) and tooth dentin (x-axis) from KA-5

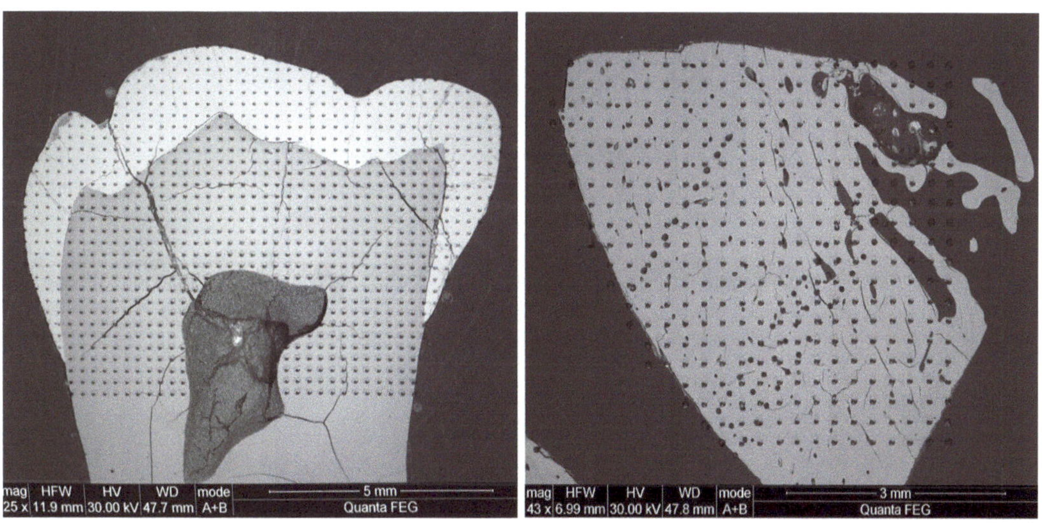

Fig. 18.3 Archaeological tooth (*left*) and femoral cortical bone (*right*) cross sections (post-LA-ICP-MS), showing grid of 100 µm diameter spots at 250 µm intervals

we are able to construct surface plots of element distributions to investigate the diffusion process across different biomineral classes (enamel, dentin and bone) for elements manganese, iron, arsenic, barium, cerium and uranium to evaluate the direction and to infer the rate of uptake of metals into the hydroxyapatite matrix in each specimen (Fig. 18.4).

18.4.3 Post-Analysis Calibration

All unknown sample concentrations were determined using the method described above, employing the liquid internal standard and SEM-EDS calcium concentration to correct for variable calcium concentration and differential laser coupling and density variation between

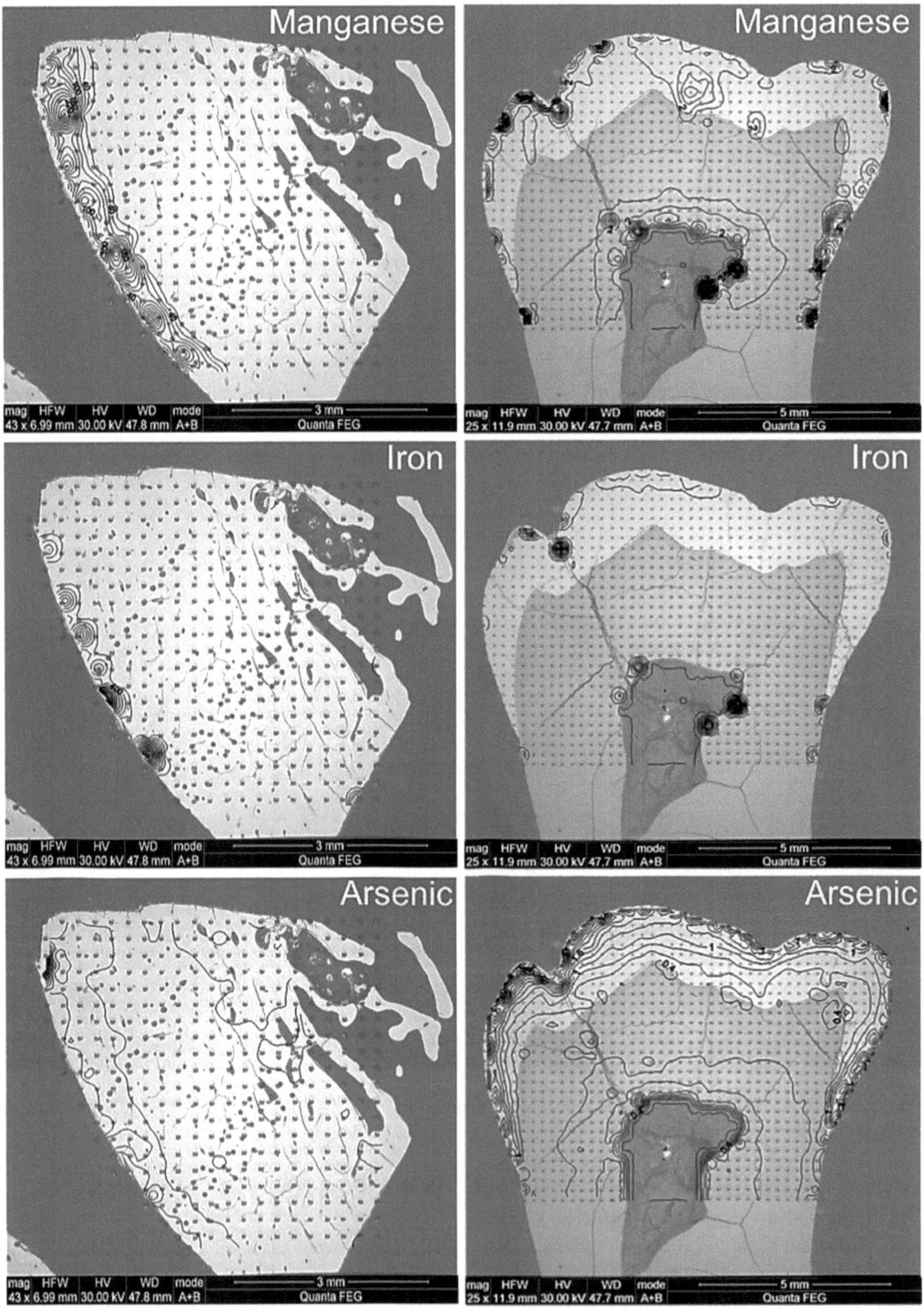

Fig. 18.4 Surface contour maps of trace element distribution for paired (single individual) human cortical bone and tooth samples

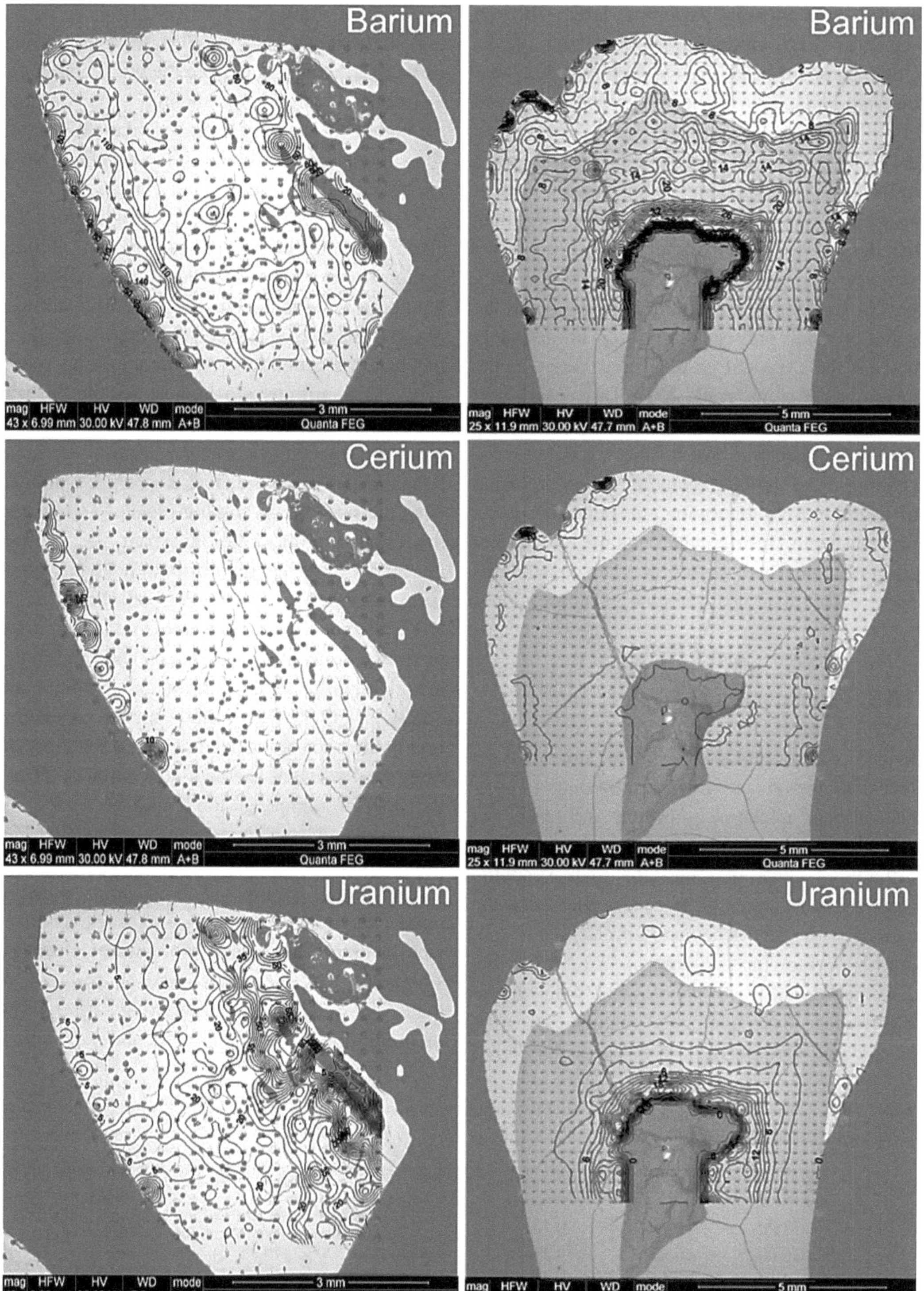

Fig. 18.4 (continued)

enamel and dentin. External laser ablation standards were composed of members of the USGS standard series and synthetic hydroxyapatite standards. This combined approach proved useful because several of the analytes of interest are either not present in consensus concentrations in the USGS standards (e.g., Hg), or are present at one or more orders of magnitude greater than the desired calibration range on the unknown samples (e.g., Na, Mg, Al, Si, K, Ti, Cr, Mn, Fe, Ni, Ba, Pb). Several authors have shown the value of using only standards that immediately bracket the expected or measured range of the unknowns (Cucina et al. 2007; Duwe and Neff 2007; Scharlotta et al. 2011). Closely bracketing standard concentrations to the expected concentrations of the unknowns increases precision of estimation by removing the potential effect of non-linear instrument response at higher concentrations of analytes in graded standards under matrix load or space charge suppression.

18.5 Discussion

In our paired-sample experiment, our analyses revealed significant diffusion-mediated post-depositional uptake in both bone and tooth, but the diffusion patterns indicate histologically-differential absorption into the hydroxyapatite matrix. In this series of plots it appears that the cations with electrochemical similarities to calcium (barium and uranium) are following a generally different histological path of incorporation from elements manganese, iron, arsenic and cerium. Barium in particular shows evidence of multiple pathways of incorporation in both tooth and bone samples, from the cortical surface of the bone and from the pulp chamber of the tooth. Alternatively, uranium concentrations are highest in the pulp chamber, like barium, but show considerably higher incorporation along the trabecular margin of inner cortical bone.

Generally, the opposite pattern obtains for elements manganese, iron, arsenic and cerium, demonstrating weakly developed diffusion gradients along the occlusal and ventral surfaces of tooth enamel, with secondary ingression along the dentin-enamel junction. In the bone specimen, these profiles are highest on the cortical surface, with secondary diffusion from arsenic along the trabecular-cortical margin of the inner bone. Based on the diffusion profile estimated from the surface and contour plots, we compared the element concentrations from the central, minimally-altered region of both bone and tooth specimens (see Fig. 18.2). For both the bone and tooth, elements iron, manganese and cerium are of similar concentration between bone and tooth, but elements arsenic, barium and uranium are nearly an order of magnitude higher concentration in the bone specimen compared to the tooth. Summary averaged element concentrations from the interior region of tooth and bone, including estimated soil concentrations, are listed in Table 18.3.

Our initial results of bone trace element uptake thought to occur from adsorptive diffusion at the soil—bone pore water interface suggested that the outer cortical surfaces of bone and teeth were substantially altered by incorporation of elements Mn and Fe, with a

Table 18.3 Concentration average and %RSD for sediment, tooth and bone, KA-5 specimens (average concentration values in ppm, parts per million; μg/g)

	Soil/sediment concentration		Sub-crown dentin concentration		Central cortical bone concentration	
	Ave	%RSD	Ave	%RSD	Ave	%RSD
^{55}Mn	917.70	19.88	0.55	14.16	0.58	52.57
^{57}Fe	35,510.00	16.12	10.21	30.44	15.15	17.79
^{75}As	9.10	21.64	0.45	4.58	2.56	5.59
^{137}Ba	171.00	21.96	12.54	11.52	88.68	3.36
^{140}Ce	19.00	16.64	0.01	27.96	0.01	38.36
^{238}U	10.00	–	0.96	14.57	10.66	35.84

Soil/sediment values represent adjacent grave soils and are compiled from unpublished data minor component of As, and the inner surface at the trabecular-cortical interface by uptake of U (Fig. 18.5). We surmised that this spatial pattern was evidence of minimal post-depositional incorporation of As, and would likely have been removed by the surface contamination procedures suggested by previous authors (Oakberg et al. 2000; Özdemir et al. 2010). The distribution of U followed a nearly opposite pattern of incorporation, with the highest concentration (up to 250 μg/g) on the trabecular-cortical interface progressing inwards, hinting at a substantially different process of diffusion-mediated uptake of water soluble elements into the bone matrix.

As noted by Pike and Richards (2002), understanding the diffusivity of trace metals into bone or tooth requires not only fine-scale data on the chemical characteristics of soil, groundwater and mineralized tissue (including pH, redox state, soil moisture and temperature), but also a mathematical model for the partition coefficient (K_d), the electrochemical potential of ionic transmission between different matrices at the pore water interface. Recent work by Kohn and colleagues (Hinz and Kohn 2010; Kohn 2008) explore the mechanisms of diffusion under fossilization, using LA-ICP-MS to measure the uptake of trace elements to understand the complex interactions of soil and mineralized tissue pore water in burial matrices. They discovered a variety of unpredictable modes of trace element incorporation in fossilized material that suggests the path of diffusion, adsorption and recrystallization in biominerals is element specific and can provide information on the specific context of burial. For instance, elements strontium, barium and uranium should diffuse and adsorb similarly, but with decreasing effective rate into bone matrices, as a function of K_d, where $K_d = \text{Sr} < \text{Ba} < \text{U}$. Practically speaking, diffusion profiles of barium should show lower spatial heterogeneity than uranium, and strontium should be lower than either barium or uranium. In the case of fossilized bone from a 25 kya Pleistocene site in southwestern Montana, diffusion of strontium,

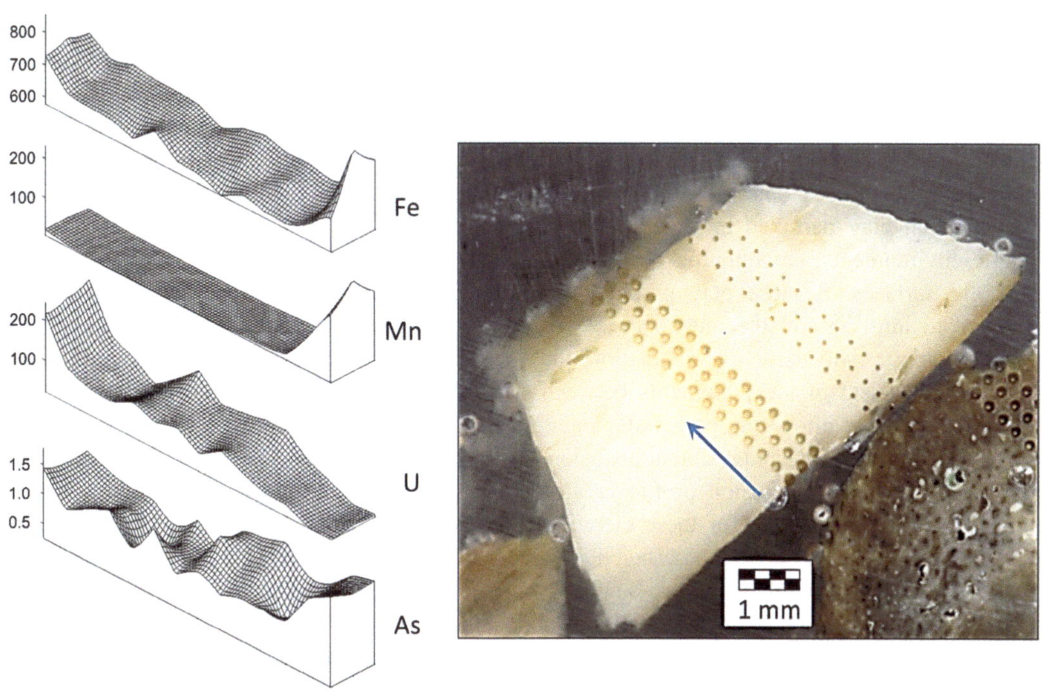

Fig. 18.5 KA-5 human bone specimen with a 3 × 15 grid of 100 μm laser spots at 250 μm intervals (*right*). Wireframe contour profiles for four elements (iron, manganese, uranium and arsenic; *left*). Concentration reported in parts per million (ppm, μg/g)

barium and uranium is complete, suggesting equilibration with local soils was complete at the time of collection (Hinz and Kohn 2010). In samples that exhibit poorly developed diffusion gradients, some have suggested that retention of a biogenic signature may be possible (Koenig et al. 2009:513), although many others are more pessimistic about recovering the living signature of incorporation (Hinz and Kohn 2010; Kohn 2008; Pike and Richards 2002; see also Berna et al. 2004).

Our results suggest that diffusion-adsorption reactions are taking place within both of the paired skeletal elements from the KA-5 cemetery, but that the rate and direction of these reactions are not consistent between material type or by element. Similar to our preliminary study of the KA-5 materials, surface enrichment of several elements (manganese, iron, cerium, arsenic) is consistent with respect to bone and tooth specimens, and indicates a mode of uptake and incorporation that is mediated by direct contact with soil adhering to the tissue surface. The observation of enrichment, relative to sediment concentration in the first row transition metals (e.g., titanium, vanadium, chromium, manganese, iron, cobalt, nickel and copper), and rare earth elements has been modeled for rock varnish systems by Thiagarajan and Lee (2004), who proposed that these elements are physically leached from clay particles in aqueous atmospheric environments, precipitating on underlying solid surfaces. Because clay particles are expected to infill around the outer surfaces of the skeleton after soft tissue decomposition, the partition coefficient of these elements between clay and hydroxyapatite at the pore water interface controls their uptake and resultant diffusion. The poorly-developed diffusion profile created from the laser ablation data for these elements suggests that either the partition coefficients for these elements are high, relative to As, Ba and U, or that pore water-mediated diffusion is less effective at transmitting these elements into the hydroxyapatite matrix. The latter case seems plausible, given the observation that seasonal variation in soil moisture and freeze-thaw cycles may contribute to localized soil moisture reduction at the pore water interface, resulting in decreased diffusion probability across the soil—bone contact area.

In contrast, the relative efficiency of transport into hydroxyapatite tissues of elements arsenic, barium and uranium—all producing water soluble compounds (e.g., H_3AsO_4, $BaCO_3$, UO_2SO_4)—may lie in their ability to remain as free ions in pore water solutions. In this situation, pore water solutions containing arsenic, barium and uranium can be delivered via capillary action into the pulp chambers of teeth and marrow cavities of bones, and proceed by diffusive uptake and incorporation in the hydroxyapatite matrix (see Millard and Hedges 1996 for a detailed discussion of the kinetics of uranium uptake). On the other hand, elements contributing to the outer surface diffusion profile reported here (Mn, Fe, As, Ce) have affinities for clay particles, and in complexation with each other (Hsia et al. 1994). Lin and Plus (2000) showed that in laboratory experiments, As(III) and As(V) were both adsorbed onto various clay particles and that over time, the desorption of these arsenic species decreased, likely owing to time-dependent diffusion and incorporation into the crystal structure of the clay particles. Therefore, arsenic adsorbed onto clay particles and complexed with iron in the local environment may eventually come into pore water contact with the surfaces of bone and teeth as adhering soils, transmitting by diffusion into the apatite mineral.

The direction of incorporation of a suite of trace elements in enamel, dentin and bone appears to be driven primarily by the element's propensity to disassociate from sediment particles and become soluble in pore water. Transition metals and rare earth elements persist as oxides on clays and other sediment particles, and produce weakly developed diffusion gradients corresponding to their partition coefficients at the pore water interface. Alternately, elements with electrochemical similarity to calcium (strontium and barium), or with phosphate affinity (uranium) may enter the larger bone cavities as aqueous ions and replace calcium in the phosphate lattice until apatite mineral equilibrium is attained (Koenig et al. 2009; Pike and Richards 2002). Viewing the laser ablation diffusion profiles of several elements in our bone and tooth specimens (Fig. 18.4), we see that barium has diffused from both cortical trabecular surfaces of the bone section and predominantly from the inner pulp chamber of the tooth specimen. We argue that this pattern is consistent with capillary uptake of aqueous barium, rather than through diffusion at the pore water interface, due to the histological microstructure revealed through SEM imaging (Fig. 18.6). These small orifices on the outer surface of compact bone (Haversian canals) are connected with other canaliculi, lacunae and voids in bone and as dentinal tubules within the pulp chamber (Kohn et al. 1999). These are contrasted with the more impermeable mineralized surfaces of enamel and cementum (covering the dentin) which, unlike the periosteal membrane, persist in the post-depositional context.

Conversely, arsenic follows the uptake profile of manganese, iron and cerium, with a strong diffusion gradient terminating about 500 μm into the cortical and trabecular surfaces of the bone specimen and about 250 μm into the occlusal enamel and pulp chamber dentin surfaces (Fig. 18.4). This pattern is more consistent with

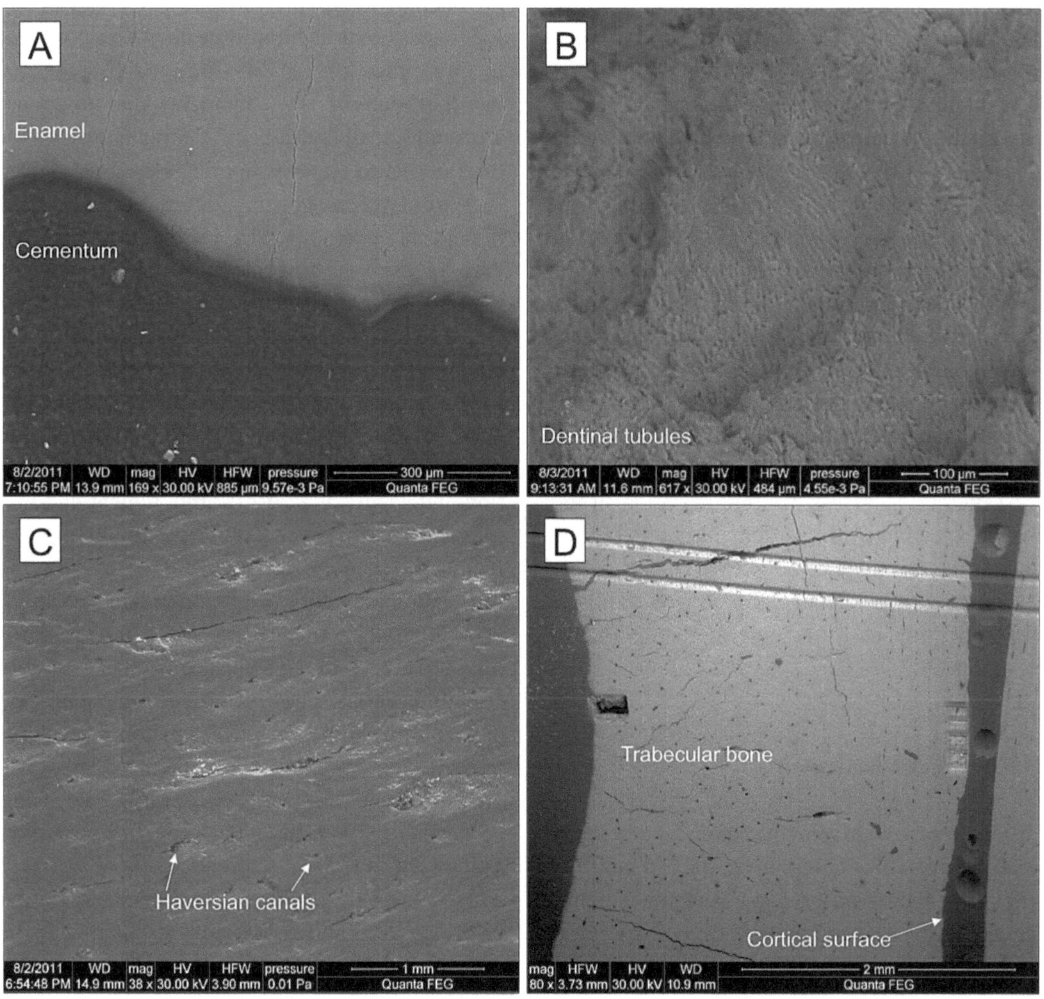

Fig. 18.6 SEM images of histological structures hypothesized to be the source of direct ionic uptake and incorporation. (**a**) Enamel—cementum interface (enamel *above*, cementum *below*); (**b**) dentin from pulp chamber; (**c**) cortical bone surface; (**d**) cross-section of cortical bone

partition coefficient-mediated diffusion from adhering clay particles at the pore water interface suggested by others (Koenig et al. 2009; Hinz and Kohn 2010), and can be discretely visualized in the laser ablation surface maps. While it may be argued that some elements—notably strontium and barium—have diffused so completely in the bone and dentin that no biogenic signature remains (what Hinz and Kohn term "diagenetic overprinting" [2010:3228]), we believe that arsenic retains evidence of biogenic uptake in both sub crown dentin and enamel, owing to the lower apparent partition coefficient of arsenic, compared to manganese, and the substantial agreement of concentrations of manganese, iron and cerium recovered from the inner portions of the bone and tooth sections (Fig. 18.2; Table 18.3). The almost 6× increase in bone versus tooth arsenic in the paired sample comparison remains a concern for future bone studies, however. While we believe it is possible that the lower values in the tooth sample reflect exposure during development of the dentition, with higher bone values representative of lifetime exposure, a conservative approach would apply only ratio-level comparisons between the permanent teeth of individuals interred in the same cemetery in the same depositional context. Using our laser ablation spot mode mapping method gives us the best method to assess discrete variation in individual specimens in histological context, and to determine the mode of uptake and the possibility of post-mortem damage which could obscure the collection of a biogenic signature (Koenig et al. 2009).

18.6 Conclusion

We have presented a methodological and analytical strategy for measuring the spatial and (by inference) temporal aspects of trace element uptake in mineralized tissues. Demonstrating the sensitivity of LA-ICP-MS for investigating the dual component of biogenic and diagenetic incorporation, we addressed one particularly thorny question in archaeological science: can we identify aspects of archaeometallurgy using trace elements specific to certain methods of manufacture (Oakberg et al. 2000; Özdemir et al. 2010; Pike and Richards 2002)? Our answer to this question is a resounding *maybe*. While we believe that our method provides the best approach to identify, separate and quantify components of biogenic incorporation and diagenetic alteration, further work is required to assess the replicability of these results in experiment situations. These include sorption experiments in modern biominerals and their analogs, interspecific comparisons of bioavailable trace metals in tissues and examination of bioarchaeometallurgy in skeletal assemblages with variable post-depositional environments. Our LA-ICP-MS experiments demonstrate that it is possible, in principle, to identify the mechanisms of incorporation of trace elements and to create archaeologically-sufficient explanations for their observed patterns in mineralized tissues.

While we have firmly internalized the cautionary tales of overreliance on trace element data as proof of biogenic signature in archaeological bones (Radosevich 1993; Sandford 1993), we show that utilizing the technological advances incorporating high-sensitivity ICP-MS trace element analysis with well-controlled UV laser ablation systems can generate spatial data on element concentration with excellent resolution and accuracy across morphological and histological structures. Combined with matrix-matched standard arrangements that provide fine-scale bracketing of specimen concentration, and internal standard routines that monitor instrumental drift and correct for laser-sample coupling variation in heterogeneous matrices, our multi-modal analytical methodology represents incremental improvements over previous strategies, with an eye towards increasing precision of measurement and repeatability. Somewhat different than other strategies that employ laser raster patterns and long acquisition times on verifiably homogeneous matrices, the procedures outlined here are crucially important for employing high-density spot analysis to understand the additive components of biogenic and diagenetic incorporation, often at sub-µg/g variation in concentration.

The use of SEM-EDS to visualize micromorphological and major chemical structural variation add to our assessment of post-depositional change and have proven useful in creating independent, internal standard calibration data to ensure comparable data across biomineralized tissues of different densities, crystal orientation and ablation response. In combination, this methodological strategy is profoundly useful for evaluating the process of biogenic and diagenetic uptake of trace elements in the archaeological skeleton, providing new empirical evidence for evaluating the presence and impact of specific patterns of human activity (Sanford and Weaver 2000).

By comparing models and empirical evidence for the process of fossilization (Hinz and Kohn 2010; Koenig et al. 2009; Kohn 2008) with our own case studies of post-depositional alteration on much shorter archaeological timescales (c. 4500 years), we have shown that well-developed diffusion gradients persist in samples of archaeological bone and tooth. Even though certain diagenetic "overprinting" indicator elements (Sr, Ba and U) demonstrate rapid diffusion and absorption in the hydroxyapatite matrix, elements (As) with explanatory potential for archaeological activities like metallurgy do not show the same pattern of uptake and incorporation (Fig. 18.4), leading us to posit a different mechanism for their occurrence. Our evidence demonstrates the presence of histologically-mediated uptake of free metal ions in pore water solutions through innervation and blood supply conduits on both the cortical and medullary aspects of long bones and within the pulp chamber of the dentinal root (Fig. 18.6). In contrast, transition metals and rare earth elements are adsorbed onto clay particles that eventually make pore water contact with biomineral surfaces such as enamel, cementum and cortical bone. Diffusive uptake of these elements is therefore guided largely by partition coefficients between clay particles and hydroxyapatite at the pore water interface, rather than by ionic solubility and electrochemical similarity to calcium in the bone mineral.

In addition to our primary aim to evaluate the biogenic versus diagenetic signatures of archaeologically important trace elements, this methodological approach has important implications for dietary stable isotope studies. A recent study suggests that collagen, in contrast to DNA and other organic proteins in biomineralized tissue, maintains a stable amino acid profile until the weight percent of collagen falls below 1.0 % (Dobberstein et al. 2009). Using our method of topographic mapping of mineralized tissues, measuring the post-depositional uptake of trace elements may yield proxy data on secondary carbonate (CO_3^{2-}) incorporation, likelihood of surviving collagen and the best locations for recovering biogenic stable isotope samples (see Trueman and Tuross 2002). Our preliminary results indicate that biogenic signatures can persist in archaeological skeletal material on the order of 4500 years old, but discerning this signature requires careful selection of specimen, sampling location within the specimen, analytical technique, and a great deal of caution in the interpretation of the proximate mechanisms producing the observed elemental distributions.

Acknowledgements The authors wish to acknowledge the support of Whitnie Rauh and Nicholas Holmer, who performed many of the analyses we report in this chapter. This material is based upon work supported by the National Science Foundation under Grant No. BCS 0821783, BCS 1024674 and OPP 0722771 and the Wenner–Gren Foundation for Anthropological Research Grant No. 7552. Support was also received from the Idaho State University Office of Research and the University of Pittsburgh. Special thanks to Chelyabinsk State University, Russia, Wenner-Gren Foundation and University of Sheffield, UK.

References

Abdullah MM, Ly AR, Goldberg WA, Clarke-Stewart KA, Dudgeon JV, Mull CG, Chan TJ, Kent EE, Mason AZ, Ericson JE (2012) Heavy metal in children's tooth enamel: related to autism and disruptive behaviors? J Autism Dev Disord 42:929–936

Aeschliman DB, Bajie SJ, Baldwin DP, Houk RS (2003) Spatially-resolved analysis of solids by laser ablation-inductively coupled plasma-mass spectrometry: trace elemental quantification without matrix-matched solid standards. J Anal Atom Spectrom 18:872–877

Arai T, Hirata T, Takagi Y (2007) Application of laser ablation ICPMS to trace the environmental history of chum salmon Oncorhynchus keta. Mar Environ Res 63:55–66

Beard BL, Johnson CM (2000) Strontium isotope composition of skeletal material can determine the birth place and geographic mobility of humans and animals. J Forensic Sci 45:1049–1061

Berna F, Matthews A, Weiner S (2004) Solubilities of bone mineral from archaeological sites: the recrystallization window. J Archaeol Sci 31:867–882

Bertini M, Shortland A, Milek K, Krupp EM (2011) Investigation of Iron Age north-eastern Scottish glass beads using element analysis with LA-ICP-MS. J Archaeol Sci 38:2750–2766

Budd P, Montgomery J, Cox A, Krause P, Barreiro B, Thomas RG (1998) The distribution of lead within ancient and modern human teeth: implications for long-term and historical exposure monitoring. Sci Total Environ 220:121–136

Budd P, Montgomery J, Evans JE, Barreiro B (2000) Human tooth enamel as a record of the comparative lead exposure of prehistoric and modern people. Sci Total Environ 263:1–10

Burton JH, Price TD, Middleton WD (1999) Correlation of bone Ba/Ca and Sr/Ca due to biological purification of calcium. J Archaeol Sci 26:609–616

Castro W, Hoogewerff J, Latkoczy C, Almirall JR (2010) Application of laser ablation (LA-ICP-SF-MS) for the elemental analysis of bone and teeth samples for discrimination purposes. Forensic Sci Int 195:17–27

Cochrane EE, Neff H (2006) Investigating compositional diversity among Fijian ceramics with laser ablation-inductively coupled plasma-mass spectrometry (LA-ICP-MS): implications for interaction studies on geologically similar islands. J Archaeol Sci 33:378–390

Craig C-A, Jarvis KE, Clark LJ (2000) An assessment of calibration strategies for the quantitative and semi-quantitative analysis of calcium carbonate matrices by laser ablation-inductively coupled plasma-mass spectrometry (LA-ICP-MS). J Anal Atom Spectrom 15:1001–1008

Cucina A, Neff H, Tiesler V (2005) Provenance of African-born individuals from the colonial cemetary of Campeche (Mexico) by means of trace elements analysis. Dent Anth 17:65–69

Cucina A, Dudgeon JV, Neff H (2007) Methodological strategy for the analysis of human dental enamel by LA-ICP-MS. J Archaeol Sci 34:1884–1888

Cucina A, Tiesler V, Sosa TS, Neff H (2011) Trace-element evidence for foreigners at a Maya port in Northern Yucatan. J Archaeol Sci 38:1878–1885

Dobberstein RC, Collins MJ, Craig OE, Taylor G, Penkman KEH, Ritz-Timme S (2009) Archaeological collagen: why worry about collagen diagenesis? Arch Anth Sci 1:31–42

Dolphin AE, Goodman AH, Amarasiriwardena DD (2005) Variation in enamel intensities among teeth and between pre- and postnatal regions of enamel. Am J Phys Anthropol 128:878–888

Dudgeon JV (2008) The genetic architecture of the late prehistoric and proto historic Rapa Nui (Easter Islanders). Ph.D. Dissertation, University of Hawaii, Honolulu

Durrant SF, Ward NI (2005) Recent biological and environmental applications of laser ablation inductively coupled plasma mass spectrometry (LA-ICP-MS). J Anal Atom Spectrom 20:821–829

Duwe S, Neff H (2007) Glaze and slip pigment analyses of Pueblo IV period ceramics from east-central Arizona using time of flight-laser ablation-inductively coupled plasma-mass spectrometry (TOF-LA-ICP-MS). J Archaeol Sci 34:403–414

Eckert SL, James WD (2011) Investigating the production and distribution of plain ware pottery in the Samoan archipelago with laser ablation-inductively coupled plasma-mass spectrometry (LA-ICP-MS). J Archaeol Sci 38:2155–2170

Gage JP, Francis MJO, Triffit JT (1989) Collagen and dental matrices. Wright, London

Golitko M, Dudgeon JV, Neff H, Terrell JE (2012) Identification of post-depositional chemical alteration of ceramics from the North Coast of New Guinea (Sanduan Province) by time of flight-laser-ablation-inductively coupled plasma-mass spectrometry (TOF-LA-ICP-MS). Archaeometry 54:80–100

Gratuze B (1999) Obsidian characterization by LA-ICP-MS and its application to prehistoric trade in the Mediterranean and the Near East: sources and distribution of obsidian within the Aegean and Anatolia. J Archaeol Sci 26:869–881

Gratuze B, Blet-Lamarquard M, Barradon JN (2001) Mass spectrometry with laser sampling: a tool to characterize archaeological materials. J Radioanal Nucl Chem 26:645–656

Hanks B (2010) Archaeology of the Eurasian Steppes and Mongolia. Annu Rev Anthropol 39:469–486

Hanks B, Doonan R (2009) From scale to practice: a new agenda for the study of early metallurgy on the Eurasian Steppe. J World Prehist 22:329–356

Hedges REM, van Klinken GJ (1992) A review of current approaches in the pretreatment of bone for radiocarbon dating by AMS. In: Long A, Kra RS (eds) Proceedings of the 14th international 14C conference. Radiocarbon 34:279–291

Hillson S (1996) Dental anthropology. Cambridge University Press, Cambridge

Hinz EA, Kohn MJ (2010) The effect of tissue structure and soil chemistry on trace element uptake in fossils. Geochim Cosmochim Acta 74:3213–3231

Horn I, Rudnick RL, McDonough W (2000) Precise elemental and isotope ratio determination by simultaneous solution nebulization and laser ablation-ICP-MS: application to U-Pb geochronology. Chem Geol 167:405–425

Horstwood MSA, Evans JA, Montgomery J (2008) Determination of Sr isotopes in calcium phosphates using laser ablation inductively coupled plasma mass spectrometry and their application to

archaeological tooth enamel. Geochim Cosmochim Acta 72:5659–5674
Hsia TH, Lo SL, Lin CF, Lee DY (1994) Characterization of arsenate adsorption on hydrous iron oxide using chemical and physical methods. Colloid Surf A 85:1–7
Kabata-Pendias A, Mukherjee AB (2007) Trace elements from soil to human. Springer, Berlin
Koenig AE, Rogers RR, Trueman CN (2009) Visualizing fossilization using laser ablation-inductively coupled plasma-mass spectrometry maps of trace elements in Late Cretaceous bones. Geology 37:511–514
Kohn MJ (2008) Models of diffusion-limited uptake of trace elements in fossils and rates of fossilization. Geochim Cosmochim Acta 72:3758–3770
Kohn MJ, Schoeninger MJ, Barker WW (1999) Altered states: effects of diagenesis on fossil tooth chemistry. Geochim Cosmochim Acta 63:2737–2747
Leach AM, Hieftje GM (2001) Standardless semi-quantitative analysis of metals using single-shot laser ablation inductively coupled plasma time-of-flight mass spectrometry. Anal Chem 73:2959–2967
Lin Z, Plus RW (2000) Adsorption, desorption and oxidation of arsenic affected by clay minerals and aging process. Environ Geol 39:753–759
Mahoney J, Langmuir D, Gosselin N, Rowson J (2005) Arsenic readily released to pore waters from buried mill tailings. Appl Geochem 20:947–959
Martíneaz-García MJ, Moreno JM, Moreno-Clavel J, Vergara N, García-Sánchez A, Guillamón A (2005) Heavy metals in human bones in different historical epochs. Sci Total Environ 348:51–72
Millard AR, Hedges REM (1996) A diffusion-adsorption model of uranium uptake by archaeological bone. Geochim Cosmochim Acta 60:2139–2152
Millard A (2006) Comment on Martínez-García et al. Heavy metals in human bones in different historical epochs. Sci Total Environ 354:295–297
Nanci A (2008) Ten Cate's oral histology: development, structure, and function, 7th edn. Mosby, St. Louis, MO
Neff H (2003) Analysis of Mesoamerican plumbate pottery surfaces by laser ablation-inductively coupled plasma-mass spectrometry (LA-ICP-MS). J Archaeol Sci 30:21–35
Neff H, Dudgeon JV (2006) LA-ICP-MS Analysis of Ceramics and Ceramic Raw Materials from the Gila River Indian Community, Arizona. Research Report: Institute for Integrated Research in Materials, Environments, and Society, California State University, Long Beach
Oakberg K, Levy T, Smith P (2000) A method for skeletal arsenic analysis, applied to the Chalcolithic copper smelting site of Shiqmim, Israel. J Archaeol Sci 27:895–901
Özdemir K, Erdala YS, Demircib Ş (2010) Arsenic accumulation on the bones in the Early Bronze Age İkiztepe Population, Turkey. J Archaeol Sci 37:1033–1041
Pike AWG, Richards MP (2002) Diagenetic arsenic uptake in archaeological bone. Can we really identify copper smelters? J Archaeol Sci 29:607–611
Radosevich SC (1993) The six deadly sins of trace element analysis: a case of wishful thinking in science. In: Sandford MK (ed) Investigations of Ancient human tissue: chemical analyses in anthropology. Gordon and Breach, Philadelphia, PA, pp 269–332
Rasmussen KL, Bjerregaard P, Gommesen PH, Jensen OL (2009) Arsenic in Danish and Swedish Mesolithic and Neolithic human bones – diet or diagenesis? J Archaeol Sci 36:2826–2834
Sandford MK (1993) Understanding the biogenic–diagenetic continuum: interpreting elemental concentrations of archaeological bone. In: Sandford MK (ed) Investigations of Ancient human tissue: chemical analyses in anthropology. Gordon and Breach, Philadelphia, PA, pp 11–15
Sanford MK, Weaver DS (2000) Trace element research in anthropology: new perspectives and challenges. In: Katzenberg MA, Saunders SR (eds) Biological anthropology of the human skeleton. Wiley-Liss, New York, NY, pp 329–350
Scharlotta I, Gilstrap W, Neff H (2011) No stone unburned: a compositional analysis of obsidian microdebitage by laser ablation TOF-ICP-MS. Archaeometry 53:873–889
Scharlotta I, Goriunova OI, Weber A (2013) Micro-sampling of human bones for mobility studies: diagenetic impacts and potentials forlemental and isotopic research. J Archaeol Sci 40:4509–4527
Speakman RJ, Neff H (eds) (2005) Laser ablation ICP-MS in archaeological research. University of New Mexico Press, Albuquerque, NM
Thiagarajan N, Lee C-TA (2004) Trace-element evidence for the origin of desert varnish by direct aqueous atmospheric deposition. Earth Planet Sci 224:131–141
Trueman CN, Tuross N (2002) Trace elements in recent and fossil bone apatite. Rev Mineral Geochem 48:498–521
Uryu T, Yoshinaga J, Yanagisawa Y, Endo M, Takahashi J (2003) Analysis of lead in tooth enamel by laser ablation-inductively coupled plasma-mass spectrometry. Anal Sci 19:1413–1416
van Elteren JT, Tennent NH, Selih VS (2009) Multi-element quantification of ancient/historic glasses by laser ablation inductively coupled plasma mass spectrometry using sum normalization calibration. Anal Chim Acta 644:1–9
Wopenka B, Pasteris JD (2005) A mineralogical perspective on the apatite in bone. Mater Sci Eng C 25:131–143
Yu Q, Huang R, Li L, Lin L, Hang W, He J, Huang B (2009) Applicability of standardless semi-quantitative analysis of solids by high-irradiance laser ionization orthogonal time-of-fight mass spectrometry. Anal Chem 81:4343–4348
Zaichick S, Zaichick V, Karandashev V, Nosenko S (2011) Accumulation of rare earth elements in human bone within the lifespan. Metallomics 3:186–194

19 Pb Isotopic Composition of Panamanian Colonial Majolica by LA-ICP-MS

Javier G. Iñañez, Jeremy J. Bellucci, Juan Guillermo Martín, Richard Ash, William F. McDonough, and Robert J. Speakman

Abstract

Panamá Viejo, founded in 1519 by the Spanish explorer Pedrarias Dávila, was the first permanent European settlement on the Pacific Ocean, and became a city, by royal decree, in 1521. Shortly after its creation, the city became an important base for trade with Spain. In 1671, the English pirate Henry Morgan waged an attack on Panamá Viejo, which resulted in a fire that destroyed the entire city. A new settlement built a few miles west, called Casco Antiguo or San Felipe, is now the historic district of modern Panama City. The Pb isotopic compositions of the glazes on the surface of sixteenth to seventeenth century majolica pottery sherds from Panama Viejo and Casco Antiguo (both in Panama), and Lima (Peru) were determined via non-destructive laser ablation multi-collector ICP-MS (LA-MC-ICP-MS). The contrast in Pb isotopic compositions in the glazes on ceramics recovered in different locations demonstrate that early majolica pottery production during this period used Pb obtained from the Andes. However, the Pb used in later majolica production in Panama is of Spanish origin. After Panamá Viejo was burned to the ground, Panamanian majolica production ended.

J.G. Iñañez (✉)
Grupo de Investigación en Patrimonio Construido (GPAC), Universidad del País Vasco/Euskal Herriko Unibertsitatea (UPV/EHU), CIEA Lascaray, Avda. Miguel de Unamuno, 3, 01006 Vitoria-Gasteiz, Álava, Spain

IKERBASQUE, Basque Foundation for Science, Maria Diaz de Haro 3, 6 floor, 48013 Bilbao, Bizkaia, Spain
e-mail: javier.inanez@ehu.eus

J.J. Bellucci
Department of Geology, University of Maryland, College Park, MD 20742, USA

Department of Geosciences, Swedish Museum of Natural History, Stockholm 104 05, Sweden
e-mail: Jeremy.bellucci@gmail.com

J.G. Martín
Universidad del Norte, Barranquilla, Colombia
e-mail: jgmartin@uninorte.edu.co

R. Ash • W.F. McDonough
Department of Geology, University of Maryland, College Park, MD 20742, USA
e-mail: rdash@umd.edu; mcdonough@umd.edu

R.J. Speakman
Center for Applied Isotope Studies, The University of Georgia, 120 Riverbend Rd., Athens, GA 30602, USA
e-mail: archsci@uga.edu

L. Dussubieux et al. (eds.), *Recent Advances in Laser Ablation ICP-MS for Archaeology*, Natural Science in Archaeology, DOI 10.1007/978-3-662-49894-1_19

19.1 Introduction

19.1.1 The Old City of Panama or Panamá Viejo

After the expedition of Christopher Columbus to the Caribbean side of the current Republic of Panama in 1502, the Spanish monarchy believed it was imperative to explore these new territories, and assigned this task to Alonso de Ojeda and Diego de Nicuesa. The mainland was divided from Cabo de la Vela to Uraba's Gulf, as "Nueva Andalucia," and from Uraba's Gulf to the west, as "Castilla de Oro," respectively. These explorations, little more than raids by today's standards, aimed to conquer and colonize the mainland, and led to the founding of San Sebastián de Urabá in 1509 (now called Necocl-í-Colombia). This settlement was subsequently destroyed by the indigenous people of this region and a year later the Spaniards founded Santa María la Antigua del Darién, near the Tanela river (now called Acandí-Colombia), which became the first settlement with the title of city in continental America (Martín 2009).

After his arrival as governor of "Castilla del Oro," Pedrarias decided to move Santa María la Antigua to the shores of the Pacific Ocean, a strategic location in which he could carry out a campaign of conquest (Martín 2009). Panama Viejo was founded on August 15th, 1519, in a native village under the command of Cori, and served as the first Spanish port on the Pacific coast of the Americas. Although the Spanish later established other Pacific Coast ports, Panama City remained one of the largest ports in the Pacific, in part due to the traffic of wealth looted from the Inca Empire. Likewise, all goods from Europe passed through this port for redistribution to the South American continent (Mena 1998). 152 years later, in 1671, the English pirate Henry Morgan attacked the city, leading to its destruction and final abandonment. This attack prompted the relocation of the city to what is now known as Casco Antiguo or San Felipe (Fig. 19.1). The new city of Panama, founded in 1673, was partly a reflection of the destroyed city. The layout—*traza*—of the new city followed the traditional rules of Spanish urban design in which the Plaza Mayor served as a point of reference for the central distribution and location of buildings within the city (Castillero 1994, 2004a, b; Mena 1984, 1992). However, "the *traza* of San Felipe is unthinkable without its walls, the need was a crucial aspect of the move" (Tejeira 2001:87).

19.1.2 Majolica and Spanish Production

Majolica is an earthenware ceramic generally characterized by a creamy light-buff colored ceramic paste and an opaque white tin-lead glaze covering the entire outer surface of the vessel. Perhaps the most characteristic feature of majolica pottery lies in the metallic oxide decorations that were applied on top of the tin-lead white glaze coat. The opaque white glaze is usually achieved after dipping the bisque ceramic into a soupy suspension made out of sand (e.g., quartz), tin and lead to the ceramic biscuit, and then fired again in the kiln. Lead plays an important role during the glaze maturation since it acts as a flux, decreasing the temperature needed for melting SiO_2, resulting in a bright and transparent/translucent glaze (Tite et al. 1998).

According to the historical majolica making tradition and extant written sources and current scientific literature, this opaque glaze is commonly achieved by the addition of a fine fraction of tin oxide (SnO_2) particles, likely cassiterite, the most common mineral source of Sn in nature. Thus, Sn might had been incorporated into the glaze mixture suspension by two different processes: the most common process was likely a frit, which is a raw mixture of Sn, Si and Pb minerals that had to be fused and then quenched, forming a glassy fine-grained compound, which was ground afterwards and added to water in order to form the glaze suspension; or as finely ground particulates added to the glaze suspension. During the cooling stage in the kiln after firing, cassiterite crystals grow within the glaze

Fig. 19.1 Map of the new and old cities of Panama

into micrometric crystals and small crystalline aggregates. The appropriate small size of newly formed cassiterite particulates, along with extant small quartz and feldspar inclusions, as well as any bubbles that may result from the firing process, absorb, scatter, and/or reflect incident light, thereby giving the transparent glaze a whitish appearance. This white opacity makes a perfect canvas on which to apply chromatic decoration, which is normally applied to the outer surfaces of the glaze coat (Iñañez 2007; Molera et al. 1999; Tite et al. 2008).

Although the first evidence of opacified glazed pottery can be traced to the Middle East as early as the fifth century BC, evidence for ceramic production showing the general features described above remain unclear until the ninth century AD (Hill et al. 2004; Mason and Tite 1997). Following the known historical occurrences, majolica technology shows a clear link to the Islamic *Al-Andalus* (the Cordoba Caliphate, and subsequently *taifas* or Islamic petty kingdoms) during the medieval period on the Iberian Peninsula. It is generally considered that from the tenth century AD onwards majolica technology became widespread throughout the entire Iberian Peninsula, even in the New Christian kingdoms and principalities of the North and Northeast, and reached the rest of Western Europe soon after. By the sixteenth century, Spanish majolica production flourished as Italian-influenced decorative styles diffused into the Iberian Peninsula, incorporating new chromatic choices to the potter's palette such as yellow, usually combined with the more traditional blue, black and green colors (e.g. Iñañez 2007 and references therein).

Many majolica production centers were fully functional in the Castile and Aragon kingdoms during the period of Spanish colonial presence in the Americas. The workshops from cities like Seville, Talavera, Manises, Muel or Barcelona, just to mention a few, might be considered as representative of a genuine proto-industrial activity, which supplied not only their immediate hinterland, but also reaching distant markets. Regarding the colonial trade towards the American market, there is one production center that stood above the rest in terms of quantity and importance—the city of Seville. This city, occupying the riverbanks of the Guadalquivir River on the southern Iberian Peninsula, served as both the departure point and the final destination for most of the Spanish galleons that traded with the Americas in the so called "*Carrera de Indias*," the official (and only allowed) armored convoy of ships from Castile to the Americas. For more than 200 years, the vast traffic of commodities that resulted from the emergence of the new colonial markets was supervised by *Casa de la Contratación*, a bureau of trade established in Seville in 1503. It is, therefore, not surprising that both the importance of, and eventual decline, of Sevillian ceramic manufactures are directly linked to the endurance of the rigid monopolistic trade established by the Castilian crown and the prevalence of the *Casa de la Contratación* in the city.

19.1.3 Archaeology of Panamanian Colonial Majolica

Panama was a territory of the Viceroyalty of Peru, and due to the rigid protectionist economy established by the Castilian Crown, it was able to trade only within this colonial administrative region but generally not to other political administrative regions, such as the Viceroyalty of New Spain, although illicit trade was regularly present, as evidenced in written sources (Glave 2000; Jamieson 2001; Stein and Stein 2002). Panamanian ceramic workshops took advantage of this legal situation and traded their products southwards following the coastline and into the nearby hinterland with cities in Ecuador (Cuenca), Colombia (Popayán) (Therrien et al. 2002) and most importantly Peru (e.g., Lima, the capital of the Viceroyalty of Peru, and Moquegua) (Iñañez et al. 2012; Rice 1997).

The development of historical archaeology in Panama is relatively recent, as is the study of Panamanian majolica (Rovira and Martín 2008). The first archaeological investigations were initiated in the 1960s, and continued intermittently until the present day. From these archaeological excavations in Panama, specifically in Panama Viejo, where kiln related evidence was found (Long 1964), emerged a defined set of ceramic types referred to as "Panamanian majolica." Panamanian ceramics visually appear different because of a "brick-red paste that makes it unmistakable at first sight" (Rovira 1997). A typological classification of Panamanian majolica serves as a chronological indicator and trade marker in colonial America and is defined basically by three types: Panama Plain, Panama Blue on White, and Panama Polychrome (Goggin 1968; Long 1967; Rovira 1997).

The first type, as defined by Long (1967), and assumed to be the earliest, is Panama Plain, which is characterized by thick enamel and in some cases white or greenish tinges. Other scholars have also noted possible technological influence from the Hispano-Moresque tradition, such the marks on the surface of plates that resulted from the use of tripods during firing, and the occurrence of flat-bottomed dishes without borders (Rovira 1997). Panama Blue on White has similar characteristics to a Talavera pottery tradition—Ichtuknee Blue on White—with some obvious American-influenced designs such as corn plant motifs, as well as motifs related to the Chinese porcelain tradition (Deagan 1987; Long 1967; Rovira 1997). Panama Polychrome occurs infrequently in excavated archaeological contexts at Panama Viejo (Goggin 1968). This type has different designs using brown, blue and green colors. In some cases it is possible to find variants that have yellow on their enamel (Rovira 1997).

According to Rovira (1997), Panama Plain ceramics occur earliest in the Panamanian majolica production sequence and have features that are reminiscent of "archaic" majolica and manufacturing techniques similar to those found in the Hispano-Moresque tradition. It is in the mid-seventeenth century when the production of

Panamanian Blue on White and Polychrome majolica purportedly started; related in turn with the decline of European types in the studied archaeological contexts at Panama Viejo.

19.1.4 Pb Isotopic Studies of Colonial Ceramics and Provenance Studies of Panamanian Majolica

Until multi-collector-inductively coupled plasma-mass spectrometry (MC-ICP-MS) gained popularity for Pb isotopic analyses, the traditional method for the measurement of Pb isotope ratios was via thermal ionization mass spectrometry (TIMS). TIMS provides high analytical accuracy and precision, although at the cost of relatively slow and arduous sample preparation (for further discussion see Stos-Gale and Gale (2009), and references therein). Recently, other laboratories have developed novel approaches to Pb isotope analysis seeking alternatives to TIMS, such as EDTA (Ethylenediaminetetracetic acid) extraction and different ICP-MS configurations (see Reslewic and Burton 2002). The use of magnetic sector ICP-MS (Woolard et al. 1998), or quadrupole ICP-Q-MS (Marzo et al. 2007) are among the different alternatives chosen by researchers. Unfortunately, none of these techniques can achieve the analytical precision that TIMS or MC-ICP-MS have demonstrated for Pb isotopic measurements. Currently, studies utilizing high-precision Pb isotope ratios characterization conducted by MC-ICP-MS report excellent agreement with data acquired by TIMS (Baker et al. 2006). Moreover, MC-ICP-MS coupled with laser ablation offers the ability to efficiently generate a large, statistically significant data set more quickly than solution analyses.

In recent years, Pb isotopic analysis of glazed ceramics has gained popularity in archaeometrical studies, although still not as recurrent as its use for analysis of artifacts made of bronze, copper, silver, or glass (e.g. Degryse et al. 2009; Ponting et al. 2003; Shortland 2006; Stos-Gale and Gale 2009; Stos-Gale et al. 1997; Thibodeau et al. 2007; Yener et al. 1991). Among the first works of this nature one can cite those published by Brill et al. (Brill and Wampler 1967; Brill et al. 1987), and, especially relevant to the present study, Joel et al. (1988), an early work focused on the Pb isotopic fingerprinting of majolica pottery in the Americas. Since these initial publications, several more projects on Pb isotope provenance of glazed ceramics have been conducted. For example, it is worth noting the analyses of Islamic glazes by Mason et al. (1992) and Islamic and Hispano-Moresque ceramics by Resano et al. (2008), ceramics from the El Paso area by Pingitore et al. (1997), Rio Grande glazed ceramics by Habicht-Mauche et al. (2000, 2002), majolica from eighteenth century New Spain presidios by Reslewic and Burton (2002), and Mexican and Spanish colonial ceramics by Iñañez et al. (2010).

The first provenance studies of Panamanian majolica were conducted by Vaz and Cruxent (1975) employing thermoluminescence to discriminate between different Spanish colonial production centers in the Caribbean. Olin et al. (1978) included three ceramics found in Panama Viejo in their large chemical study of Spanish and Colonial Spanish majolica conducted by instrumental neutron activation analysis (INAA), which showed a different chemical composition than those studied from other areas. More recently, Jamieson and Hancock (2004) conducted chemical analyses by INAA on a set of ceramics collected in Cuenca, Ecuador, including a set of ceramics found in the same site and archaeologically identified as Panamanian. Soon after, Rovira et al. (2006) reported the chemical characterization by INAA of Panamanian majolica and other ceramic types unearthed at the site of Panama Viejo and two clay samples. Recently, Iñañez et al. (Iñañez and Martín 2011; Iñañez et al. 2012) reported the chemical and technological characterization by INAA and scanning electron microscopy (SEM) of Panamanian ceramics unearthed during recent archaeological excavations at the sites of Panamá Viejo and Casco Antiguo, as well as the convent of Santo Domingo in Lima, Peru. This study included over-fired ceramics and kiln related materials, such as clay spurs, confirming the local origin of

Panamanian ceramics, in agreement with Rovira et al. (2006).

19.2 Goals and Sample

The Pb isotopic composition of the glaze coating of majolica pottery can provide constrains on its source. The main goal of this study is to establish the origin of the Pb used in the manufacturing of Panamanian majolica, taking into account available archaeological evidence and the particular historical circumstances at Panamá Viejo and Casco Antiguo. Majolica technology required significant amounts of Pb and Sn in order to obtain the white glazed enamel characteristic of this ware. However, although quite common in the Earth's crust, Pb is not ubiquitous, so it has to be mined and traded from where Pb minerals are found abundantly. During the early stages of Spanish settlement in the Americas, almost every artifact and commodity was imported from the Iberian Peninsula, including Pb. However, around the mid-sixteenth century, the occupation of the Americas by Spaniards was extensive, and many important Au and Ag mines were being exploited. Therefore, the supply of Pb to Panamanian workshops was oriented towards the southern areas of the viceroyalty of Peru instead of Spain. In addition, Pb is often related to Ag in sulfide ore deposits, so ancient miners used to extract galena (PbS) and other sulfides to obtain Ag in significant quantities. Thus, studying the origin of Pb in American metallic and glazed artifacts can provide significant information regarding trade within the colonial market. In addition, assessing the use of Pb isotopic analysis by LA-MC-ICP-MS as a tool to study the provenance of colonial archaeological material in a nearly non-destructive fashion, consequently providing a powerful technique for cultural heritage studies, is also among the goals of the present work.

In order to achieve these objectives, 30 majolica Pb-glazed ceramics from Panama and Lima, Peru, were analyzed by laser ablation multi-collector inductively coupled plasma mass spectrometry (LA-MC-ICP-MS) (Table 19.1). As seen in Fig. 19.2, typical Panamanian majolica decorations range from plain white glazed coats to various geometric motifs produced in blue on white, green and black on white, or even polychrome patterns. The samples investigated in this study include (1) 15 previously studied ceramics from Panama Viejo, which have been identified as of Panamanian origin according to the chemical composition of their clay pastes and archaeological evidence that included studies of kiln related materials (Iñañez et al. 2012; Rovira et al. 2006); (2) five ceramics excavated at Casco Antiguo, the new city built after the destruction and abandonment of Panamá Viejo in 1671; and (3) ten majolica ceramics that date to the sixteenth to seventeenth centuries and recovered at the Convento de Santo Domingo, Lima, which have been recently identified as Panamanian, likely from Panamá Viejo, according to the chemical analysis of their clay pastes (Iñañez et al. 2012). Additionally, this study also incorporates extant majolica and non-tin-lead glazed Pb isotopic data from sixteenth to eighteenth centuries Spanish ceramic production centers (Iñañez et al. 2010 and references therein; Joel et al. 1988), as well as Andean Pb ores (Gunnesch and Baumann 1984; Gunnesch et al. 1990; Kontak et al. 1990; Mukasa et al. 1990; Sangster et al. 2000; Tilton et al. 1981), and Spanish Pb ores (Arribas and Tosdal 1994; Canals and Cardellach 1997; Hunt 2003; Santos Zalduegui et al. 2004; Tornos and Chiaradia 2004; Velasco et al. 1996). These studies, combined with the data obtained by this study, will put Panamanian majolica Pb isotopic data into an interpretable context, and allow us to determine whether or not Panamanian majolicas used American or Spanish Pb for their glazed coatings.

Mexican Pb analyses have not been included in this study because of historical and geological reasons. The fact that some Pb ores from the Andes and Central Mexico share similar metallogenetic ages provides some overlap in their Pb isotopic signatures. Additionally, for historical reasons, one has to bear in mind that the main Mexican Ag and Au mines during Spanish colonial times, like Zacatecas, Guanajuato,

Table 19.1 Pb isotopic values of the Majolica ware analyzed from Panama and Peru

Sample	Decorations	Origin location	Chronology	Pb provenance	$^{208}Pb/^{204}Pb$	$2\sigma_{(mean)}$	$^{207}Pb/^{204}Pb$	$2\sigma_{(mean)}$	$^{206}Pb/^{204}Pb$	$2\sigma_{(mean)}$	$^{207}Pb/^{206}Pb$	$2\sigma_{(mean)}$	$^{208}Pb/^{206}Pb$	$2\sigma_{(mean)}$
OP0012	Green and black on white	Panama Viejo	1519–1671	Andes	38.776	0.0108	15.679	0.0030	18.756	0.0034	0.836	0.00013	2.068	0.00059
OP0018	Blue, green and black on white	Panama Viejo	1519–1671	Andes	38.837	0.0057	15.690	0.0016	18.796	0.0016	0.835	0.00007	2.067	0.00048
OP0019	Blue, green and black on white	Panama Viejo	1519–1671	Andes	38.836	0.0075	15.686	0.0026	18.788	0.0024	0.835	0.00006	2.068	0.00018
PVM002	n/a	Panama Viejo	1519–1671	Andes	38.828	0.0483	15.680	0.0027	18.801	0.0104	0.834	0.00029	2.065	0.00181
PVM003	n/a	Panama Viejo	1519–1671	Andes	38.971	0.0440	15.670	0.0084	18.841	0.0062	0.832	0.00068	2.068	0.00325
PVM006	n/a	Panama Viejo	1519–1671	Andes	38.763	0.1258	15.656	0.0388	18.789	0.0381	0.833	0.00028	2.063	0.00265
PVM008	n/a	Panama Viejo	1519–1671	Andes	38.853	0.0108	15.684	0.0049	18.802	0.0111	0.834	0.00033	2.066	0.00067
PVM013	n/a	Panama Viejo	1519–1671	Andes	38.807	0.1088	15.667	0.0411	18.777	0.0513	0.834	0.00014	2.066	0.00030
PVM020	Plain white	Panama Viejo	1519–1671	Andes	38.943	0.0188	15.712	0.0102	18.846	0.0091	0.834	0.00016	2.067	0.00041
PVM026	Blue on white	Panama Viejo	1519–1671	Andes	38.945	0.0091	15.718	0.0045	18.841	0.0061	0.834	0.00007	2.067	0.00047
PVM031	Blue on white	Panama Viejo	1519–1671	Andes	38.827	0.0291	15.676	0.0107	18.799	0.0100	0.834	0.00018	2.066	0.00073
PVM032	Blue on white	Panama Viejo	1519–1671	Andes	38.810	0.0180	15.671	0.0068	18.795	0.0102	0.834	0.00017	2.065	0.00075
PVM033	Blue on white	Panama Viejo	1519–1671	Andes	38.923	0.0253	15.714	0.0067	18.840	0.0053	0.834	0.00014	2.066	0.00089
PVM042	Polychrome	Panama Viejo	1519–1671	Andes	38.916	0.0272	15.718	0.0051	18.810	0.0039	0.836	0.00032	2.069	0.00169
PVM043	Polychrome	Panama Viejo	1519–1671	Andes	38.929	0.0103	15.725	0.0035	18.816	0.0039	0.836	0.00005	2.069	0.00030
PVM045	Plain white	Panama Casco Antiguo	1673–1750	Spain	38.606	0.0310	15.663	0.0076	18.503	0.0079	0.847	0.00040	2.087	0.00175
PVM046	Blue on white	Panama Casco Antiguo	1673–1750	Andes	38.661	0.0099	15.670	0.0029	18.654	0.0051	0.840	0.00009	2.072	0.00043
PVM050	Blue on white	Panama Casco Antiguo	1673–1750	Andes	38.838	0.0186	15.677	0.0066	18.786	0.0081	0.835	0.00007	2.068	0.00051

(continued)

Table 19.1 (continued)

Sample	Decorations	Origin location	Chronology	Pb provenance	$^{208}Pb/^{204}Pb$	$2\sigma_{(mean)}$	$^{207}Pb/^{204}Pb$	$2\sigma_{(mean)}$	$^{206}Pb/^{204}Pb$	$2\sigma_{(mean)}$	$^{207}Pb/^{206}Pb$	$2\sigma_{(mean)}$	$^{208}Pb/^{206}Pb$	$2\sigma_{(mean)}$
PVM052	Polychrome	Panama Casco Antiguo	1673–1750	Spain	38.615	0.0427	15.657	0.0082	18.483	0.0083	0.847	0.00056	2.090	0.00236
PVM056	Polychrome	Panama Casco Antiguo	1673–1750	Spain	38.597	0.0213	15.662	0.0072	18.502	0.0077	0.847	0.00009	2.087	0.00045
LIM002	Plain white	Lima, Peru	16th–17th cent.	Andes	39.035	0.0679	15.694	0.0274	18.868	0.0315	0.832	0.00016	2.069	0.00032
LIM024	Green and yellow on white	Lima, Peru	16th–17th cent.	Spain	38.576	0.0107	15.647	0.0031	18.496	0.0065	0.846	0.00019	2.086	0.00098
LIM037	Green on white	Lima, Peru	16th–17th cent.	Andes	38.811	0.0128	15.670	0.0023	18.791	0.0024	0.834	0.00022	2.066	0.00091
LIM038	Green on white	Lima, Peru	16th–17th cent.	Andes	38.671	0.0233	15.626	0.0082	18.686	0.0065	0.836	0.00024	2.069	0.00104
LIM040	Green and yellow black on white	Lima, Peru	16th–17th cent.	Andes	38.737	0.0124	15.669	0.0049	18.663	0.0063	0.840	0.00018	2.076	0.00079
LIM041	Green and yellow black on white	Lima, Peru	16th–17th cent.	Andes	38.825	0.0178	15.671	0.0072	18.800	0.0079	0.834	0.00017	2.066	0.00059
LIM054	Green on white	Lima, Peru	16th–17th cent.	Andes	38.802	0.0213	15.660	0.0070	18.925	0.0071	0.827	0.00011	2.050	0.00045
LIM056	Plain white	Lima, Peru	16th–17th cent.	Andes	38.880	0.0252	15.642	0.0102	18.814	0.0106	0.831	0.00011	2.066	0.00063
LIM062	Blue on white	Lima, Peru	16th–17th cent.	Spain	38.551	0.0076	15.638	0.0043	18.487	0.0059	0.846	0.00013	2.086	0.00047
LIM063	Blue on white	Lima, Peru	16th–17th cent.	Andes	38.778	0.0213	15.659	0.0064	18.787	0.0059	0.834	0.00020	2.065	0.00108

Fig. 19.2 Examples of plain *white*, *blue* on *white* and polychrome majolica found in Panama (from *left* to *right* and from *top* to *bottom*: PVM020, PVM006, PVM031, PVM042)

Pachuca and Sombrerete, are located in Central and North Mexico. Because of their geographic location and the rigid monopolistic control exhibit by Spaniards towards trade between colonial viceroyalties, metals mined in Mexico were generally carried by ground transport to the ports of Veracruz and Acapulco to be shipped to Spain or to the Spanish colonies in Asia (Castillo and Lang 1995; Lacueva Muñoz 2010). Therefore, Mexican influence on Pb supply in Panama had little historical significance and, given the geological constraints, consequently Mexican Pb isotopic signatures have been ruled out of the examples presented here in order to provide less cluttered representations of isotopic data.

19.3 Analytical Methods

Lead has four isotopes, ^{208}Pb, ^{207}Pb, ^{206}Pb, and ^{204}Pb; ^{204}Pb is invariant in nature, whereas ^{208}Pb, ^{207}Pb, ^{206}Pb are daughter products of the decay of ^{232}Th, ^{235}U, and ^{238}U, respectively. Therefore, variation in the Pb isotopic composition of a material is a function of its initial U, Th and Pb concentrations, the starting Pb isotopic composition, and the time-integrated growth of radiogenic Pb. Due to dissimilarity in the chemical behavior of U, Th, and Pb, the Pb isotopic composition of different materials can vary widely in nature. These natural variations, therefore, make the Pb isotopic system an ideal candidate for constraining the potential provenance of geologic materials and the archaeological materials derived from them (e.g. Brill and Wampler 1967; Pollard et al. 2007; Pollard 2009; Shortland 2006; Stos-Gale and Gale 2009 and references therein).

All analyses were conducted at the University of Maryland College Park Plasma Laboratory following the methodology reported by Iñañez et al. (2010). Pb isotopic compositions were determined *in situ* via LA-MC-ICP-MS employing a New Wave UP-213 laser system and a Cetac Aridus desolvating nebulizer system coupled to a Nu Plasma multiple-collector ICP-MS. Before reaching the plasma torch, the He gas from the laser ablation cell was combined with an Ar and N_2 gas flow from the Aridus nebulizer via a T-junction. During each analytical session, ultra-pure 18 MΩ (milli-Q) water was flushed through the Aridus ensuring only Ar and N_2 reached the plasma (see Iñañez et al. 2010 for gas flow settings). Laser ablation MC-ICP-MS analysis of majolica glazes has several benefits when compared to traditional thermal ionization mass spectrometry (TIMS) or MC-ICP-MS analyses. These benefits are: (1) minimally destructive analysis, which preserves samples for future investigations; (2) rapidity—each analysis takes ~2 min, which allows the collection of far more data and in turn generates statistically significant datasets; and (3) the precision of LA-MC-ICP-MS on glazes that contain wt% Pb (as the ceramics in this study), approaching that of TIMS or solution MC-ICP-MS.

Before each sample analysis, an on-peak background was taken for 45 s with the laser on and shuttered. The Nu Plasma time-resolved software was used to establish the average of the background for each analysis and to calculate each ratio using the background corrected signals for each time-resolved measurement. Typical ablation spectra were collected for ~60 s. The isobaric interference of ^{204}Hg on ^{204}Pb was monitored by measuring the background-corrected ^{202}Hg signal and corrected for using the natural isotopic abundances of each Hg isotope, ^{202}Hg/^{204}Hg $= 0.2299$ (de Laeter et al. 2003). The total Hg interference was insignificant (<1 part per 10,000) during our analyses due to the large amount of Pb ($\geq$30 wt% in all cases) and negligible amount of Hg ($\leq$µg/g) in the analyzed ceramic glazes. In addition, matrix effects are limited when using laser ablation systems with wavelengths in the deep UV (213 and 193 nm) and with nanosecond pulse durations.

Isotopic fractionation corrections were performed using the Exponential Law and NIST SRM610 values from Baker et al. (2004) by means of standard-sample bracketing (e.g., Jochum et al. 2006; Kent 2008; Paul et al. 2005; Simon et al. 2007). The fractionation factor was determined using the ^{208}Pb/^{206}Pb ratio measured in NIST SRM 610. Block analyses consisted of two standard measurements (NIST SRM610) before and after six sample measurements.

Replicate analyses of NIST 610 during each day of analyses yielded an external precision of 0.4 % on ^{20x}Pb/^{204}Pb ratios and 0.2 % on ^{20x}Pb/^{206}Pb (with x being 6, 7 or 8, as appropriate). Typical internal and external precisions for ^{20x}Pb/^{204}Pb of ceramic glazes, based on triplicate analyses of an individual sample, are 0.05 % and <0.1 %, respectively (Iñañez et al. 2010). All errors in this study are reported as internal $2\sigma_{mean}$ on an individual analysis. Iñañez et al. (2010) performed triplicate analyses on similar samples and showed that the internal precision is roughly the same as the external precision due to the high concentrations (30–50 wt%) of Pb in the glazes. All analyses were conducted on untreated ceramics on the white glazed area of the coated vessel. Small sherds of around 1 cm^2 of each sample were gently wiped with ethanol and mounted together on an aluminum microscopy wafer using a double-side sticker. This setup was used to avoid repeated manipulation of the laser chamber to introduce new samples. Reference materials were also placed inside the laser chamber at the same time. No other pre-treatment was necessary, as laser pre-ablation removed any possible superficial contamination on the samples.

19.4 Results and Discussion

The Pb isotopic compositions for the Panamanian ceramics discussed above are illustrated in Fig. 19.3 and reported in Table 19.1. Andean Pb isotopic compositions are clearly distinct from that of European produced Spanish majolica—a consequence of the different geological sources of Spanish and American materials, albeit there is some limited compositional

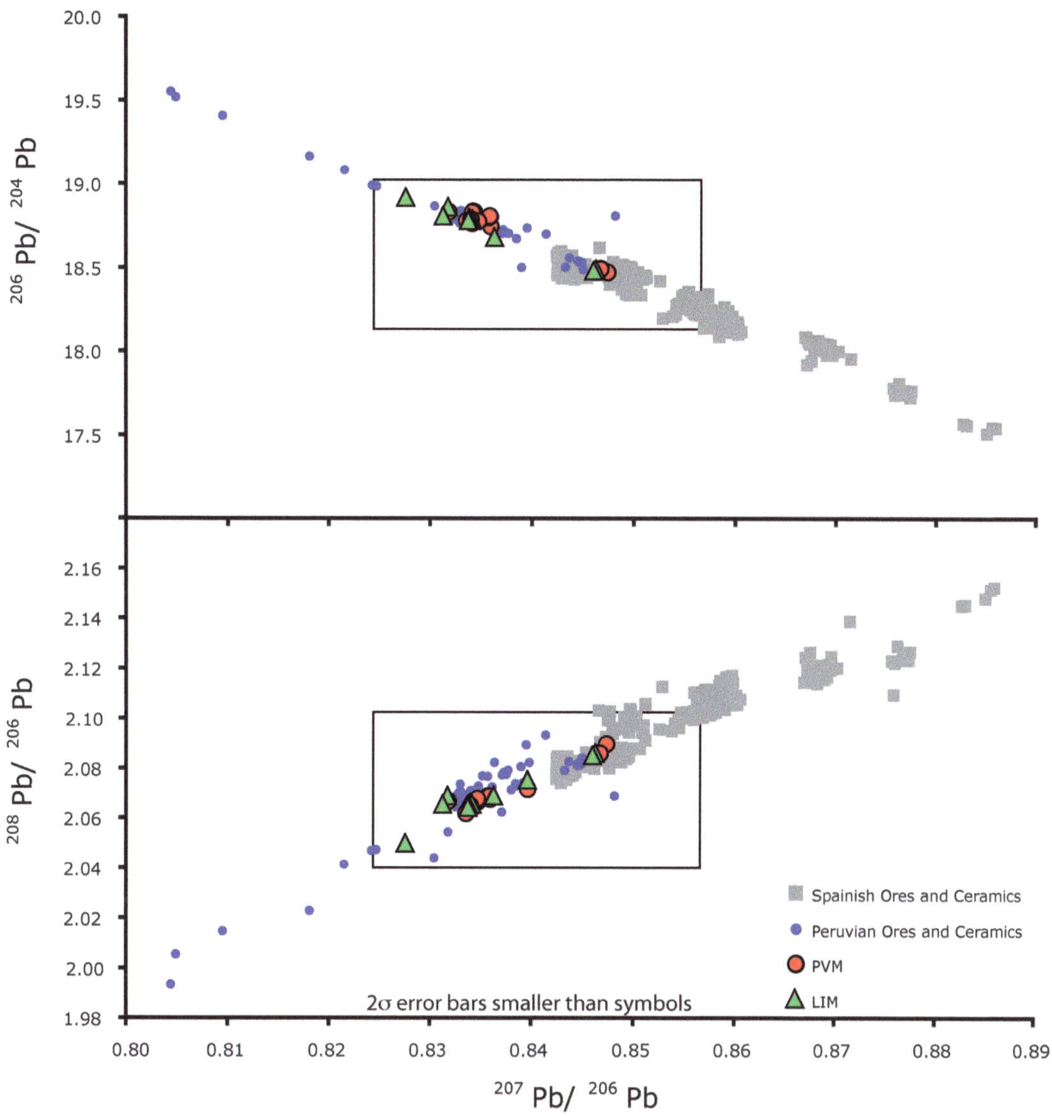

Fig. 19.3 Pb isotopic ratios of Panamanian, Mexican and Spanish ceramics, and lead deposits

overlap. Figure 19.4 represents an enlarged view of the measured Pb isotopic compositions, shading the composition for each historical ore area and projecting the Panamanian ceramics onto these shaded regions.

The Panamanian ceramics found in Panama and in Lima have similar Pb isotopic compositions, which corroborates further observations (Iñañez et al. 2012), that the ceramics found in Lima (PVM) were Panamanian products. Pb isotopic compositions for most of the ceramics found in Panama and Lima (PVM and LIM) overlap with Andean Pb isotopic ratios measured in ore samples. Furthermore, the Pb ratios of three ceramics from Panama Casco Antiguo (PVM045, PVM052 and PVM056) show higher ^{208}Pb/^{206}Pb and ^{207}Pb/^{206}Pb ratios and lower ^{206}Pb/^{204}Pb ratios (Fig. 19.4), exhibiting similar Pb isotopic compositions to Spanish ores and artifacts. Two of the Panamanian ceramics found in Lima (LIM) show Pb isotopic compositions similar to the Panamanian outliers and likely were produced

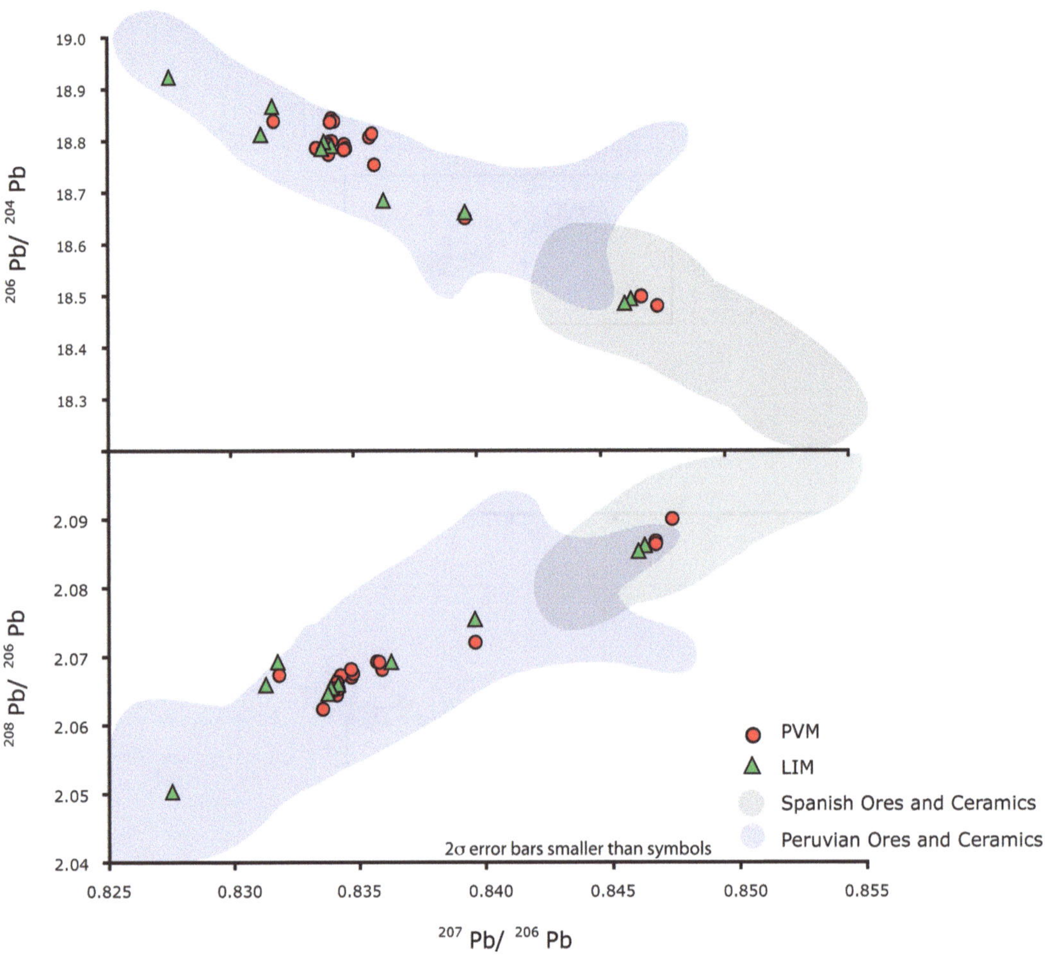

Fig. 19.4 Pb isotopic ratios of Panamanian ceramics. *Grey symbols* correspond to Spanish and Peruvian lead deposits and ceramics

from Spanish Pb. Interestingly, the outlier ceramics from Panama, with white and polychrome decorations respectively, were unearthed at the Casco Antiguo site, chronologically the later of the contexts studied. Unfortunately, the two ceramics traded to Lima, one with blue on white and the other with green and yellow on white decoration, can only be dated approximately to the seventeenth century, which may coincide with the earlier dates attributed to Casco Antiguo.

According to the data presented here, sixteenth and early seventeenth century majolica from Panama Viejo was manufactured using Pb that was most likely obtained from the Andes. This conclusion is in harmony with the idea that Panama Viejo played a leading role during the early colonial period and was considered a commercial hub in the Spanish colonial market due to its strategic location. At that time, Panama Viejo was considered one of the most important ports on the Pacific Ocean, serving as a pivotal connecting point from the territories recently conquered by the Spaniards in South America—the subsequent Viceroyalty of Peru and their important supplies of metals—to Mexico, the Caribbean, and Spain.

Nonetheless, three out of five ceramics analyzed from Casco Antiguo and two out of the ten Panamanian ceramics from the archaeological site of convent of Santo Domingo in Lima, the new capital of the Viceroyalty of Peru, have Pb isotopic compositions compatible with Spanish, not Andean, sources. Unfortunately, the chronological attribution for these ceramics has not been established accurately. However, it is possible to argue a *post quem* date of the first decade of the seventeenth century for these ceramics, when written sources provide data about tiles ordered and traded from Spain to Lima for decorating the convent, with the supply ship stopping by in Panama on the way (Rovira 2002).

One possible interpretation of our results is that as a result of the attack and sack of the city by the English pirate Morgan, which resulted in the relocation of the city 8 km southwest of its original location, the provisioning of Pb was largely supplied by new sources. Thus, at least during the first years of life of the new city, the utilization of Spanish Pb by the pottery industry may imply a shift of the Pb supply and/or a recycling process, perhaps by scavenging objects found in the ruins of or in the vicinity of the destroyed city, instead of relying on the provisioning by trade from South America. However, the archaeological surveys conducted so far at Casco Antiguo do not provide evidence regarding majolica production after the relocation of the city in 1673.

Historical written documentation provides another piece of evidence to help solve this archaeological puzzle. The government of Felipe II (second half of the sixteenth century) was known to have continued the economic and fiscal reforms established by his father, Carlos I. Among the fiscal reforms implemented by the king, the so called "*Siete Rentillas*" (seven little charges) duty is of great importance for studying Pb supply during the sixteenth to eighteenth centuries in Spain and the colonial market. This is because this duty taxed seven products of little profit to the state: black powder, sulfur, mercury, sealing wax, playing cards, iron-based pigments and, importantly, Pb (Torrente 1835). Some of these products were likely exported from the Viceroyalty of Peru. It is no coincidence that the trade of Pb from South America is linked to the beginning of extensive mining activity in Peru around 1545. Thus, Andean Pb easily arrived to Panama, becoming the primary ore used in the Panama Viejo majolica workshops, in agreement with the isotopic data from this study.

However, around the mid-seventeenth century, a deep economic crisis hit the Viceroyalty of Peru. This crisis had multiple parallel causes—perhaps the most relevant to this study is the decline in the amount of precious metals mined at Peruvian mines, which includes indirectly the quantity of Pb extracted and later traded to Panama. The purported decline in the amount of metals mined at the Andean mines resulted from several causes—the depletion of the main metallic ore veins that had been mined since the mid-sixteenth century, and the demographic decline of the native Andean population were perhaps the most significant. The native population formed the main workforce in the mine districts at that time, working under two similar types of exploitive labor conditions: forced labor (*encomienda*) or very low waged labor (*repartimiento*), both resulting in very strenuous work conditions, which ultimately greatly contributed to the decimation of the native population. In addition, a new fiscal reform that increased tax pressure fostered illicit trade by colonial market agents, who constantly broke the once rigid monopoly established by the Spanish crown and often traded with the British, French, and Dutch. During this crisis situation, the precious metals and Pb trade from Peru northwards to Panama significantly reduced in volume, and Spanish imports may have increased in importance as a result (Andrien 2011). These historical data are in agreement with the results presented here.

Furthermore, the Panamanian majolica found in Lima show a *post quem* dating to the first decades of the seventeenth century. Thus, a relative chronological assignment for Lima ceramics having a Pb isotopic signature compatible with Spanish Pb origin can be demonstrated. Based on the Pb isotopic compositions measured here and the historical and archaeological evidence known to date, these ceramics were likely manufactured in Panama after the mid-seventeenth century and transported to Peru. Consequently, and according to the current archaeological knowledge of

Panamanian majolica production, we can argue that the *ante quem* chronology for these ceramics must be around the date of 1671, when the city, and likely also its ceramic industry, was destroyed. As a final remark, we attribute the Pb used in the sixteenth and early seventeenth century Panamanian majolica to an Andean Pb ore source. However, around the second third of the seventeenth century, Panamanian ceramics were mainly manufactured with Spanish Pb.

19.5 Final Remarks

Although we acknowledge the limitations of this work because of the relatively limited number of samples analyzed, it is important to highlight implications of this study for future research. This is one of the few studies to date that successfully combines Pb isotopic analysis and available chemical data of ceramic pastes, with archaeological and historical data to reconstruct the production chronology of one of the most important majolica production centers during colonial times. In addition, we have demonstrated that integrated studies, such as what we have described above, are critical for assessing the provenance and technology of colonial glazed ceramics in order to avoid erroneous conclusions about ceramic provenance and to provide additional insight into colonial technologies.

Acknowledgements This work was supported in part by the Marie Curie International Outgoing Fellowships program, endorsed by the European Commission ("ARCHSYMB", PIOF-GA-2008-223319) and IKERBASQUE, Basque Foundation for Science. Lead isotope analyses were carried out at the Plasma Laboratory of the University of Maryland. Samples were provided by Patronato de Panamá Viejo (PVM), R. L. Bishop and J. Blackman from the Smithsonian Institution (OP), and A. Coello from the Universidad San Marcos la Mayor (LIM).

References

Andrien KJ (2011) Crisis y decadencia. El virreinato del Perú en el siglo XVII, Banco Central de Reserva del Perú. Instituto de Estudios Peruanos, Lima

Arribas A, Tosdal RM (1994) Isotopic composition of Pb in ore deposits of the Betic Cordillera, Spain; origin and relationship to other European deposits. Econ Geol 89:1074–1093

Baker J, Peate D, Waight T, Meyzen C (2004) Pb isotopic analysis of standards and samples using a Pb-207-Pb-204 double spike and thallium to correct for mass bias with a double-focusing MC-ICP-MS. Chem Geol 211:275–303

Baker J, Stos S, Waight T (2006) Lead isotope analysis of archaeological metals by multiple-collector inductively coupled plasma mass spectrometry. Archaeometry 48:45–56

Brill RH, Wampler JM (1967) Isotope studies of ancient lead. Am J Archaeol 71:63–77

Brill RH, Barnes IL, Tong SSC, Joel EC, Murtagh MJ (1987) Laboratory studies of some European artifacts excavated on San Salvador Island. In: Gerace DT (ed) First San Salvador conference: Columbus and his world. CCFL Field Station, Fort Lauderdale, pp 247–292

Canals A, Cardellach E (1997) Ore lead and sulphur isotope pattern from the low-temperature veins of the Catalonian Coastal Ranges (NE Spain). Miner Deposita 32:243–249

Castillero A (1994) La Vivienda Colonial en Panamá. Historia de un Sueño, Fondo de Cultura. SHELL, Panamá

Castillero A (2004a) Los Edificios Religiosos en la Nueva Panamá. In: Castillero A (ed) Historia General de Panamá. Comité Nacional del Centenario de la República, Panamá, pp 375–398

Castillero A (2004b) Las Fortificaciones. In: Castillero A (ed) Historia General de Panamá. Comité Nacional del Centenario de la República, Panamá, pp 27–51

Castillo M, Lang MF (1995) Metales preciosos: Unión de dos mundos. Tecnología, comercio y política de la minería y metalurgia Iberoamericana. Muñoz Moya y Montraveta editores, Sevilla

de Laeter JR, Bohlke JK, Bievre PD, Hidaka H, Peiser HS, Rosman KJR, Taylor PDP (2003) Atomic weights of the elements: review 2000. Pure Appl Chem 75:683–800

Deagan KA (1987) Artifacts of the Spanish Colonies of Florida and the Caribbean, 1500–1800. Ceramics, glassware and beads. Smithsonian Institution Press, Washington, DC

Degryse P, Henderson J, Hodgins G (2009) Isotopes in vitreous materials, Studies in archaeological sciences. Leuven University Press, Leuven

Glave LM (2000) Intercambio y productos de comercio. In: Pease FGY, Moya Pons F (eds) El primer contacto y la formación de nuevas sociedades. Ediciones UNESCO, Paris, pp 391–404

Goggin JM (1968) Spanish majolica in the New World. Types of the sixteenth to eighteenth centuries. Department of Anthropology, Yale University, New Haven, CT

Gunnesch KA, Baumann A (1984) The Atacocha district, central Peru: some metallogenetic aspects.

In: Wauschkuhn A, Kluth C, Zimmermann RA (eds) Syngenesis and epigenesis in the formation of mineral deposits. Springer, Berlin, pp 448–456

Gunnesch KA, Baumann A, Gunnesch M (1990) Lead isotope variations across the central Peruvian Andes. Econ Geol 85:1384–1401

Habicht-Mauche JA, Glenn ST, Milford H, Flegal AR (2000) Isotopic tracing of prehistoric Rio Grande glaze-paint production and trade. J Archaeol Sci 27:709–713

Habicht-Mauche JA, Glenn ST, Schmidt MP, Franks R, Milford H, Flegal AR (2002) Stable lead isotope analysis of Rio Grande glaze paints and ores using ICP-MS: a comparison of acid dissolution and laser ablation techniques. J Archaeol Sci 29:1043–1053

Hill DV, Speakman RJ, Glascock MD (2004) Chemical and mineralogical characterization of Sasanian and Early Islamic glazed ceramics from the Deh Luran plain, Southwest Iran. Archaeometry 46:585–605

Hunt MA (2003) Prehistoric mining and metallurgy in South West Iberian Peninsula. Archaeopress – Publishers of British Archaeological Reports, Oxford

Iñañez JG (2007) Caracterització arqueomètrica de la ceràmica vidrada decorada de la Baixa Edat Mitjana al Renaixement dels principals centres productors de la Península Ibèrica. Dissertation, Universitat de Barcelona

Iñañez JG, Martín JG (2011) Old city, new city: majolica production strategies in colonial Panamá. Paper presented at the 11th European meeting on ancient ceramics, Museum of Natural History, Vienna, 29 September–1 October 2011

Iñañez JG, Bellucci JJ, Rodríguez-Alegría E, Ash R, McDonough W, Speakman RJ (2010) Romita pottery revisited: a reassessment of the provenance of ceramics from Colonial Mexico by LA-MC-ICP-MS. J Archaeol Sci 37:2698–2704

Iñañez JG, Martín JG, Coello A (2012) La mayólica del convento de Santo Domingo (siglos XVI-XVII), Lima (Perú). La evidencia arqueométrica. In: Teixeira A (ed) Arqueologia Moderna. Centro de História de Além-Mar (CHAM), Lisboa, pp 837–846

Jamieson RW (2001) Majolica in the early colonial Andes: the role of Panamanian wares. Lat Am Antiq 12:45–58

Jamieson RW, Hancock RGV (2004) Neutron activation analysis of colonial ceramics from Southern Highland Ecuador. Archaeometry 46:569–583

Jochum KP, Stoll B, Herwig K (2006) Improved in situ Pb isotope analysis of low Pb samples by LA-ICP-MS using a solid-state 193 nm Nd:YAG laser. Geochim Cosmochim Acta 70:A294

Joel EC, Olin JS, Blackman MJ, Barnes IL (1988) Lead isotope studies of Spanish, Spanish-Colonial and Mexican Majolica. In: Farquhar RM, Hancock RGV, Pavlish LA (eds) Archaeometry 88, Proceedings of the 26th international archaeometry symposium. University of Toronto, Toronto, pp 188–195

Kent AJR (2008) In-situ analysis of Pb isotope ratios using laser ablation MC-ICP-MS: controls on precision and accuracy and comparison between Faraday cup and ion counting systems. J Anal Atom Spectrom 23:968–975

Kontak DJ, Cumming GL, Krstic D, Clark AH, Farrar E (1990) Isotopic composition of lead in ore deposits of the Cordillera Oriental, southeastern Peru. Econ Geol 85:1584–1603

Lacueva Muñoz JJ (2010) La plata del rey y sus vasallos: minería y metalurgia en México (siglos XVI y XVII). Universidad de Sevilla, Diputación de Sevilla, Sevilla

Long G (1964) Excavations at Panama Vieja. Fla Anth 17:104–109

Long G (1967) Archaeological investigations at Panamá Vieja. University of Florida, Gainesville

Martín JG (2009) Arqueología de Panamá La Vieja: del asentamiento prehispánico a la ciudad colonial. Dissertation, Universidad de Huelva

Marzo P, Laborda F, Pérez-Arantegui J (2007) A simple method for the determination of lead isotope ratios in ancient glazed ceramics using inductively coupled plasma-quadrupole mass spectrometry. Atom Spectrosc 28:195–201

Mason RB, Tite MS (1997) The beginnings of tin-opacification of pottery glazes. Archaeometry 39 (1):41–58

Mason RB, Farquhar RM, Smith PE (1992) Lead-isotope analysis of Islamic glazes: an exploratory study. Muqarnas 9:67–71

Mena MdC (1984) La Sociedad de Panamá en el siglo XVI, Excma. Diputación Provincial de Sevilla. Sección Historia, Sevilla

Mena MdC (1992) La ciudad en un cruce de caminos: Panamá y sus orígenes urbanos, Consejo Superior de Investigaciones Científicas-Junta de Andalucía. Consejería de Cultura y Medio Ambiente, Sevilla

Mena MdC (1998) Sevilla y las Flotas de Indias: La Gran Armada de Castilla del Oro (1513–1514). Universidad de Sevilla, Sevilla

Molera J, Pradell T, Salvado N, Vendrell-Saz M (1999) Evidence of tin oxide recrystallization in opacified lead glazes. J Am Ceram Soc 82:2871–2875

Mukasa SB, Vidal CE, Injoque-Espinoza J (1990) Pb isotope bearing on the metallogenesis of sulfide ore deposits in central and southern Peru. Econ Geol 85:1438–1446

Olin JS, Harbottle G, Sayre EV (1978) Elemental compositions of Spanish and Spanish-Colonial Majolica ceramics in the identification of provenience. In: Carter GF (ed) Archaeological chemistry II. American Chemical Society, Washington, DC, pp 200–229

Paul B, Woodhead JD, Hergt J (2005) Improved in situ isotope analysis of low-Pb materials using LA-MC-ICP-MS with parallel ion counter and Faraday detection. J Anal Atom Spectrom 20:1350–1357

Pingitore NE Jr, Leach JD, Villalobos J, Peterson JA, Hill DJ (1997) Provenance determination from ICP-MS elemental and isotopic compositions of El Paso area ceramics. In: Vandiver PB, Druzik JR, Merkel JF,

Stewart J (eds) Material issue in art and archaeology V. Materials Research Society, Pittsburgh, PA, pp 59–70
Pollard M (2009) What a long strange trip it's been: lead isotopes and archaeology. In: Shortland AJ, Rehren T, Freestone IC (eds) From mine to microscope. Oxbow Books, Oxford, pp 181–189
Pollard M, Batt C, Stern B, Young SMM (2007) Analytical chemistry in archaeology. Cambridge University Press, Cambridge
Ponting M, Evans JA, Pashley V (2003) Fingerpriting of Roman mints using Laser-Ablation MC-ICP-MS Lead isotope analysis. Archaeometry 45:591–597
Resano M, Marzo P, Perez-Arantegui J, Aramendia M, Cloquet C, Vanhaecke F (2008) Laser ablation-inductively coupled plasma-dynamic reaction cell-mass spectrometry for the determination of lead isotope ratios in ancient glazed ceramics for discriminating purposes. J Anal Atom Spectrom 23:1182–1191
Reslewic S, Burton JH (2002) Measuring lead isotope ratios in Majolica from New Spain using a nondestructive technique. In: Jakes KA (ed) Archaeological chemistry: materials, methods, and meaning. American Chemical Society, Washington, DC, pp 36–47
Rice PM (1997) Tin-enameled wares of Moquegua, Perú. In: Gasco J, Smith GC, Fournier P (eds) Approaches to the historical archaeology of México, Central and South America. The Institute of Archaeology, University of California, Los Angeles, p 173–180
Rovira B (1997) Hecho en Panamá: la manufactura colonial de mayólicas. Revista Nacional de Cultura 27:67–85
Rovira B (2002) Paredes no tan desnudas. La muestra de azulejos sevillanos del sitio de Panamá La Vieja. Arqueología de Panamá La Vieja Avances de Investigación 2:166–182
Rovira B, Martín JG (2008) Arqueología histórica de Panamá. La experiencia en las ruinas de Panamá Viejo. Vestigios 2:7–34
Rovira B, Blackman J, van Zelst L, Bishop R, Rodríguez CC, Sánchez D (2006) Caracterización química de cerámicas coloniales del sitio de Panamá Viejo. Resultados preliminares de la aplicación de activación neutrónica instrumental. Canto Rodado 1:101–131
Sangster DF, Outridge PM, Davis WJ (2000) Stable lead isotope characteristics of lead ore deposits of environmental significance. Environ Rev 8:115–147
Santos Zalduegui JF, Garcia de Madinabeitia S, Gil Ibarguchi JI, Palero F (2004) A lead isotope database: the Los Pedroches – Alcudia area (Spain); implications for archaeometallurgical connections across southwestern and southeastern Iberia. Archaeometry 46:625–634
Shortland AJ (2006) Application of lead isotope analysis to a wide range of late Bronze Age Egyptian materials. Archaeometry 48:657–669
Simon JI, Reid MR, Young ED (2007) Lead isotopes by LA-MC-ICPMS: tracking the emergence of mantle signatures in an evolving silicic magma system. Geochim Cosmochim Acta 71:2014–2035
Stein SJ, Stein BH (2002) Plata, comercio y guerra. España y América en la formación de la Europa Moderna. Crítica, Barcelona
Stos-Gale Z, Gale H (2009) Metal provenancing using isotopes and the Oxford archaeological lead isotope database (OXALID). J Archaeol Anthropol Sci 1:195–213
Stos-Gale ZA, Maliotis G, Gale NH, Annetts N (1997) Lead isotope characteristics of the Cyprus copper ore deposits applied to provenance studies of copper oxhide ingots. Archaeometry 39:83–123
Tejeira E (2001) La ciudad, sus habitantes y su arquitectura, El Casco Antiguo de la Ciudad de Panamá. Publicaciones de Cultura Metropolitana, Panamá
Therrien M, Uprimny E, Lobo-Guerrero J, Salamanca M, Gaitán F, Fandiño M (2002) Catálogo de cerámica colonial y republicana de la Nueva Granada: producción local, materiales foráneos (Costa Caribe, Altiplano Cundiboyacense-Colombia). FIAN-Banco de la República, Bogotá
Thibodeau AM, Killick DJ, Ruiz J, Chesley JT, Deagan K, Cruxent JM, Lyman W (2007) The strange case of the earliest silver extraction by European colonists in the New World. Proc Natl Acad Sci USA 104:3663–3666
Tilton GR, Pollak RJ, Clark AH, Robertson RCR (1981) Isotopic composition of lead in Central Andean ore deposits. Geol Soc Am Mem 154:791–816
Tite MS, Freestone IC, Mason I, Molera J, Vendrell M, Wood N (1998) Lead glazes in antiquity-methods of production and reasons for use. Archaeometry 40:241–260
Tite M, Pradell T, Shortland A (2008) Discovery, production and use of tin-based opacifiers in glasses, enamels and glazes from the late Iron Age onwards: a reassessment. Archaeometry 50:67–84
Tornos F, Chiaradia M (2004) Plumbotectonic evolution of the Ossa Morena Zone, Iberian Peninsula: tracing the influence of mantle-crust interaction in ore-forming processes. Econ Geol 99:965–985
Torrente M (1835) Revista General de la Economía Política. Imprenta Jordan, La Habana
Vaz E, Cruxent JM (1975) Determination of the provenience of majolica pottery found in the Caribbean area using its gamma-ray induced thermoluminesce. Am Antiq 40:71–82
Velasco F, Pesquera A, Herrero JM (1996) Lead isotope study of Zn-Pb ore deposits associated with the Basque-Cantabrian basin and Paleozoic basement, Northern Spain. Miner Deposita 31:84–92
Woolard D, Franks R, Smith D (1998) Inductively coupled plasma magnetic sector mass spectrometry method for stable lead isotope tracer studies. J Anal Atom Spectrom 13:1015–1019
Yener KA, Sayre EV, Joel EC, Özbal H, Barnes IL, Brill RH (1991) Stable lead isotope studies of central Taurus ore sources and related artifacts from eastern Mediterranean Chalcolithic and Bronze Age sites. J Archaeol Sci 18:541–577

MIX
Papier aus verantwortungsvollen Quellen
Paper from responsible sources
FSC® C105338

If you have any concerns about our products,
you can contact us on
ProductSafety@springernature.com

In case Publisher is established outside the EU,
the EU authorized representative is:
Springer Nature Customer Service Center GmbH
Europaplatz 3, 69115 Heidelberg, Germany

Printed by Libri Plureos GmbH
in Hamburg, Germany